TIDAL SIGNATURES IN MODERN AND ANCIENT SEDIMENTS

SPECIAL PUBLICATION NUMBER 24 OF THE
INTERNATIONAL ASSOCIATION OF SEDIMENTOLOGISTS

Tidal Signatures in Modern and Ancient Sediments

EDITED BY B. W. FLEMMING
AND A. BARTHOLOMÄ

Blackwell
Science

and published for them by
Blackwell Science Ltd
Editorial Offices:
Osney Mead, Oxford OX2 0EL
25 John Street, London WC1N 2BL
23 Ainslie Place, Edinburgh EH3 6AJ
238 Main Street, Cambridge
Massachusetts 02142, USA
54 University Street, Carlton
Victoria 3053, Australia

Other Editorial Offices:
Arnette Blackwell SA
1, rue de Lille, 75007 Paris
France

Blackwell Wissenschafts-Verlag GmbH
Kurfürstendamm 57
10707 Berlin, Germany

Feldgasse 13, A-1238 Wien
Austria

First published 1995

Set by Semantic Graphics, Singapore
Printed and bound in Great Britain
at the Alden Press Limited,
Oxford and Northampton

A catalogue record for this title
is available from the British Library

ISBN 0-86542-978-2

DISTRIBUTORS

Marston Book Services Ltd
PO Box 87
Oxford OX2 0DT
(*Orders*: Tel: 01865 791155
Fax: 01865 791927
Telex: 837515)

USA
Blackwell Science, Inc.
238 Main Street
Cambridge, MA 02142
(*Orders*: Tel: 800 215-1000
617 876-7000
Fax: 617 492-5263)

Canada
Oxford University Press
70 Wynford Drive
Don Mills
Ontario M3C 1J9
(*Orders*: Tel: 416 441-2941)

Australia
Blackwell Science Pty Ltd
54 University Street
Carlton, Victoria 3053
(*Orders*: Tel: 03 347-0300
Fax: 03 349-3016)

Library of Congress
Cataloging-in-Publication Data

Tidal signatures in modern and ancient sediments /
edited by B.W. Flemming and A. Bartholomä.
p. cm. — (Special publication number 24 of the International Association of Sedimentologists)
"Papers presented at the 3rd International Research Symposium on Modern and Ancient Clastic Tidal Deposits, TIDAL CLASTICS 92, held in Wilhelmshaven, Germany, from 25–28 August 1992 ... host to the Conference was the Marine Science Division of the Senckenberg Institute ... making it the 6th International Senckenberg Conference" — Pref.
Includes bibliographical references (p. –) and index.
ISBN 0-86542-978-2
1. Sediments (Geology) — Congresses. 2. Marine sediments — Congresses. 3. Tidal currents — Congresses. I. Flemming, B. W. II. Bartholomä, A. III. Forschungsinstitut Senckenberg. Abt. für Meeresforschung. IV. International Research Symposium on Modern and Ancient Clastic Tidal Deposits (3rd: 1992: Wilhelmshaven, Germany) V. International Senckenberg Conference (6th: 1992: Wilhelmshaven, Germany) VI. Series: Special publication ... of the International Association of Sedimentologists; no. 24.
QE471.2.T53 1995
551.3′6 — dc20 95-4005
CIP

Contents

Preface

This Special Publication contains 23 papers presented at the 3rd International Research Symposium on Modern and Ancient Clastic Tidal Deposits, **Tidal Clastics 92**, held in Wilhelmshaven, Germany, from 25 to 28 August 1992. In all, 54 papers and 26 posters were presented by delegates representing 20 countries from five continents (Flemming, 1992). Host to the conference was the Marine Science Division of the Senckenberg Institute in Wilhelmshaven, home of many early pioneers in clastic tidal research in Germany, e.g. Richter (1922), Häntzschel (1936), Schäfer (1956, 1962), Reineck (1958). Indeed, as has become tradition, the *Senckenbergische Naturforschende Gesellschaft*, seated in Frankfurt, acted as generous sponsor to this memorable occasion, making it the 6th International Senckenberg Conference.

The conference was preceded and followed by a number of field trips into some of the most prominent tidal flat and barrier island systems of continental Europe. It was a stroke of luck that all the field trip organizers agreed to rewrite their guide books into overview papers summarizing the current state of knowledge about the tidal regions visited in the course of the various field trips (Flemming & Hertweck, 1994). Included in the volume are the upper macrotidal Bay of Mont St-Michel in France (Larsonneur, 1994), the lower mesotidal West Frisian Wadden Sea in The Netherlands (Oost & de Boer, 1994), the upper meso- to lower macrotidal East Frisian Wadden Sea in Germany (Flemming & Davis, 1994; Hertweck, 1994; Irion, 1994), and the micro- to lower mesotidal North Frisian Wadden Sea in Denmark (Bartholdy & Pejrup, 1994).

As pointed out above, Tidal Clastics 92 was the third international conference focusing attention on modern and ancient clastic tidal and tide-influenced deposits. As such, this series of specialized conferences has now established a sound international reputation and will in future be held at 4-year intervals, the next one being scheduled for Savannah, USA in 1996. It is appropriate to mention here that the initiator of this conference series was Poppe L. de Boer of Utrecht University (The Netherlands), who held the inaugural meeting in 1985 (de Boer *et al.*, 1988). The second meeting followed in 1989, being held in Calgary, Canada (Smith *et al.*, 1991). Although the formal start of the tidal clastics conference series was in 1985, it should be pointed out that it does have a number of spiritual forerunners, e.g. Lauff (1967), Ginsburg (1975) and Hobday & Eriksson (1977).

The editors of this volume wish to thank the numerous referees for their conscientious and constructive perusal of the manuscripts. Special thanks are due to those who have contributed in one way or another to the tedious editorial process, in particular Astrid Raschke for her scanning and typing work and Dr Monique Delafontaine for her language corrections and sharp eye in repeated proofreading.

B.W. Flemming
A. Bartholomä

REFERENCES

Bartholdy, J. & Pejrup, M. (1994) Holocene evolution of the Danish Wadden Sea. In: *Tidal Flats and Barrier Systems of Continental Europe: A Selected Overview* (Eds Flemming, B.W. & Hertweck, G.). *Senckenbergiana marit.* **24**, 187–209.

de Boer, P.L., Gelder, A. van & Nio, S.D. (Eds) (1988) *Tide-influenced Sedimentary Environments and Facies.* Reidel, Dordrecht, 530 pp.

Flemming, B.W. (Ed.) (1992) *Tidal Clastics 92 – Abstract Volume.* Cour. Forsch.-Inst. Senckenberg **151**, 105 pp.

Flemming, B.W. & Davis, R.A., Jr (1994) Holocene evolution, morphodynamics and sedimentology of the Spiekeroog barrier island system (southern North Sea). In: *Tidal Flats and Barrier Systems of Continental Europe: A Selected Overview* (Eds Flemming, B.W. & Hertweck, G.). *Senckenbergiana marit.* **24**, 117–155.

Flemming, B.W. & Hertweck, G. (Eds) (1994) *Tidal Flats and Barrier Systems of Continental Europe: A Selected Overview. Senckenbergiana marit.* **24**, 209 pp.

Ginsburg, R.N. (Ed.) (1975) *Tidal Deposits – A Casebook of Recent Examples and Fossil Counterparts.* Springer, New York, 428 pp.

Häntzschel, W. (1936) Die Schichtungsformen rezenter Flachmeer-Ablagerungen. *Senckenbergiana* **18**, 316–356.

Hertweck, G. (1994) Zonation of benthos and lebensspuren in the tidal flats of the Jade Bay, southern North Sea. In: *Tidal Flats and Barrier Systems of Continental*

Europe: A Selected Overview (Eds Flemming, B.W. & Hertweck, G.). *Senckenbergiana marit.* **24**, 157–170.

IRION, G. (1994) Morphological, sedimentological and historical evolution of Jade Bay, southern North Sea. In: *Tidal Flats and Barrier Systems of Continental Europe: A Selected Overview* (Eds Flemming, B.W. & Hertweck, G.). *Senckenbergiana marit.* **24**, 171–186.

LARSONNEUR, C. (1994) The Bay of Mont-Saint-Michel: A sedimentation model in a temperate macrotidal environment. In: *Tidal Flats and Barrier Systems of Continental Europe: A Selected Overview* (Eds Flemming, B.W. & Hertweck, G.). *Senckenbergiana marit.* **24**, 3–63.

LAUFF, G.H. (Ed.) (1967) *Estuaries*. Am. Assoc. Adv. Sci. Publ. **83**, 757 pp.

OOST, A.P. & BOER, P.L. DE (1994) Sedimentology and development of barrier islands, ebb-tidal deltas, inlets and backbarrier areas of the Dutch Wadden Sea. In: *Tidal Flats and Barrier Systems of Continental Europe: A Selected Overview* (Eds Flemming, B.W. & Hertweck, G.). *Senckenbergiana marit.* **24**, 65–115.

REINECK, H.-E. (1958) Wühlbau-Gefüge in Abhängigkeit von Sediment-Umlagerung. *Senckenbergiana Lethaea* **39**, 1–24.

RICHTER, R. (1922) Flachseebeobachtungen zur Paläontologie und Geologie, III–IV. *Senckenbergiana* **4**, 103–141.

SCHÄFER, W. (1956) Wirkungen der Benthos-Organismen auf den jungen Schichtverband. *Senckenbergiana Lethaea* **37**, 183–263.

SCHÄFER, W. (1962) *Aktuo-Paläontologie nach Studien in der Nordsee*. Kramer, Frankfurt, 666 pp.

SMITH, D.G., REINSON, G.E., ZAITLIN, B.A. & RAHMANI, R.A. (Eds) (1991) *Clastic Tidal Sedimentology*. Mem. Can. Soc. petrol. Geol. **16**, 387 pp.

HOBDAY, D.K. & ERIKSSON, K.A. (Eds) (1977) *Tidal Sedimentation with Special Reference to South African Examples*. Special Issue, *Sediment. Geol.* **18**, 356 pp.

Modern Tidal Processes and Sediment Dynamics

Spec. Publs int. Ass. Sediment. (1995) **24**, 3–18

What is a bedload parting?

P.T. HARRIS*¶, C.B. PATTIARATCHI†, M.B. COLLINS‡ *and* R.W. DALRYMPLE§

**Ocean Sciences Institute, University of Sydney, Sydney NSW 2006, Australia;*
†Centre for Water Research, University of Western Australia, Nedlands WA 6009, Australia;
‡Department of Oceanography, University of Southampton, Southampton SO9 5NH, UK; and
§Department of Geological Sciences, Queen's University, Kingston, Ontario K7L 3N6, Canada

ABSTRACT

Net sediment transport paths and their associated bedload partings (BLPs) are important elements of sediment dispersion on tidally dominated continental shelves. The facies distribution associated with these paths is believed to reflect primarily a downcurrent decrease in tidal bottom stress. With decreasing tidal bottom stress, scour zones characterized by lag gravel and/or bedrock exposures give way to a mobile sand sheet facies, ending with a muddy sand facies at the lowest tidal energy. A review of available data on the distribution of these facies shows that they are developed at the location of local maximum bed stress, induced by either standing-wave nodal points (type A) or geomorphic constrictions (type B). Around the western European continental shelf, 22 different locations having a length scale exceeding 10 km are associated with local bottom-stress maxima (type A or B).

A model is proposed for the development of scour zones as a function of sediment supply, with consideration given to the movement of facies boundaries through time in relation to the bottom-stress maxima. Where sand is plentiful, the transport pattern is dominated by linear sand banks and associated dunes; these delimit a mutually evasive transport system. Incipient scour zones are those that occupy a significant (~50%) portion of the channel width. Partial scour zones occupy nearly the entire channel, except for deposits along the margins which link the flanking depositional areas. Complete scour zones separate totally the mobile sand deposits located on opposite flanks of a channel, with no significant marginal sand deposits occurring. Only 11 of the 22 zones of local maximum bottom stress correlate with 'complete' scour zones.

Bedload partings are classified as a special case of scour zone, in which local bottom-stress maxima coincide with divergent patterns in sand transport (inferred from bedform asymmetries). A minimum scale of about 10 km is proposed for a 'complete' BLP which separates flanking depositional areas on the basis of the tidal excursion length of a sand grain. The model proposed for scour zone development applies also to BLPs, suggesting that only three of the six BLPs previously identified for the west European shelf are at a 'complete' stage of development.

INTRODUCTION

The currently accepted conceptual model for sand dispersal on a tidally dominated continental shelf evolved from concepts originally presented by Stride (1963, 1973), Belderson & Stride (1966) and Kenyon & Stride (1970). The model incorporates *tidal current transport paths*, along which sand particles move. A spatial plot of such paths demonstrates regions of convergence and divergence, i.e. *bedload partings* (BLPs; see Fig. 1A). This model has gained wide acceptance and is cited in many sedimentology texts as the 'type case' for sediment dispersal on tidally dominated continental shelves (e.g. Allen, 1970, p. 175; Leeder, 1982, p. 204; Johnson & Baldwin, 1986, p. 218, p. 241; Dalrymple, 1992, p. 209).

¶*Correspondence address*: Australian Geological Survey Organisation, Antarctic CRC, University of Tasmania, GPO Box 252C, Hobart, Tasmania, 7001, Australia.

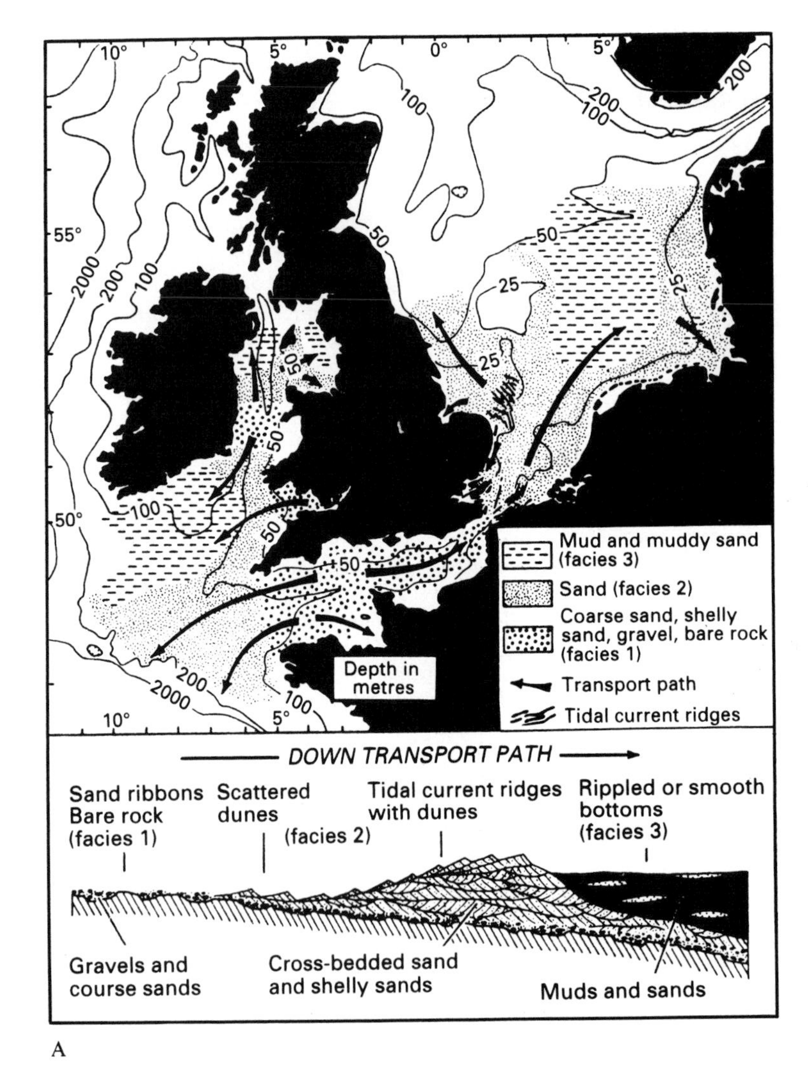

A

Fig. 1. (A) Tidal current sand streams on the western European continental shelf, with an idealized cross-section showing distribution of sea-bed facies (1, 2 and 3) along a velocity gradient (after Allen, 1970). The actual position of sand banks along the transport path is shown by Johnson *et al.* (1982) to be located towards the high-velocity end of the sand sheet facies. (B) Location of tidal current sand streams and the six bedload parting zones (BLPs) identified by Johnson *et al.* (1982) on the bases of bedform morphologies, current meter data and other data concerning net directions of bedload transport: (1) Pentland Firth; (2) North Channel of the Irish Sea; (3) St George's Channel; (4) Bristol Channel; (5) English Channel; and (6) Southern Bight of the North Sea.

A key aspect of the model is that BLPs form the 'heads' of individual transport paths; they are characterized by an eroded 'basal' substrate (Belderson & Stride, 1966), from which surficial sediments may be winnowed to a lag gravel (armoured bed) or, in some cases, exposed bedrock; such areas will be referred to as 'scour zones' in the present paper. With increasing distance from the scour zone, scattered dunes (definition of Ashley 1990) and sand ribbons are found; these grade into large linear sand-bank (tidal current ridge) fields and finally to muddy sand deposits towards the end of the transport path (see facies 1, 2 & 3 defined in Fig. 1A).

This concept of BLPs was extended by Kenyon *et al.* (1981) and by Johnson *et al.* (1982), to include a divergent pattern in inferred sand bank migration directions in the southern North Sea (Fig. 1B). Thus, in this extended definition, a BLP may be any divergent pattern in bedload transport, not necessarily linked with any specific (scour zone) facies. The definition of a BLP suggested by Johnson *et al.* (1982, p. 81) is concerned with regional divergences in net bedload transport vectors, such that BLPs can be thought of as 'S' or bow-shaped lines or 'more or less narrow zones' (Fig. 1B).

Application of this conceptual model to locations outside of the western European shelf has been limited (e.g. Belderson *et al.*, 1978; Kenyon *et al.*, 1981). This is due partly to the limited global occurrence of tidally dominated shelves and partly, we suggest, to the inconsistent way in which the concept of BLPs has been developed and applied. If

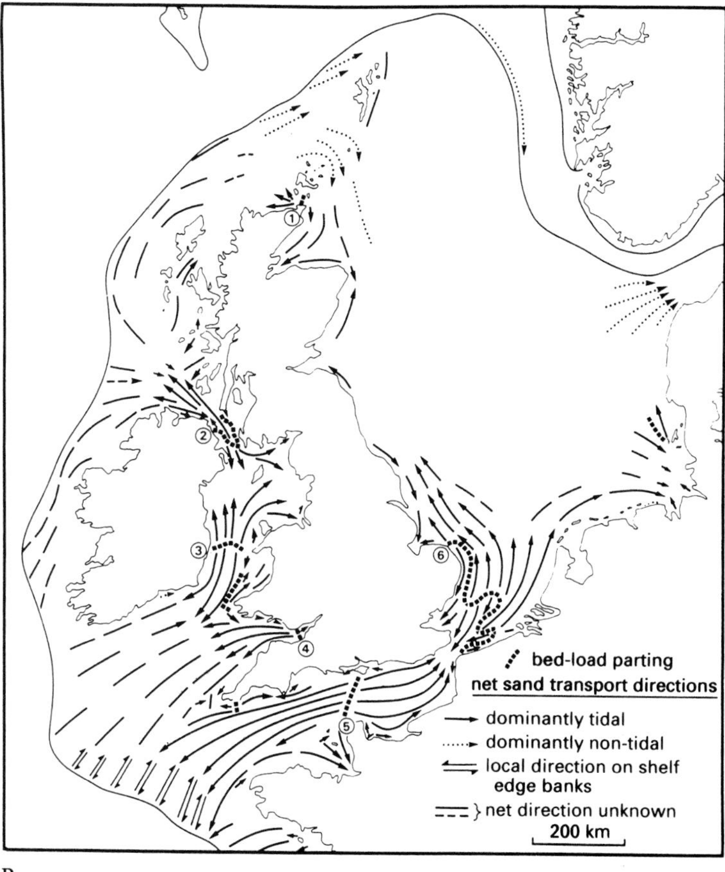

Fig. 1. *(Continued.)* B

BLPs are defined geologically as being associated with the scour zone facies of a (grain-size) sequence ranging from gravel through sandy bedforms to fine-grained mud deposits (Fig. 1A), then the occurrence of such scour zones, extending for many tens of kilometres on either side of the English Channel BLP, could hardly be termed a 'more or less narrow zone'. If BLPs are defined oceanographically as a divergence in transport paths, there is the problem of scale: are BLPs shelf-scale systems or do small-scale bedload divergences, such as occur in association with tidal inlets, qualify as BLPs? It has also been debated whether the simple divergent transport pattern defining a BLP can be applied appropriately to the understanding of the complex sedimentation patterns associated with BLPs in large macrotidal estuaries (Harris & Collins, 1991; Stride & Belderson, 1991).

The main objective of the present study is to derive a conceptual model that is useful in explaining, and ultimately in predicting, the occurrence and distribution of sedimentary facies on a tidally dominated continental shelf. In order to clarify the meaning of the terminology used and to facilitate the application of BLP theory to other shelves, two important questions need to be answered: (i) What physical processes control the occurrence and distribution of sedimentary facies in relation to tidal bedload transport processes? (ii) What criteria (e.g. facies, scale, transport pattern, etc.) must be met for an area to be considered a BLP? In short, the question posed is: *What is a bedload parting?*

PHYSICS OF SCOUR ZONES AND BEDLOAD PARTINGS

The early discussions of Stride (1963), Kenyon & Stride (1970) and Stride (1973) noted that sand transport was generally in the direction of the 'peak' (maximum) tidal current, where differences between ebb and flood current speed maxima of only about 0.1 knot (5 cm s^{-1}) appear to be sufficient to induce a net sand transport vector. The locations of BLPs around the western European shelf were identified by these investigators on the basis of inferred transport paths, determined from bedform morphology and from available tidal current data (abstracted mainly from nautical charts). No specific theoretical or physical bases were suggested as to why BLPs should exist. However, it was (and still is) considered that the main physical parameter controlling the locations of BLPs is tidal current flow. Other possible factors influencing the net movement of sand include: (i) the effect of Stoke's drift; (ii) frictional effects; and (iii) the superposition of non-tidal currents (e.g. wind-driven or density currents). Although such currents may determine patterns of net sand movement in some areas (even resulting in non-tidal BLPs as described by Johnson *et al.* (1982) and Flemming (1988)), they will not be considered further in the present discussion.

Major progress in understanding the physics of BLPs was made by Pingree & Griffiths (1979), who noted that BLPs correlate with areas where the peak bottom stress (average over many tidal cycles) reaches a local maximum, such as those found in association with M2 tidal amphidromic points (cf. Figs 1B & 2). At such points, the change in surface elevation of the M2 tide is theoretically zero (i.e. a node in a non-rotating, standing-wave case) and horizontal currents and bottom stress are at a maximum (amphidromes must be located near coastlines to coincide with a current maximum; those located away from coasts, as, for example, that in the centre of the North Sea, will coincide with weak currents as the tidal wave will be rotational in nature). Regions of maximum current speed (and bottom stress) occur also where flow is accelerated by tidal forcing through coastal constrictions (Johnson *et al.*, 1982) as in tidal inlets, embayments with narrow entrances and tidal channels (e.g. Pentland Firth, Bristol Channel, English Channel and Dover Strait) and around headlands (e.g. off Anglesey; Figs 1 & 2).

Hence there are two types of tidal-current, bottom-stress maxima: (i) those related to flow acceleration under amphidromic points (standing-wave nodes); and (ii) those related to flow acceleration induced by local constrictions, related to coastal geometry. Both types (A and B) exhibit a bed shear stress gradient, centred around a zone where tidal current speed and bed shear stress are at a maximum. The main difference between these two types of bottom-stress maxima is their scale. Type 'A' maxima can occur only in shelf seas that are large enough to contain M2 amphidromic systems; their spacing is a function of tidal wavelength (λ), which is given by:

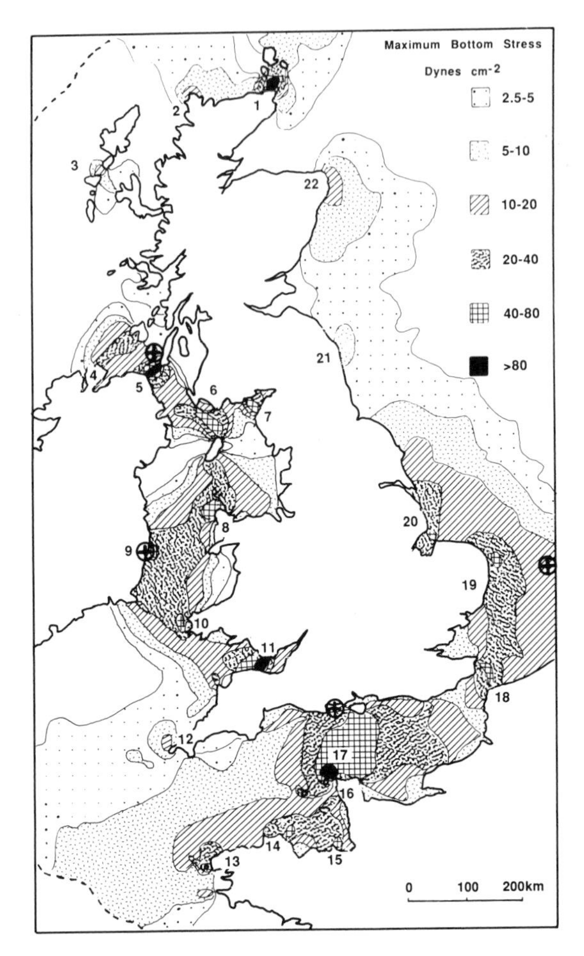

Fig. 2. Distribution of numerically estimated maximum bottom stress due to M2 and M4 currents and locations of M2 amphidromic points ⊕ (after Pingree & Griffiths, 1979). Locations of bottom-stress maxima are numbered 1 to 22 (refer to Table 1 for location names). Note that the model was run using a grid spacing of 5 nautical miles (9.26 km).

$$\lambda = \sqrt{ghT}$$

where h is the mean water depth, g is the acceleration due to gravity and T is the tidal period. Thus, a

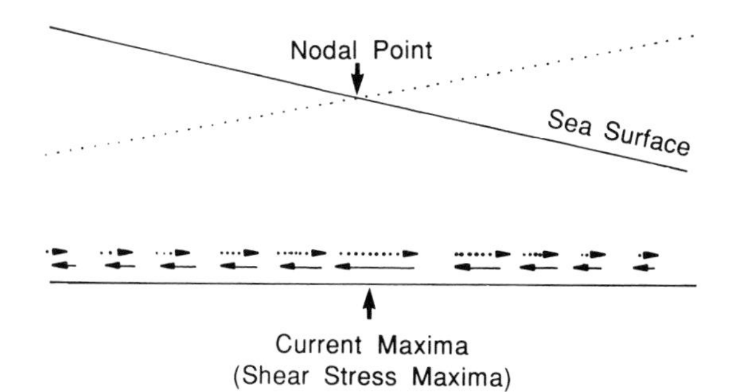

Fig. 3. Sketch showing the sea surface variation and bottom current speeds associated with a standing wave nodal point. Sea surface with a solid line corresponds with solid arrows and sea surface with dotted line corresponds with dotted arrows.

shelf with a mean depth of 30 m would need dimensions of approximately 740 km to contain a single M2 wavelength.

Type 'B' bottom-stress maxima, on the other hand, can occur at much smaller scales. Thus, the divergent transport patterns associated with tidal inlets (with their associated ebb- and flood-tidal deltas) correlate with type 'B' bottom-stress maxima, even though their cross-channel widths may be only a few 10s of metres.

The occurrence of *scour zones* under bottom-stress maxima is explained in terms of diffusion down the current gradient. Since there is a decrease in current amplitude away from the nodal point (Fig. 3), there is an associated bedload transport gradient, with erosion occurring on the accelerating flank and deposition occurring on the decelerating flank, the two sides reversing sign with each change of the tide. The movement of sand away from the zone of maximum bottom stress is thus a *diffusive* response, resulting from that part of the transport process which is random and isotropic (i.e. the paths of sand grains integrated over many tidal cycles). Given that such a process operates over a long time interval (100s to 1000s of years?) sand-sized grains will be winnowed away, moving down the transport gradient, eventually resulting in the development of a scour zone.

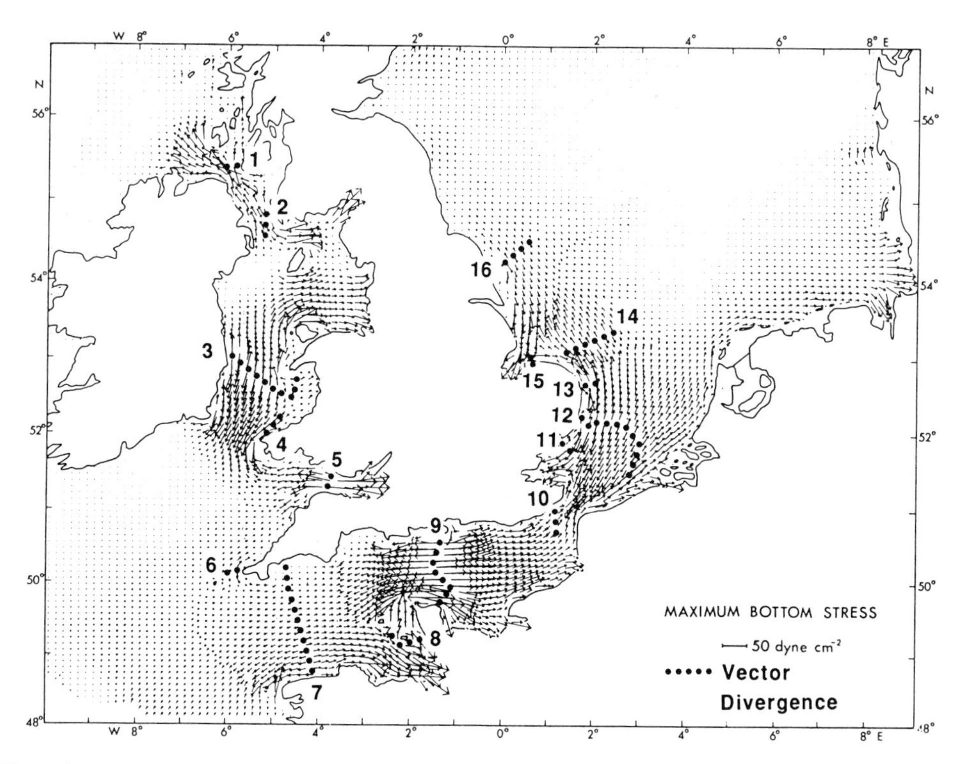

Fig. 4. Numerically modelled, maximum bottom-stress vectors due to M2 and M4 currents (after Pingree & Griffiths, 1979). Vector divergences have been numbered 1 to 16 for comparison with bottom-stress maxima and facies (see Table 2). Note that the model was run using a grid spacing of 5 nautical miles (9.26 km).

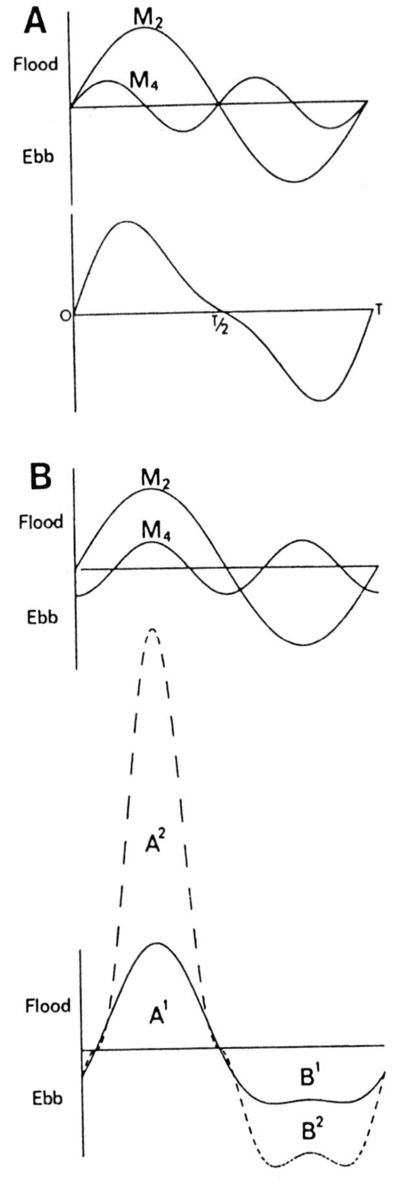

Fig. 5. Curves of current speed versus time over one (12-h) semi-diurnal tidal cycle, showing the effects of the combination of M2 and M4 tidal constituents. (A) When the phase of the M2 is equal to one half that of the M4 + 90°, the resulting curves are symmetrical and maximum currents are equal. (B) For any angle less than 90°, however, interference results in an asymmetry where the maximum current in one direction exceeds that in the other; the maximum effect occurs when the phase of the M2 is equal to one half that of the M4 + 0° (after Pingree & Griffiths, 1979).

The development of BLPs, however, requires the existence of a bedload transport divergence (and not necessarily a scour zone). Pingree & Griffiths (1979) showed that the divergent patterns in vectors of numerically-modelled maximum bottom stress correlate with the BLPs identified by Kenyon & Stride (1970; Figs 1B & 4). Such a divergent pattern in bottom-stress vectors is explained by inequalities between maximum ebb and flood tidal currents that occur in response to the 'interference' between the M2 semi-diurnal tidal currents and the M4 harmonic (Fig. 5A & B). Since sand transport is a non-linear process (being a function of current speed cubed (cf. Hardisty, 1983)), this inequality in ebb and flood velocities is enhanced further, resulting in the advective transport of sand in a particular ebb or flood direction.

It can be seen that the sand transport paths mapped by Johnson *et al.* (1982) do not exhibit a divergent pattern at every bottom-stress maximum (compare Figs 1B & 2). Hence, the simple correlation between the occurrence of scour zones (facies 1) and the strength of tidal flow (bottom stress; e.g. Stride, 1963; Belderson & Stride, 1966) cannot be extended to include BLPs, which are defined on the basis of a divergent transport pattern.

SCOUR ZONES AND BLPs ON THE WESTERN EUROPEAN SHELF

Scour zones and bottom-stress maxima

As noted above, an area of localized maximum bottom stress may evolve into a scour zone over a long time period (100s or 1000s of years). The terms 'bottom-stress maxima' and 'scour zone' should not be used interchangeably since one is a process and the other is an evolutionary (geological) product. In this section, we review the occurrence of bottom-stress maxima around the western European shelf in order to determine the extent to which scour zones have developed at each location.

The output of Pingree & Griffiths' (1979) model suggests that there are 22 locations around the west European shelf with an area of locally elevated bottom stress having a horizontal scale in excess of the model's grid spacing (i.e. 10 km; see Fig. 2). Each of these locations exhibits a bottom-stress gradient (given in Table 1), determined as the maximum rate of change in modelled bottom shear stress per kilometre length of sea-bed as measured from Fig. 2. The areas of sea-bed shaded in Fig. 2 all experience a modelled, maximum bed shear stress τ_0 of over 2.5 dynes cm^{-2}, this value is equivalent to a current speed at 100 cm above the bed (U_{100}) of 28 cm s^{-1} based on the Quadratic Stress Law: $\tau_0 = C_{100}\ U^2{}_{100}$, where C_{100} is a drag coefficient which has been taken as 3.1×10^{-3} (Sternberg, 1972). At four locations on Fig. 2 the maximum bottom stress (due to M2/M4 current interactions) exceeds 80 dynes cm^{-2} which is equivalent to a U_{100} of 160 cm s^{-1}; these are numbered 1, 5, 11 and 17 (Table 1). Thus, over a spring–neap cycle, threshold velocities are exceeded and bedload transport of particles ranging from fine sand to gravel can be expected over that part of the eastern European shelf shown as shaded areas in Fig. 2.

Assessment of the sea-bed morphology and sediment distribution at each of the 22 bottom-stress maxima (Fig. 2) is facilitated by reference to published work and, in particular, the British Geological Survey Sea Bed Sediments charts (1:250 000 scale, UTM Series). An excellent review of much of this work has also been presented by Pantin (1991). A summary of each assessment is listed in Table 1 and the distribution of lag gravel and exposed bedrock (facies 1), mobile sand deposits, linear sand banks and dunes (facies 2) and of muddy deposits (facies 3) is shown in Fig. 6. The length of the scour zones (Table 1) was determined in different ways for channels and for headlands. For channels, the length was determined as the minimum distance of lag gravel and exposed bedrock (facies 1) that must be crossed between flanking sand deposits (facies 2), measured parallel to the axis of tidal flow. For headlands, the length was determined as the minimum distance of lag gravel and exposed bedrock (facies 1) that must be crossed between land and sand deposits (facies 2), measured normal to the headland. Table 1 also lists derived information, including the average maximum bottom stress for the boundaries between facies 1–2 and facies 2–3 (from Figs 2 & 6).

Comparison of transport pattern, bottom-stress maxima and facies

Within the area of the west European shelf modelled by Pingree & Griffiths (1979), there are 16 locations exhibiting local divergence in maximum bottom-stress vectors (having a horizontal scale in excess of ~10 km; Fig. 4). Examination of the relationships between the occurrence of divergences (Fig. 4), bottom-stress maxima (Fig. 2), the associated facies types (Fig. 6) and observed sand transport directions (Fig. 1B) shows that (see Table 2): (i) nine out of the 16 divergences correlate with zones of maximum bottom stress; (ii) 11 out of the 16 divergences correlate with scour zones (facies 1); and (iii) seven out of the 16 divergences correlate with observed divergences in bedload transport as deduced from bedform orientations.

Comparisons of the 22 bottom-stress maxima with observed bedload transport paths are listed in Table 1. The six previously identified BLPs (Fig. 1B) are the only locations where a modelled divergence (Fig. 7) correlates with a bottom-stress maximum (Fig. 2) and a clearly observed divergence in bedload transport direction (see also Table 2). One bottom-stress maximum (Dover Strait) corresponds with a sand stream convergence. At six bottom-stress maxima (The Minches, Isle of Man, Anglesey, Plateau des Roches Douvres, Plateau des Minquiers, The Wash) sand is transferred across the area in a constant direction.

At seven locations (North Channel of the Irish Sea #1, St David's Head, Lands End, Chanel du Four, the Channel Isles, Farne Islands and Kinnairds Head) sand streams are complex and do not clearly exhibit any specific pattern (divergent, convergent or transfer) and at two locations (Cape Wrath and Solway Firth) data on bedload transport paths are unavailable. However, in an earlier interpretation, Kenyon & Stride (1970) identified BLPs at the North Channel of the Irish Sea #1, Chanel du Four and Farne Islands and a transport convergence at Kinnairds Head. It is unclear as to why these transport patterns were not incorporated into the revised summary published by Johnson *et al.*, (1982; Fig. 1B) and hence they are listed in Table 1 with question marks.

DISCUSSION

Examination of the relationships between the sedimentary facies, the bottom-stress distribution and the distribution of the maximum stress vectors (see Figs 2, 4 & 6; Tables 1 and 2) shows that in three cases (Solway Firth, Thames Estuary and The Wash) zones of maximum bed stress are associated with areas of numerous sand banks (facies 2; see Fig. 6). In most other areas, however, the bottom-stress maxima correspond with a scour zone (facies 1). Similar variation is exhibited by the facies

Table 1. M_2 and M_4 bottom-stress maxima (see Fig. 2 for locations) with characteristic length, maximum bottom stress (from Pingree and Griffiths, 1979), surface current speed (after Dietrich, 1950), inferred type (A or B) and stage of development (as described in the text). 'Gradient' refers to the maximum change in bottom stress per km distance occurring in association with a bottom stress maximum (as per Fig. 2). Average maximum bottom stress along boundaries dividing facies 1–2 and 2–3 is derived from Figs 2 and 6. The type of bedload transport path refers to divergences (BLPs), convergences and transfer zones, determined using Johnson *et al.*'s (1982) assessment (Fig. 1B). See text for discussion of the data

	Location	Length of scoured zone (km)	Max. bottom stress (dyne cm^{-2})	Bottom stress gradient (dn cm^{-2} km^{-1})	Max. surf. current (cm/sec)	Type of stress Maxima	Facies 2-3 boundary (dyne cm^{-2})	Facies 1-2 boundary (dyne cm^{-2})	Development stage	Transport path type
1	Pentland Firth	28	>80	3.8	120	B	1.3	11.3	Complete	BLP-1
2	Cape Wrath	20	10–20	0.6	100	B	1.3	8.5	Complete	no data
3	The Minches	3	10–20	0.8	200	B	1.3	2.5	Partial	transfer
4	North Channel Irish Sea[1]	14	20–40	0.4	150	A	1.3	10	Partial	BLP?
5	North Channel Irish Sea[2]	40	>80	4.4	200	B	3.8	15	Complete	BLP-2
6	Isle of Man to Galloway	24	40–80	2.8	200	B	12.5	22.5	Complete	transfer
7	Mouth of Solway Firth	0	40–80	2.1	200	B	15	15	Mutually evasive	no data
8	Anglesey	25	40–80	1.5	120	B	6.9	15	Complete	transfer
9	St. George's Channel	8	40–80	1.1	160	A	5.6	37.5	Incipient	BLP-3
10	St. David's Head	20	40–80	0.6	150	B	2.5	13.1	Complete	?
11	Bristol Channel	10	>80	5.3	200	B	10	20	Partial	BLP-4
12	Lands End	32	10–20	0.6	100	B	3	5	Complete	?
13	Chanel du Four	48	40–80	3.3	120	B	3.8	15	Complete	BLP?
14	Plat. des Roches Douvres	15	40–80	2.2	160	B	7.5	45	Partial	transfer
15	Plateau des Minquiers	16	40–80	4.0	120	B	11.3	22.5	Complete	transfer
16	Channel Islands	12	20–40	1.5	200	B		30	Partial	?
17	English Channel	30	>80	3.3	180	A	3.8	15	Complete	BLP-5
18	Dover Strait	8	40–80	0.7	180	B	3.8	18.8	Incipient	convergence
19	Thames Estuary Mouth	0	20–40	0.7	120	A	7.5	30	Mutually evasive	BLP-6
20	Mouth of the Wash	0	40–80	4.4	120	B	16.9	60	Mutually evasive	transfer
21	Farne Islands	20	5–10	0.1	75	B	2.5	6.3	Complete	BLP?
22	Kinnairds Head	12	10–20	0.4	100	B	5	15	Partial	convergence?
Mean		17.5	50.5	2.03	148		6.0	19.7		
SD		12.9	28.9	1.61	41.1		4.7	13.7		

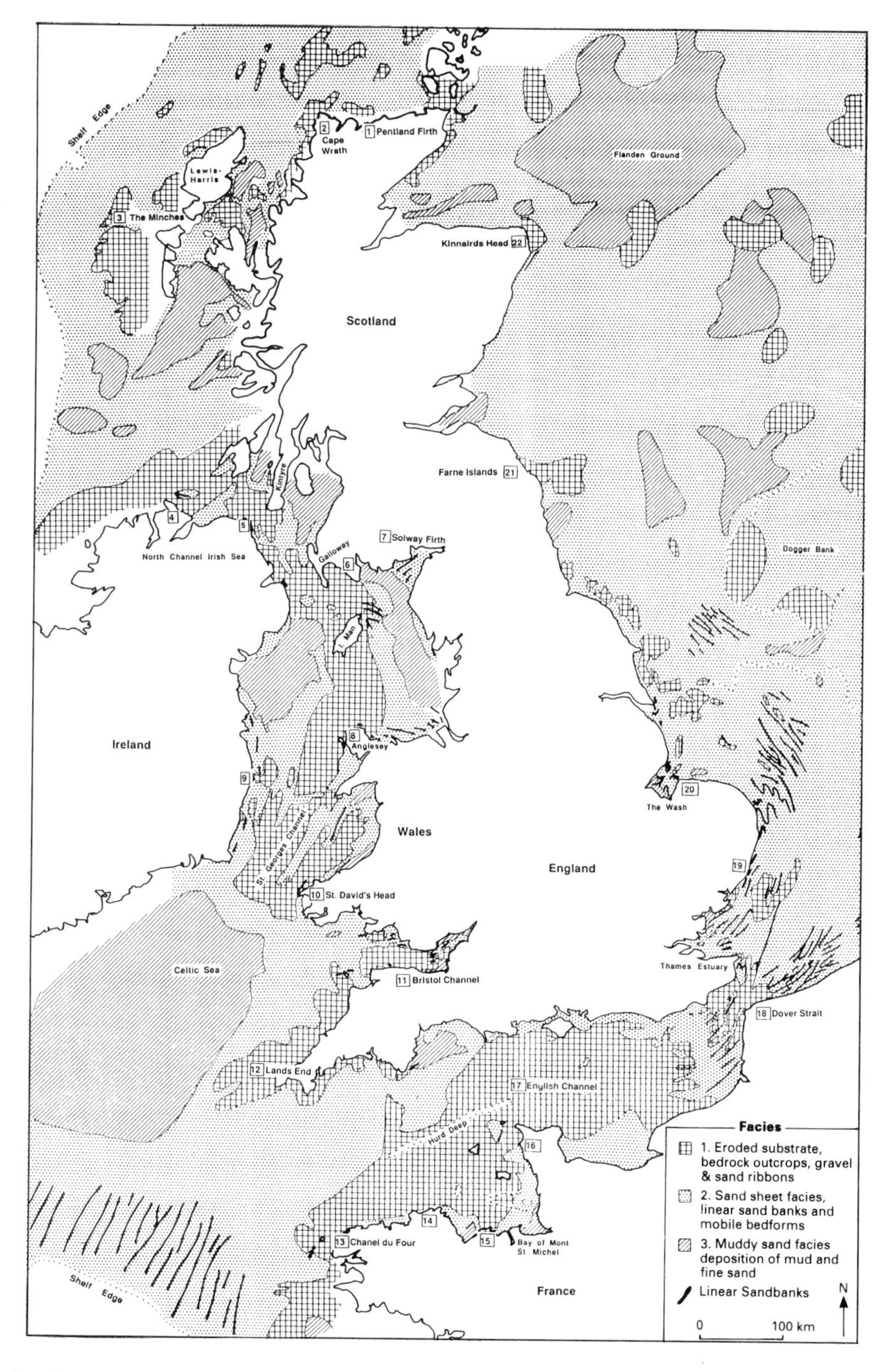

Fig. 6. Distribution of facies on the western European shelf. Data compiled from Eisma (1973; 1981), G.F. Caston (1976), Evans & Collins (1976), Vaslet *et al.* (1978), Bishop & Jones (1979), V.N.D. Caston (1979), Owens (1981), Tappin (1983), Graham (1984), Pattiaratchi & Collins (1984), Pantin & Evans (1984), Evans (1985), Wingfield (1983, 1985), Evans *et al.* (1986), Crosby *et al.* (1987), Ruckley & Chester (1987), Harris & Collins (1988, 1991), James (1988, 1990), Hamblin (1989), Harrison (1989a,b), Balson & D'Olier (1990), Pantin (1991). The numbered locations are sites of bottom-stress maxima as determined numerically by Pingree & Griffiths (1979; see Fig. 4).

Table 2. List of areas having a divergent pattern of maximum bottom stress vectors numbered as on Fig. 4, and correlation with zones of maximum bed stress (based on Fig. 2), facies (based on Fig. 6) and bedform orientations (based on Johnson *et al.*, 1982; see Fig. 1B)

Divergence no.	Correlates with a bottom stress maxima (number/location)?	Facies	Correlates with an observed divergence in bedform orientations?
1	5. North Ch. Irish Sea[2]	1	no
2	no	1	yes
3	9. St. George's Channel	1–2	yes
4	10. St. David's Head	1	yes
5	11. Bristol Channel	1	yes
6	12. Lands End	1	no
7	no	1	no
8	16. Channel Islands	1–2	no
9	17. English Channel	1	yes
10	no	2	no
11	no	2	no
12	19. Thames Estuary	2	yes
13	no	2	yes
14	no	2	no
15	20. The Wash	2	no
16	no	1	no

associated with the divergences in bottom-stress vectors (Table 2). The facies associated with the 22 different examples of bottom-stress maxima (Fig. 6) suggest a conceptual model for the development of scour zones.

Sequential development of scour zones in relation to tidal bottom-stress maxima

Consider a channel partially infilled with unconsolidated sediment and subject to tidal current flow in

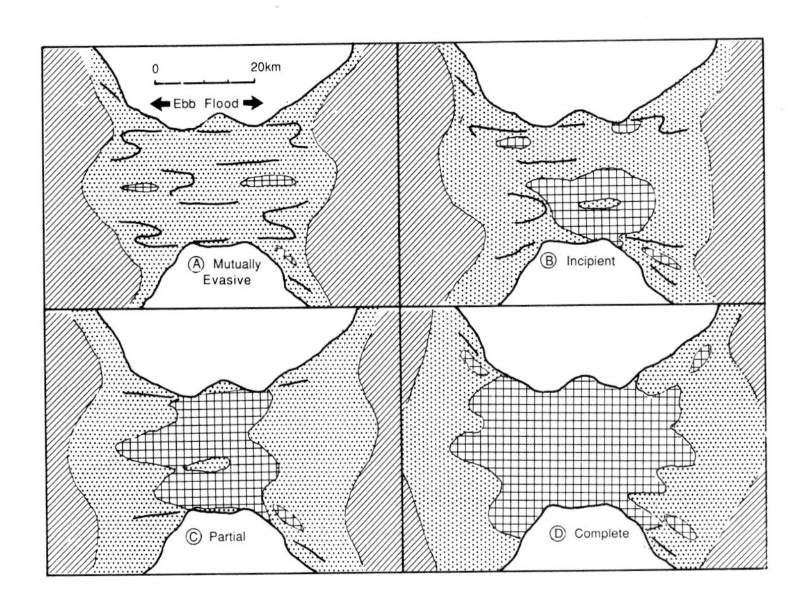

Fig. 7. Idealized diagrams showing the surficial sediment distribution patterns associated with scour zones having different relative sand supplies: (A) mutually evasive transport patterns; (B) incipient scour zone; (C) partial scour zone, and (D) a complete scour zone. Shading of facies is the same as in Fig. 6 (see text for discussion).

which a type A or type B bottom-stress maximum occupies the channel centre (Fig. 7). In channels where sand is plentiful, the deposits are dominated by linear sand banks and associated dunes, delimiting complex and *mutually evasive* transport patterns (Harris, 1988; Fig. 7A). The sea-bed may be scoured locally to a lag gravel (e.g. in the trough between sand banks) but is characterized mainly by mobile sand deposits. *Incipient* scour zones are those in which an eroded substrate occupies a significant proportion (more than 50%) of the channel width. Sand is winnowed from the scour zone and is deposited elsewhere in the channel centre and along the channel margins in the form of sand sheet and linear sand-bank deposits (Fig. 7B). *Partial* scour zones contain a well-developed eroded substrate, with a narrow sandy zone containing limited sand banks along the margins; the scoured zone separates flanking depositional areas apart from these marginal deposits (Fig. 7C). *Complete* scour zones totally separate the mobile sand deposits located on opposite flanks of the channel in the absence of any marginal sand deposits (Fig. 7D).

The stages of development represented in Fig. 7 may be viewed as a model for the sequential development of channel or headland associated tidal-shelf deposits. The sequence A–B–C–D represents the stages of an area undergoing erosion, whereas D–C–B–A represents an area being infilled with sand. The proposed model incorporates mutually evasive patterns and sediment-starved scour zones as end members of a continuum. An area at one of the intermediate stages of development (i.e. Figs 7B or C) would be expected to possess attributes of both patterns. With respect to the three main sedimentary facies associated with tidal current transport paths (Fig. 1A), the mutually evasive pattern exists when the zone of maximum bottom stress coexists with facies 2 (the mobile sand sheet facies), whereas complete scour zones correlate with facies 1 (tidally scoured sea-bed).

In some cases, the scoured zone does not extend completely across a channel but merely occupies part of a channel or area surrounding a promontory. The facies (1, 2 and 3) are arranged in a series of concentric patterns about the scour zone, denoting the fluid-power gradient. Thus, the arrangement of the facies is not always along the direction of tidal flow and may even be at right angles to the flow direction(s) (e.g. the gradient in bottom stress and facies associated with Cape Wrath, Anglesey, Farne Islands and Kinnairds Head; sites 2, 8, 21 and 22 in Figs 2 & 6). Sand is winnowed from the zone of maximum bed stress and is removed downgradient (not necessarily downstream); this is clearly a diffusive response.

From the above discussion, it appears that BLPs may be viewed as a special type of tidal scour zone. Since all BLPs identified up to the present correlate with bottom-stress maxima, their development also occurs according to the proposed model (Fig. 7). Of the six BLPs identified (Fig. 1B) one is at the 'mutually evasive' stage of development (southern North Sea), one is at the 'incipient' stage (St George's Channel), one is at the 'partial' stage (Bristol Channel) and three are at the 'complete' stage (Figs 1B & 7; Table 1). Whereas sand may continually pass through a scour zone (i.e. the transfer scour zones listed in Table 1), a 'complete' BLP should act as a barrier to transport. However, this will only occur provided a BLP has reached a size that prohibits the movement of sand through the scour zone.

Scaling of a 'complete' BLP

The maximum distance required between flanking depositional areas to prevent sand from being transported across the scoured zone is determined by the maximum distance a sand grain can travel during a single tidal cycle, i.e. the 'tidal excursion' of a sand grain. As an example, the tidal excursion may be calculated using current meter data of the type obtained from the Bristol Channel zone of maximum bottom stress (Harris & Collins, 1988). The data include current speed and direction observations obtained at 10-min intervals over a spring–neap cycle in August 1983. For the purposes of this exercise, a 24-h segment has been selected (Fig. 8) which shows rectilinear spring tidal flow, with a peak speed of about 1.55 m s^{-1} measured 1.5 m above the bed in 14 m of water depth (see Harris & Collins, 1988, 1991). Fine sand and silt-sized sediment grains, transported entirely in suspension, would move at the same speed as the water once the threshold speed has been exceeded. For grains transported in suspension, the Bristol Channel current meter data give an excursion distance of 21 km.

For coarser sediment with a higher threshold velocity and transported in bedload only, the excursion length is smaller. Wiberg & Smith (1985) have proposed a model for the prediction of the velocity of a sand grain (U_s) moving as saltating bedload.

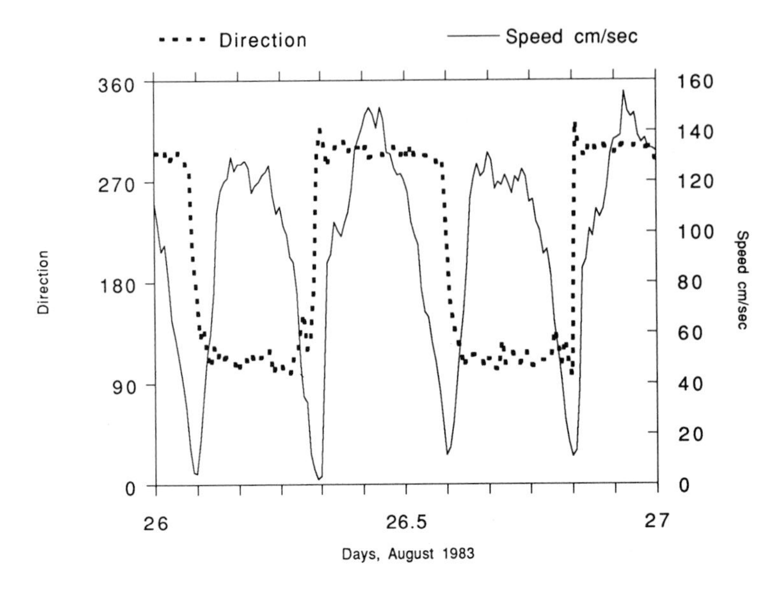

Fig. 8. Plot of tidal current speed and direction measured at 1.5 m above the bed at a site on a planar sand sheet situated marginal to the Bristol Channel scour zone (site CM2 of Harris & Collins, 1988).

Using 0.5 mm particles (medium sand) together with the current meter data (Fig. 8) yields a mean value of U_s of 0.45 m s^{-1} and an excursion length of about 10 km. Lower spring current speeds for a given area will result in a diminished excursion length, although it should be noted that spring surface current speeds averaged 1.54 ± 0.04 m s^{-1} over the 22 locations studied (Table 1). Similarly, still coarser sediments have an even shorter excursion length.

This length scale (~10 km) sets a *conservative* limit to the size which a BLP must achieve before it will act as a barrier to sand movement; BLPs which contain scour zones of this size will almost certainly act as barriers to transport, although smaller-scaled systems may still act as barriers if the current speed and grain-size conditions are appropriate. In the Bristol Channel, for example, sediment distribution maps reveal an area of exposed bedrock and gravel in the central Channel about 40–50 km in length (Fig. 6). However, along the northern margin sand accumulations are separated by bedrock exposures less than 10 km in length (Harris & Collins, 1988). Thus, the scoured zone of the Bristol Channel acts as a barrier to 0.5 mm sand moving along the central axis of the channel but not along the margins, where existing sand deposits may serve as areas of temporary deposition for a sand grain in transit up- or down-channel. Hence, we classify the Bristol Channel as a *partial* BLP (Table 1).

For many small-scaled tidal systems such as tidal inlets (<1 km in width), the dimensions of the channel are much less than the sand excursion length and the system cannot contain a complete BLP. This may indeed be the case for small-scaled systems like Solway Firth or The Wash (see Fig. 6 and Table 1). In the case of the southern North Sea (BLP No. 6 in Fig. 1B), the divergence coincides with a large area of facies 2, which exhibits a mutually evasive system of sand banks and channels, in which sand may recirculate freely within the zone of maximum bottom stress. Thus the evolution of this BLP may be arrested or slowed, depending on the rate at which sand escapes from this area.

Controls on scour zone development

Factors which might intuitively be considered to influence the evolutionary stage of a scour zone include: (i) the magnitude of the maximum bottom stress (which includes the effects of water depth, current speed and drag coefficients); (ii) the magnitude of the bottom-stress gradient; (iii) the size of the scour zone; and (iv) the available sand supply. The information compiled in Table 1 shows that no correlation is apparent between the stage of development and any one of the first three of these factors. For the 22 areas examined, the boundary between facies 1 and 2 experiences an average maximum bottom stress of 19.7 ± 13.7 dynes cm^{-2}, whereas the boundary between facies 2 and 3 experiences an average maximum bottom stress of

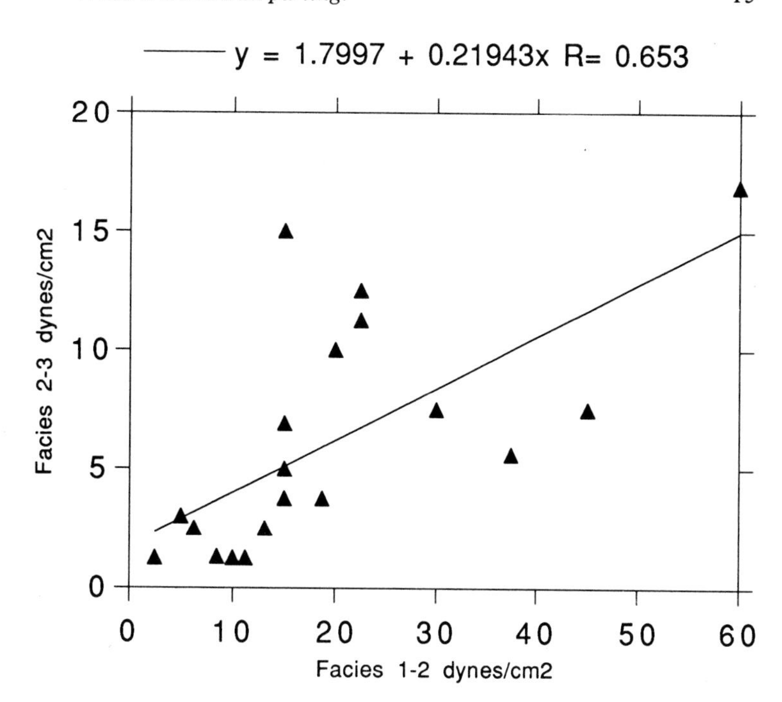

Fig. 9. Plot of maximum bottom stress due to M2/M4 current interaction averaged along facies boundaries 1–2 and 2–3 for 22 different locations (based on the data presented in Figs 2 & 6 and Table 1).

6.0 ± 4.7 dynes cm^{-2}. These mean values are not statistically different (Table 1); however, the magnitude of the maximum bottom stress associated with the boundary between facies 1 and 2 is about five times larger than that for the boundary between facies 2 and 3 in the examples studied (Fig. 9).

If the scour zones are evolving as suggested in Fig. 7, one would expect the facies boundaries to shift with time. In that case there should be no fixed value of shear stress associated with each facies boundary. Instead, the maximum bottom stress associated with a facies boundary should decrease as the evolutionary stage increases (Fig. 10). Such a trend is in fact suggested by the data (Fig. 10), the 1–2 facies boundary occurring at a higher bottom stress for 'immature' mutually evasive scour zones than for 'mature' (complete) scour zones.

Characteristics of scour zones and BLPs

It is worth reviewing the differences between scour zones and BLPs in terms of their governing processes and recognition in both modern and ancient deposits. Scour zones are produced by advective and/or diffusive processes in association with bottom-stress maxima. BLPs, on the other hand, arise only from the advective processes related to asymmetries in the tidal wave (Fig. 5). Scour zones correlate specifically with the lag-gravel/exposed-bedrock endmember of a facies succession. In contrast, BLPs do not correspond to any specific facies and may, in fact, cross facies boundaries as in the case of the Southern Bight of the North Sea, which includes facies 1 and facies 2 along its length (cf. Figs 1B & 6). Recognition of the facies succession in the geological record should enable one to identify scour zones; identification of BLPs requires palaeocurrent data.

The distribution of facies as shown in Fig. 6 is easily mapped from surficial sediment and side-scan sonar data. The facies distribution is related to the distribution of bottom stress; scour zones may be developed in relation to bottom-stress maxima and will evolve through time (e.g. Figs 7 & 10). The distribution of BLPs, on the other hand, is related to the complex interactions of M2 and M4 tidal constituents (Fig. 5). Divergences in maximum bottom-stress vectors produced by one numerical model (Fig. 4) does not permit prediction of BLP occurrence, since divergences in bottom-stress vectors correlate with four locations where an observed divergence in bedform orientation is absent (i.e. North Channel of the Irish Sea #2, Lands End, Channel Islands and The Wash; see Table 2). Such

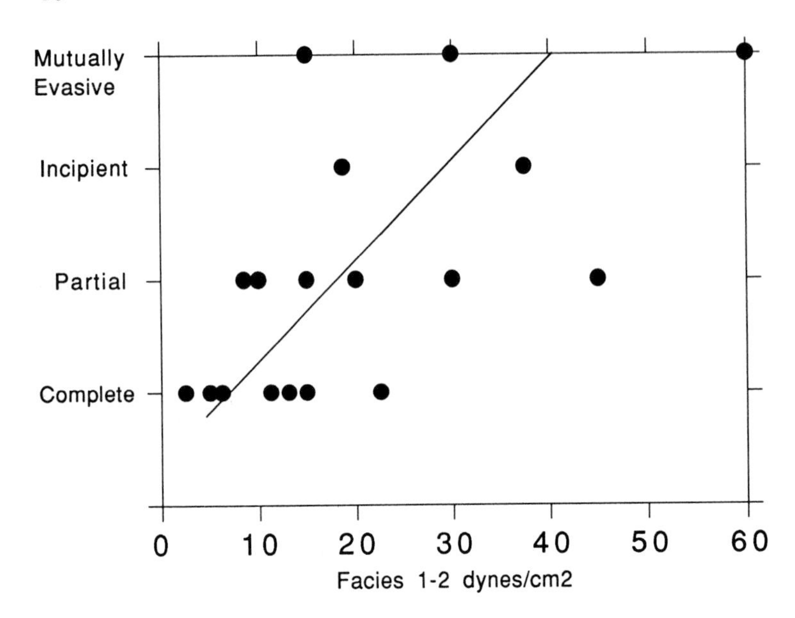

Fig. 10. Plot of maximum bottom stress due to M2–M4 current interaction averaged along facies boundaries 1–2 in relation to development stage. The trend line suggests that the facies boundary shifts towards a lower maximum bottom stress as the evolution of scour zones progresses.

an apparent discrepancy between the direction of maximum stress vectors (Fig. 4) and that of observed transport (Fig. 1B) may be explained by a number of factors: (i) invalid input parameters were used in the model (see Johnson *et al.*, 1982, p. 78); (ii) the model does not include tidally induced residuals caused by coastline irregularities and sand banks (Pattiaratchi & Collins, 1987; Harris & Collins, 1991); (iii) the model does not include the effects of non-tidal currents; and (iv) the directions of observed transport as shown schematically in Fig. 1B are incorrect locally, due to insufficient bedform data. Thus, in the case of the west European continental shelf, there is insufficient data to map the distribution of BLPs conclusively, even though this shelf belongs to the best studied marine environments in the world.

CONCLUSIONS

The tidally dominated, western European continental shelf is characterized by sedimentary facies, the organization of which is controlled by the distribution of bed shear-stress maxima and/or the net sediment transport pathways, acting independently or together. Bottom-stress maxima may be induced by either standing wave nodal points (type A) or geomorphic constrictions (type B). Information on surficial sediment deposits associated with 22 different zones of maximum bottom stress suggests a conceptual model, in which evolutionary stages of scour zones and BLPs are linked to relative sand supply. The proposed model represents the sequential development of deposits with mutually evasive transport patterns and extensive scour zones as end members. Hence, facies boundaries do not occur at fixed values of shear stress, but instead evolve with time. The facies distribution is controlled by the local distribution of shear stress and by the stage of evolution.

All 22 of the areas studied show a spatial succession of facies as originally outlined by Stride (1963), in which grain size decreases as the maximum tidal bed shear stress decreases. However, the facies gradient is not necessarily aligned parallel to the axis of tidal flow, thus indicating that diffusive processes may supplement or even replace advection in the development of scour zones. Of the 22 stress maxima examined, 11 are considered to be complete scour zones in terms of their evolutionary stage, based on the facies distribution pattern.

Bedload partings are a special case of scour zone, characterized by a local, tidally-induced bottom shear-stress maxima which coincides with a divergent pattern in observed sand transport paths. In order for a BLP to be classified as 'complete' in terms of its evolutionary stage, it must reach a scale related to the tidal excursion of a sand grain, whereby a conservative length scale of 10 km is suggested. It is hoped that the criteria and definitions of terminology proposed here will facilitate the distinction of BLPs and associated deposits from scour zones and mutually evasive systems in other tidally dominated shelf environments.

ACKNOWLEDGEMENTS

This contribution was prepared as part of a project supported by a grant from the Australian Research Council (Ref. A38830391) awarded to two of us (PTH and CBP). Burg Flemming and two anonymous reviewers are thanked for comments and suggestions on an earlier version of the manuscript.

REFERENCES

ALLEN, J.R.L. (1970) *Physical Processes of Sedimentation.* Allen and Unwin, London, 248 pp.

ASHLEY, G.M. (1990) Classification of large-scale subaqueous bedforms: a new look at an old problem. *J. sediment. Petrol.* **60**, 160–172.

BALSON, P.S. & D'OLIER, B. (1990) Thames Estuary Sheet 49°N–04°W, British Geological Survey, 1:250,000 Series, Sea Bed Sediments and Quaternary Geology. Ordnance Survey, Southampton, UK.

BELDERSON, R.H. & STRIDE, A.H. (1966) Tidal current fashioning of a basal bed. *Mar. Geol.* **4**, 237–257.

BELDERSON, R.H., JOHNSON, M.A. & KENYON, N.H. (1982) Bedforms. In: *Offshore Tidal Sands, Processes and Deposits* (Ed. Stride, A.H.), pp. 27–57. Chapman & Hall, London.

BELDERSON, R.H., JOHNSON, M.A. & STRIDE, A.H. (1978) Bed-load partings and convergences at the entrance to the White Sea, USSR, and between Cape Cod and George's Bank, USA. *Mar. Geol.* **28**, 65–75.

BISHOP, P. & JONES, J.W. (1979) Patterns of glacial and post-glacial sedimentation in the Minches. North-West Scotland. In: *The North-West European Shelf Seas: the Sea Bed and the Sea in Motion I. Geology and Sedimentology* (Eds Banner, F.T., Collins, M.B. & Massie, K.S.), pp. 89–194. Elsevier, Amsterdam.

CASTON, G.F. (1976) The floor of the North Channel, Irish Sea, a side-scan sonar survey. *Inst. Geol. Sci.* (U.K.) Rept **76–77.**

CASTON, V.N.D. (1979) The Quarternary sediments of the North Sea. In: *The Northwest and European Shelf Seas: The Sea Bed and the Sea in Motion.* (Eds Banner, F.T., Collins, M.B. & Massie, K.S.), pp. 195–267. Elsevier, Amsterdam.

CROSBY, A. (1983) Portland Sheet Sheet 50°N–02°W, British Geological Survey, 1:250,000 Series, Sea Bed Sediments and Quarternary Geology. Ordnance Survey, Southampton, UK.

CROSBY, A., GRAHAM, C.C., RUCKLEY, N.A. & PANTIN, H.M. (1987) Sea Bed Sediments Around the United Kingdom (North and South Sheets) British Geological Survey, Nat. Environ. Res. Council.

DALRYMPLE, R.W. (1992) Tidal depositional systems. In: *Facies Models* (Eds Walker, R.G. & James, N.P.), pp. 195–218. Geol. Ass. Canada, Toronto, 409 pp.

DIETRICH, G. (1950) Die natürliche Region der Nord- und Ostsee auf hydrographischer Grundlage. *Kieler Meeresforsch.* **7**, 35–69.

EISMA, D. (1973) Sediment distribution in the North Sea in relation to marine pollution. In: *North Sea Science* (Ed. Goldberg, E.D.), pp 131–150. MIT Press, New York.

EISMA, D. (1981) Supply and deposition of suspended matter in the North Sea. *Spec. Publs. int. Ass. Sediment.* **5**, 415–428.

EVANS, D. (1985) Clyde Sheet 55°N–06°W, British Geological Survey, 1:250,000 Series, Sea Bed Sediments and Quaternary Geology. Ordinance Survey, Southampton, UK.

EVANS, D., RUCKLEY, N.A., MCELVANNEY, E.P. *et al.* (1986) Malin Sheet 55°N–08°W, British Geological Survey, 1:250,000 Series, Sea Bed Sediments and Quaternary Geology. Ordnance Survey, Southampton, UK.

EVANS, G. & COLLINS, M.B. (1976) The transportation and deposition of suspended sediments over the intertidal flats of the Wash. In: *Nearshore Sediment Dynamics and Sedimentation* (Eds Hails, J.B. & Carr, A.P.), pp. 273–306. John Wiley, London.

FLEMMING, B.W. (1988) Pseudo-tidal sedimentation in a non-tidal shelf environment (southeast African continental margin). In: *Tide-influenced Sedimentary Environments and Facies* (Eds De Boer, P.L., Van Gelder, A. & Nio, S.D.), pp. 167–180. Reidel, Dordrecht.

GRAHAM, C. (1984) Peterhead Sheet 57°N–2°W, British Geological Survey, 1:250,000 Series, Sea Bed Sediments and Quaternary Geology. Ordnance Survey, Southampton, UK.

HAMBLIN, R.J.O. (1989) Dungeness–Boulogne Sheet 50°N–00°W, British Geological Survey, 1:250,000 Series, Sea Bed Sediments and Quarternary Geology. Ordnance Survey, Southampton, UK.

HARDISTY, J. (1983) An assessment and calibration of formulations for Bagnold's bedload equation. *J. sediment. Petrol.* **53,** 1007–1010.

HARRIS, P.T. (1988) Large scale bedforms as indicators of mutually evasive sand transport and the sequential infilling of wide-mouthed estuaries. *Sediment. Geol.* **57**, 273–298.

HARRIS, P.T. & COLLINS, M.B. (1988) Estimation of annual bedload flux in a macrotidal estuary, Bristol Channel, UK. *Mar. Geol.* **83,** 237–252.

HARRIS, P.T. & COLLINS, M.B. (1991) Sand transport in the Bristol Channel: bedload parting zone or mutually evasive transport? *Mar. Geol.* **101**, 209–216.

Harrison, D.J. (1989a) Guernsey Sheet 49°N–04°W, British Geological Survey, 1:250,000 Series, Sea Bed Sediments and Quaternary Geology. Ordnance Survey, Southampton, UK.

HARRISON, D.J. (1989b) Farne Sheet 55°N–02°, British Geological Survey, 1:250,000 Series, Sea Bed Sediments and Quaternary Geology. Ordnance Survey, Southampton, UK.

JAMES, J.W.C. (1988) Cardigan Bay Sheet 52°N–06°W, British Geological Survey, 1:250,000 Series, Sea Bed

Sediments and Quaternary Geology. Ordnance Survey, Southampton, UK.

JAMES, J.W.C. (1990) Anglesey Sheet 53°N–08°W, British Geological Survey, 1:250,000 Series, Sea Bed Sediments and Quaternary Geology. Ordnance Survey, Southampton, UK.

JOHNSON, H.D. & BALDWIN, C.T. (1986) Shallow siliciclastic seas. In: *Sedimentary Environments and Facies*, 2nd edn. (Ed. Reading, H.G.), pp. 229–282. Elsevier, New York.

JOHNSON, M.A., KENYON, N.H., BELDERSON, R.H. & STRIDE, A.H. (1982) Sand transport. In: *Offshore Tidal Sands, Processors and Deposits* (Ed. Stride, A.H.), pp. 58–94. Chapman & Hall, London.

KENYON, N. & STRIDE, A.H. (1970) The tide-swept continental shelf sediments between the Shetland Islands and France. *Sedimentology* **14**, 159–173.

KENYON, N.H., BELDERSON, R.H., STRIDE, A.H. & JOHNSON, M.A. (1981) Offshore tidal sandbanks as indicators of net sand transport and as potential deposits. *Spec. Publs. int. Ass. Sediment.* **5**, 257–268.

LARSONNEUR, C., BOUYSSE, P. & AUFFRET, J.P. (1982) The superficial sediments of the English Channel and its western approaches. *Sedimentology* **29**, 851–864.

LAWSON, M.J. & HAMBLIN, R.J.O. (1989) Wight Sheet 50°N–02°W, British Geological Survey, 1:50,000 Series, Sea Bed Sediments and Quarternary Geology. Ordinance Survey, Southampton, UK.

LEEDER, M.R. (1982) *Sedimentology: Process and Product.* George Allen and Unwin, London, 344 pp.

MALIKIDES, M., HARRIS, P.T. & TATE, P.M. (1989) Sediment transport and flow over sandwaves in a nonrectilinear tidal environment: Bass Strait, Australia. *Continent. Shelf Res.* **9**, 203–221.

MILLER, M.C., MCCAVE, I.N. & KOMAR, P.D. (1977) Threshold of sediment motion under unidirectional currents. *Sedimentology* **24**, 507–527.

OWENS, R. (1981) Holocene sedimentation in the northwestern North Sea. *Spec. Publs int. Ass. Sediment.* **5**, 303–322.

PANTIN, H.M. (1991) The sea-bed sediments around the United Kingdom. *British Geol. Surv. Res. Rept* **SB/90/1**, 47 pp.

PANTIN, H.M. & EVANS, C.D.R. (1984) The Quaternary history of the central and southwestern Celtic Sea. *Mar. Geol.* **57**, 259–293.

PATTIARATCHI, C.B. & COLLINS, M.B. (1984) Sediment transport under waves and tidal currents: a case study from the northern Bristol Channel, U.K. *Mar. Geol.* **56**, 27–40.

PATTIARATCHI, C.B. & COLLINS, M.B. (1987) Mechanisms for linear sandbank formation and maintenance in relation to dynamical oceanographic observations. *Progr. Oceanogr.* **19**, 117–176.

PINGREE, R.D. & GRIFFITHS, D.K. (1979) Sand transport paths around the British Isles resulting from M2 and M4 tidal interactions. *J. Mar. Biol. Assoc. U.K.* **59**, 497–513.

Ruckley, N.A. & Chesher, J.A. (1987) Caithness Sheet 58°N–04°W, British Geological Survey, 1:250,000 Series, Sea Bed Sediments and Quaternary Geology. Ordnance Survey, Southampton, UK.

STERNBERG, R.W. (1972) Predicting initial motion and bedload transport of sediment particles in the shallow marine environment. In: *Shelf Sediment Transport* (Eds Swift, D.J.P., Duane, D.B. & Pilkey, O.H.), pp. 61–82. Dowden, Hutchinson & Ross, Stroudsburg.

STRIDE, A.H. (1963) Current swept sea floors near the southern half of Great Britain. *Q. J. Geol. Soc. London* **119**, 175–199.

STRIDE, A.H. (1973) Sediment transport by the North Sea. In: *North Sea Science* (Ed. Goldberg, E.D.), MIT Press, Cambridge, MA.

STRIDE, A.H. & BELDERSON, R.H. (1991) Sand transport in the Bristol Channel east of Bull Point and Worms Head: a bedload parting model with some indications of mutually evasive sand transport paths. *Mar. Geol.* **101**, 204–208.

TAPPIN, D.R. (1983) Lundy Sheet 52°N–06°W, British Geological Survey, 1:250,000 Series, Sea Bed Sediments and Quaternary Geology. Ordnance Survey, Southampton, UK.

VASLET, D., LARSONNEUR, C. & AUFFRET, J.P. (1978) Map of the surficial sediments of the English Channel, 1:500,000. BRGM, Orleans, France.

WIBERG, P.L. & SMITH, J.D. (1985) A theoretical model for saltating grains in water. *J. geophys. Res.* **90 C4**, 7341–7354.

WINGFIELD, R.T.R. (1983) Lake District Sheet 54°N–04°W, British Geological Survey, 1:250,000 Series, Sea Bed Sediments and Quaternary Geology. Ordnance Survey, Southampton, UK.

WINGFIELD, R.T.R. (1985) Isle of Man Sheet 54°N–08°W, British Geological Survey, 1:250,000 Series, Sea Bed Sediments and Quaternary Geology. Ordnance Survey, Southampton, UK.

Spec. Publs int. Ass. Sediment. (1995) **24**, 19–32

Hydraulic roughness of tidal channel bedforms, Westerschelde estuary, The Netherlands

J.H. VAN DEN BERG, N.E.M. ASSELMAN *and* B.G. RUESSINK

Institute for Marine and Atmospheric Research, Utrecht University, 3508 TC Utrecht, The Netherlands

ABSTRACT

Resistance to flow was studied in a 6–7-m deep straight tidal channel of the Westerschelde estuary. For this purpose long-term current velocity and water level measurements were carried out at three locations. Vertical current and sand concentration profiles were also measured. Bed morphology was characterized by 0.2–0.4-m high dunes which showed complete reversal of asymmetrical shape during a spring tidal cycle.

Hydraulic roughness was calculated in three ways: (i) by solving the one-dimensional momentum equation; (ii) by fitting log-normal regressions to vertical current velocity profiles; and (iii) by fitting regressions to best-fit Rouse distributions of vertical sand concentration profiles.

Consistent values of hydraulic roughness were obtained only by the momentum equation. During a tidal cycle the values of hydraulic roughness varied by one order of magnitude. The data indicate an inverse relationship between the Nikuradse roughness parameter k_s and the water depth. This conforms to findings of an earlier study in the Conwy estuary, North Wales. Also, an inverse relationship between k_s and flow velocity is evident. At maximum tidal flow the k_s-value is equal to about half the dune height.

INTRODUCTION

Hydraulic roughness is one of the key parameters in physical models that simulate the velocity and direction of currents in rivers, estuaries and shallow seas. In the case of sand as bed material, the roughness is caused by turbulence behind bedforms. The roughness value is generally considered to be a function of the amplitude, shape and frequency of vortices near the bed behind dunes and ripples. If bedforms are adapted to the flow, the roughness properties of these wakes may be converted to bedform parameters such as bedform height (H), bedform length (λ), and bedform steepness (H/λ). Several functions are proposed that predict the bedform roughness as a function of such bedform parameters (e.g. Vanoni & Hwang, 1967; Wooding *et al.*, 1973; Engelund, 1977; van Rijn, 1984, 1989).

The various hydraulic roughness parameters found in the literature can in most cases be converted into one another. In the present study the well-known Nikuradse roughness length, k_s, is used. This is a convenient measure, as it is expressed in metres and can therefore be easily related to bedform dimensions.

In mathematical models simulating natural flows, a constant value of bedform roughness is generally assumed. Such a constant value may have some physical justification in the case of rivers, in which bedform morphology has reached a stable dynamic equilibrium in the course of prolonged periods of relatively steady flow. However, any change in the flow conditions will automatically result in bedform adaptation and thereby induce a change in roughness. Furthermore, substantial time lags between shape characteristics of (large) bedforms and instantaneous flow conditions may occur (Allen, 1976). Thus, even at times of steady discharge, significant changes in bed roughness may occur.

In tidal flows the relationship between bedform shape and roughness becomes much more complicated. The constant, rapid change in velocity and direction of the flow results in a continuous lag in the adjustment of dune shape to the instantaneous flow conditions. The shape of the wakes behind dunes will therefeore not be a simple function of just one or two bedform parameters. As a result of this, the predictors of bedform roughness outlined above will not be expected to perform well.

In this paper the results of a measurement campaign devoted to the study of hydraulic roughness in a tidal channel characterized by dune bedforms are presented. The study area is located along an almost straight channel section of the Westerschelde estuary (Fig. 1). The tidal range at this location averages about 4.3 m. Spring and neap tidal ranges average 5.2 m and 3.3 m respectively. The bed material consists of well sorted, fine siliciclastic sand with median diameters of 200–300 μm.

METHODS

Background information

There are several ways to calculate the hydraulic roughness in natural flows. The most straightforward method is the direct measurement of roughness by solving the momentum equation. In this case the study reach consisted of a straight section of a natural channel with a very large width–depth ratio and a very small bottom slope in the flow direction. Density gradients were virtually absent. Under such conditions, and during periods of low to moderate wind velocity, the momentum equation may be simplified to the one-dimensional function

$$\frac{\delta u}{\delta t} + u\left(\frac{\delta u}{\delta x}\right) + \frac{g}{hC^2}(u\,|\,u\,|) - gi = 0 \qquad (1)$$

where u = depth-averaged flow velocity (m s^{-1}), x = longitudinal distance (m), h = mean water depth of the channel section (m), g = acceleration of gravity (m s^{-2}), i = downcurrent water surface slope (–), and C = Chézy roughness coefficient (m$^{1/2}$ s^{-1}) which can be converted to the Nikuradse roughness length parameter, k_s, according to the formula of White–Colebrook and by implicitly assuming the validity of the mixing length theory of Prandtl:

$$C = 18 \log \frac{12h}{k_s} \qquad (2)$$

In the case of steady and uniform river flow, equation 1 further reduces to the well-known Chézy function:

$$i = \frac{u^2}{hC^2} \qquad (3)$$

Apart from this direct calculation, the hydraulic roughness can also be estimated by two indirect procedures. These comprise regression analyses of either the vertical current velocity profile or the vertical distribution of suspended bed material, assuming the validity of theoretical flow or concentration profiles.

The velocity profile method is based on the validity of the Kármàn–Prandtl expression for hydraulically rough flow:

$$u_z = \frac{u_*}{\kappa} \ln \frac{z}{z_0} \qquad (4)$$

where u_z = velocity at height z above the bed, u_* = shear velocity, κ = Kármàn constant, z_0 = roughness length (height z above the bed at which $u_z = 0$); z is converted to k_s by

$$z_0 = 0.033k_s \qquad (5)$$

The concentration profile method is based on the validity of the theoretical distribution proposed by Rouse (1937):

$$\frac{C_z}{C_a} = \left[\frac{h-z}{z}\,\frac{a}{h-a}\right]^Z \qquad (6)$$

with

$$Z = \frac{w_s}{\beta\kappa u_*} \qquad (7)$$

where

$$\beta = 1 + 2\left[\frac{w_s}{u_*}\right]^2 \qquad (8)$$

and, according to Zanke (1977)

$$w_s = 10\frac{\nu}{D_s}\left[\left(1 + \frac{0.01\,(\rho_s - \rho)gD_s^3}{\rho\nu^2}\right)^{0.5} - 1\right] \qquad (9)$$

in which C_z, C_a = suspended sand concentration at distance z, above the bed (kg m^{-3}), a = reference height above the bed (m), Z = suspension number (–), w_s = particle fall velocity (m s^{-1}), u_* = bed shear velocity (m s^{-1}), β = sediment mixing coefficient (–), D_s = representative grain size of the suspended sand (m), ρ = the fluid density, and g = acceleration due to gravity.

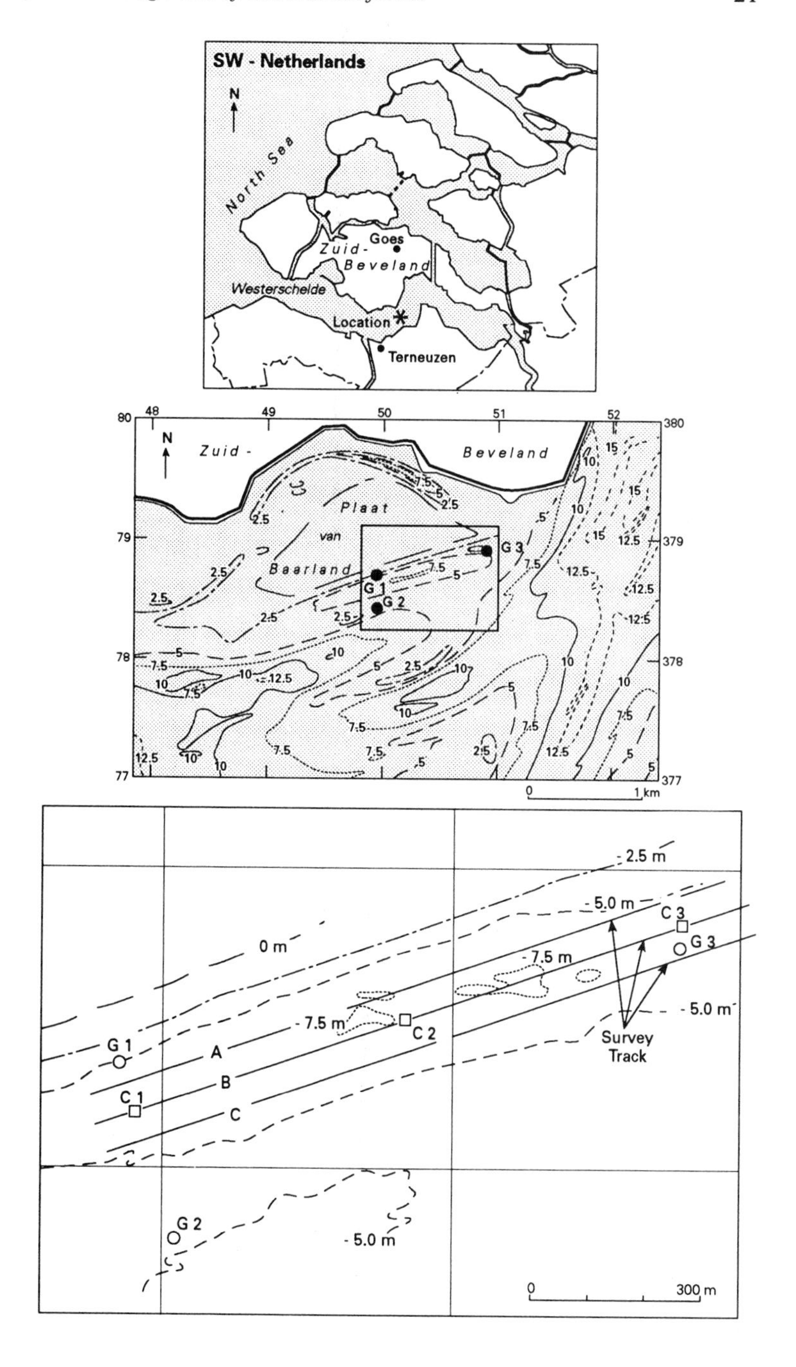

Fig. 1. Location of study area, measuring points and sounding tracks.

The literature contains a number of predictors by which bedform roughness can be deduced from geometrical properties of the bedforms. In most of the available equations the properties involved are the height and length of the bedforms. In the case of tidal environments, the bedforms continuously adapt to the velocity and direction of the tidal flow. This may eventually result in a complete reversal of bedform orientation during a flood or ebb current phase. As a result, the shape of tidal bedforms may vary between strongly asymmetrical to more or less symmetrical forms. In the present analysis, a bedform roughness predictor proposed by Van Rijn (1989) is used which is the only predictor that to some extent considers the influence of bedform asymmetry on roughness. In the case of subaqueous dunes, the Van Rijn roughness predictor reads as follows:

$$k_s = 1.1\gamma\Delta(1 - e^{-25(\Delta/\lambda)}) \quad (10)$$

in which Δ = dune height (m), γ = form factor (–) and λ = dune length (m). The form factor is a function of a leeside slope parameter $\gamma = f(\cot g\phi)$, with ϕ = average leeside slope angle (measured from the base of the trough to the summit of the dune), based on flume experiments by Ogink (1989). In these experiments, both the roughness of artificial triangular dunes with a frontal slope as steep as the natural avalanche leeside slope as well as the roughness of dunes with a much smaller inclination of the frontal slope was studied. The effective form roughness of the dunes with the small frontal slope appeared relatively small. According to Klaassen *et al.* (1986), this effect is related to the gradual disappearance of the wake behind the dunes as the angle of inclination decreases.

Field measurements

To obtain a good data base for calculating the hydraulic roughness, two field campaigns were carried out, the first from 6 to 29 May 1991, the second from 1 to 15 May 1992. The same instrument configuration was used in both cases.

Water levels were measured at three tide gauges in the tidal channel (G1, G2 and G3; see Fig. 1), and used to calculate the water surface slope. Current velocities and current directions were obtained at three locations using three FLACHSEE self-recording instruments (C1, C2 and C3; see Fig. 1) located 2.5 m above the bed. The tide gauges and the current meters measured every 10 min. All instruments functioned well during the first field campaign, but the tide gauge G1 and the current meter C3 were inoperative during the second field campaign.

The bedforms in the channel were measured by means of echo-soundings along survey tracks as well as from a fixed position. Surveys were executed every 3 or 4 days along predefined sounding tracks (Fig. 1), using an automated TRIDENT positioning system (accuracy 1–3 m). In the fixed station mode, a rotating echosounder (profiler) attached to the tide gauge G3 was used to obtain a more continuous record of the bedforms and their temporal behaviour. The bed was scanned over a distance of about 25 m along a line parallel to the axis of the channel. The profiler unfortunately did not produce the desired results for two reasons. One, the scanner was mounted 1.5 m away from the tide gauge pole but this was apparently not far enough to scan beyond the local erosional scour hole around the pole; two, the signal of the profiler was quite sensitive to the presence of suspended sand particles. As a result, clear registrations of the bed profile were obtained only during slack water periods (see also Fig. 7).

Flow velocity and sand concentration profiles were measured near each FLACHSEE current meter position for 13 h on 8 May (neap tide), 16 May (spring tide) and 24 May 1991 (mean tide) using ship-mounted measuring instruments. The current velocities and current directions were measured by means of an OTT-propeller current meter in combination with a giro-compass mounted in the same streamlined carrier. The sand concentration profiles were obtained by using an acoustic sand transport meter (Van den Berg, 1984). The flow velocity and sand concentration profiles were obtained within 12 min by measuring successively over 2 min at five different elevations above the bed, the lowest elevation being 0.2 m above the bed. The data were not corrected for flow accelerations occurring during the measuring period. However, the analysis of other measurements taken in the area indicated that, except for the period around slack water when flow acceleration may be relatively large, errors in the calculation of the depth-averaged flow velocity and, therefore, of k_s remained rather small. The current profiles were used to convert the point measurements of the FLACHSEE instruments to depth-averaged values. A comparison of the depth-averaged flow data with the results of the FLACHSEE instruments showed that deviations were generally small and random (Fig. 2). No corrections were therefore made to the depth-averaged flow data.

RESULTS

The data collected in the course of the two field campaigns allowed the calculation of k_s-values according to the three methods outlined above. The hydraulic roughness values obtained by log-normal regression of the vertical current velocity profiles are shown in Fig. 3. Correlations between measured and fitted velocity profiles are moderate for meas-

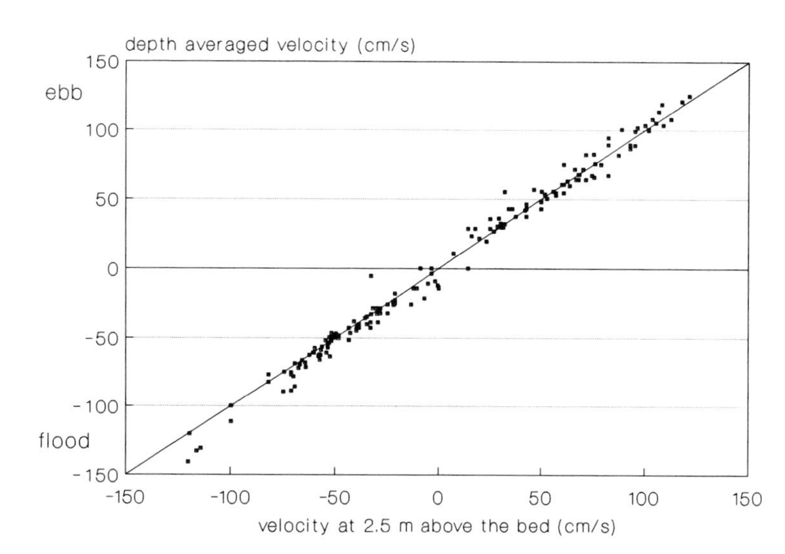

Fig. 2. Current measurements at 2.5 m above the bed compared to depth-averaged flow velocities at the same time and location.

urements performed around slack water, but good for measurements made at maximum flow. The calculated roughness values vary over a tidal cycle. During the ebb phase, k_s-values were found to be larger than during the flood phase. Fluctuations of the k_s-value in the course of one ebb or flood period are the result of local differences in velocity distributions over the bed topography.

In calculating the hydraulic roughness from vertical sand concentration profiles, the concentration at the lowest sampling point (0.15–0.20 m above the bed) was taken as the reference concentration (C_a). At spring tide, C_a-values ranged from zero at slack water to 0.6 kg m^{-3} at the maximum flood current. The method proved very sensitive at the sediment fall velocities used in this study. Unfortunately, it was not possible to measure directly the size of the suspended sediment in the field. Instead, the particle fall velocity was estimated at 0.02 m s^{-1} using equation 9, which corresponds to the diameter of the bed material (200–300 μm). A value of 1 was taken for the β-coefficient. The hydraulic roughness values calculated in this way are presented in Fig. 4. The vertical concentration profiles correspond reasonably well to the Rouse distribution, i.e. the correlation coefficients for the measured and fitted profiles range between 0.8 and 0.95. Again the calculated roughness values were found to fluctuate strongly in the course of a tidal cycle, with values ranging between 1 and 50 m. The k_s-values are lower during the ebb phase (positive values of the current velocity) than during the flood phase, the roughness values being 5–10 times larger than those obtained by log-normal regression of current velocity profiles.

The hydraulic roughness values calculated from the momentum equation are shown in Fig. 5. During both measurement campaigns, the hydraulic roughness values show the same general development over a tidal cycle. They are smaller during the ebb than during the flood. After high water slack, the hydraulic roughness decreases from > 5 m to about 0.1–0.25 m in 1991 and to 0.45 m in 1992. About 2 h before low water slack, the hydraulic roughness slowly increases again. After the turn of the tide it remains at a relatively high value of 0.8–1.0 m at first. Two hours before high water slack, k_s decreases again to reach a minimum value of 0.05 m 1 h before the high water flow reversal. This lowest roughness value is reached about 20 min after maximum current velocities are reached and maximum water level slopes are recorded.

In general, roughness values at mean tide and spring tide are lower than during neap tide (Fig. 6) and observed differences are larger during the flood than during the ebb. However, during the largest part of the ebb period no systematic differences are noted.

Owing to the poor quality of the scanner data, our original aim of correlating the temporal behaviour of hydraulic roughness over a tidal cycle or over a neap–spring tidal cycle with bedform morphology was not achieved. Instead, only the results of the echo-sounding surveys were available. From these, an average height and length of the leeside slope

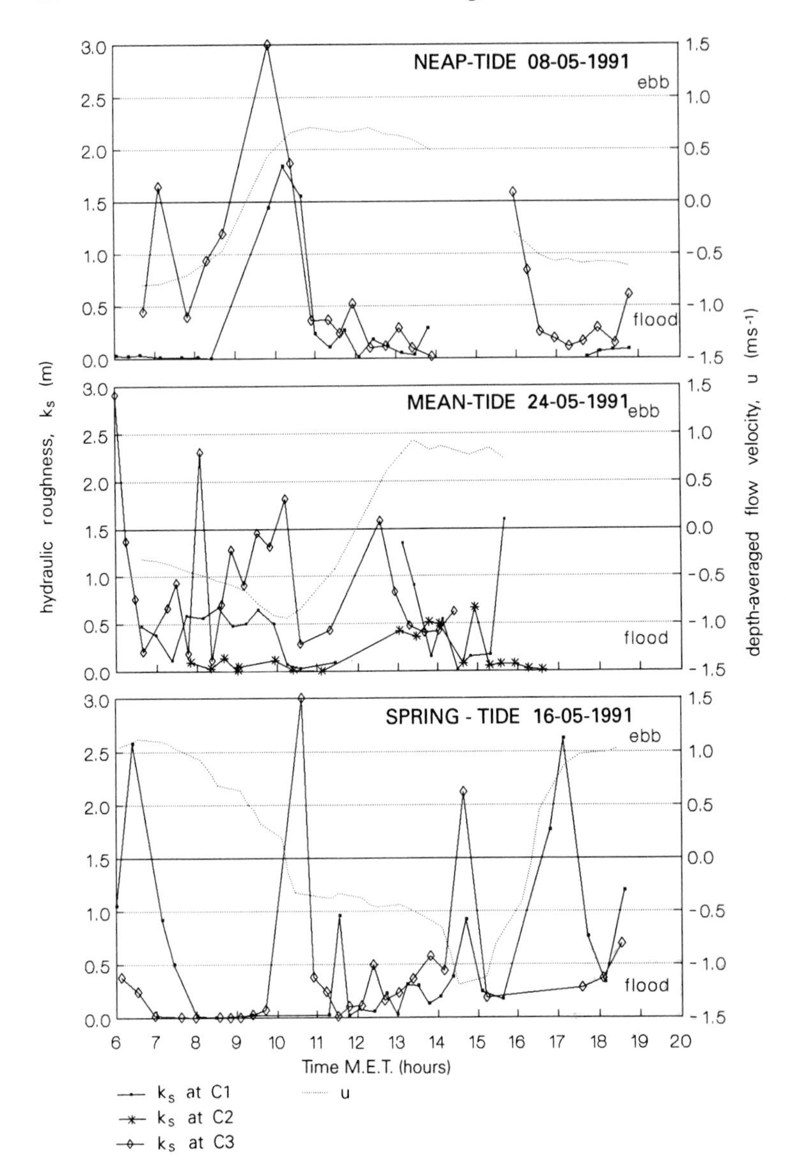

Fig. 3. Nikuradse roughness length (k_s) calculated from vertical current velocity profiles. Depth-averaged flow velocities (u) refer to location Cl (see Fig. 1).

(L1) and the stoss-side slope (L2) of about 30 dunes in each survey line were measured. The results are shown in Table 1. The predicted k_s-values range between 0.04 and 0.27 m.

DISCUSSION

Reliability of the methods

The reliability of the methods used for calculating hydraulic roughness is related to the validity of the assumptions involved and to the accuracy of the data needed for computation. As far as the assumptions are concerned, the momentum equation has the advantage of providing a direct method, the roughness being directly related to the energy gradient. Errors in k_s-values computed by this method will mainly reflect systematic errors in the calculation of a mean water depth due to pronounced irregularities in the mean bed elevation of the channel section. The influence of a systematic error in the mean water depth on the roughness values is large (an increase in water depth of about 1 m

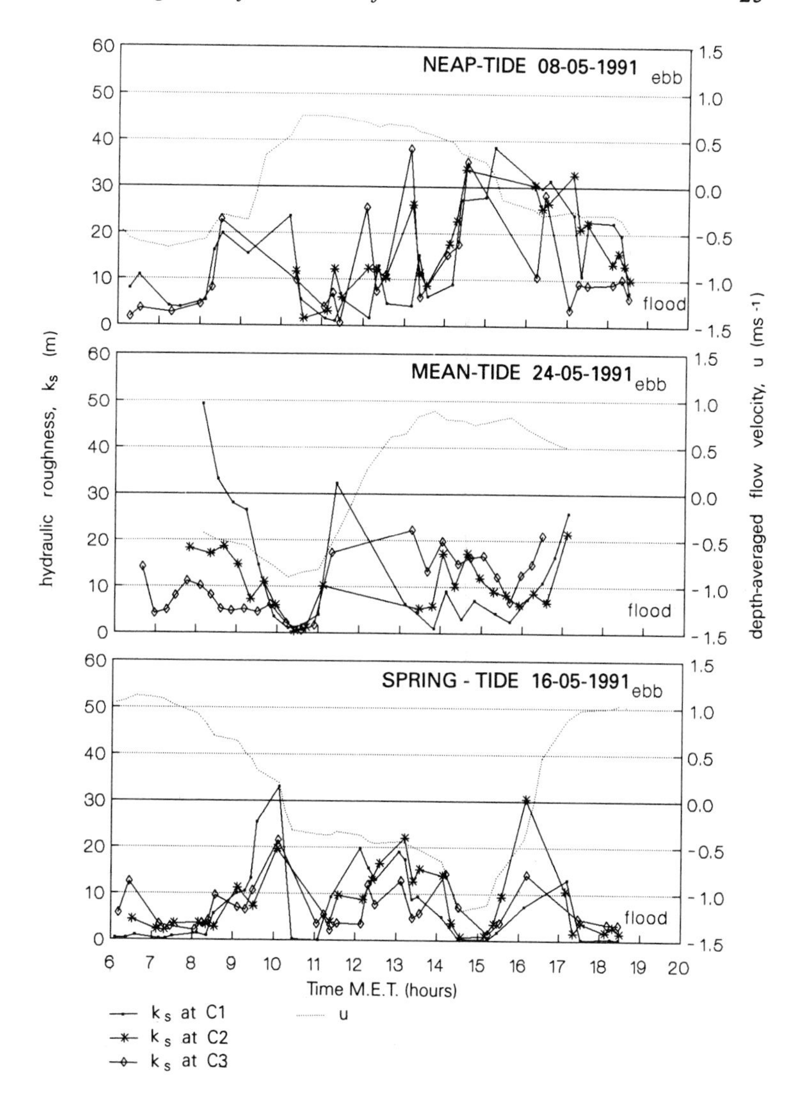

Fig. 4. Nikuradse roughness length (k_s) calculated from vertical sand concentration profiles. Depth-averaged flow velocities (u) refer to location C1 (see Fig. 1).

results in a decrease in the corresponding k_s-value by a factor of 2). However, it will not alter the temporal pattern of relative change of the roughness parameter. It may thus be concluded that the apparent differences in calculated roughness between1991 and 1992 are probably the result of errors in the calculated mean bed elevation.

In contrast to the momentum equation approach, the reliability of k_s-values obtained using vertical current velocity profile data is expected to be relatively small because the conditions at the study site were far from ideal for its application. The method requires a flat bed as well as uniform and steady flow conditions, neither condition being fulfilled in the present case. Furthermore, the shape of the current profile will depend on its position with respect to the dune bedform. Especially the shape of the lower part of the current profile is affected by the presence of bedforms. As this method considers most of all the lower part of the current profile, it is likely to produce relatively large errors in the calculation of k_s-values under the ambient conditions of this study.

The same drawbacks apply to the vertical sand concentration profiles, thus also rendering the results of the Rouse-curve-fit method unreliable. It must be emphasized that the pre-condition of steady and uniform flow applies even more strictly

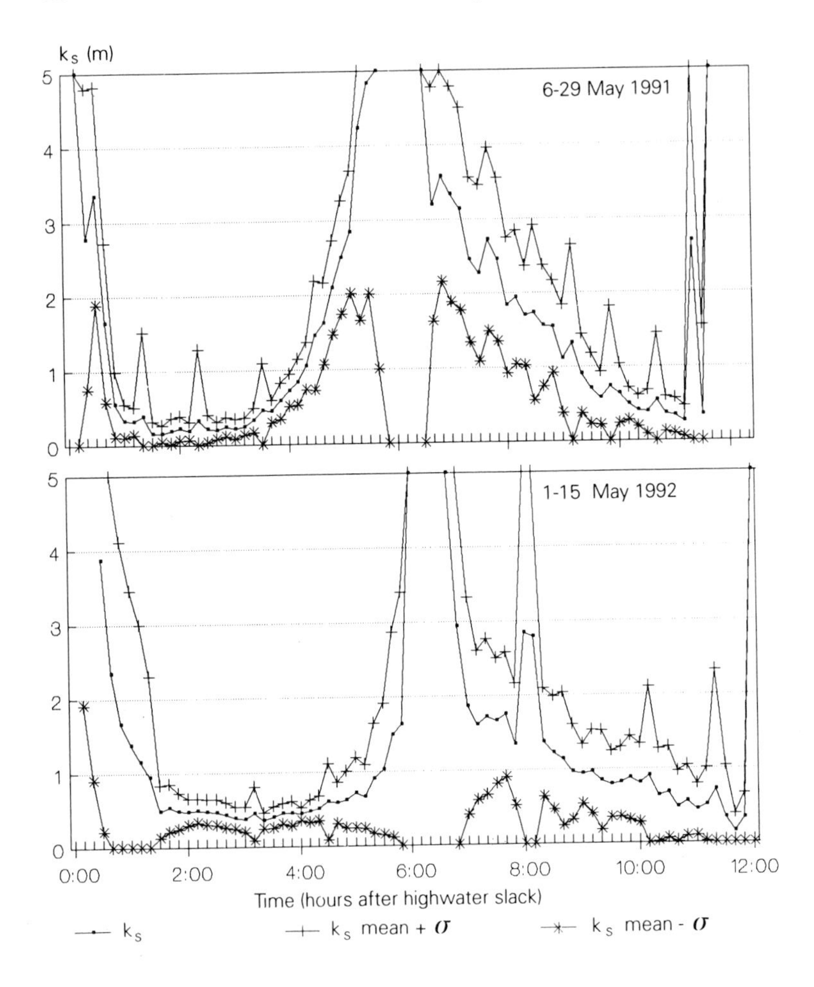

Fig. 5. Hydraulic roughness values obtained using the momentum equation averaged over the measuring periods.

here, since processes of adaptations in the suspended sand concentrations are superimposed on changes in the flow field. Several additional sources of errors, related to the size of the suspended sand, can be identified. Thus, the size of the suspended sand used in the calculation was represented by a single grain size, estimated from the grain-size distribution of the bed material. It is possible that this size deviates significantly from the mean grain size of the suspended sand. The use of a single grain size in the computation also ignores the influences of grain size segregation effects on the shape of the vertical concentration profile.

It is thus clearly evident that k_s-values obtained by the Rouse-curve-fit procedure are the least reliable and that the best results are obtained by the application of the momentum equation. This conclusion is supported by the chaotic nature of temporal changes in k_s-values obtained by the curve-fit methods (compare Figs 3 & 4 with Figs 5 & 6). The large scatter obtained by the latter methods conforms to findings of earlier attempts to calculate roughness from the shape of vertical current profiles in tidal settings with dune bedforms (Sternberg, 1972; McCave, 1971; Ludwick, 1974).

The bedform height (H) and length (λ) in the channel section investigated in this study remained rather constant, both in space and in time (see Table 1). After the turn of the tide the bedforms adapt to the opposing current and the bedform asymmetry is gradually reversed. During spring tide this process ends in a complete reversal of dune orientation during both the ebb and the flood current phases (Fig. 7). It is self-evident that application of equation 10 in the case of a gradual reversal of bedform asymmetry will result in in-

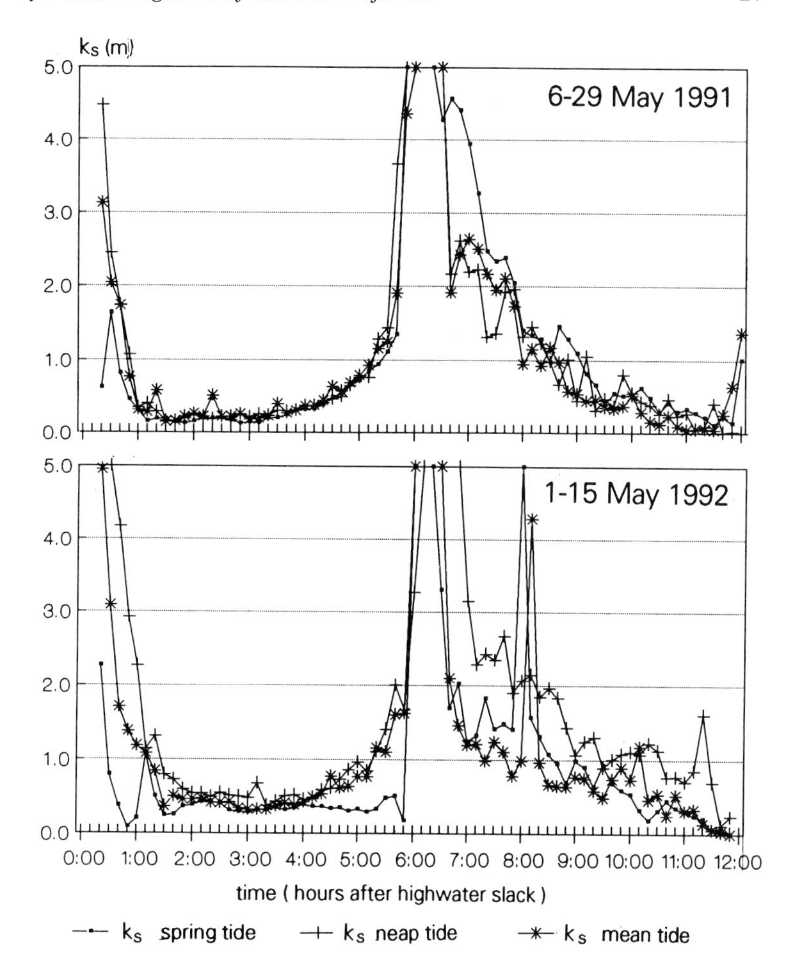

Fig. 6. Hydraulic roughness values obtained using the momentum equation during spring tide, neap tide and mean tide.

creasing hydraulic roughness during the ebb or flood stage. However, such a behaviour is not supported by the results using the momentum equation (see Fig. 6). During all the sounding surveys the dunes presented the asymmetrical shape obtained during the preceding maximum flow stage. It is thus reasonable to assume that the values of the bedform parameters used in equation 10 (as measured from the echo-sounder profiles) do not deviate significantly from the situation of the preceding maximum flow. In Fig. 8, the k_s-values listed in Table 1 are compared with roughness values obtained by the momentum equation from the corresponding maximum flow. The roughness values derived by the two methods are within the same order of magnitude but are not well correlated because the calibration of the form factor (γ) in equation 10 is as yet based on rather limited flume data derived from small and artificial triangular dunes.

Comparison with Conwy estuary data

The study of Knight (1981) in the Conwy estuary (North Wales) is, to our knowledge, the only other case in which the flow resistance of a tidal channel was analysed using the momentum equation. His findings are in agreement with those presented here (Fig. 9). Thus, the roughness parameter k_s appears to be inversely related to the water depth also in the Conwy estuary. It should be noted that, because the high roughness values computed at relatively large water depths relate to conditions of low current velocity near slack water, they are unreliable and have therefore been omitted from this comparison. It is stressed that the inverse relationship of rough-

Table 1. Average values of dune parameters in the three survey profiles A, B and C (see Fig. 1)

	Time	H (m)	L1 (m)	L2 (m)	γ (–)	k_s (m)
May 14 1991						
Line A	01h20 after LW	0.30	2.42	4.48	0.36	0.08
B	01h30 after LW	0.27	2.07	4.63	0.50	0.09
C	01h50 after LW	0.33	2.31	5.39	0.62	0.15
May 17 1991						
Line A	03h00 after HW	0.35	1.18	4.75	0.90	0.27
B	02h50 after HW	0.31	1.08	4.33	0.90	0.23
C	02h20 after HW	0.40	1.60	6.40	0.85	0.27
May 21 1991						
Line A	04h10 after HW	0.31	2.20	4.50	0.58	0.14
B	03h50 after HW	0.24	1.60	3.77	0.61	0.11
C	03h30 after HW	0.35	2.82	5.26	0.45	0.12
May 29 1991						
Line A	03h10 after LW	0.34	2.96	4.50	0.31	0.08
B	03h20 after LW	0.29	2.28	3.45	0.46	0.11
C	03h30 after LW	0.34	3.44	5.20	0.25	0.06
May 05 1992						
Line A	02h50 after LW	0.21	2.50	5.80	0.15	0.02
B	02h50 after LW	0.22	1.70	2.10	0.50	0.07
C	02h50 after LW	0.22	2.00	4.10	0.30	0.04

ness with water depth is purely empirical and lacks a theoretical explanation.

Hydraulic roughness and dune shape

All published methods for the prediction of form roughness from geometrical properties of bedforms have in principle been designed for steady flow conditions. In the present case the bedform shape lags behind the change of the tidal flow. This is obviously the case during the reversal of bedform asymmetry after the turn of the tide. The evolution of the local flow pattern over the dunes during this adaptation and its effect on the form roughness is largely unknown. One possible sequence of change in bedform shape and flow pattern has been outlined by Boersma & Terwindt (1981). It illustrates that the flow pattern and the presence of wakes behind a dune during the adaptation process deviates substantially from the dynamic equilibrium under steady flow conditions. By implication, the roughness properties of the bedforms will not be constant but will probably change dramatically during the adaptation period. Variations in hydraulic roughness during the tidal cycle recorded by our measurements must be attributed to this process of adaptation.

During a spring tidal cycle, slightly lower values of the roughness coefficient are reached in comparison with neap tide (see Fig. 6). This is apparently related to higher current velocities at spring tide. The inverse relationship of flow velocity and Nikuradse roughness length is clearly demonstrated in Fig. 10. It may be explained partly by a change in the vortex movement occurring when a well-developed, steep avalanche front of the dunes is re-established. In a companion study, carried out in a flood-dominated intertidal shoal area in the eastern part of the Westerschelde estuary, morphological changes of dune bedforms were studied in greater detail (Van Gelder & Terwindt, 1992). Bedform migration was controlled by the flood current which reached depth-averaged velocities of 1.2 m s^{-1} at spring tide. Sedimentary structures of longitudinal sections of flood-oriented dunes revealed a temporal lowering of the flow separation point (brinkpoint) behind the dune summit during maximum tidal flow (Fig. 11). This is caused by a temporal steepening of the slope between the summit and the brinkpoint of the dune to a maximum

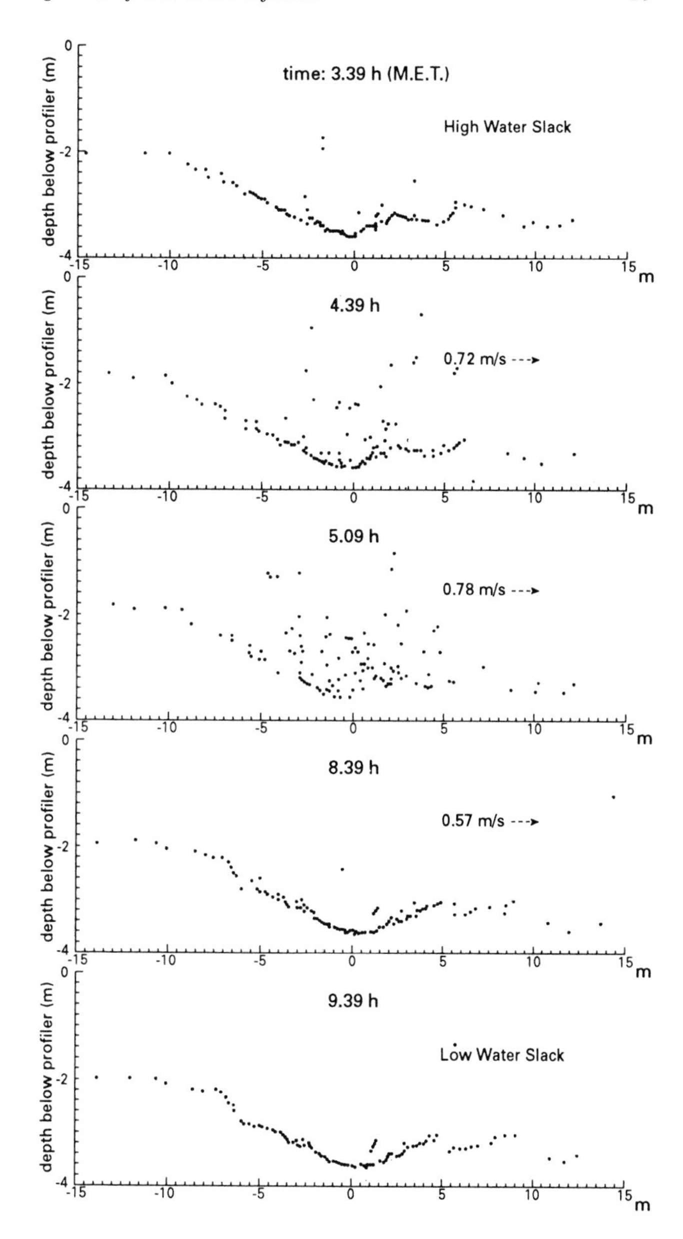

Fig. 7. Time series of echo-sounder records showing complete reversal of dune asymmetry during an ebb period.

of about 8°. This steepening generally results in a straight erosional boundary, truncating the uppermost part of the steep foreset strata deposited in front of the brinkpoint during the preceding stage of the accelerating tidal flow. During the remaining part of the flood period, the slope between the summit and the brinkpoint of the bedform diminishes to about 6°. A similar behaviour of tidal dunes can be observed on illustrations presented by Kohsiek & Terwindt (1981) and Van den Berg (1982). The apparent link between variations in slope angle and flow intensity appears to be analogous to the findings of Achenbach (1968) in his study of wind flow past a circular cylinder. The experiments of the latter investigation indicate that the position of the separation point behind a cylinder oriented transverse to the flow is a function of a cylinder Reynolds number. Whether the slope angle

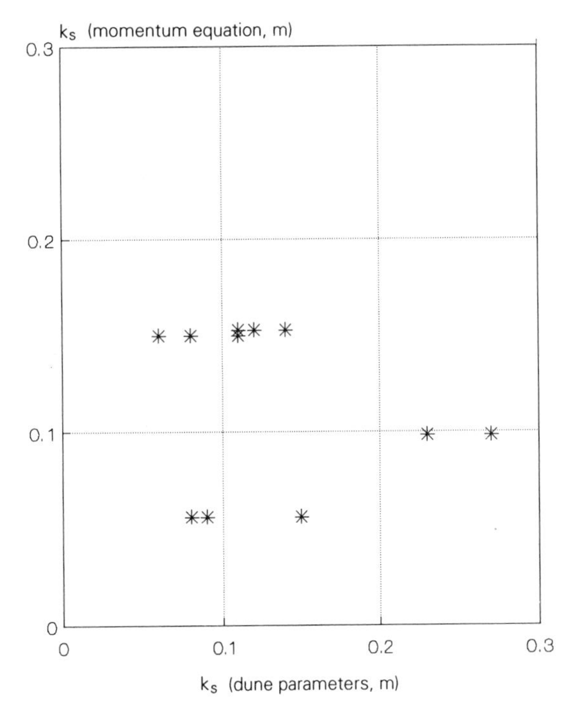

Fig. 8. Comparison of roughness values obtained from the momentum equation and by applying a bedform roughness predictor (Eq. 10).

between the summit and the brinkpoint of dunes in water flows can similarly be characterized by a dune Reynolds number remains to be investigated. If this were confirmed, dune Reynolds numbers could be used to determine hydraulic roughness, since the angle of flow separation significantly influences the form and extension of the wake behind the dune.

CONCLUSIONS

The objective of the present study was to investigate the hydraulic roughness of a straight tidal channel with dune bedforms. On the basis of a large number of measurements, which allowed the computation of roughness from the one-dimensional momentum equation and curve fitting of the vertical current and sand concentration distributions, the following conclusions are drawn:

1 Reliable values of hydraulic roughness are obtained only by solving the momentum equation.

2 The poor performance (large scatter) of the curve-fitting methods is caused by the unsteady conditions of the tidal flow and by the influence of the dunes on the shape of the lower part of the current and sand concentration profiles.

3 The Nikuradse roughness length (k_s) was found to vary by one order of magnitude during a tidal cycle. It appears to be inversely related to flow velocity. Furthermore, as in the case of the Conwy estuary (Knight, 1981), k_s was also found to be inversely related to the flow depth. However, the latter relation is purely empirical and lacks a theoretical basis. The change in roughness must be attributed to changes in the effective roughness of the dune bedforms. These changes cannot be estimated satisfactorily by using the presently available bedform roughness predictors.

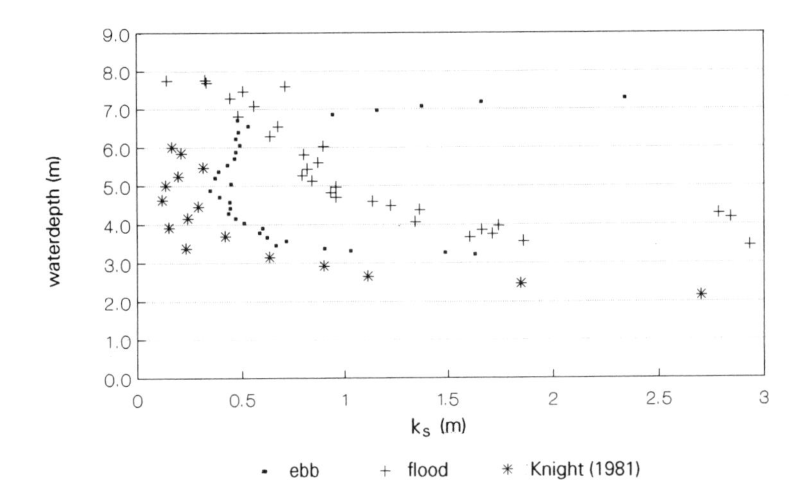

Fig. 9. Hydraulic roughness (k_s) as a function of water depth: a comparison with data of the Conwy estuary. The k_s data refer to values from the 1992 measuring period as used in Fig. 5.

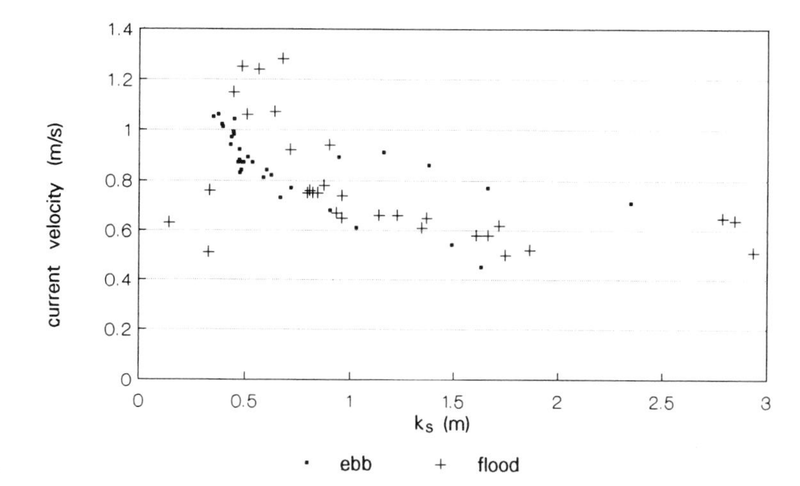

Fig. 10. Hydraulic roughness as related to flow velocity. The k_s data refer to values from the 1992 measuring period as used in Fig. 5.

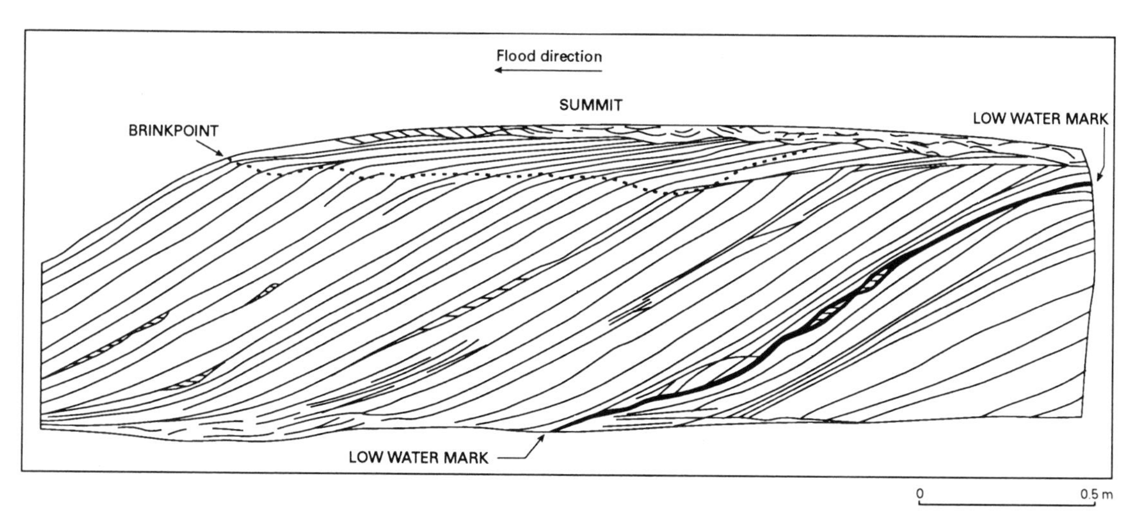

Fig. 11. Change in frontal shape of a tidal dune: lacquer peel of a longitudinal section through an intertidal dune on a shoal of the Westerschelde estuary.

4 Evidence from sedimentary structures of preserved dune foreset stratification of tidal dunes indicates that the shape and amplitude of the wake behind dune bedforms is related to flow intensity. This suggests that the effective hydraulic roughness of tidal dunes is related not only to their shape but also to a dune Reynolds number.

ACKNOWLEDGEMENTS

This work was part of a joint project of the Ministry of Transport and Public Works, Tidal Waters Division, and the Physical Geography Department, Utrecht University, dealing with tidal bedforms and hydraulic roughness in the Westerschelde. We gratefully acknowledge the technical support of the Measuring Department of the Rijkswaterstaat Zeeland Directorate, Middelburg. The manuscript benefited from the comments of Jesper Bartholdy and an anonymous reviewer.

REFERENCES

Achenbach, E. (1968) Distribution of local pressure and skin friction around a circular cylinder in cross-flow up to Re = 5×10^6. *J. Fluid Mech.* **34**, 625–639.

Allen, J.R.L. (1976) Time-lag of dunes in unsteady flows: an analysis of Nasner's data from the river Weser, Germany. *Sediment. Geol.* **15**, 309–321.

Boersma, J.R. & Terwindt, J.H.J. (1981) Neap–spring tide sequences of intertidal shoal deposits in a mesotidal estuary. *Sedimentology* **28**, 151–170.

Engelund, F. (1977) Hydraulic resistance for flow over dunes. Inst. Hydrodyn. Hydraul. Eng., Techn. Univ. Denmark, *Progr. Rept* **44**, 19–20.

Klaassen, G.J., Ogink, H.J.M. & van Rijn, L.C. (1986) DHL-research on bed forms, resistance to flow and sediment transport. In: *Proc. 3rd Internat. Symp. River Sediment* (Eds Wang, S.Y., Shen, H.W. & Ding), University of Mississippi, Jackson, Mississippi, USA.

Knight, D.W. (1981) Some field measurements concerned with the behaviour of resistance coefficients in a tidal channel. *Est. coast Shelf Sci.* **12**, 303–322.

Kohsiek, L.H.M. & Terwindt, J.H.J. (1981) Characteristics of foreset and topset bedding in megaripples related to hydrodynamic conditions on an intertidal shoal. In: *Holocene Marine Sedimentation in the North Sea Basin* (Eds Nio, S.D., Schüttenhelm, R.T.E. & Van Weering, T.C.E.) *Spec. Publs int. Ass. Sediment.* **5**, 22–37.

Ludwick, J.C. (1974) Tidal currents and zig-zag sand shoals in a wide estuary entrance. *Geol. Soc. Am. Bull.* **85**, 717–726.

McCave, I.N. (1971) Sand waves in the North Sea off the coast of Holland. *Mar. Geol.* **10**, 199–225.

Ogink, H.J.M. (1989) Hydraulic roughness of single and compound bed forms. *Delft Hydraulics Rept* **Q 786**, 80 pp.

Rouse, H. (1937) Modern conceptions of the mechanics of turbulence. *Trans. Am. Soc. Civil Eng.* **102**, 463–543.

Sternberg, R.W. (1972) Predicting initial motion and bedload transport of sediment particles in the shallow marine environment. In: *Shelf Sediment Transport: Process and Pattern* (Eds Swift, J.P, Duane, D.B. & Pilkey, O.H.), pp. 61–82. Dowden, Hutchinson & Ross, Stroudsburg.

Van den Berg, J.H. (1982) Migration of large-scale bedforms and preservation of crossbedded sets in highly accretional parts of tidal channels in the Oosterschelde, SW Netherlands. *Geol. Mijnb.* **61,** 253–263.

Van den Berg, J.H. (1984) The determination of the suspended sand concentration. In: *The Closure of Tidal Basins* (Eds Huis in't Veld, J.C., Stuip, J., Walther, A.W. & van Westen, J.M.), pp. 208–211. Delft University Press, Delft.

Van Gelder, A. & Terwindt, J.H.J. (1992) Variability of bedforms and deposited sedimentary structures on an intertidal floodshield, Westerschelde estuary, The Netherlands. In: *Tidal Clastics 92, Abstr. Vol.* (Ed. Flemming, B.W.,), pp. 32–33. Cour. Forsch.-Inst. Senckenberg **151**.

Van Rijn, L.C. (1984) Sediment transport, III: Bed forms and alluvial roughness. *J. Hydraul. Eng., ASCE* **110**, 1733–1754.

Van Rijn, L.C. (1989) Handbook of sediment transport in currents and waves. *Delft Hydraulics Rept* **H 461**, 500 pp.

Vanoni, U.A. & Hwang, L.S. (1967) Relation between bed-forms and friction in streams. *Proc. ASCE* **93** (HY3, paper 5242), 121–144.

Wooding, R.A., Bradley, E.F. & Marshal, J.K. (1973) Drag due to regular arrays of roughness elements of varying geometry. *Boundary-Layer Meteorol.* **5**, 285–308.

Zanke, U. (1977) Berechnung der Sinkgeschwindigkeiten von Sedimenten. *Mitt. Franzius-Inst. Wasserb. Küsten-ing. Univ. Hannover* **46**, 231–265.

Spec. Publs int. Ass. Sediment. (1995) **24**, 33–51

Bedforms on the Middelkerke Bank, southern North Sea

J. LANCKNEUS *and* G. DE MOOR
Marine and Coastal Geomorphology Research Unit, Geological Institute, University of Gent, Krijgslaan 281, B-9000 Gent, Belgium

ABSTRACT

Large dunes cover the flanks and summit of the Middelkerke Bank. They have constant orientation, heights ranging from 0.5 to 5 m, wavelengths of 75 to 150 m and in most cases asymmetrical profiles. No relationships could be established between the heights of the large dunes and the grain sizes of the sediment or the water depth in which they are developed. The slopes of both lee and stoss flanks are very low. Almost every large dune on the bank can assume both ebb and flood orientations through time. Two cases are presented in which particular swell and wind conditions induced common orientations to nearly all the large dunes. No significant migration was observed for six large dunes on the northern end of the bank over a time span of 13 months during which five surveys were carried out. Fields of small and medium dunes occur on the whole bank and in the adjacent swales. They are two-dimensional and have wavelengths ranging from 1 to 15 m. They are oriented at an oblique angle to the large dunes with a counter-clockwise offset of approximately 20°, their alignment being perpendicular to the direction of the peak tidal currents. Flood- and ebb-oriented small dunes cover the western and eastern flanks of the bank respectively. The boundary line between the dune fields having opposing asymmetries more or less follows the axis of the bank and often coincides with the crest lines of the larger dunes.

INTRODUCTION

The Middelkerke Bank, a tidal sand bank located on the Belgian Continental Platform (Fig. 1), was the object of a multidisciplinary research project carried out by several European universities and one private company. The project was partly financed by the European Commission within the framework of its MAST I (Marine Science and Technology) programme and is known under the acronym RESECUSED (Relationship between Sea Floor Currents and Sediment Mobility in the Southern North Sea) (Lanckneus *et al.*, 1991; De Moor *et al.*, 1993; De Moor & Lanckneus, 1993). Major objectives were the detailed analysis of a sand bank and that of its sediments and bedforms in a macrotidal offshore environment over short and medium time scales and an assessment of interactions between water movement, sediment transport and bedform mobility.

The residual dynamics of the bedforms were studied by repeated bathymetric and side-scan sonar surveys. Their analysis allowed an assessment of the short- (time scale of a few weeks) and medium-term (time scale of seasons) evolution of (i) the volume, position and morphology of the sand bank; (ii) the residual sediment transport paths; and (iii) the movement of major bedforms such as large dunes.

This paper discusses the geographical distribution and characteristics of bedforms on the Middelkerke Bank and their residual morphodynamic behaviour.

MORPHOLOGY AND HYDRODYNAMICS

The Middelkerke Bank has a length of 12 km, a mean width of 1.5 km and a height varying from 8 m in the north-east to 15 m in the south-west. The axis of the bank runs oblique to the coastline and

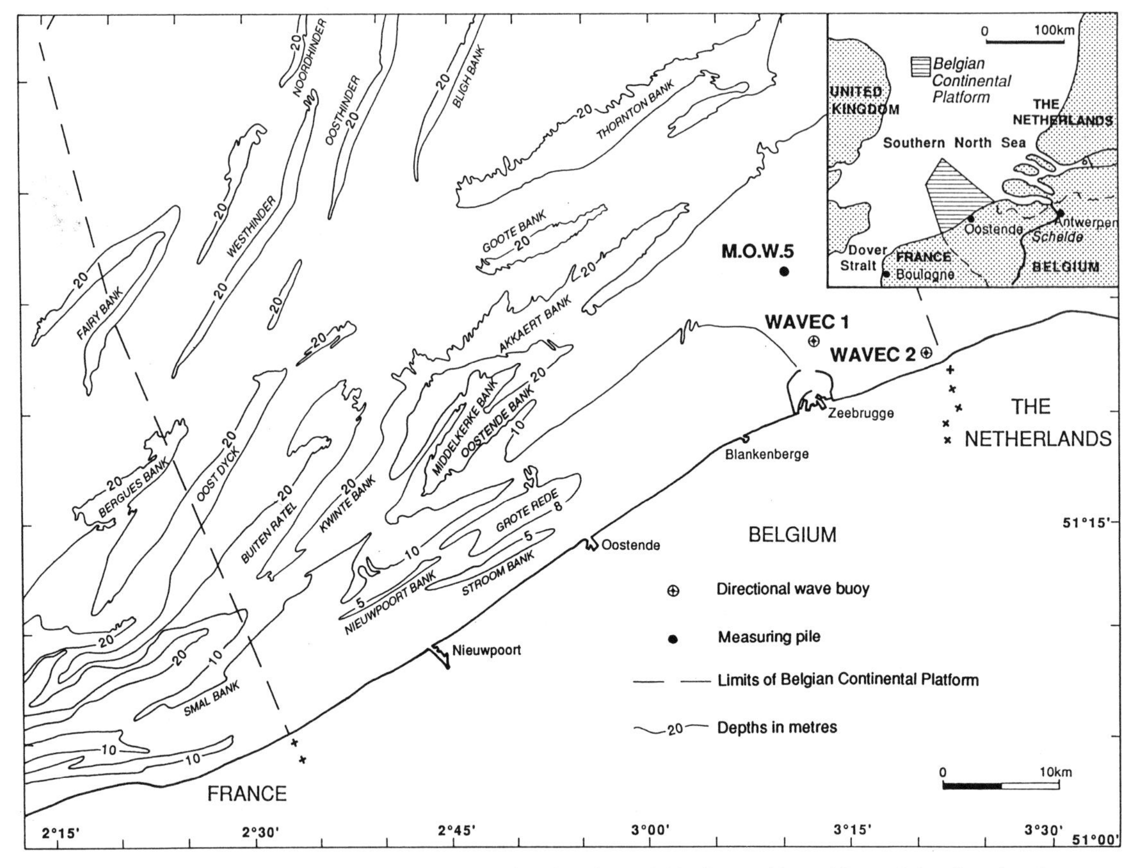

Fig. 1. Location of the Middelkerke Bank on the Belgian Continental Platform with positions of the directional wave buoys and of the measuring pile on which meteorological data were acquired.

has a SW–NE orientation. The depth at low water varies from 4 m in the south-west to 11 m in the north-east. The southern edge of the bank is relatively wide while the northern extremity tends to be rather narrow. The two adjacent swales of the Middelkerke Bank are known as the Negenvaam (north-west) and the Uitdiep (south-east). The Negenvaam swale is 2–3 km wide and 12–20 m deep. The Uitdiep swale has a width of 1–3 km and a depth of 12–20 m. Both swales become narrower and shallower landwards where they merge into a coast-parallel swale system. The bank shows a pronounced asymmetrical cross-section with the steeper side to the north-west and the more gentle slope to the south-east.

The hydrodynamics around the Middelkerke Bank are characterized by semi-diurnal tides of meso- to macrotidal range, the tidal amplitude being 4.8 m at spring tide and 2.7 m at neap tide. The hydrodynamics on the Middelkerke Bank were studied in detail by Stolk (1993). Measurements over full tidal cycles were carried out with propeller-type Elmar current meters in seven different locations. The current was measured at six or seven points along a vertical profile over a tidal cycle of 13 h. The discrete monitoring of wave and tidal movement, current characteristics and suspended sediment concentration was carried out by means of sea-bed-mounted measuring frames (Stolk, 1993). They were deployed twice on both bank flanks (Fig. 2). Velocities and directions of tidal currents vary at different locations on the bank. An upslope increase in maximum speed of the ebb and the flood currents was detected close to the sea-bed; it increases from 0.6 m s^{-1} in the swales to 0.8 m s^{-1} at the top. The velocity of the surface peak currents attains values of 1 m s^{-1} (Van Cauwenberghe, 1992). The bank shows an anti-clockwise offset with respect to the peak tidal currents, as do most of the linear sand banks in the southern North Sea (Kenyon, 1981). The angle between the direction of the peak flood and ebb currents and the overall axis of the bank/swale system varies from 10° in the swales to 28° on the top of the bank (Stolk, 1993).

DATA ACQUISITION AND PROCESSING

Data acquisition

A preliminary reconnaissance of the bank was carried out on 16 May 1990. Sonograms and bathymetric profiles were recorded along profiles across the bank. Track lines were spaced such as to obtain a sonar mosaic. An Atlas Deso XX echosounder and a 500-kHz Klein two-channel side-scan sonar were used to obtain bathymetric recordings and sonograms respectively (De Moor & Lanckneus, 1988). A ship speed of 4 knots was maintained during all surveys involving side-scan sonar operations.

Repeated bathymetric recordings were carried out three to five times per year between 1987 and 1992. Surveys were carried out at speeds of 10 knots along a number of reference tracks defined by fixed start and end points situated on the red Decca lines of the 5B English chain (Fig. 2). In all cases, the positioning and navigation made use of a Syledis UHF-radio positioning system. The precision of the Syledis system (variation of the measured values in a single location caused by noise in the signal) was calculated on the basis of strict statistical norms by means of the POSAN software package of the Dutch Rijkswaterstaat (Van Cauwenberghe & Den Duyver, 1993). Areas of similar precision were drawn as a function of the existing array of the Syledis beacons, the antenna heights and the standard deviation of the master and slave signals. The calculations demonstrated that the Syledis system had a precision of better than 10 m over the entire survey area.

Owing to the lack of rock outcrops, attempts were made to calibrate the accuracy of the positioning system with the help of a shipwreck located on one of the reference lines. The position of the Sigurd Faulbaums, which was sunk on 29 May 1940, was checked during two of the surveys. Variations of 1 m and 4 m were found along Easting and Northing directions respectively. However, these measurements have only a limited significance for the determination of the accuracy of the positioning system as the wreck has a length of 110 m. It is thus possible to record different 'correct' positions of the wreck while sailing along survey tracks with different offsets to the reference line.

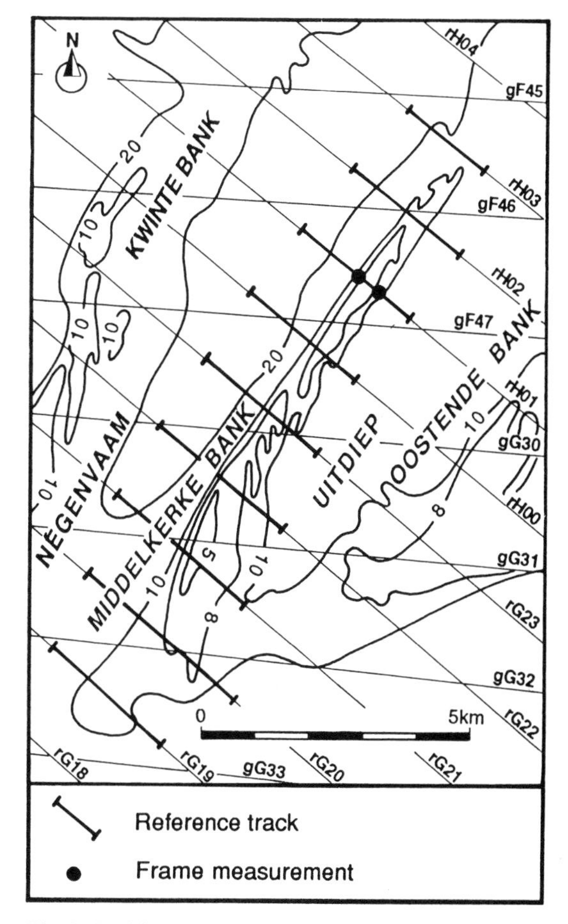

Fig. 2. Position of bathymetric survey lines across the Middelkerke Bank. The survey lines are situated on the red (r) Decca-lines of the British 5B chain; green Decca-lines on the figure are preceded by the letter g. Hydrodynamic data were collected on both flanks of the sand bank by means of sea-bed-mounted measuring frames (bold points).

As measurements of the crest positions of large dunes were based on bathymetric profiling, errors of navigation must in such cases also be taken into account. Slight deviations from the fixed track are unavoidable as current speeds transverse to the ship's course will differ in the swales and on the bank. This means that successive echosounder recordings along the same reference track can produce slightly different cross-sections. As the orientation of the large dunes was mostly oblique to the ship's heading, different XY positions of the crests were obtained on several adjacent survey lines. These positions were plotted on a digitized crest-line map of the bank obtained from a sonogram mosaic. To check the feasibility of this method, a particular reference track was surveyed 22 times on a single day. The positions of the crest along a straight section of a large dune were calculated and plotted for all 22 tracks. Figure 3 shows the results of this operation. Each track had a slightly different offset from the reference line, thus causing the intersection points between crest line and survey track to have different positions. The maximum offset from the base track was 13 m to the south and 7 m to the north. If a straight crest is drawn on the basis of the 22 recordings by linear regression, an estimation of the accuracy of the positioning system can be obtained. The majority of the crest positions deviate 0–2 m from the inferred crest line, although two values of

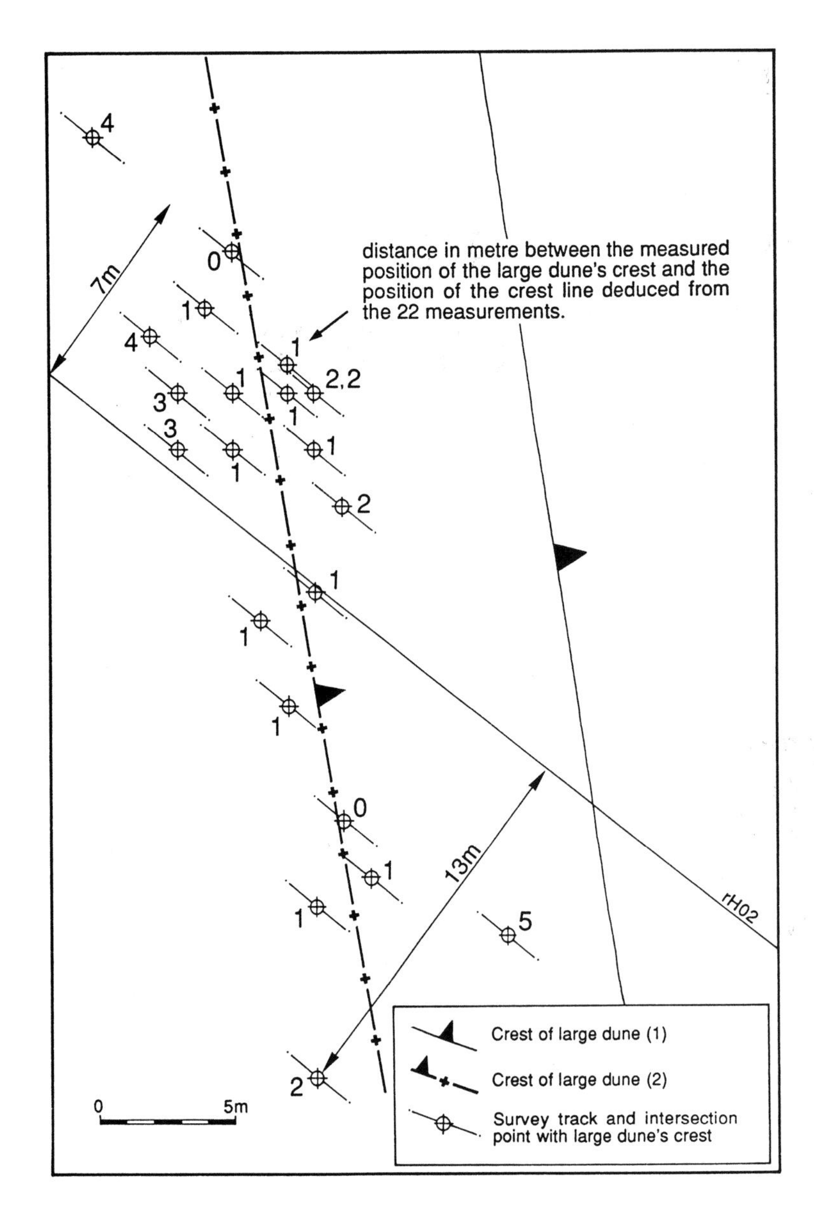

Fig. 3. Accuracy control of the positioning system. The reference line rH02 was sailed 22 times on a single day with calm weather conditions. Sections of all 22 survey tracks and their intersection points with the crest of a large dune are indicated. A straight line drawn through the 22 points represents the crest of large dune No 2. The crest position of dune No 1 was deduced from the sonograms recorded in May 1990. The numbers 1 to 5 represent the distance in metres between the calculated intersection point and the assumed location of the crest line.

4 and 5 m were also found. The inferred crest line of this particular dune ran parallel to the position of a large dune recorded by side-scan sonar almost 2 years earlier. Sonograms recorded between the two dates indicate that both features relate to the same dune. This example and similar studies dealing with the migration of large dunes on adjacent sand banks (Lanckneus & De Moor, 1991) suggest that the displacement of crest lines is mostly uniform over considerable distances. Undulations of crest lines do not interfere with such analyses as the large dunes in the study area can be considered as straight-crested over a distance of 10–20 m. This example shows that the accuracy of the positioning system is probably better than 5 m. Apparent migrations of bedforms over distances smaller than 10 m were therefore not considered as significant.

Meteo-marine observations

Registrations of meteo-marine parameters are carried out throughout the year by the Belgian Service of Coastal Harbours. Wind directions and velocities are measured on an observation pile (named MOW 5) located offshore of the harbour of Zeebrugge (Fig. 1). Wave data are produced by directional wave buoys (Wavec buoys measuring wave height and wave direction). Wave characteristics are measured continuously and the data are transmitted by radio to an acquisition station on land. Such information was obtained from a buoy located in front of the Harbour of Zeebrugge (Bol van Heist; 25 km eastwards of the research area) and from another one near the Belgian–Dutch border (Zwin; 40 km eastwards of the research area) (Fig. 1).

Data processing

All sonograms were processed into isometric representations of the sea-floor following Flemming (1976). The crests and the leeside troughs of the large dunes were mapped on a scale of 1:5000. Both the asymmetries and the heights of the large dunes were determined from the echosounder records. The small- and medium-sized dune fields were also mapped, dune orientation being determined from the sonograms.

The processing of the raw bathymetric data comprised tidal reduction (Van Cauwenberghe, 1977) and corrections for variations in the ship's velocity and heading. Net bathymetric profiles were then plotted in relation to a zero level which corresponds to the local MLWS.

The flow-transverse bedforms described below can be classified according to Ashley (1990) as large to very large dunes and as small to medium dunes.

GEOMETRY OF LARGE AND VERY LARGE DUNES

Lateral extension

The positions of the crest lines of the large and very large dunes (both defined hereafter as large dunes), deduced from the side-scan mosaic, are presented in Fig. 4. Large dunes occur on both the flanks and the summit of the Middelkerke Bank but are completely absent in the two adjacent swales.

The strike of the crest lines is quite uniform and varies between N–S and N 15° W on the whole bank except on the southern edge where it distinctly bends westwards. In this part, several large dunes have a strike of N 30° W. The crest lines of most large dunes are nearly straight to slightly sinuous. They display a good lateral continuity and are traceable in most cases from one flank of the bank to the other over a distance up to 3 km. Shorter large dunes with a spacing of 200–300 m occur frequently between the longer ones. Branching of crest lines was observed occasionally.

Height

Heights of the large dunes range from 0.5 m to some exceptional values of 4–5 m (Fig. 5), the most common height being 1.5–2 m. The height of an individual dune can vary between 1 and 4 m along its crest.

The existence of possible relationships between the height and other morphological or sedimentological parameters of large dunes was investigated by means of a data set exclusively comprising the observed maximum heights of all 95 large dunes.

Dune height versus water depth

Allen (1984) proposed a relationship between the heights of large dunes (H, in metres) and the water depth (d, in metres):

$$H = 0.086\, d^{1.19}$$

Figure 6 illustrates the relationship between observed maximum heights of the large dunes and

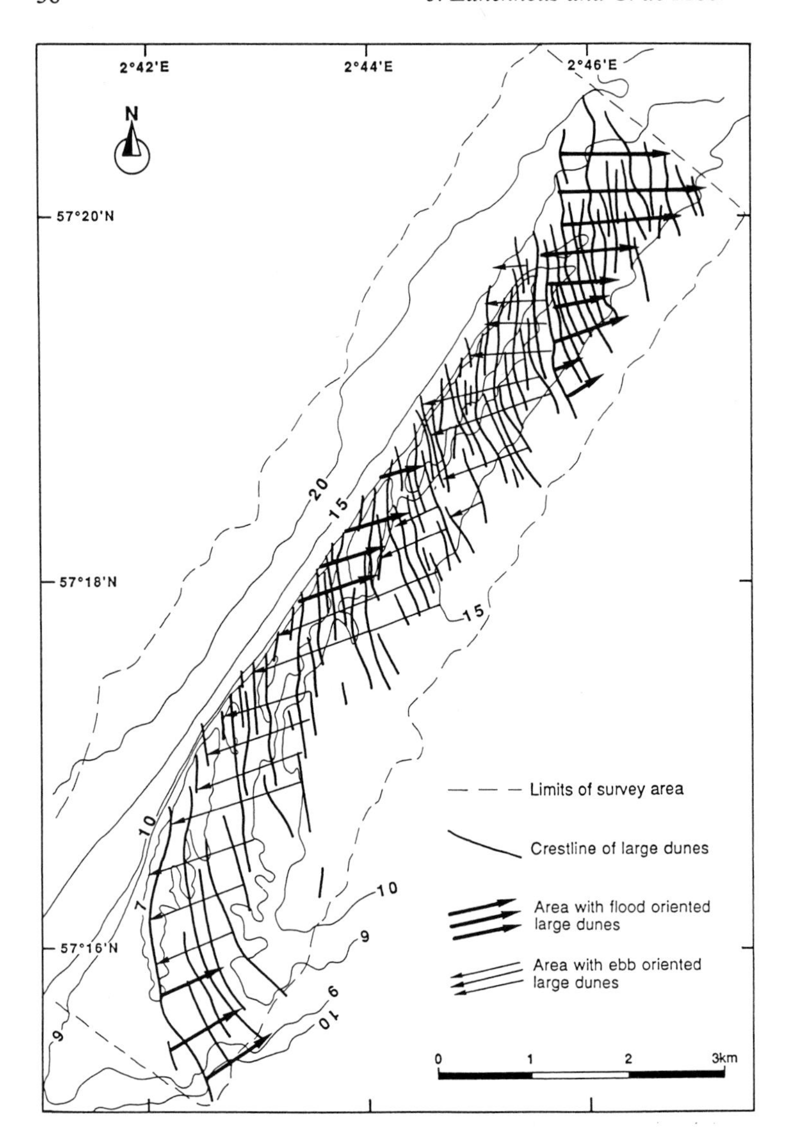

Fig. 4. Crest positions and orientation of the large dunes with respect to the dominant tidal flow direction on the Middelkerke Bank based on the recordings of May 1990.

water depths corresponding to the tide-corrected depths at the base of the dunes, both parameter sets being plotted on a logarithmic scale. No correlation between the two parameters is observed. Large dunes with heights of 2 m are, for instance, found in water depths ranging from 7 to 18 m. It is clear from this example that theoretical equations, which are commonly used in mathematical models, have to be used cautiously.

Dune height versus grain size

Several authors point to the existence of a relationship between dune height and mean grain size of the sediment in which the dunes develop. Laboratory experiments and observations in nature indicate an increase of the height of the dunes (H) with the diameter of the grains (D). Yalin (1964) proposed the following linear relationship:

$$H = 1000\,D$$

Figure 7 illustrates the relationship between the maximum heights of the 95 large dunes and the corresponding mean grain sizes (mean calculated according to Folk & Ward, 1957) of the sediment. Mean grain sizes were derived from a set of 84

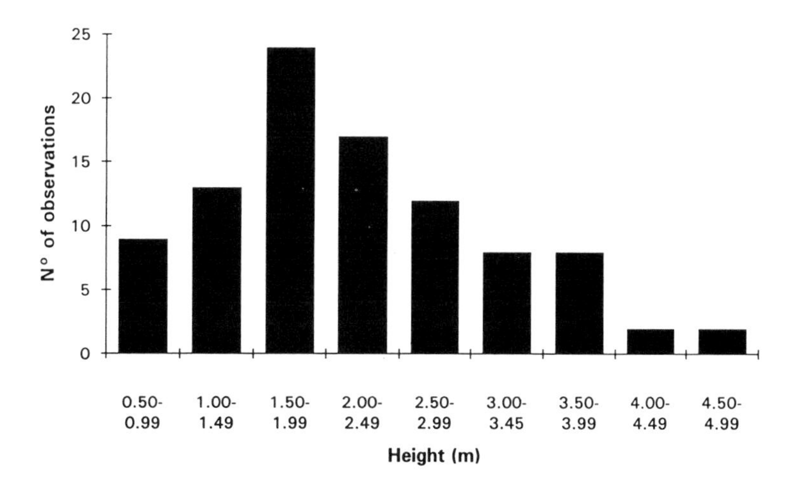

Fig. 5. Height frequency distribution of the large dunes on the Middelkerke Bank.

samples taken immediately after the side-scan sonar and bathymetric surveys of May 1990. Samples were taken on a rectangular grid of 400 m (transverse to the bank's axis) * 700 m (parallel to the bank's axis). Their positions were pre-determined and the exact topographic locations of the samples were thus not taken into consideration, since the aim was to obtain a general grain-size trend only. A mean grain-size map was constructed at intervals of 50 μm. The observed grain-size trends were very similar to the ones found on the adjacent Kwintebank. On this bank, grain size also increases towards the northern edge of the bank, the western flank being coarser-grained than the eastern flank (Lanckneus, 1989). Figure 7 demonstrates that no correlation exists between height and grain size of dunes on the Middelkerke Bank. Large dunes with heights ranging from 0.5 to 4.5 m develop in sediments with mean grain sizes of 250–300 μm. Points a, b and c on Fig. 7 represent large dunes detected along the western flank of the northern edge of the bank which is characterized by very coarse sands.

The lack of correlation between dune height and grain size could be due to the fact that large dunes do not always develop fully (Berné, 1991). The growth of large dunes on the Middelkerke Bank may be obstructed by the erosional effect of storms or by insufficient sediment input. Whatever the case may be, the data clearly contradict the relationship proposed by Yalin (1964).

Spacing

The spacing of dunes is the distance measured between two adjacent crest lines on the sonogram. The large dunes on the Middelkerke Bank have a rather constant spacing ranging, in most cases, from 75 to 150 m.

If sufficient sediment is available to allow full bedform development, a correlation between the

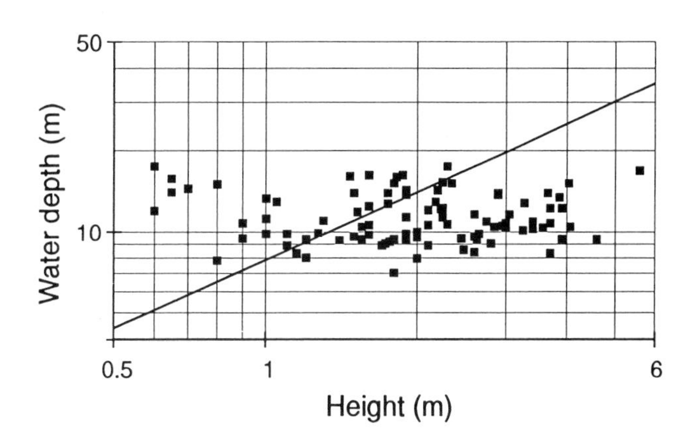

Fig. 6. Relationship between dune heights and water depth on Middelkerke Bank. The diagonal line represents the relationship between the two parameters as proposed by Allen (1984).

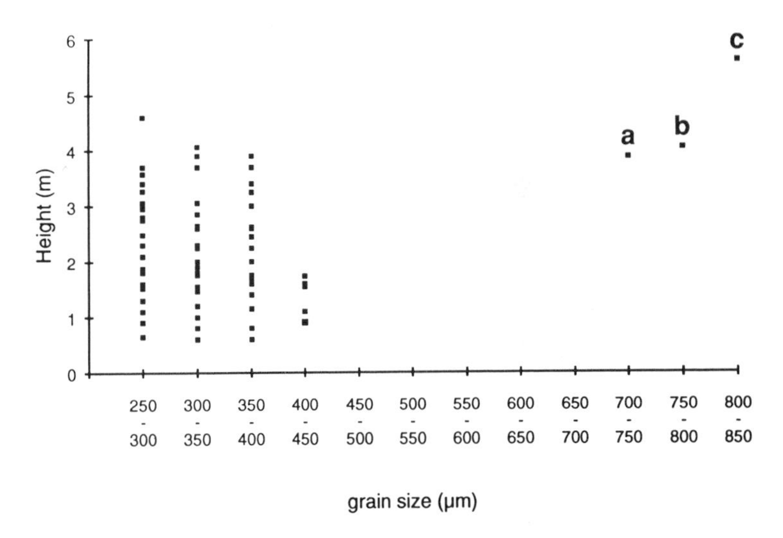

Fig. 7. Relationship between dune height and mean grain size. Cases a, b and c correspond to large dunes developed in very coarse sands.

height and spacing of large dunes is observed. Flemming (1988) established a simple relationship between dune height (H, in metres) and spacing (L, in metres):

$$H = 0.0677\, L^{0.8098}$$

The height-spacing relationship of the 95 dunes of the Middelkerke Bank illustrated in Fig. 8 is characterized by a high degree of scatter. Nevertheless, a large amount of points fall close to the mean trend of Flemming (1988). Most of the points, however, are situated below the theoretical mean. This means that for a given spacing the corresponding height is smaller than the one predicted. Again, this could be explained by a lack of full bedform development.

Slopes

Figure 9 shows the ratios between the angles of the lee and stoss slopes of all 95 dunes of the Middelkerke Bank. Only one measurement per dune, corresponding to the value of the observed maximum height, is included.

The angle of the lee slopes varies between 1° and 10°, the most common value being 2–3°. Stoss slopes are generally in the range of 1–7°. Modern tide-generated dunes are known to have slopes with small angles, although values of up to 10° are quite common (McCave, 1985). Berné *et al.* (1989) even mention values of 30–35° measured in several dune fields of the continental shelf around France. The slopes of the large dunes on the

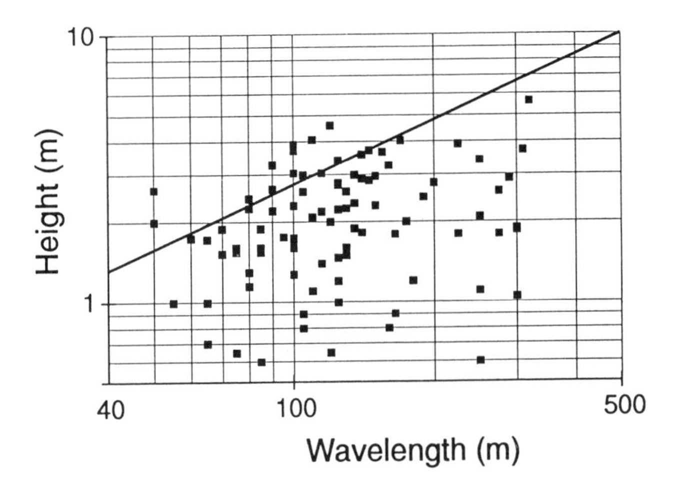

Fig. 8. Relationship between height and wavelength of the large dunes on the Middelkerke Bank. The diagonal line represents the mean relationship between the two parameters as proposed by Flemming (1988).

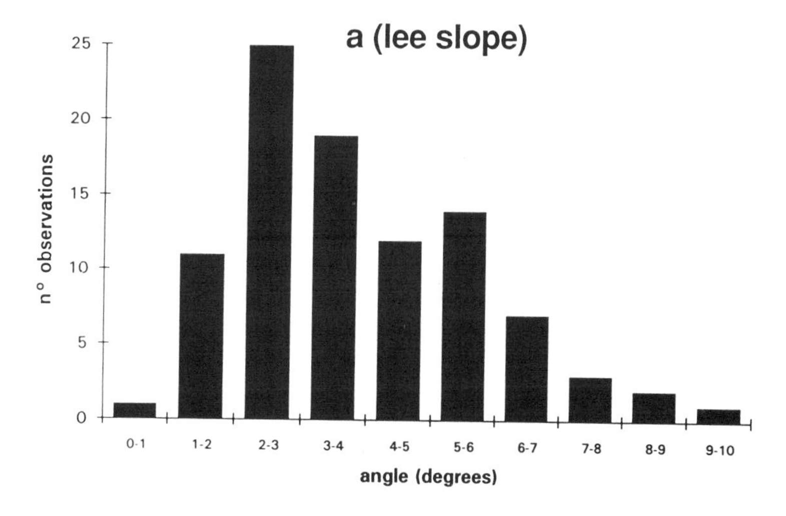

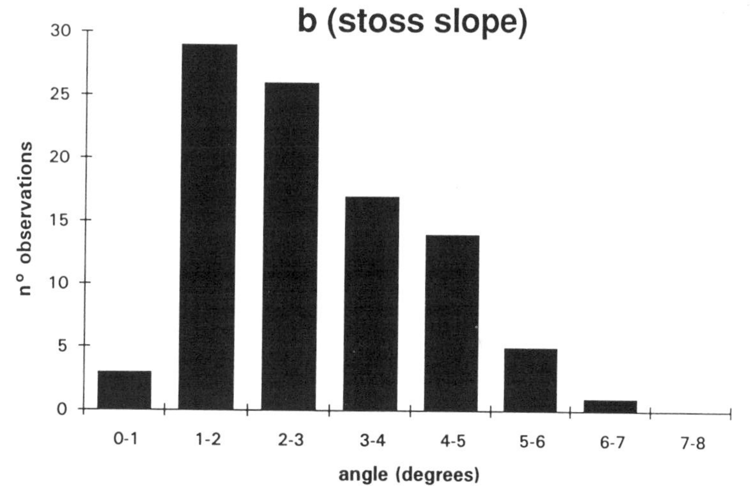

Fig. 9. Dip angle frequencies of the lee (a) and the stoss (b) slopes of the large dunes.

Middelkerke Bank can thus be considered as quite low.

Asymmetry

Most large dunes have a pronounced asymmetrical profile. Many authors (e.g. Caston, 1972) describe asymmetrical large dunes migrating along both flanks of similar sand banks, generally moving towards their crest. In this case, the asymmetries of the large dunes face opposite directions on both flanks, suggesting a convergence of sand streams towards the crest line of the bank (De Moor, 1985). Symmetrical large dunes can occur in the transition area between dunes of opposite asymmetry (Berné, 1991).

The records of May 1990 presented in Fig. 4 allowed a detailed analysis of the asymmetry of the dunes along the whole bank. All large dunes at both extremities of the bank have their lee slopes dipping towards the north-east. The central area of the bank is characterized by large dunes with lee slopes dipping in opposite directions on both flanks, facing towards the axis of the bank. Two areas with large dunes facing towards the south-west were observed between the central area and the two edges of the bank.

The asymmetry of a large dune is linked to the long-term hydrodynamic environment and in many cases it can be considered as permanent (Dalrymple, 1984). This 'permanent' character of the asymmetry may explain the fact that few authors comment on medium-term reversal of the asymmetry of large dunes. Figure 10 presents corrected bathymetric records across the Middelkerke Bank along the same reference line (red Decca line rH00) recorded in the period November 1987 to March 1992. Several changes in asymmetry can clearly be observed. Furthermore, the analysis showed that some large dunes reverse their asymmetry within a single month and that individual dunes can disappear completely, as observed on 23/03/1990. The bank, however, always seems to recover from such abrupt changes, the restored cross-section in this case being characterized by four large dunes.

The records of Fig. 10 were selected from a large data set to illustrate changes in asymmetry. Such reversals, however, do not occur along all reference lines. All available bathymetric data between 1987 and 1992 along the reference line rG20 (Fig. 11) reveal a permanent asymmetry of the large dunes with lee slopes dipping seawards. Other reference lines, such as rG21 (Fig. 12), have characteristic

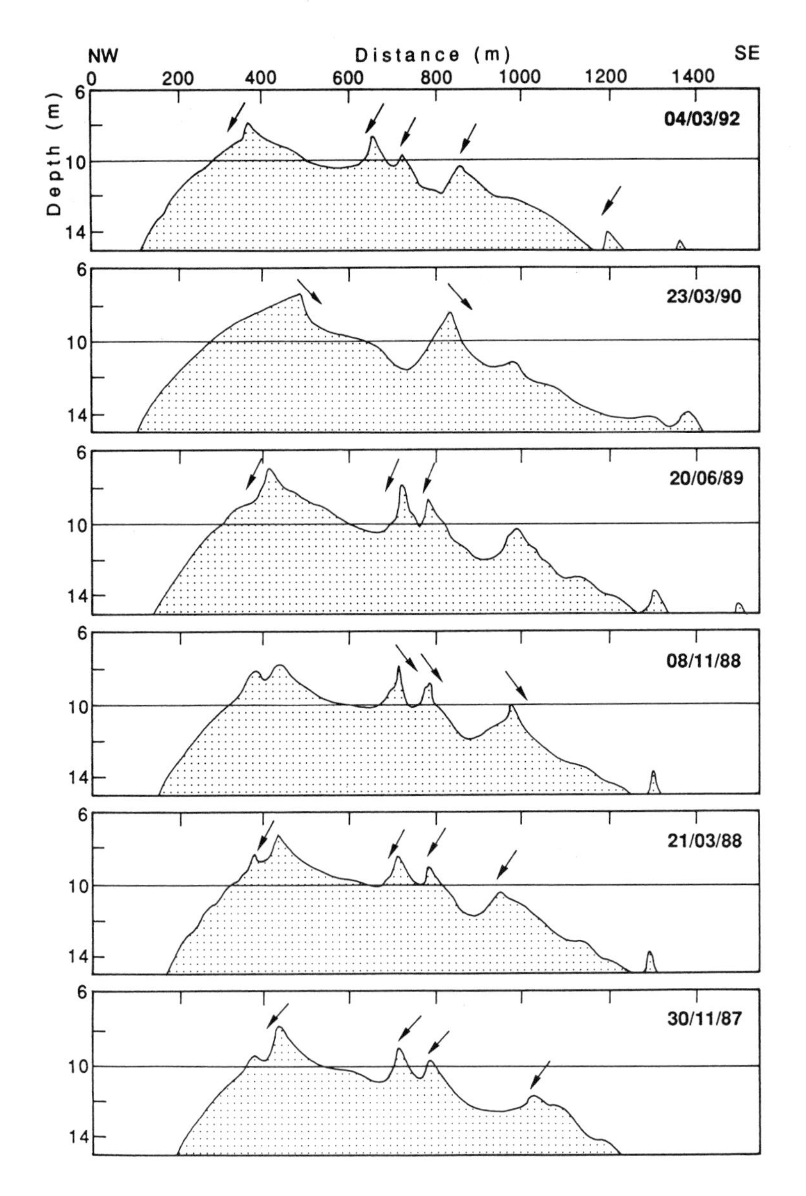

Fig. 10. Corrected bathymetric records along reference line rH00 obtained in the course of repeated surveys between 1987 and 1992.

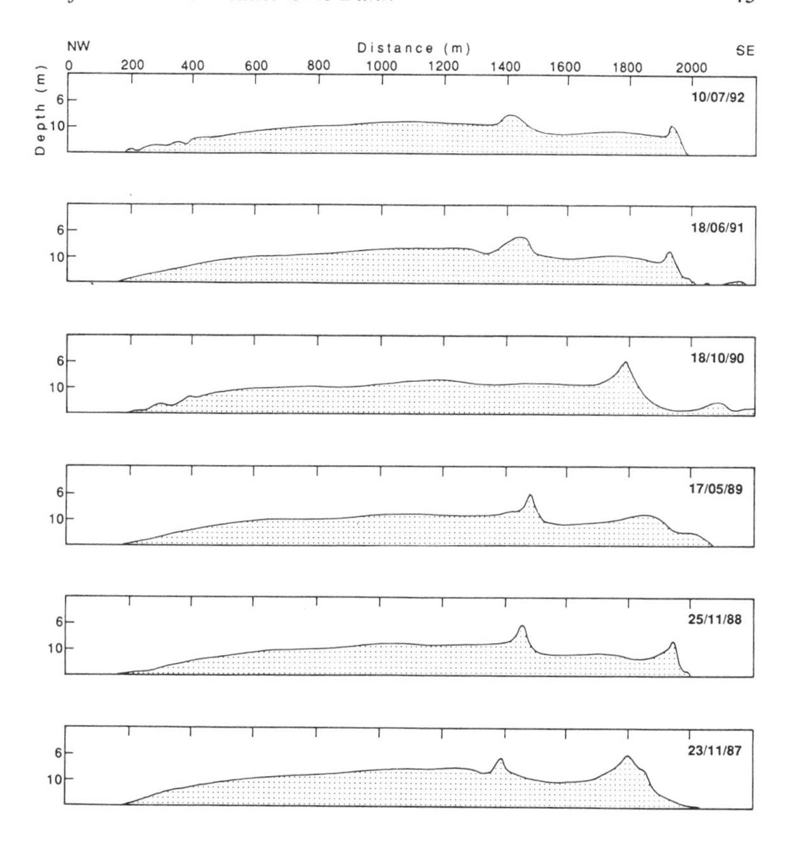

Fig. 11. Corrected bathymetric records along reference line rG20 obtained in the course of repeated surveys between 1987 and 1992.

profiles which may temporarily be disturbed by some reversals in asymmetry or by the appearance of an individual large dune.

The asymmetries of large dunes detected in the course of a particular survey are also of interest. Figure 13 illustrates conditions of bedform asymmetry in May 1990. The majority of the large dunes are ebb-oriented (lee sides dipping to the south-west) with the exception of some flood-dominated dunes (lee sides dipping to the north-east) located at the two extremities and in a small western section of the central part of the bank. The records of September 1992 show a different picture (Fig. 14), all large dunes with the exception of one feature along the red Decca line rG21 now being flood-oriented. Similar maps were compiled for each survey period and the results show that practically all large dunes can change from flood- to ebb-dominance (or vice versa) through time. Strangely, no apparent correlation between this phenomenon and the characteristics of the tidal currents could be found.

Finally, the impact of meteo-marine conditions on the asymmetry of the large dunes was analysed. Figure 15a displays the wind and wave directions from 12 to 22 May 1990. Information from the closest Wavec buoy, 'Bol van Heist', was unfortunately only available from 17 May onwards. The data from 12 to 17 May came from the more distant Wavec buoy located at the 'Zwin'. This buoy registered a north-westerly–northerly swell over this period. Thereafter the 'Bol van Heist' buoy recorded a north–north-easterly swell. The corresponding wind directions over this period were dominated by north–north-easterly winds. Swells and winds from this direction seemed to have amplified the ebb current, causing the large dunes to adjust their asymmetry towards the south-west. Similarly, Fig. 15b shows that during the survey period in September 1992, swells and winds from the south-west–west were dominant over a long period and probably induced the flood-asymmetry of the large dunes. Berné (1991) was also not able to explain the asymmetry of his large dunes by tidal dynamics alone and thus also invoked other processes. Langhorne (1982) showed that swells and winds were able to modify the duration and inten-

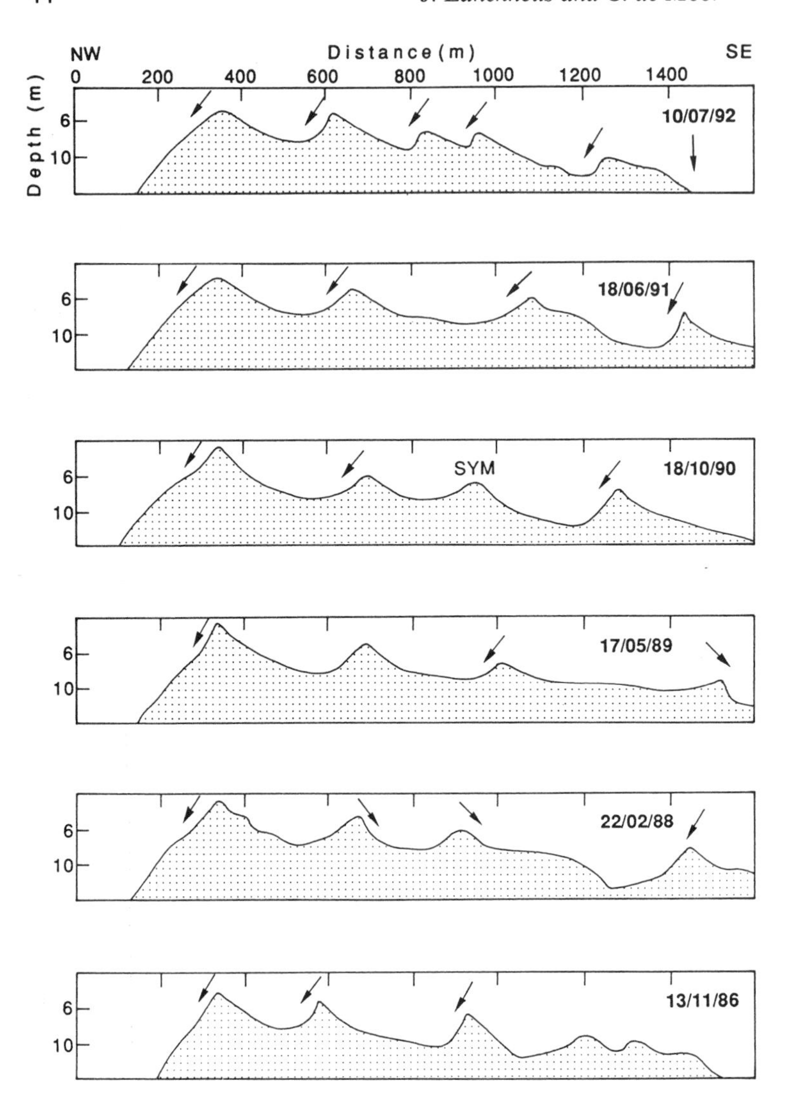

Fig. 12. Corrected bathymetric records along reference line rG21 obtained in the course of repeated surveys between 1987 and 1992.

sity of tidal currents in Start Bay (UK). Such modifications can thus increase or decrease the strength of either flood or ebb currents and thereby induce a particular asymmetry in the large dunes.

Winds blowing for an extended time from the same direction are rather rare in the North Sea. As demonstrated by the two examples above, it is unlikely that the majority of large dunes should have their lee slopes dipping in the same direction at any particular time. Instead, as shown in Fig. 4, ebb-dominated dunes are present in more or less the same proportion as flood-dominated ones.

DISPLACEMENT OF LARGE DUNES

Measuring technique

Bathymetric data were acquired along a number of reference lines on 5 May 1992 and 23 February, 12 May, 18 June and 15 July 1993. As positioning and depth data were digitally stored two times per second, it was possible to infer the positions of the crest lines of the large dunes along the cruise track. The positions of each survey were plotted on to the digitized crest line map of the Middelkerke Bank.

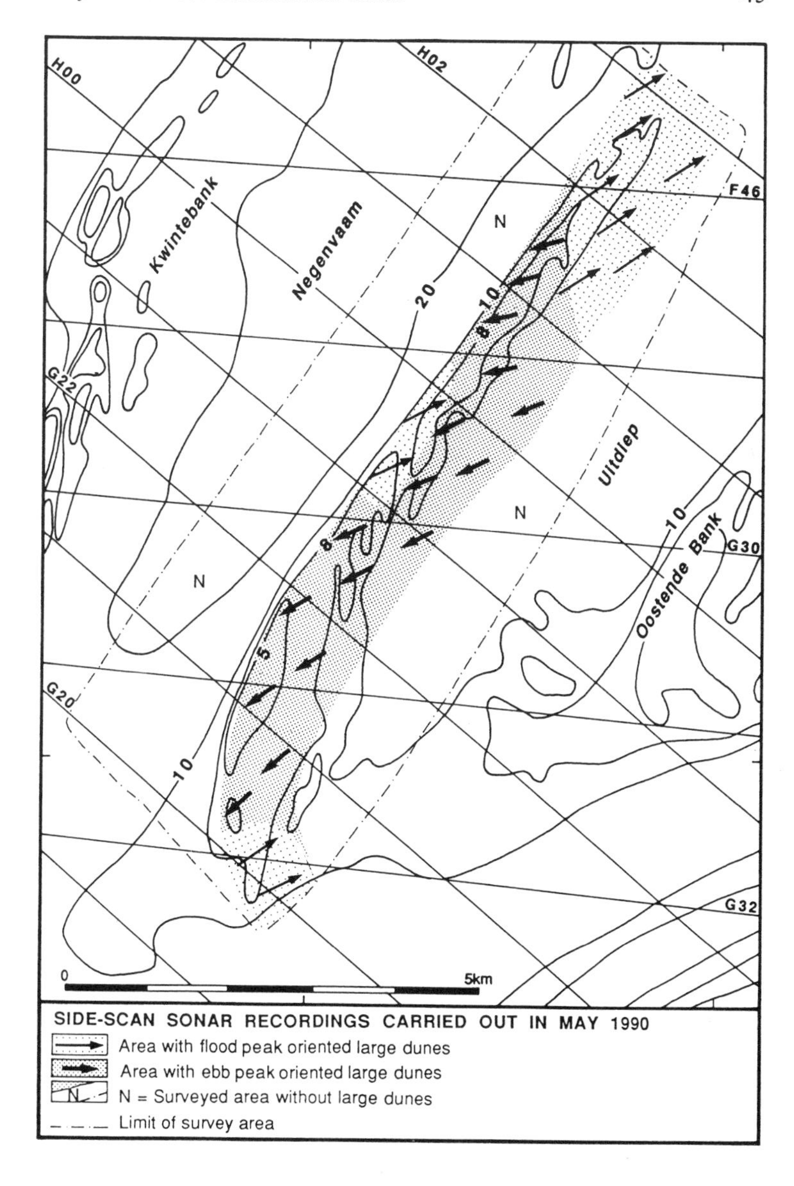

Fig. 13. Orientation of large dunes based on surveys carried out in May 1990.

An example of this procedure is illustrated in Fig. 16, in which the crest locations of two large dunes (recorded on 15 May 1990) are compared to the inferred positions of the crest points of these same dunes (based on the bathymetric profiling of May 1992 and February, May, June and July 1993).

Results

Assuming that all crests moved parallel to the mapped crest lines of May 1990, the net displacements of the crest lines of large dunes Nos 2 and 6 over the five observation periods were inferred and are listed in Table 1. Most values are extremely low (0–2 m), only one value of 7 m being obtained for the displacement of both large dunes between June and July 1993. All the values fall into the range of positional error. As a result, no net movement can be inferred over the different observation periods which cover a total time span of 13 months. Similar values are obtained for the five remaining large dunes along reference line rH02.

If the positions of the large dunes determined in the five surveys are compared to the ones inferred

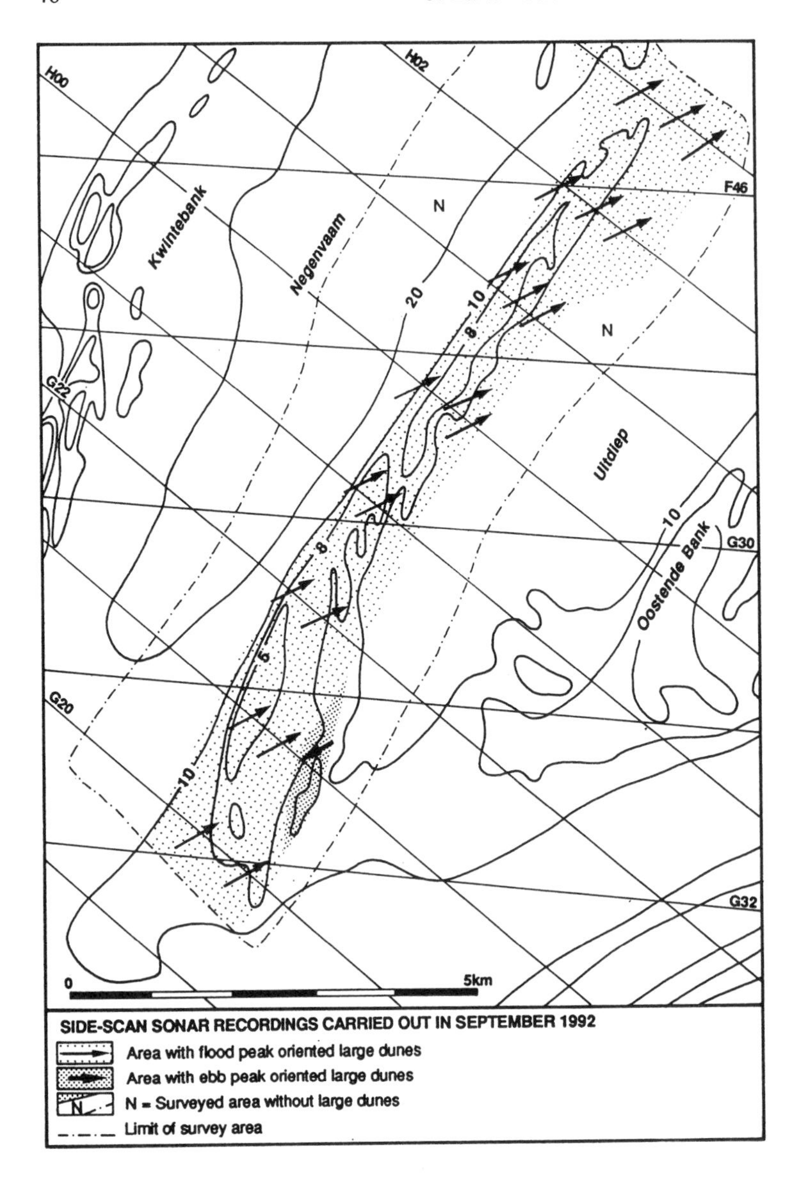

Fig. 14 Orientation of large dunes based on surveys carried out in September 1992.

for the crest lines from the sonograms of May 1990, again no net movement can be postulated, as all calculated values are smaller than 10 m. This means that the positions of both large dunes were extremely stable and that possible net movements over a 3-year period were all smaller than 10 m.

Discussion

Observations of the migration rates of the large dunes indicate that they are essentially stable in the short term (13 months). Three hypotheses can be proposed to explain the low migration rates of the large dunes on the Middelkerke Bank: (i) large dunes migrate at rates smaller in value than the assumed accuracy of the positioning system over the time span of this study; (ii) large dunes are only influenced by storm events strong enough to induce bedform migration; and (iii) large dunes are subject to oscillatory movements resulting in negligible net migration.

Fenster *et al.* (1990) studied the migrational trends of large dunes in a high-energy tidal environment in eastern Long Island Sound, USA. The

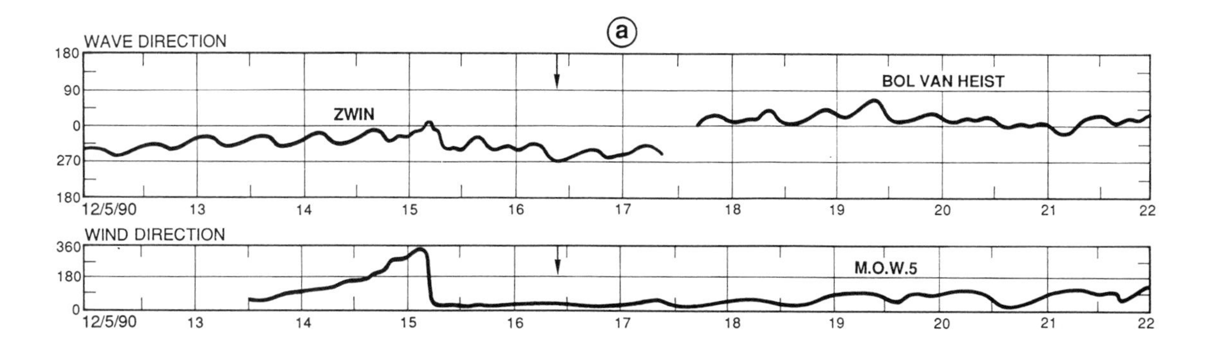

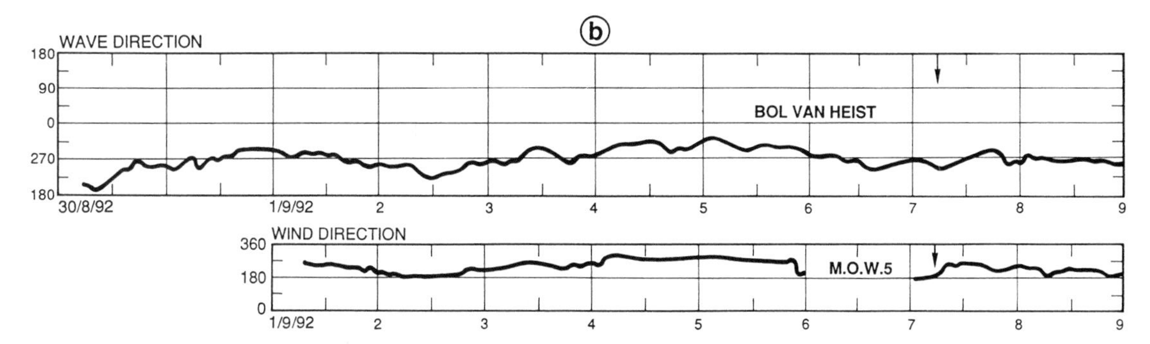

Fig. 15. Wave directions recorded with directional wave buoys and wind direction data from the measuring pile MOW 5 (for location see Fig. 1) for the periods 12–22 May 1990 (a) and 30 August–9 September 1992 (b). Arrows indicate the days of the bathymetric surveys.

results of their study demonstrate that storms apparently played a major role in the dynamics of large dunes. The impact of storm events on the morphology and movement of large dunes on the northern Middelkerke Bank was studied by Houthuys (1993). Surveys were carried out by hovercraft, the accuracy of positioning being estimated at 2 m. A first survey was carried out on 2 September 1991 in a period of fair weather. A second survey took place on 27 November 1991, i.e. the first day that sea conditions were favourable after several weeks of stormy weather. A comparison of the pre- and post-storm results showed that crest lines of the large dunes were lowered by up to 1.2 m. Crest lines migrated a maximum distance of 5 m to the west in the direction of their lee slopes. This storm-induced migration of 5 m is evidently a significant event in the dynamics of the bedforms but this does not exclude the potential importance of longer-term processes.

Table 1. Net displacement of the crest lines of the large dunes Nos 2 and 6 measured along the reference line rH02 (for location of large dunes see Fig. 16). An asterisk indicates that the displacement is smaller than the assumed accuracy of the positioning system. The measured amount can thus not be considered as significant

Displacement between	Large dune No 2	Large dune No 6
	N° 2	N° 6
05/05/92 and 23/02/93	0 m*	2 m*
23/02/93 and 12/05/93	2 m*	2 m*
12/05/93 and 18/06/93	1 m*	0 m*
18/06/93 and 15/07/93	7 m*	7 m*

Large dunes on the adjacent Kwintebank have been studied for many years (De Moor, 1985) but no net movement has ever been detected. The migrational behaviour of the large dunes on the bank was studied in a similar manner as that on the Middelkerke Bank (Lanckneus & De Moor, 1991).

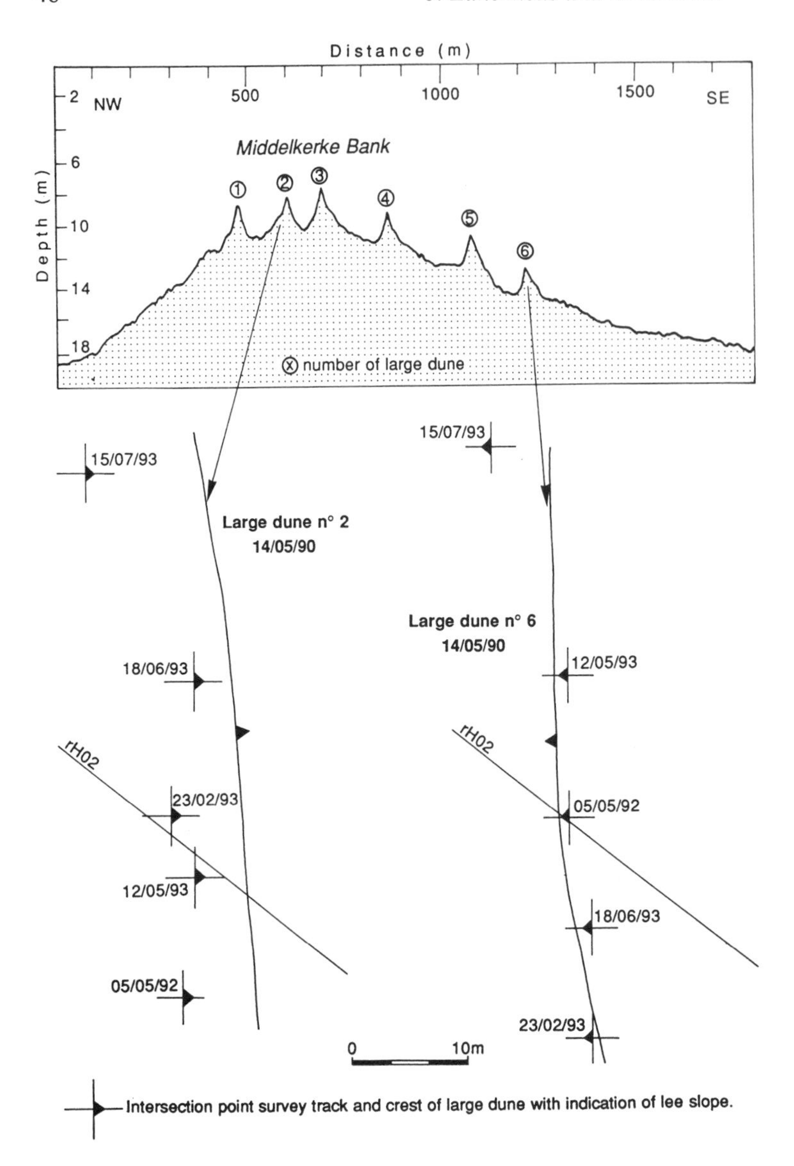

Fig. 16. Migration of crest lines of the large dunes Nos 2 and 6 along reference line rH02. The locations of the crest lines are based on the sonograms recorded in May 1990. The positions of the crests, defined as the intersection points between the survey tracks and crests of the large dunes, are shown for the survey days 5 May 1992, 23 February 1993, 12 May 1993, 18 June 1993 and 15 July 1993.

A comparison of three sonogram mosaics recorded in February, June and November 1989 on the northern part of the Kwintebank showed that strike, asymmetry and height remained unchanged over the observation periods. However, the positions of the crest lines of the large dunes changed. Between the first two recordings a net movement of 30 m towards the west was detected. The sonograms of November 1989 showed a similar migration but now towards the east, causing the crests to return to roughly the same positions they had 11 months earlier. In a general way, the direction of migration was similar for the entire crest line. However, flexing (movement of adjacent sections of the crest line in opposite directions; Langhorne, 1973; Fenster *et al.*, 1990) occured in some cases. The orientation of the dunes on the northern Kwintebank, which have heights of up to 8.6 m, can be considered as permanent as no reversal was observed in the course of 40 surveys between 1984 and 1992. The movement detected in June 1989 took place in the direction of the stoss slope. Fenster

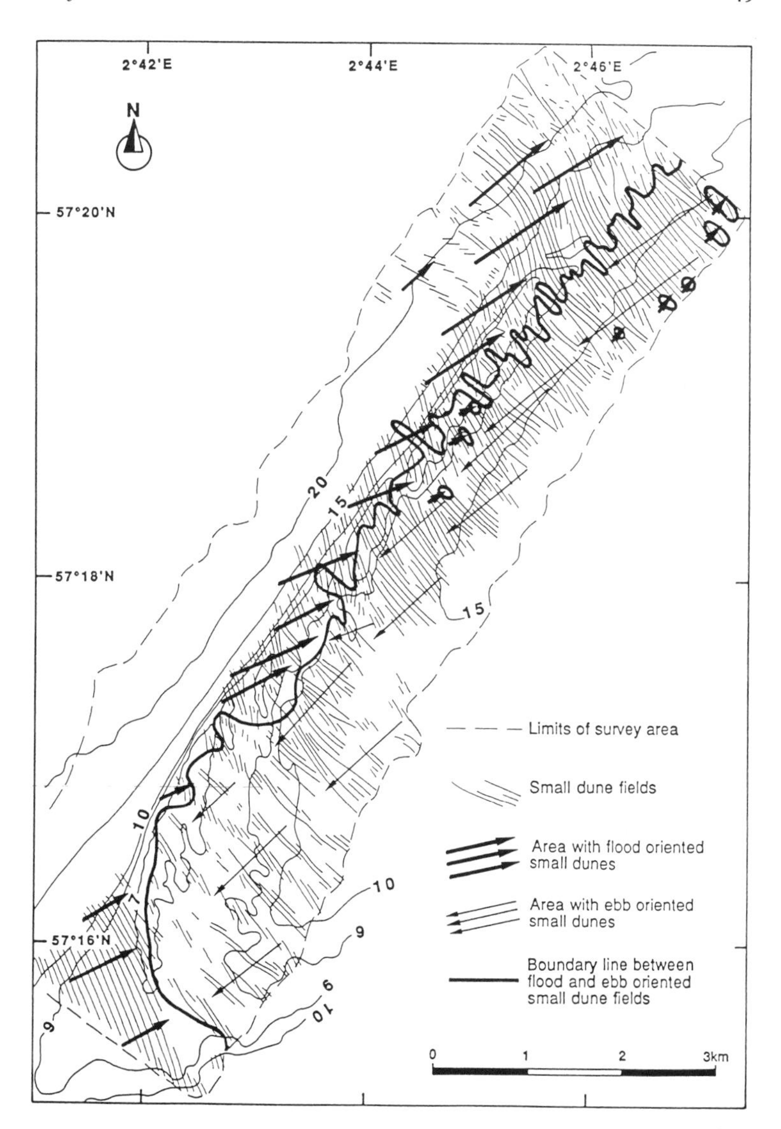

Fig. 17. Extent and orientation of small dunes on the Middelkerke Bank. The separation between flood- and ebb-oriented dunes is indicated by a thick line.

et al., (1990) found a similar behaviour in large dunes and stressed the fact that asymmetrical large dunes do not always move in the direction of their lee slopes.

To conclude, it can be said that the large dunes on the Middelkerke Bank appear as stable features. No significant migration was detected on the northern end of the bank over a 13-month period. If an oscillatory movement of their crest lines exists, as on the adjacent Kwintebank, it is probable that wind- and swell-induced processes determine the net direction of migration.

GEOMETRY OF SMALL AND MEDIUM DUNES

The distribution of small to medium dunes (defined hereafter as small dunes) was also mapped, their asymmetry being determined from the sonograms (Fig. 17). Small dunes are found over the whole bank and in a large part of the adjacent swales. Often they superpose both flanks of the large dunes. They are two-dimensional with wavelengths ranging in most cases from 1 to 15 m. The strike of the small dunes ranges between N 10° and N 30° W.

This means that they are generally oriented at an oblique angle to the large dunes with a counter-clockwise angle-offset of approximately 20°. Current measurements carried out 1.3 m above the bottom on both flanks of the Middelkerke Bank show a direction of N 65° E to the peak flood current (Stolk, 1993). Taking the strike of both small and large dunes into consideration, it is clear that the small dunes are oriented perpendicular to the peak currents.

The lee slopes of the small dunes dip in opposite directions on either side of the bank. In the western swale and western section of the bank they dip to the north-east, whereas in the eastern swale and eastern section of the bank they dip towards the south-west. The division between the ebb- and flood-oriented small dune fields is sharp and a line can be drawn from north-east to south-west, more or less parallel to the axis of the bank. This line often coincides with sections of the crest lines of the larger dunes which, in these cases, have small dunes with opposite asymmetry on both flanks. Nevertheless, eight small areas with flood-oriented small dunes occur within the area of the ebb-asymmetric features. In the southern part of the survey area, where the bank is hardly detectable, all small dunes dip to the north-east.

GENERAL CONCLUSIONS

A very detailed description of the very large, large, medium and small dunes of the Middelkerke Bank was facilitated by a side-scan sonar mosaic. No correlation was found between the maximum heights of large dunes and water depth. A clear relationship between the heights of large dunes and grain size could not be established.

Time-series analyses of cross-sections across large dunes showed that practically all large dunes on the Middelkerke Bank can be both flood- and ebb-oriented through time. A correlation was found between the asymmetry of the large bedforms and prevailing meteo-marine conditions. The results suggest that dominant north-north–easterly swells and winds can induce an ebb-asymmetry in the large dunes. Similarly, swells and winds from the south-west–west can cause the large dunes to be flood-asymmetrical. No significant migration could be established for six large dunes on the northern end of the Middelkerke Bank over a 13-month period in the course of which five surveys were carried out.

ACKNOWLEDGEMENTS

This research was partly financed by the European Commission within the framework of its MAST programme. We are grateful to the Management Unit of the Mathematical Model North Sea and to the Belgian National Fund for Scientific Research for their support. We also wish to thank the captain and crew of the oceanographic vessel Belgica for their constant help and co-operation during the surveys.

REFERENCES

ALLEN, J.R.L. (1984) *Principles of Physical Sedimentology.* George Allen & Unwin, London, 272 pp.

ASHLEY, G.H. (1990) Classification of large-scale subaqueous bedforms: a new look at an old problem. *J. sediment. Petrol.* **60**, 160–172.

BERNÉ, S. (1991) *Architecture et dynamique des dunes tidales. Exemples de la marge atlantique française.* Thèse Univ. Sci. Tech. Lille Flandres-Artois.

BERNÉ, S., ALLEN, G., AUFFRET, J.P., CHAMLEY, H., DURAND, J. & WEBER, O. (1989) Eassia de synthèse sur les dunes hydrauliques géantes tidales actuelles. *Bull. Soc. Géol. France* **V6**, 1145–1160.

CASTON, V.N.D. (1972) Linear sandbanks in the southern North Sea. *Sedimentology* **18**, 63–78.

DALRYMPLE, R.W. (1984) Morphology and internal structure of sandwaves in the Bay of Fundy. *Sedimentology* **31**, 365–382.

DE MOOR, G. (1985) Shelf bank morphology off the Belgian Coast. Recent methodological and scientific developments. In: *Recent Trends in Physical Geography.* Liber Amicorum L. Peeters, Brussels, VUB Study Ser. **20**, 47–90.

DE MOOR, G. & LANCKNEUS, J. (1988) Acoustic teledetection of seabottom structures in the Southern Bight. *Bull. Soc. belge Géol.* **97**, 199–210.

DE MOOR, G. & LANCKNEUS, J. (Eds) (1993) *Sediment Mobility and Morphodynamics of the Middelkerke Bank.* Final Rept MAST-0025, Gent, 255 pp.

DE MOOR, G., LANCKNEUS, J., BERNÉ, S. *et al.* (1993) Relationship between sea floor currents and sediment mobility in the Southern North Sea (RESECUSED, MAST Contract 25-C). *Proc. MAST Days Symp., Comm. Europ. Commun., Direct. Gener. Sci.*, 193–207.

FENSTER, M.S., FITZGERALD, D.M., BOHLEN, W.F., LEWIS, R.S. & BALDWIN, C.T. (1990) Stability of giant sandwaves in eastern Long Island Sound, U.S.A. *Mar. Geol.* **91**, 207–225.

FLEMMING, B.W. (1976) Side-scan sonar, a practical guide. *Int. Hydrog. Rev.* **LIII 1**, 65–92.

FLEMMING, B.W. (1988) Zur Klassifikation subaquatischer strömungstransversaler Transportkörper. *Boch geol. u. geotechn. Arb.* **29**, 44–47.

FOLK, R.L. & WARD, W.C. (1957) Brazos river bar: a study

in the significance of grain-size parameters. *J. sediment. Petrol.* **27,** 3–26.

HOUTHUYS, R. (1993) Impact of a storm period on the morphology of the Middelkerke Bank. In: *Sediment Mobility and Morphodynamics of the Middelkerke Bank* (Eds De Moor, G. & Lanckneus, J.), pp. 18–46. Final Rept MAST-0025, Gent.

KENYON, N.H. (1981) Offshore tidal sandbanks as indicators of net sand transport and as potential deposits. *Spec. Publ. int. Ass. Sediment.* **5**, 257–268.

LANCKNEUS, J. (1989) A comparative study of some characteristics of superficial sediments on the Flemish banks. In: *The Quarternary and Tertiary Geology of the Southern Bight, North Sea* (Eds Henriet, J.-P. & De Moor, G.). Min. Economic Affairs and Belgian Geological Service, pp. 229–241.

LANCKNEUS, J. & DE MOOR, G. (1991) Present-day evolution of sandwaves on a sandy shelf bank in the Southern Bight. *Oceanologica Acta, Spec. Vol.* **11**, 123–127.

LANCKNEUS, J., DE MOOR, G., BERNÉ, S. *et al.*, (1991) Cartographie du Middelkerke Bank: dynamique sédimentaire, structure géologique, faciès sédimentaires. *Int. Symp. OSATES (Ocean Space Advanced Technologies European Show), Brest*, 11 pp.

LANGHORNE, D.N. (1973) A sandwave field in the Outer Thames Estuary. *Mar. Geol.* **14**, 129–143.

LANGHORNE, D.N. (1982) A study of the dynamics of a marine sand wave. *Sedimentology.* **29**, 571–594.

MCCAVE, I.N. (1985) Recent shelf clastic sediments. In: *Sedimentology: Recent Developments and Applied Aspects* (Eds Brenchley, P.J. & Williams, B.P.J.), Spec. Publ. Geol. Soc., London, pp. 49–65.

STOLK, A. (1993) Hydrodynamics and suspended load; shipborne and stand-alone frame measurements. In: *Sediment Mobility and Morphodynamics of the Middelkerke Bank* (Eds De Moor, G. & Lanckneus, J.), pp. 194–210. Final Rept MAST-0025, Gent.

VAN CAUWENBERGHE, C.G. (1977) Overzicht van de tijwaarnemingen langs de Belgische Kust. Periode 1941–1970 voor Oostende, 1959–1970 voor Zeebrugge en Nieuwpoort. *Tijds. Openbare Werken Belgie* **4**, 15 pp.

VAN CAUWENBERGHE, C.G. (1992) *Stroomatlas 1992 Noordzee VlaamseBanken.* Dienst Kusthavens, Hydrografie Oostende, Ministerie Vlaamse Gemeenschap, Departement Leefmilieu Infrastructuur, 26 pp.

VAN CAUWENBERGHE, C.G. & DEN DUYVER, D. (1993) *Het radioplaatsbepalingssysteem Syledis langs de Belgische Kust en aangrenzend gebied.* Rapport **41** Hydrografische Dienst Kust, Ministerie Vlaamse Gemeenschap, Departement Leefmilieu Infrastructuur, 22 pp.

YALIN, M.S. (1964) Geometrical properties of sandwaves. *ASCE, J. hydraul. Div.* **90 (HY5)**, 105–119.

Spec. Publs int. Ass. Sediment. (1995) **24**, 53–68

Storm-enhanced sand transport in a macrotidal setting, Queen Charlotte Islands, British Columbia, Canada

C.L. AMOS*, J.V. BARRIE† *and* J.T. JUDGE‡

**Geological Survey of Canada, Bedford Institute of Oceanography, PO Box 1006, Dartmouth, Nova Scotia, Canada;*
†Geological Survey of Canada, Institute of Ocean Sciences, PO Box 6000, Sidney, British Columbia, Canada; and
‡Martec Limited, 1888 Brunswick Street, Halifax, Nova Scotia, Canada

ABSTRACT

This paper presents a method for the evaluation of regional sand transport patterns in a continental shelf setting, the inner Queen Charlotte Island shelf, western Canada. The method uses hydrodynamic and geological data sets of the region in a numerical model of sand transport (SEDTRANS). The model uses as input spatially variable bathymetry and grain size with time-variable surface gravity waves, tidal currents and wind-driven currents for a severe winter storm that took place on 26 February, 1984 (1 : 5 year event). Bed shear stress is computed following the methods of Grant & Madsen (1979) and Grant & Glenn (1983). This shear stress is then used to estimate sand transport rate using thresholds of sand motion outlined in Amos *et al.* (1988) with a modified algorithm from Engelund & Hansen (1967). Results conform with geophysical and geological observations on sand transport pathways. The tidal contribution to sand transport for the storm period accounts for only a small portion of the net sand transport in the region. Most of the predicted transport takes place as a consequence of the storm. As storms from the south-east typify this region, we conclude that such storms dominate the signature of sand transport. Under such storm conditions, coastal erosion of eastern Graham Island appears stimulated by high rates of coast-parallel sand transport. When the storm coincides with an ebbing tide, sand is predicted to move around Rose Spit and westwards into MacIntyre Bay, where it is available for subsequent fairweather beach progradation. We interpret the observed decrease in coastal erosion (cf. Barrie & Tucker, 1992) in the south of the study region as largely due to the presence of an amphidromic point in the M2 tide, which induces beach accretion and the formation of log barricades.

Rose Bar, located at the northern edge of Dogfish Bank, is interpreted to have formed as a result of storm-driven sand motion on the shelf, coupled with topographic steerage of both tidal and wind-driven flows and an associated sharp gradient in the sand transport rate at the bank margin. Thus, Rose Bar is the result of what we term a *hydraulic fence* to storm sand transport and is, therefore, a relatively stable feature. We conclude that Rose Bar is a nearshore storm ridge *sensu* Swift *et al.* (1981) which is the result of storm-induced continental shelf sand transport similar to that described by Amos & Judge (1991) for the east coast of Canada. In so far as these processes also led to the formation of Sable Island (Scotian Shelf), the development of Rose Bar may offer insight into the genesis of this engimatic east coast feature.

INTRODUCTION

The north-eastern tip of Graham Island, Queen Charlotte Islands, British Columbia (Fig. 1) is noted for its intensive and dynamic shore processes (Harper, 1980). This is evident in the high rates of coastal retreat along the eastern margin of Graham Island (>12 m yr^{-1}), where Pleistocene fine sand of the Queen Charlotte Lowlands crops out in its eroding cliffs (Barrie *et al.*, 1991). The rate of erosion decreases to the south where a wide intertidal region of intertwined tree trunks (log barricades,

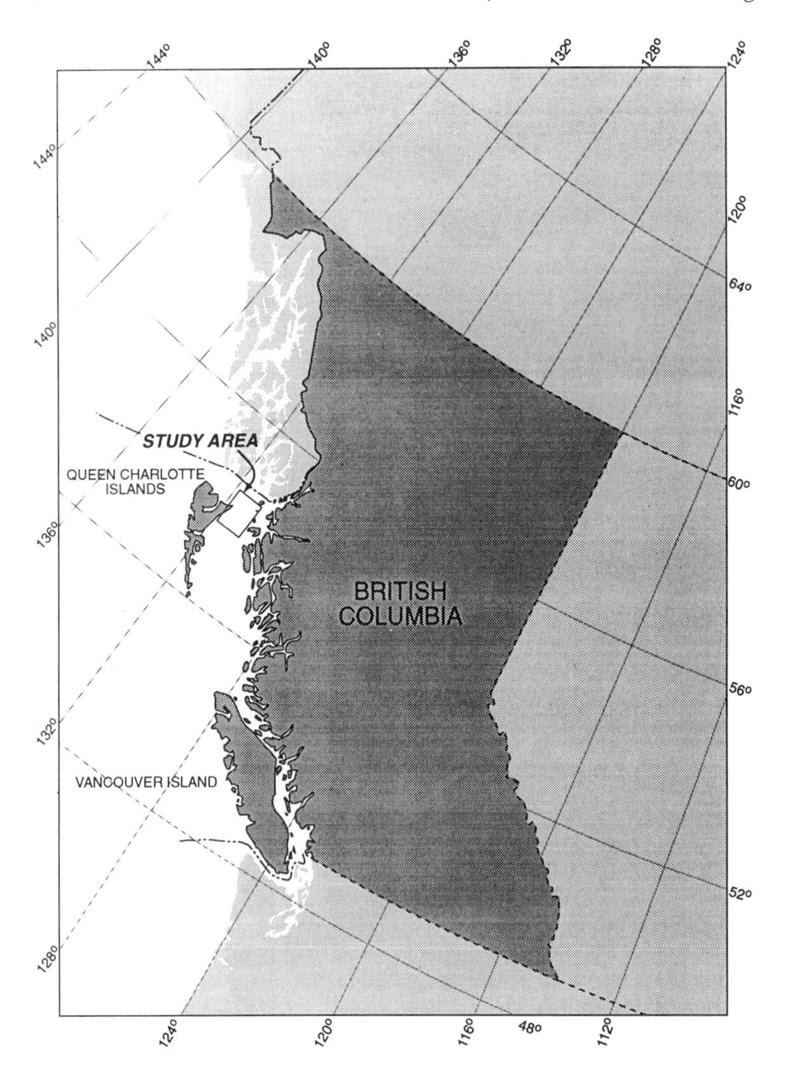

Fig. 1. A map of the study region, situated off the north-eastern tip of Graham Island (the northermost island of the Queen Charlotte Islands). The setting is within glaciated coastal channels that are subject to strong tidal flows and periodic winter storms.

Fig. 2) and a wide sandy beach has formed. What were the processes that led to the formation of the log barricades? Has the barricade trapped the beach sand and inhibited coastal erosion, or is the reduction in coastal erosion and the formation of the barricade and the associated beach merely the result of the same physical processes?

Clague & Bornhold (1980) and Harper (1980) showed that the sandy coastline of MacIntyre Bay is prograding approximately 0.3–0.4 m yr^{-1} through input of 140 000 m^3 yr^{-1} of sand (Harper, 1987). Harper (1980) suggested that the source of the sand was offshore, yet recent surveys of this region show it to be predominantly underlain by bedrock (Barrie *et al.*, 1990). Furthermore, there is no evidence for a source of sand in the west, where again bedrock prevails (Barrie & Tucker, 1992). Could the sand be derived from coastal erosion of eastern Graham Island as suggested by Bornhold & Barrie (1991)? Harper (1980) thinks not, stating. '...there is no evidence...that sediment moves from Rose Spit westward...' Indeed, a residual current flow to the east prevails in the region (Crawford & Thomson, 1991). So what is the source of the large supply of fine sand that feeds MacIntyre Bay? If the fine sand released from the eroding cliffs of eastern Graham Island does not enter MacIntyre Bay, then is it 'lost offshore' as suggested by Harper (1980)? If so, then where is it to be found?

Rose Bar is precariously perched on the north-

Fig. 2. The log barricade located on the southern coastal margin of eastern Graham Island.

western extremity of Dogfish Bank and forms the upper part of the steeply-dipping south flank of Dixon Entrance (Fig. 3). Nevertheless, there is no evidence for spill-over of the Bar down this south flank. Given the observed high degree of mobility of Rose Bar, why is this the case? Also, the sand bodies of Sable Island, Sable Island Bank and Eastern Shoal, Banquereau are situated in similar bathymetric settings to Rose Bar (Amos, 1989), so are there overriding bathymetric controls to shelf sand transport that leads to the development of these large-scale features?

The purpose of this paper is to examine the processes and events that lead to sand transport in the Dixon Entrance/Hecate Strait region, to determine the transport pathways of its sand, and to determine the regions of net sea-bed accretion and erosion. Thereafter, we will attempt to answer fundamental questions on the evolution of the Holocene sand cover in relation to: (i) the origin and stability of log barricades; (ii) the sources and transport paths of the fine sand in the region; and (iii) the origin and stability of Rose Bar.

BACKGROUND AND GEOGRAPHICAL SETTING

Bathymetry and surficial sediments

The study area is situated at the northern limit of the British Columbian continental shelf, western Canada. It is bound to the west by Graham Island (the northernmost of the Queen Charlotte Islands) and to the east by the mainland of British Columbia (Fig. 1). It encompasses Dixon Entrance to the north, a glacially excavated deep trough (>200 m) that trends west–east, and Hecate Strait, a shallow sheltered shelf that separates the Queen Charlotte Islands from the mainland.

The regional surficial geology has been described by Bornhold & Barrie (1991) and Moslow *et al.*, (1991), while detailed maps of the study region are provided by Barrie *et al.* (1990, 1991). The study region can be subdivided into five major surficial geological terrains diagnostic of differing modes of evolution and transport of surficial sediment. (1) Dixon Entrance deep water (depths >100 m); (2) Dixon Entrance south flank (depths <100 m); (3) the Rose Bar complex; (4) western Dogfish Bank, Hecate Strait (depths <30 m); and (5) eastern Dogfish Bank, Hecate Strait (depths >30 m).

1 The deeper regions of Dixon Entrance form the northernmost part of this study region and are characterized by a moribund or relict sea-bed. Relict iceberg scouring and glacial boulder fields crossed by modern trawl marks are prevalent (Fig. 4). The dominance of ponded fine-grained sediment in its basins is diagnostic of an absence of reworking, a prevalence of quiescent deposition and a general lack of sand transport.

2 The south flank of Dixon Entrance forms the boundary between the quiescent setting to the north and the more active shelf regions to the south. The

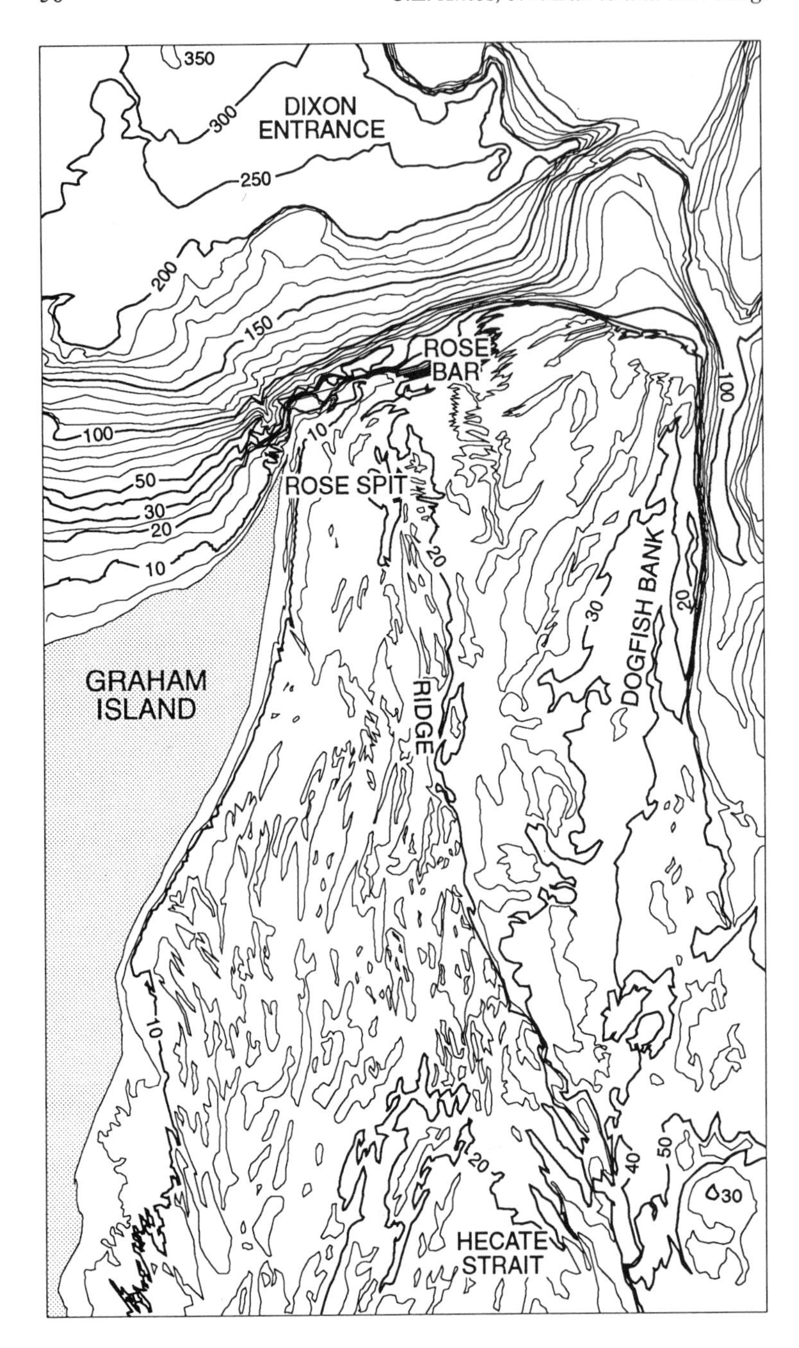

Fig. 3. The bathymetry of the study region and the domain of the numerical simulation of sand transport. The major morphological features are the glaciated channels of Dixon Entrance and Hecate Strait. Dogfish Bank, located in Hecate Strait, accommodates the sedimentary features that are the focus of this study. These are: Rose Spit, Rose Bar, the nearshore channel and the shore-parallel sand ridge.

steep slope is devoid of any notable features, except at the base where a belt of well-developed 2D megaripples are found on a thin veneer of mobile medium sand (Fig. 4B). These megaripples are probably modern and are oriented eastward, parallel to the local isobaths, being diagnostic of a net transport of sand in that direction.

3 The Rose Bar complex is a sand body of late Holocene age that is located at the northern extremity of Dogfish Bank, Hecate Stait. It occurs in

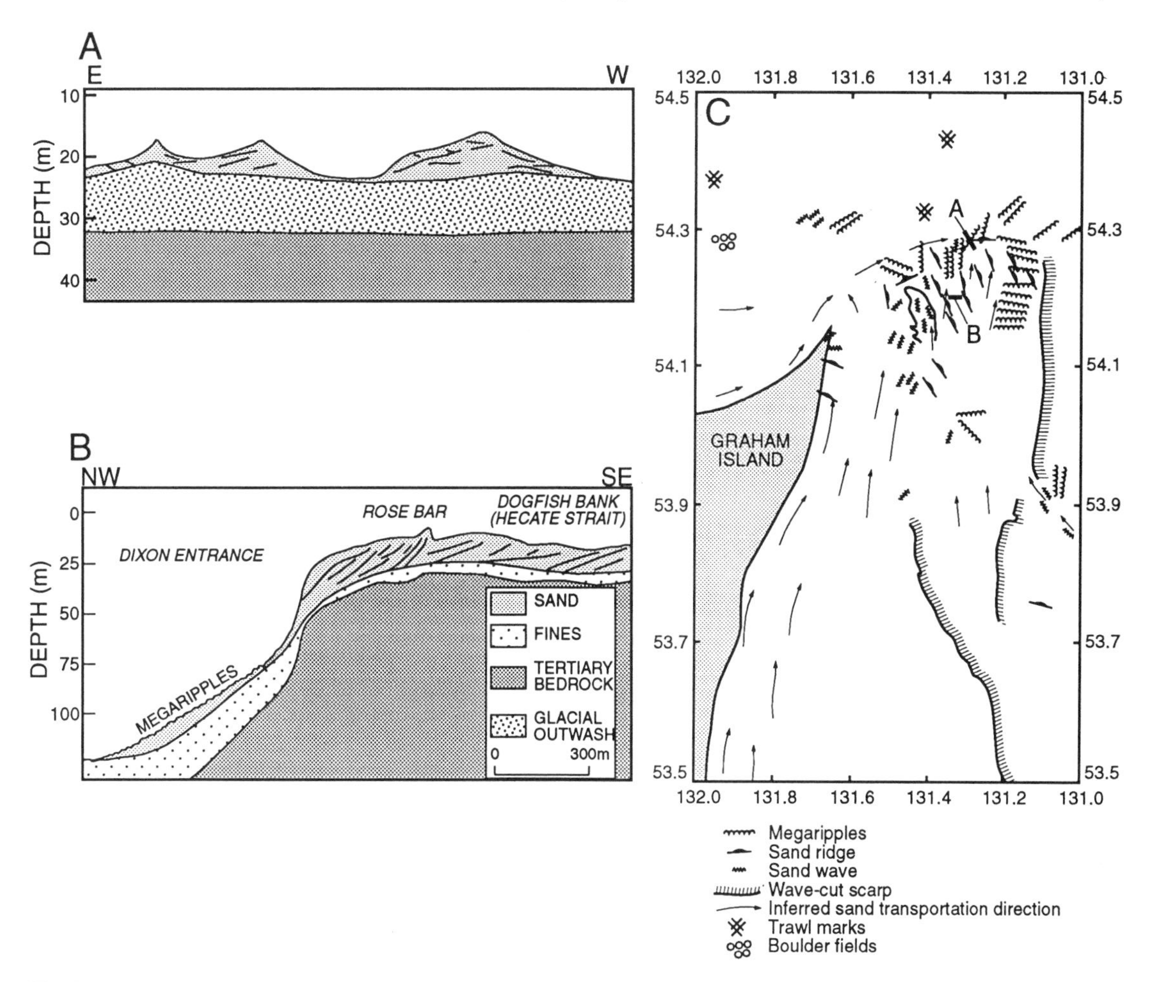

Fig. 4. The surficial features of the study region synthesized from Barrie *et al.* (1990). The direction of sand transport inferred from bedform orientation, grain size patterns and internal cross-stratification (see insert sections A and B) is also shown. In general, net transport is to the north in Hecate Strait and to the east in Dixon Entrance.

10–30 m of water. It is formed of medium and coarse sand and is moulded into a rich suite of sand waves, megaripples and sand ridges (Fig. 4A, C). Rose Bar is known locally for the rapidity of bathymetric changes during storms and is thus considered a possible hazard to shipping. The internal bedding planes dip to the north, indicating a progradation in this direction (Fig. 4B). The dip angles of the beds increase as the south flank of Dixon Entrance is approached, indicating a steepening of the bar slip face.

The top of the bar is today in morphological continuity with this south flank, yet the medium-coarse sand of which it is composed appears not to be spilling over into deeper water. The general orientation of the 2D megaripples is diagnostic of a northward and north-eastward net transport of sand.

4 Western Dogfish Bank, Hecate Strait, is largely a shoreface and inner shelf environment, characterized by a well-defined suite of shoreface-connected ridges and shore parallel sand ridges that are evident in the bathymetry (Fig. 3). It lies above the late Pleistocene wave-cut scarps mapped by Barrie *et al.* (1991), thus having been transgressed in the Holocene period. The region is largely composed of fine sand in the northernmost nearshore (0–15 m depths) and medium sand elsewhere. The direction of closure of the shoreface-connected ridges with the shoreline is thought to be diagnostic of the

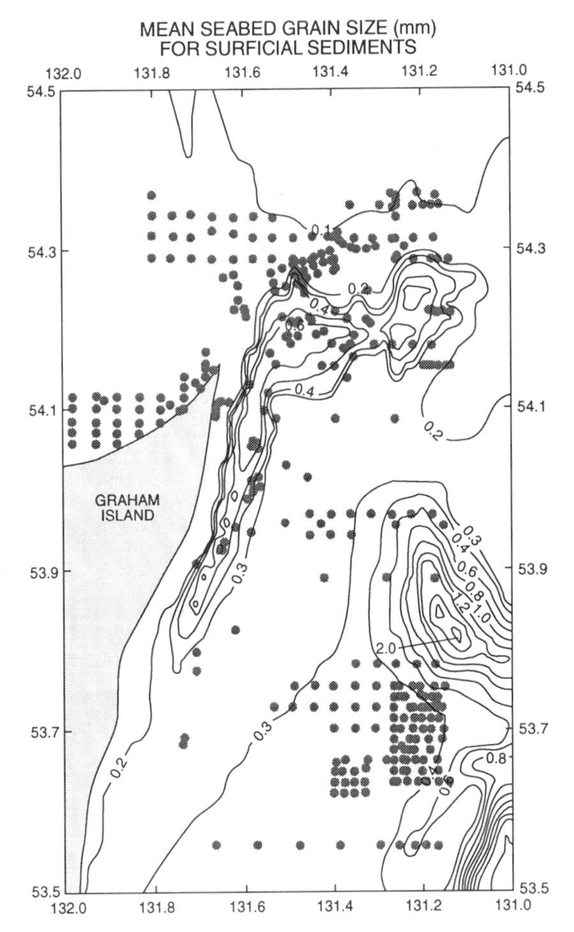

Fig. 5. The mean grain diameter of sea-bed sediments. Notice that the size of sediments comprising Rose Bar (d_{50} = >500 μm) is larger than that of the shoreface sand of eastern Graham Island (d_{50} = 125 μm). Also note the similarity in grain size between the shoreface of Graham Island and MacIntyre Bay. The dots denote sample locations.

longshore sand transport direction (Duane *et al.*, 1972), which in our case is to the north. The orientation of low-amplitude megaripples also indicates a transport in this direction. The layer of mobile sand is only about 10 m thick and overlies glacial outwash sand and gravel or Tertiary bedrock (Fig. 4A, B).

5 Eastern Dogfish Bank, Hecate Strait largely comprises a region below the wave-cut scarp (Fig. 4C). It shows an irregular bathymetry and is virtually devoid of large-scale bedforms. It is underlain by glacial sands and gravels that have been reworked in parts to form starved, low-amplitude megaripples. The morphology of these bedforms indicates a net sand transport to the north.

The fine sand (d_{50} = 160 μm) found within the shoreface of eastern Graham Island, in the greater part of MacIntyre Bay and in parts of Hecate Strait (Fig. 5) is considered to be derived from the eroding cliffs of eastern Graham Island (Barrie *et al.*, 1991). Given this source of fine sand, the longshore sand transport must also be northwards on to Rose Spit and thereafter into MacIntyre Bay, as no other areas are underlain by sands of this grain size. Although the northward development of Rose Spit supports this transport direction, there is no obvious mechanism by which the sand could be moved around the Spit to be pushed westwards into MacIntyre Bay against the prevailing sand transport direction (mapped in Fig. 4C) and against the residual circulation patterns (see below). From Fig. 5 it is evident that the grain-size trends that encompass Rose Bar extend southwards immediately below the shoreface (*c.* 15 m) of eastern Graham Island. The contrast in sediment size between the shoreface and inner shelf suggests that longshore rather than cross-shore transport of sand is taking place. It also suggests that Rose Bar is not derived from sediment supplied from the shoreface of eastern Graham Island.

Tides

The physical environment of the Hecate Strait region has been described by Chevron Canada Resources Limited (1982), Langford (1983) and Petro-Canada Limited (1983). Tides in the region are mixed diurnal/semi-diurnal. Tidal range varies from 3 m in the south to 5 m in the north. The tidal wave propagates eastwards into Dixon Entrance and northwards through Hecate Strait. The two branches of the flood tide meet near Rose Bar approximately 1 h into the flood tide, resulting in a complex flow pattern over Dogfish Bank. The shallow, depth-averaged tidal flow in the region is strongly semi-diurnal with a moderate tidal asymmetry and strong diurnal inequality. Plate 1 (facing p. 62) shows the spatial distribution of tidal currents at three different times of the day during a simulated storm (see below): (a) at peak flood (1300 GMT); (b) at high tide (1700 GMT); and (c) at peak ebb (2000 GMT). The strong flood current over Rose Bar (*c.* 0.50 m s^{-1}; Plate 1a), the clockwise veering of the currents over Dogfish Bank, and the

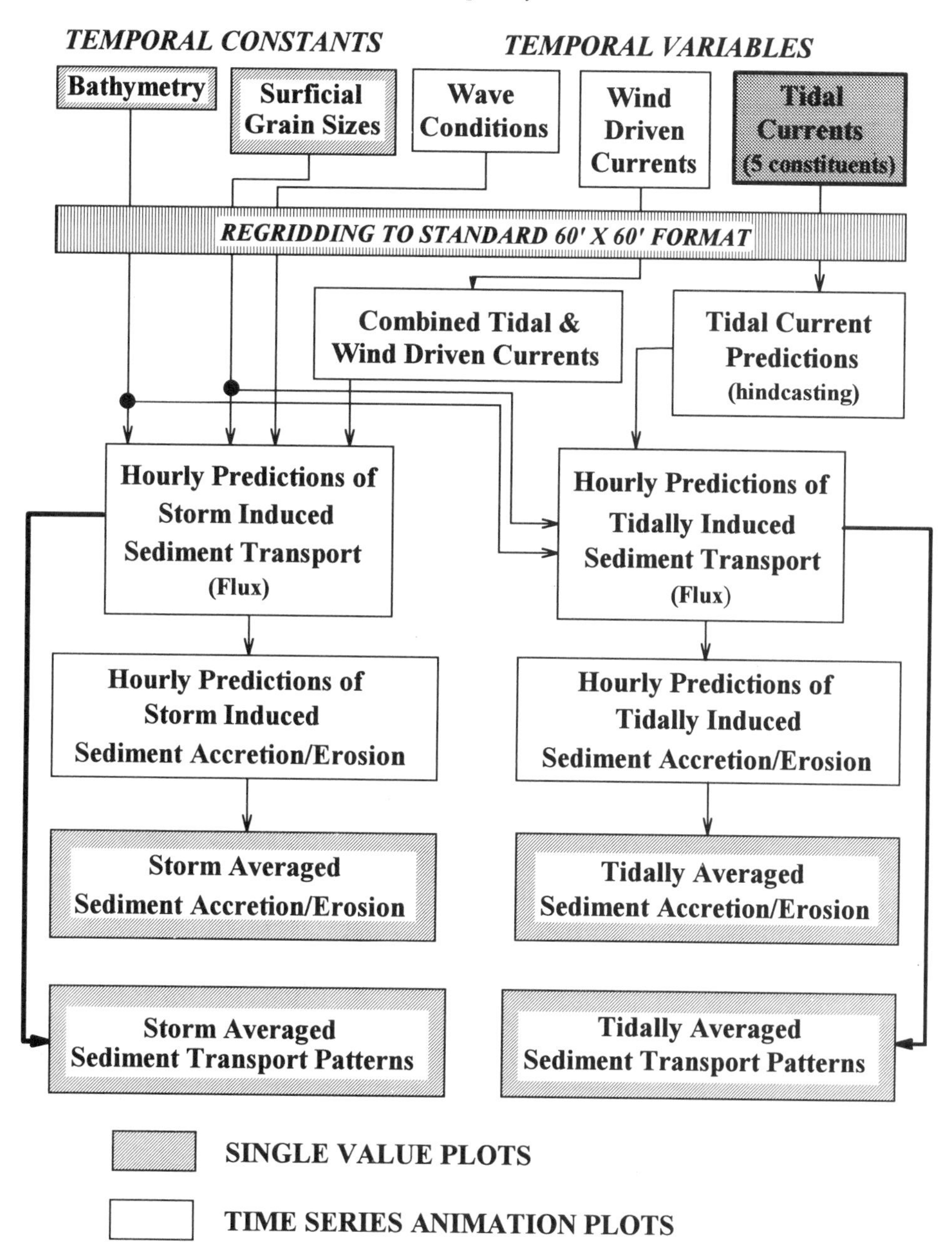

Fig. 6. A flow chart of the numerical simulation of sand transport undertaken in this study.

tidal convergence at the south coast of eastern Graham Island should be noted. In Plate 1b, two gyres signify the change in the tide from flood to ebb: a clockwise gyre centred immediately north of Rose Bar and an anti-clockwise one over southern Dogfish Bank. This pattern results in a divergence of flow along the coastline of eastern Graham Island that is enhanced during the peak ebb shown in Plate 1c. Notice in this latter figure that the coastal currents are greatest ($0.15\ \mathrm{m\ s^{-1}}$) along the north-eastern coastline and diminish southwards where they are directed offshore.

The residual tidal flow for the region is strongly rectified due to bottom frictional effects. This rectification results in an anti-clockwise gyre (the Rose Spit Eddy) within Dixon Entrance. It is evident along the southern flank of Dixon Entrance at depths <100 m (Crawford & Greisman, 1987) as a strong (0.06–0.11 m s^{-1}) eastward flow (Flather, 1987). In Hecate Strait, a residual current of about 0.10 m s^{-1} flows in a northward direction. These patterns are unaffected by river discharge but are modified by storms.

Temperature and salinity

Temperature and salinity are uniform with depth during winter months due to mixing. Typical values are 6°C and 32‰ respectively (Crawford *et al.*, 1988). In summer, the topmost 20 m are diluted to 23‰ by freshwater inflow (10^4 m^3 s^{-1}) and are heated to 14°C (Chevron Canada Resources Limited, 1982). Observations by Dodimead (1980) show that a strong thermocline is only developed in the months of June and July, salinities increasing gradually with depth over the remainder of the year. The water column is mixed during storms. This justifies the use of depth-averaged currents to simulate winter storm conditions within our model.

Weather patterns and sea states

The study region experiences intense storms that track northwards through the region, often deepening explosively. Lewis & Moran (1985) found that over the 25-year period between 1957 and 1983, at least two storms with maximum winds in excess of 60 knots occurred in every month except July. Nevertheless, there was a marked seasonal variation in the frequency of severe storms, with highest frequencies being recorded during late autumn and early winter. During these periods, wind-induced inertial oscillations resulted in currents exceeding 0.75 m s^{-1} (Petro-Canada Limited, 1983). Wind-driven currents principally flow along bathymetric contours due to strong topographic steering, such as that described by Loder & Wright (1985), and may even flow into the wind. Hannah *et al.* (1991) have modelled these currents for Hecate Strait. They show depth-averaged currents for a typical winter storm to be in excess of 0.30 m s^{-1} over much of the region. These currents flow north-eastwards along bathymetric contours. Major changes in both magnitude and direction of the currents take place along the break in slope immediately to the north of Rose Bar. Here the currents are much weaker and they flow westwards in the opposite direction to the residual tidal flow. This westward flow is also evident in the coastal region of MacIntyre Bay.

Though sheltered from Pacific Swell, the study region is strongly affected by local wind waves. Conditions are most severe when winds blow from the south-east. Thomson (1981) shows that significant wave heights >3.5 m prevail for 20–30% of the time during the winter months. The development and distribution of surface wave conditions in the Hecate Strait region have been examined and modelled by Hodgins & Nikleva (1986). They concluded that the strongest winds and the highest waves would be caused by south-easterly gales under pre-frontal conditions of a broad pressure gradient, the centre of which would be located to the west. Because such gales predominate in the study region, a storm of this type was chosen for the modelling approach in this study.

METHODS

Predictions of sand transport were made using the Geological Survey of Canada sediment transport numerical model—SEDTRANS. The structure of this model is shown in Fig. 6. SEDTRANS integrates geological and hydrodynamical data that were digitized and gridded on a 1 * 1 minute 2D matrix. Geological data include: bathymetry (Fig. 3), mean grain size (Fig. 5), sediment sorting and density. Bathymetry was derived from Canadian Hydrographic Chart 3802. The bathymetry was reduced to lowest low water. Water level was held constant in the model in view of the uncertainties over wind-induced water level fluctuations. Also, the CHS bathymetry data set was chosen in preference to the commonly used ETOPO5 world bathymetry data file, as the latter is highly inaccurate in the study region. The mean grain size of the surficial sediment was derived from samples collected by Barrie *et al.*, (1990, 1991; Fig. 5). Samples were analysed in a settling column and grain sizes thus reflect mean sedimentation diameters. Hydrodynamic data include: tidal current speed and direction, wind-driven current speed and direction, significant wave height, wave period and direction of propagation. Tidal currents were reconstructed from the five largest tidal constituents: M2, S2, N2, K2 and K1. These constituents were derived from a finite-element, two-dimensional depth-averaged model developed by Foreman & Walters (1990) which is accurate to within 3% in amplitude and 2° in phase (Flather, 1987). In our model, each constituent was defined by the amplitude of the semi-major axis, the amplitude of the semi-minor axis, the direction of the semi-major axis, and the Greenwich phase lag. The five constituents comprised a minimum of 72% of the tidal current amplitude (de Lange Boom, personal communication, 1990).

The wind-driven currents were reconstructed on the basis of a finite-differencing, two-dimensional depth-averaged numerical model developed and calibrated by Hannah *et al.* (1991). From an analysis of current records from several sites in the region, they found that the wind-driven current speed for a south-easterly storm varied more or less linearly with wind stress, with no significant change in the current direction. The predicted directions appear accurate to within ±20°, whereas the current magnitude is accurate to ±50% at depths <100 m. Our model uses nominal vectors from the Hannah *et al.* (1991) model which are modulated as a function of the wind speed (from Hodgins & Nikleva, 1986). Wind speed was converted to wind stress following the method of Smith (1988).

The wave climate was abstracted from plots of a full spectral wave hindcasting model undertaken by Hodgins & Nikleva (1986). In our case, wave height was digitized at 3 h intervals over the storm, while wave period and direction of propagation were held constant at 11 s and 330°T (true) respectively. The predicted wave climate for the storm under consideration was calibrated by Hodgins & Nikleva (1986) against two wave rider buoys in the region and was found to be accurate to ±15%.

The grid size, grid projections and positions of grid points varied between the three models described above. All data were therefore regridded to a standard 60 * 60 matrix between latitudes 53.5° N and 54.5° N and longitudes 131° W and 132° W. The resulting grid size was 1 * 1.5 nautical miles. Bathymetry and grain size were also digitized on this grid system so that trends in depth related to large-scale bedforms were largely smoothed. Interpolation and extrapolation of input data were based on an inverse distance method, wherein 10 adjacent data points were used in the definition of the surface trend. This was undertaken using SURFER 4 (Golden Software Inc., 1990) and the results were exported in the form of 2D arrays. For the period under consideration (0000Z/26 February 1984 to 0800Z/27 February 1984), tidal current constituents and the wind-driven currents were convolved to produce a combined mean flow for each grid point on a 12 min time step (Fig. 7). Thereafter, current flow, wave parameters, grain size and bathymetry were read by SEDTRANS to solve the magnitude and direction of sand transport. We adopted the method of Madsen & Grant (1976), Grant & Madsen (1979) and Grant & Glenn (1983) to evaluate the combined flow (wave-induced oscillatory flow and steady currents) total bed shear stress (skin friction plus form drag). Bed stress was transformed to sand transport, adopting a modified form of the total load (bedload and suspension) algorithm of Engelund & Hansen (1967). This method, when used with the total bed shear stress (skin friction and form drag), produced a good fit to observed sand transport rates on the Scotian Shelf (Fig. 8). The details of the model and the calibration data set are described by Amos (1987) and Amos *et al.* (1988) respectively.

THE MODEL STORM

The storm simulated in this study occurred on 26 February 1984. It was chosen on the basis of an analysis of 32 winter storms undertaken by Hodgins & Nikleva (1986). It was one of the most severe storms measured in the region. The peak wind speed at Sand Spit, Queen Charlotte Island (50 km south of the study region), correspond to the 1:5 year event (Brown *et al.*, 1986). The storm was the result of a large cyclone, the centre of which was situated west of the Queen Charlotte Islands in the North Pacific. The pressure field of the storm created northward-directed kinematic winds over the entire study region that blew consistently for 48 h at speeds up to 50 knots. Although the magnitude of the storm was exceptional, the spatial character of this south-easterly storm was typical of the region. Wave-rider measurements in western Dixon Entrance (50 km west of the study region) showed peak wave conditions to be 7–9 m. Wave heights varied significantly across our study region during the storm. According to Hodgins & Nikleva (1986), waves were highest over southern Dogfish Bank (H_s = 5.5 m) and lowest in MacIntyre Bay (Plate 2a). The direction of wave propagation was consistently 330°T.

The tides on 26 February were at a neap stage (see Fig. 7A). Thus the errors associated with using a

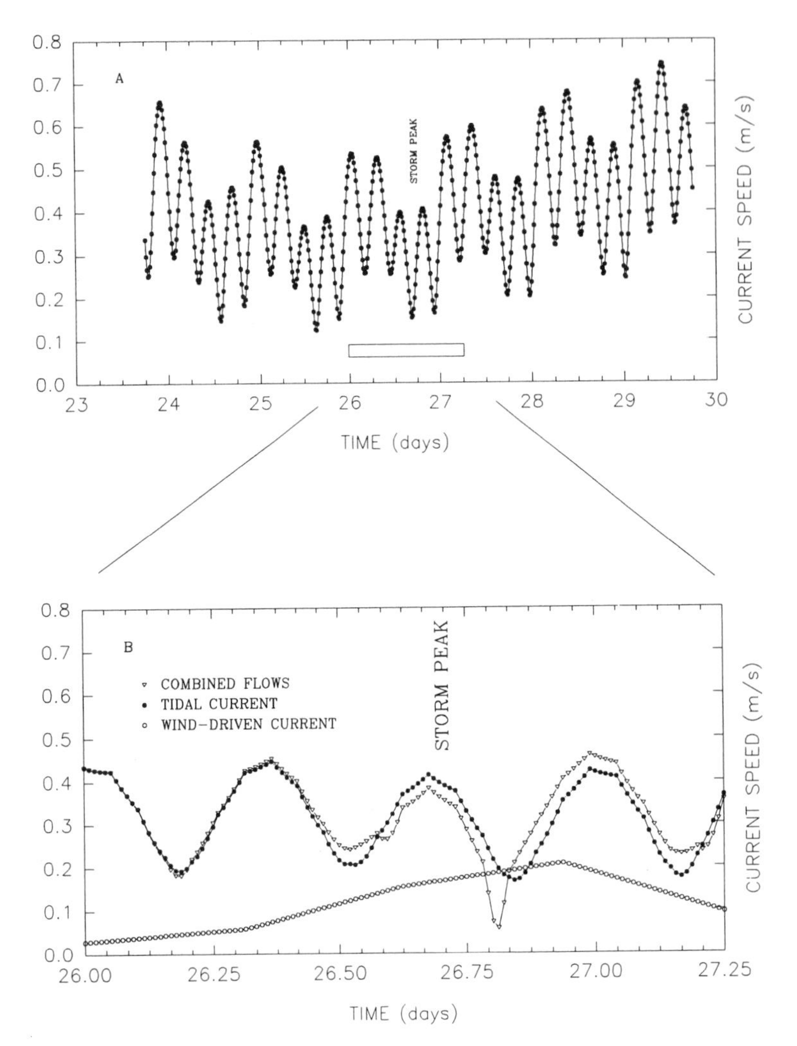

Fig. 7. Time-series of depth-averaged currents from a site on Rose Bar (latitude: 54°11′ N; longitude: 131°29′ W). (A) Predicted current speeds for the week centred about the simulated storm from the tidal model of Foreman & Walters (1990). Notice that the storm occurred during the neap stage of the tide. (B) Predicted currents for the duration of the storm: (i) wind-driven currents were derived from the model of Hannah *et al.* (1991); (ii) the regridded (60 * 60) tidal currents interpolated from Foreman & Walters (1990); differences between the peaks of these currents are the result of the interpolation procedure and slightly differing locations in the nodes of the two predictions; and (iii) the combined (wind-driven and tide) flows. Notice that the combined flow magnitude is not the sum of the two constituents. It is the equivalent velocity of the sum of the constituent bed shear stresses.

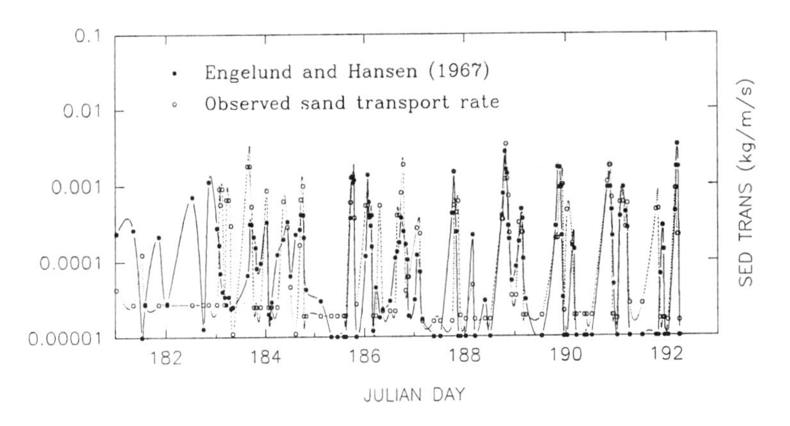

Fig. 8. The calibration data of the sediment transport model SEDTRANS. It is based on medium sand (d_{50} = 230 μm) transport measures made on Sable Island Bank and reported in Amos *et al.* (1988). In general, good results from SEDTRANS were obtained that fitted the observations to within ± a factor of five. Details of the model are given in Amos & Judge (1991) and Amos *et al.* (unpublished).

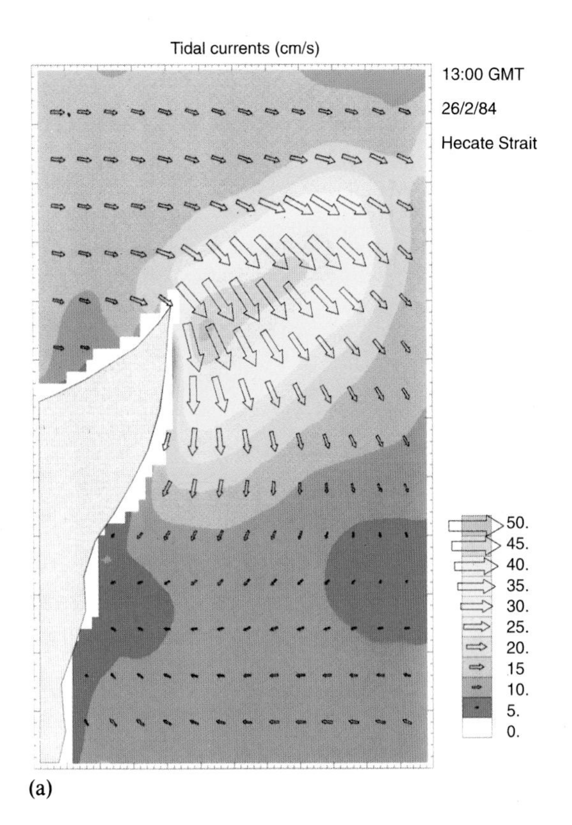

(a)

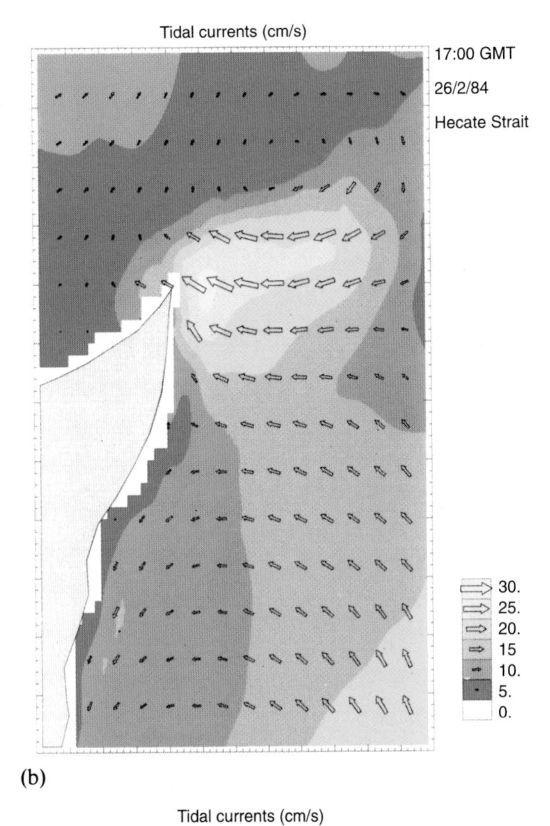

(b)

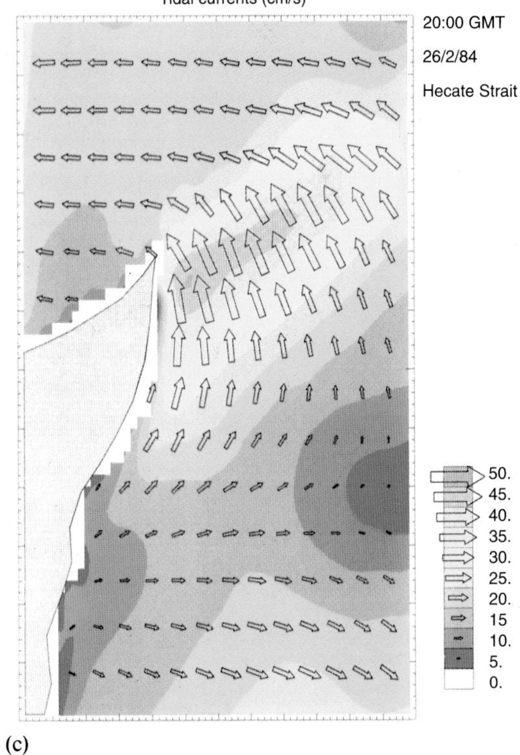

(c)

Plate 1 Predicted tidal current vectors for the day of the storm simulated in this study (26 February 1984): (a) peak flood (1300 GMT); (b) high tide (1700 GMT); and (c) peak ebb (2000 GMT). Notice the two tidally driven gyres evident at high tide and the divergence of longshore currents along the coast of eastern Graham Island during the ebb tide.

(*Facing page 62*)

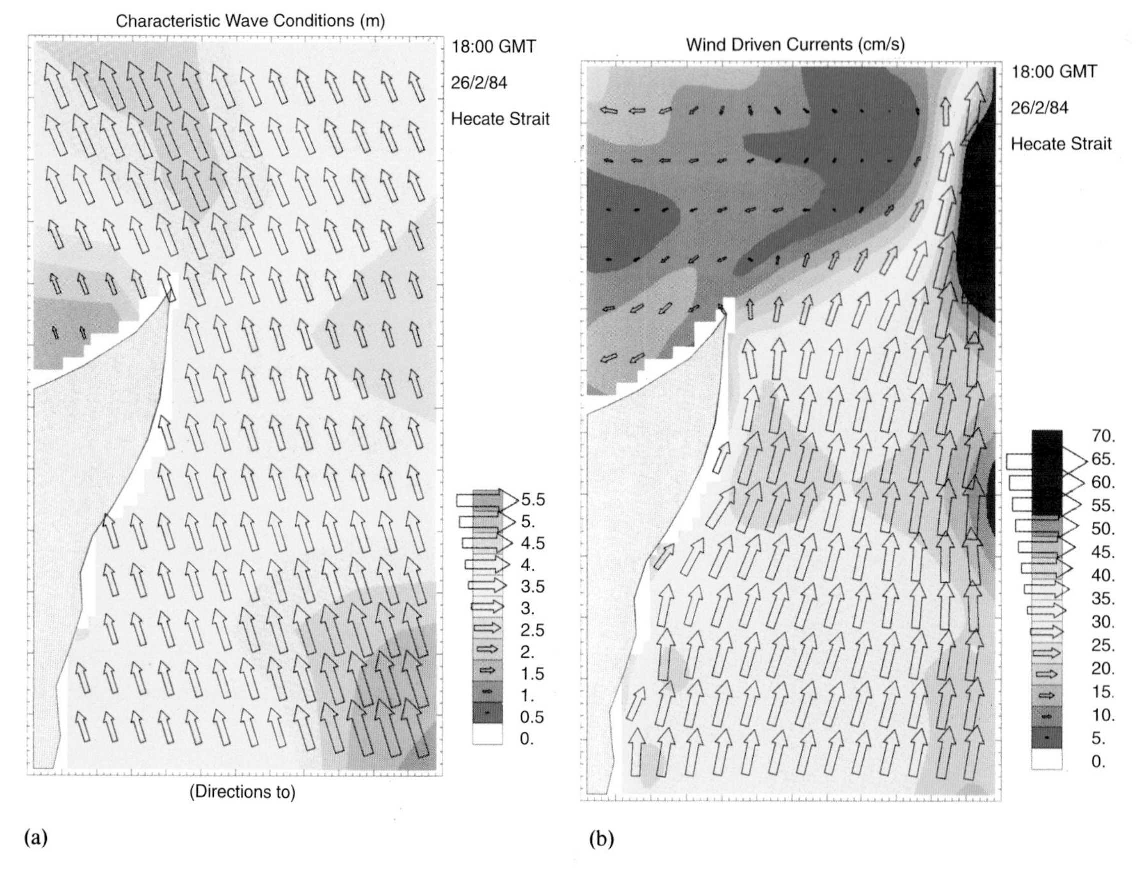

(a) (b)

Plate 2 The conditions during the storm peak for: (a) significant wave heights and directions of propagation taken from Hodgins & Nikleva (1986); and (b) depth-averaged wind-driven currents adapted from Hannah *et al.* (1991). Notice the rapid reduction in velocity over the southern flank of Dixon entrance and the backing of vectors to a south-westerly direction. The flow around Rose Spit and into MacIntyre Bay is particularly prominent.

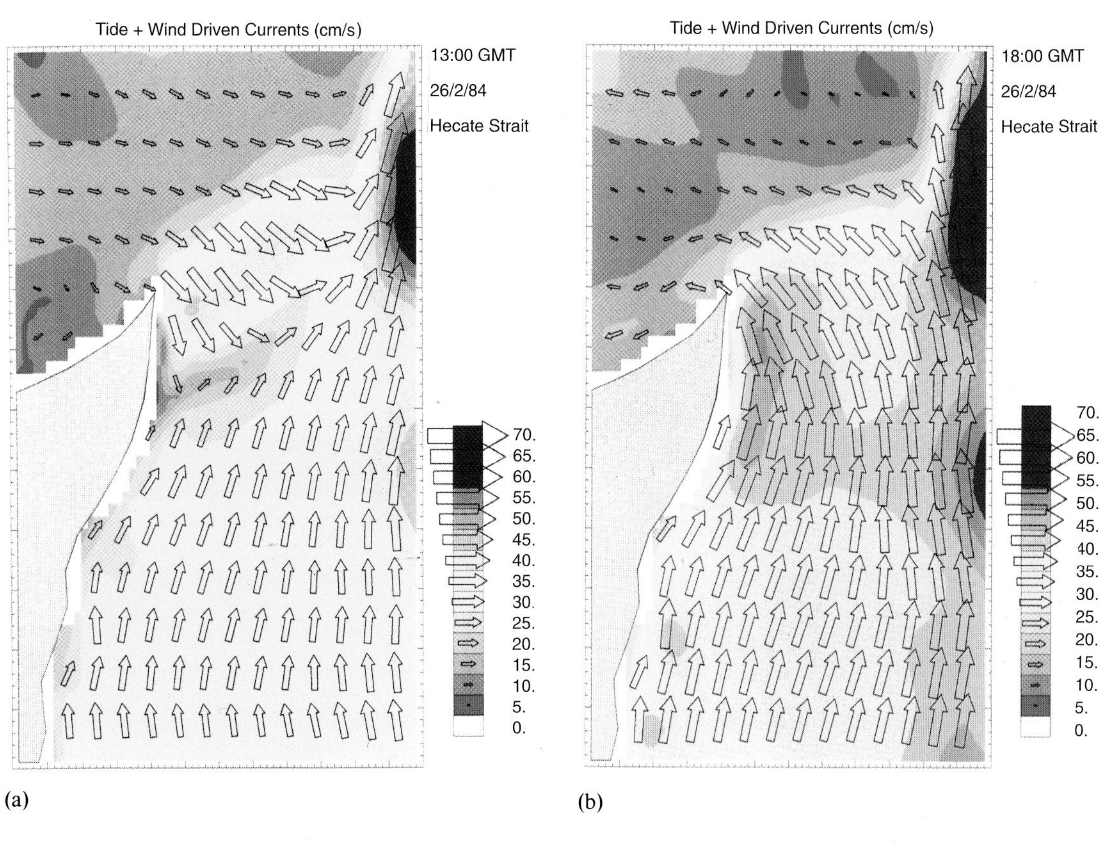

(a)

(b)

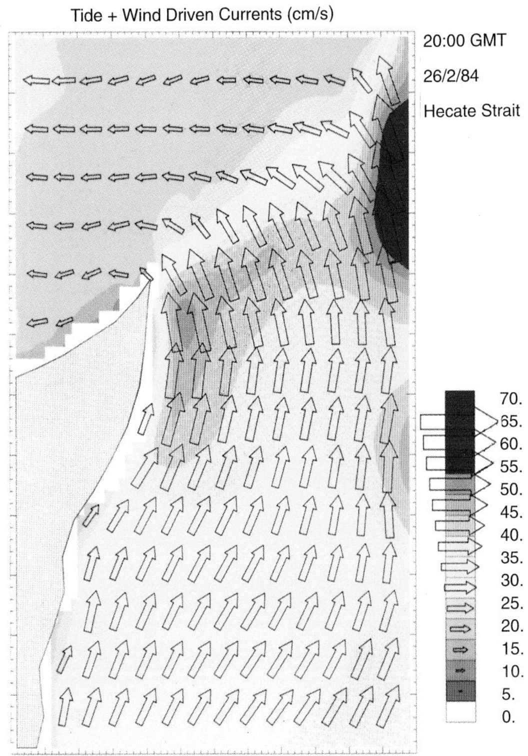

(c)

Plate 3 Predicted depth-averaged combined flow patterns for the day of the storm during: (a) peak flood (1300 GMT); (b) storm peak (1800 GMT); and (c) peak ebb (2000 GMT).

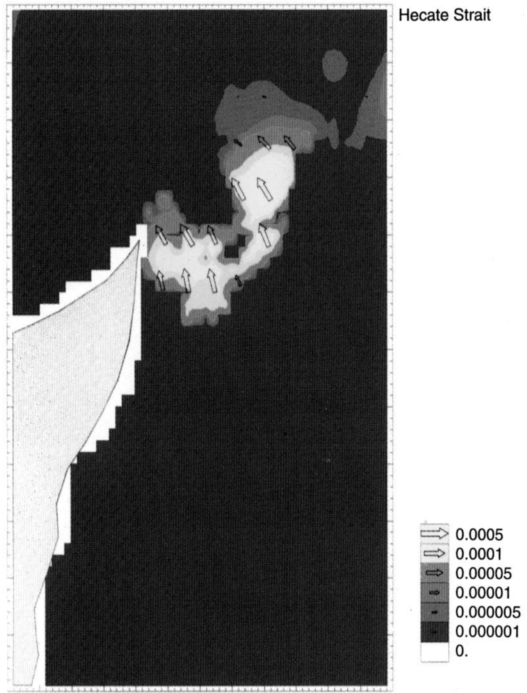

Plate 4 The tidal-averaged predicted sand transport rates (in kg m^{-1} s^{-1}) for the day of the storm. Notice that transport is restricted to the vicinity of Rose Bar and is to the north-northwest.

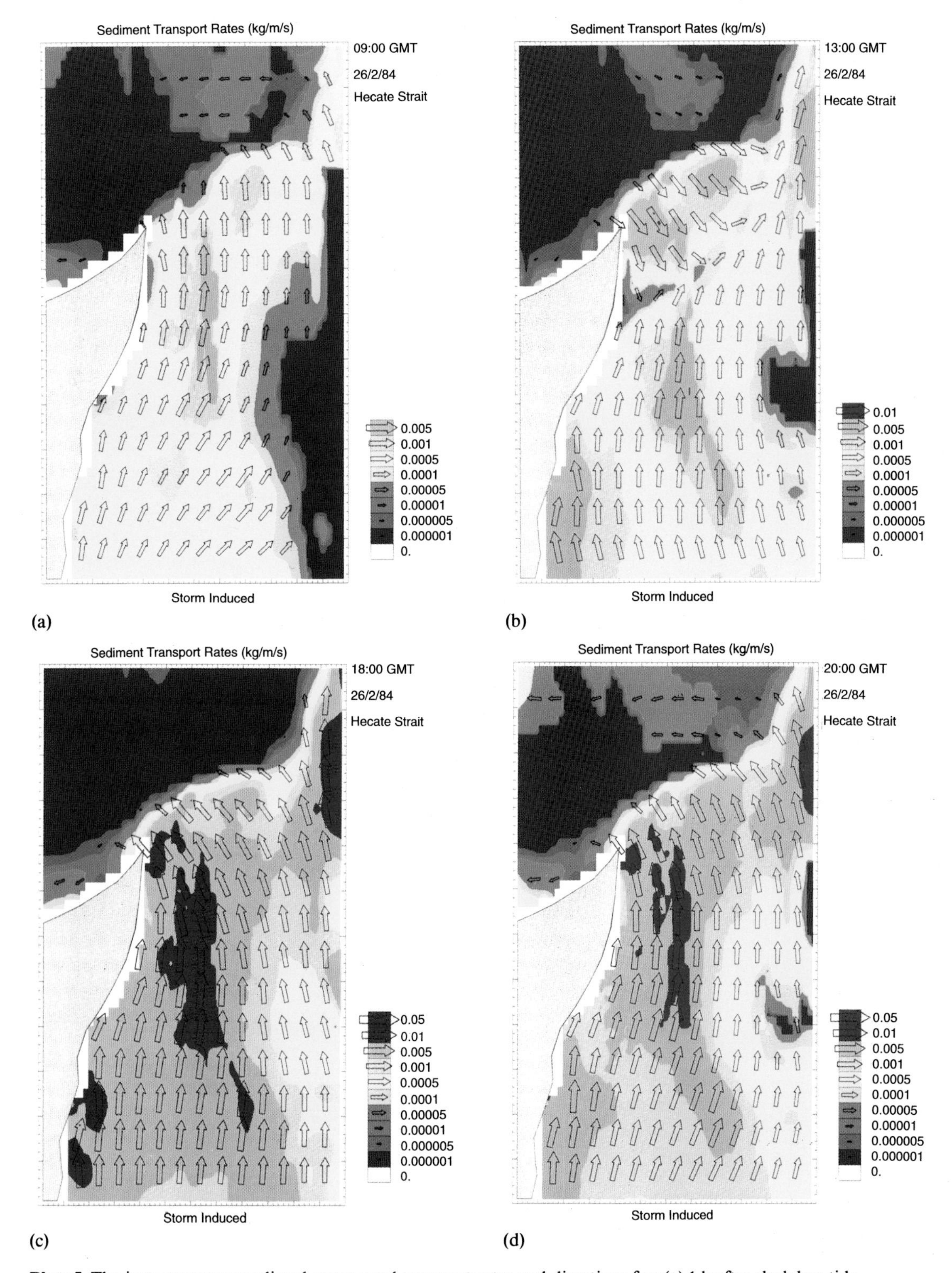

Plate 5 The instantaneous predicted storm sand transport rates and directions for: (a) 1 h after slack low tide (0900 GMT); (b) strong flood (1300 GMT); (c) storm peak (1800 GMT); and (d) peak ebb (2000 GMT). Notice that most of the sea-bed of Dogfish Bank and MacIntyre Bay is mobilized by the storm and that the sand moves predominantly northwards. The greatest transport rates coincide with the crest of the shore-parallel ridge and the nearshore channel. On the ebbing tide, sand is moved around Rose Spit and into MacIntyre Bay. There is no evidence for sand spill-over into Dixon entrance.

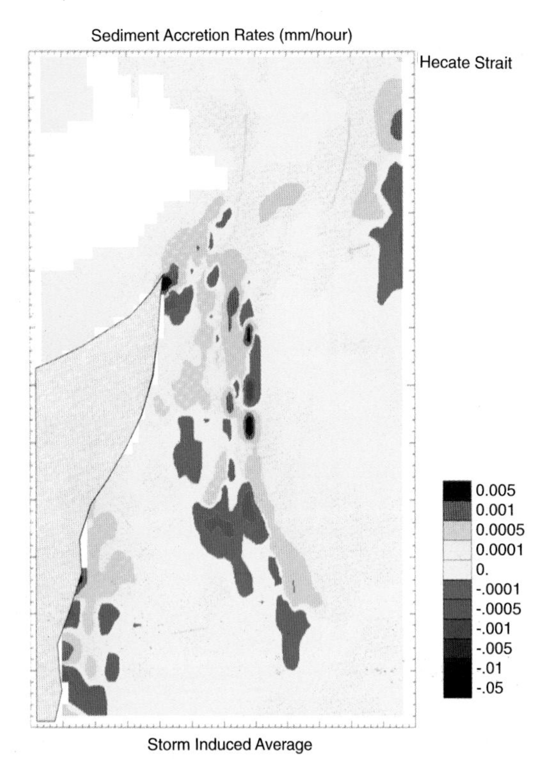

Plate 6 The predicted net sand accretion/erosion rates for the storm. Notice that Rose Bar and the shore-parallel sand ridge are largely accretional, whereas southern Dogfish Bank is largely erosional.

time-constant bathymetry are minimized to ± 1 m. High tide on Rose Bar was predicted at 1620Z and low tides at 1120Z and 2335Z (see Fig. 7B). Tidal currents in the region were strongest over Rose Bar. Peak predicted flood tides over Rose Bar were about 0.40 m s^{-1} (Plate 1a) and peak ebb tides were about 0.50 m s^{-1} (Plate 1c). Had the storm coincided with spring tides, the associated peak tidal currents over Rose Bar would have exceeded 0.70 m s^{-1} (Fig. 7A). Wind-driven currents (depth-averaged) over Rose Bar (latitude: 54°11′ N; longitude: 131°29′ W) for the duration of the storm are shown in Fig. 7B. The currents accelerated throughout the storm and were strongest well after the storm peak. The predicted spatial distribution of these currents for the storm peak is shown in Plate 2. It can be seen that the maximum current speed of 0.70 m s^{-1} was predicted for the extreme east of the region, where north-flowing tidal waters in Hecate Strait were accelerated through a narrow channel east of Rose Spit. In the shallower regions of Hecate Strait, wind-driven currents were strongest in the nearshore channel, where they attained a speed of 0.50 m s^{-1}. Elsewhere on the southern Dogfish Bank, these currents decreased to *c.* 0.40 m s^{-1}. Over Rose Bar the currents were 0.15 m s^{-1}, decreasing rapidly to 0.05 m s^{-1} along its northern flank. Flows were generally northwards except in Dixon Entrance and MacIntyre Bay where a weak westward-directed flow was predicted. This westward flow was greatest close to shore.

The combined flow time-series (tides and wind-driven currents: $\bar{U}_{tot}$) is shown in Fig. 7B. It is derived from the sum of the constituent bed shear stresses:

$$\tau = 0.5 f \rho \bar{U}_{tot}^2 \quad (1)$$

and

$$\bar{U}_{tot} = 1.41\sqrt{(\tau_w + \tau_t)/f\rho} \quad (2)$$

where τ_w is the average wave-induced total bed shear stress, τ_t is the combined steady flow total bed shear stress, and f is the friction factor derived from

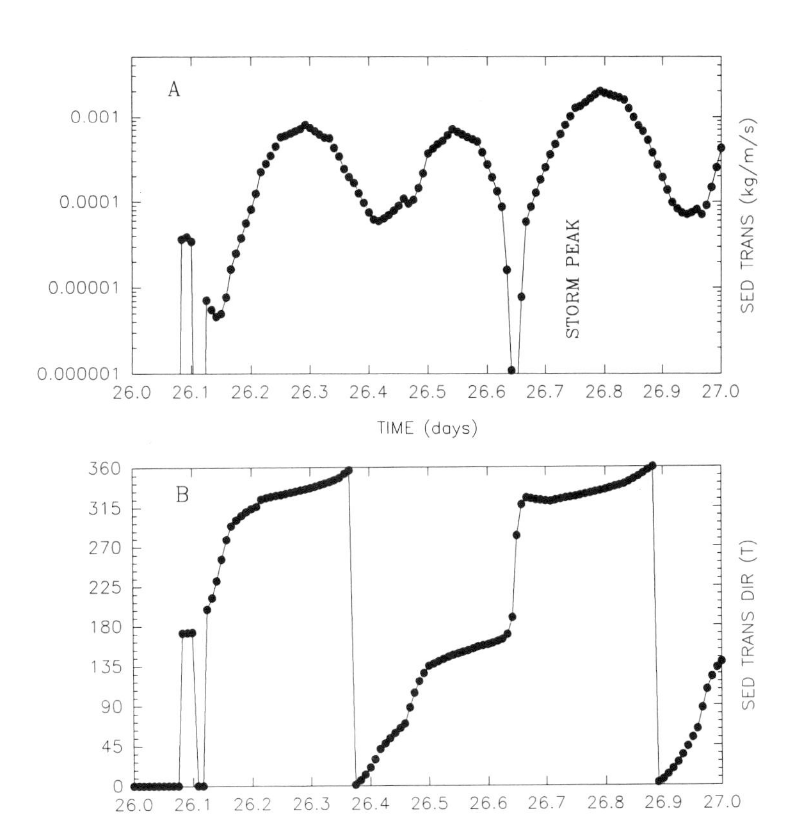

Fig. 9. A time-series of (A) the predicted storm sand transport rate and (B) the direction for a site on Rose Bar (latitude: 54°11′ N; longitude: 131°29′ W). Notice the strong tidal modulation in the rate and the two complete rotations in direction during the storm. The predicted sand transport rate approaches zero about 2 h before the peak of the storm, presumably because of a short-lived balance between the flooding tide and the wind-driven current.

Grant & Madsen (1979), but modified from its original use in that it is scaled to bed roughness, not skin friction. This roughness was equated with ripple height (H) on the basis of the relationship $H \approx 100d_{50}$ (where d_{50} is the median grain diameter). It is derived from the relationship between ripple wavelength and grain diameter proposed by Allen (1970) and assumes a ripple steepness of 0.1 (Nielsen, 1992). On this basis, bed roughness was assigned a mean constant value of 0.03 m. Under conditions where steady flows dominate, the combined bed shear stress is prevented from falling below the equivalent bed stress derived at by the method of Sternberg (1972). In this fashion, the addition of a small wave to a constant, steady current would not cause a decrease in sediment discharge, as was the case when only the skin friction component of f was used.

The regional flow patterns for the peak flood (1300 GMT), storm peak (1800 GMT) and peak ebb (2000 GMT) are shown in Plate 3a–c. At 1300 GMT a strong convergence is predicted immediately south of Rose Spit. Flows in Dixon Entrance are largely eastwards at 0.10 m s^{-1}, while those in Hecate Strait are to the north at 0.30 m s^{-1}. At the storm peak (Plate 3) the combined flows are greatest in the nearshore channel (0.50 m s^{-1}), in the eastern extremity of the region (0.70 m s^{-1}) and over Dogfish Bank (0.50 m s^{-1}). We predict a rapid decrease in flow strength (to 0.15 m s^{-1}) and an anti-clockwise gyre immediately north of Rose Bar. We also noted strong currents in the coastal region separating Rose Spit from Rose Bar. Peak ebb flows (at 2000 GMT, see Plate 3c) exceeded 0.50 m s^{-1} on Rose Bar and again maximize in the easternmost section of the region at 0.75 m s^{-1}. The westward flow in Dixon Entrance is at its strongest (0.20 m s^{-1}) and is apparent throughout this region.

RESULTS

Tidal sand transport for the day of the storm (26 February 1984) is predicted to be low. Sand transport occurs only at peak flows and is restricted to the tidal channel off Rose Spit and the outer part of Rose Bar. Maximum transport rates (at 2000 GMT) are estimated at $5 * 10^{-4}$ kg m^{-1} s^{-1} (see Plate 4). Notice that net transport is to the northwest, and is greatest on the southern side of Rose Bar. The north flank of the Bar is accretional because of a deceleration in transport rate across it, while the converse appears to be the case for the southern flank. Sediment accretion is predicted for a narrow band along the north side of Rose Bar only, while erosion is predicted for eastern Rose Spit. Elsewhere, tidal flow causes very little change in bed elevation.

Storm-induced bed changes were far more extensive and of greater magnitude than under tides alone. This is evident in Plate 5a–d which illustrates predicted sand transport for: (a) 0800 GMT—slack low tide on Dogfish Bank; (b) 1400 GMT—strong flood tide; (c) 1800 GMT—storm peak; and (d) 2000 GMT—peak ebb tide. By 0800 GMT on the day of the storm, most of the sand on Dogfish Bank was mobilized and was moving to the north in the direction of the wind-driven currents. This mobilization was almost entirely the result of the storm. Sand transport was greatest (*c.* $1 * 10^{-3}$ kg m^{-1} s^{-1}) over the dominant shore-parallel sand ridge (see Fig. 3 for the ridge location) and in the nearshore region between the ridge and the shoreline. More sediment was reaching Rose Bar at its southern margin than was leaving it at its northern end, indicating accretion. During the flood tide (1400 GMT), the direction of sand transport in Dixon Entrance and over Rose Bar had reversed, but it continued to be northwards over most of Dogfish Bank. Peak sand transport rate reached $1 * 10^{-3}$ kg m^{-1} s^{-1}. The area of this peak transport had grown to include Rose Bar, the coast-parallel ridge, and the southern coastal region of Graham Island. A zone of convergence was predicted to occur parallel with Rose Bar, and to the south of it. The storm peak coincided with an ebbing neap tide. The predicted peak sand transport rates were estimated to exceed $1 * 10^{-2}$ kg m^{-1} s^{-1} (Plate 5) over the shore-parallel sand ridge crest and on the western parts of Rose Bar. The nearshore channel and the southern coastline were also areas of strong northerly sand transport. The north flank of Rose Bar continued to be a site of net accretion through the deceleration of sand transport rate. There also appeared to be strong sand transport around Rose Spit and westwards into MacIntyre Bay. The pattern of sand transport predicted for the storm peak is apparent throughout the ebb (Plate 5d). The magnitude of transport was, however, less over southern Dogfish Bank and greater in northern Hecate Strait. The anti-clockwise rotation and deceleration of the sand transport vectors on the north flank of Rose Bar were still present, as was the westward transport of sand into MacIntyre Bay, albeit at very low rates. There was no predicted sand transport down the flank of Dixon Entrance at any stage of the storm.

The pattern of sand transport was in general one of a steady increase to the storm peak with strong semi-diurnal modulations caused by the changes in tidal flow magnitude and direction. This is clearly demonstrated in Fig. 9, which shows a predicted sediment transport time-series for a site on Rose Bar (latitude: 54°11′ N; longitude: 131°29′ W). Notice that the transport rate varied from zero to $2 * 10^{-3}$ kg m^{-1} s^{-1} and the direction of transport during the storm showed two clockwise rotations of 360° (i.e. two tidal cycles). Also note that the predicted sand transport rate decreased to zero 2 h before the peak of the storm, presumably because of a short-lived balance between the flooding tide and the wind-driven current. The net effect of these patterns is complex when viewed spatially. This we have expressed as storm-averaged sediment accretion or erosion in Plate 6. Peak erosion was found principally in the shallow waters south of Rose Spit, on the flanks of the shore-parallel sand ridge, and the central coastline of Graham Island. Peak deposition was predicted on the shore-parallel sand ridge crest, on Rose Spit, and over much of Rose Bar.

DISCUSSION

The patterns of sediment transport illustrated in this paper support the concept that storms are important in the distribution and transport of sand around Graham Island. The patterns of the predicted storm transport provide explanations for the three principal questions addressed in the introduction: (i) the origin of the log barricades of eastern Graham Island; (ii) the source of sand in MacIntyre Bay; and (iii) the origin of Rose Bar. The continuous coastal erosion of north-eastern Graham Island is not surprising given the high removal rates of sand from the shoreface by northward-directed longshore currents during south-easterly winter storms. Consequently, wide sand beaches or log barricades do not occur in this region. Further south, however, we note that the coastline is punctuated by zones of high deposition, separated by zones of high erosion. The wide beaches in the south of the region may be related to the predicted zones of coastal storm accretion that is influenced by the two coastal, tidally driven gyres. The region of rapid coastal retreat is influenced by the Rose Spit Gyre that enhances the northward storm-driven flow along the coastline of north-eastern Graham Island and so increases the longshore sand transport rate. The southern gyre, by contrast, flows southwards along the coast and works in opposition to the wind-driven currents, thereby reducing the longshore transport of sand. We propose that the wide beaches of the southern coastline are the consequence of this latter effect and would form regardless of the presence of the log barricades. Furthermore, it appears that the trees are trapped by the same processes that trap the sand. The natural formation of log barricades to the north is therefore unlikely and would have little influence on beach stability were such log barricades to be introduced.

The model shows that the origin of sand in MacIntyre Bay could easily be eastern Graham Island. The fine sand, transported northwards during storms, moves around Rose Spit and westwards into MacIntyre Bay when such storms are coincident with ebbing tides. We found no mechanism, either tidal or storm-related, to transport beach material eastwards across Dogfish Bank. Nor did we predict the spill-over of sand down the steep south flank of Dixon Entrance. Thus, almost all the sand eroded from the cliffs of north-eastern Graham Island during storms is predicted to end in MacIntyre Bay.

The shore-parallel sand ridge on eastern Dogfish Bank was predicted to accrete at its crest during storms. The high transport rates over the ridge and the erosion of parts of its flanks suggest that the ridge is active and migrating. We propose that the ridge is in dynamic equilibrium with storm conditions and is therefore a modern phenomenon, not a relict one. Our results show no clear direction of ridge migration. We suggest that it is modified only slightly by each storm, while maintaining its general position which is dictated by the dominance of storm-driven currents. Also, the ridge appears to be unaffected by neap tidal currents.

Rose Bar appears to be fed by sand transported from southern Dogfish Bank during storms. Little, if any, sand comes from the coastline of eastern Graham Island (this material being trapped and transported within the nearshore channel off Rose Spit). The bar crest and southern flank accreted during our simulation. This, together with the low net tidal transport rates, suggest that storms and not tidal currents are essentially responsible for its growth. Consequently, Rose Bar is not an extension of Rose Spit as it is not fed by its adjacent shoreline.

We propose that it is similar in genesis to shelf-edge sand ridges found on other storm-influenced continental shelves (Swift & Thorne, 1991). The accretion of Rose Bar suggests that it is constructional and in dynamic equilibrium with the stormy conditions of the region. The strong deceleration and rotation of sand transport vectors along Rose Bar's northern flank suggest that it is no longer migrating northwards, nor is it spilling over into Dixon Entrance. The absence of spill-over is consistent with seabed granulometry which shows a dominance of fine-grained sediments below 100 m of water depth (Fig. 5).

The physical setting and the sediment transport patterns around Rose Bar appear similar to those predicted for Sable Island, which is located on the eastern Canadian continental shelf (Amos & Judge, 1991). The bathymetric and hydrodynamic settings also appear similar. Furthermore, the internal tabular cross-stratification (dipping in the direction of major changes in slope and of storm propagation) are similar in both cases (compare Fig. 2.10 in Amos, 1989 to Fig. 4B herein). It follows that Rose Bar may be useful as a model for the evolution of this enigmatic east coast feature. Historic evidence compiled by one of the authors (J.V.B.) shows that Rose Bar has become shallower over the last 100 years through accretion, a series of isolated banks now existing above mean water level at its western extremity (adjacent to Rose Spit). It is possible that Sable Island evolved in the same way.

CONCLUSIONS

The model presented herein appears to answer many basic questions regarding the origin and stability of surficial sediments along the coastline and inner shelf of northeastern Graham Island. Although the model is strictly valid for one particular storm only, it appears to fit generalized geological interpretations based on geophysical and geomorphological evidence. A greater degree of certainty could be achieved by collecting calibration data on sediment transport in the study region. This will be the focus of future efforts. The following are the major points of conclusion of this study:

1 Nothwithstanding the meso-macrotidal nature of Hecate Strait, storm sediment transport is a dominant process in the evolution of the coastal and nearshore surficial geology.

2 A simulation of sand transport for a storm on 26 February 1984 (1 : 5 year event) showed significant increases in sand transport over purely tidally driven transport.

3 Sand transport was largely alongshore and to the north. This sand was funnelled within a nearshore channel off Rose Spit and was redirected westwards into MacIntyre Bay when the storm coincided with ebbing tides. This mechanism may be the cause of the continued erosion of eastern Graham Island and the continued growth of northern Graham Island.

4 The eastern coastal region is subject to two tidally driven gyres that significantly affect coastal stability. The Rose Spit Gyre enhances northerly directed storm-driven flows in the north, whereas the Dogfish Bank Gyre opposes such flows in the south. Consequently, accretion of wide beaches in the south takes place independently of the development of log barricades (also present in the coastal zone). We propose that the beaches and barricades are merely the result of the same oceanographic phenomenon. If this is so, then the barricades have little effect on beach development. Furthermore, wide sandy beaches are unlikely to develop further to the north under present conditions. As a result, high rates of coastal erosion are likely to continue.

5 The shore-parallel sand ridge on Dogfish Bank does not appear to be a relict submerged barrier island. Our results show it to be the most active part of the shelf during storms and to possess a rapidly accreting ridge crest. We suggest that the ridge is modern, active and in dynamic equilibrium with modern storm conditions. Its position is apparently stable, being at the margin of strongly aligned flows in the nearshore region.

6 Rose Bar is separated from Rose Spit by an active nearshore channel which prevents beach sediment from eastern Graham Island from reaching the bar. We predict accretion of the crest of the bar during storms by sand reworked from shelf sediments to the south. Also, we saw no spill-over of sand into Dixon Entrance to the north. Rose Bar may not be a true bar in terms of its evolution. It is, however, similar in genesis to sand ridges formed on stormy continental shelves.

7 Historical charts show that Rose Bar has evolved quickly. Its relatively constant position together with its recent shoaling suggest it to be in dynamic equilibrium with modern storm conditions. The physical and dynamic settings of Rose Bar are similar to those of Sable Island. More detailed studies of this Bar may thus offer insights into the processes controlling the evolution of Sable Island, whose history is much longer and less well understood.

ACKNOWLEDGEMENTS

The authors wish to acknowledge the contributions of the following persons: K. Conway, who compiled much of the sample data; S. Davidson, who assisted in adapting SEDTRANS; M. Foreman for the provision of raw tidal data; C. Hannah, who provided raw wind-driven current predictions; M. Best and the Pacific Geoscience Centre for use of their facilities; and J. Verhoef and Regional Reconnaissance subdivision for provision of colour graphic facilities. The third author of this paper (J.T. Judge) passed away during the preparation of this paper. We mourn his loss and acknowledge his excellence as a coastal engineer.

REFERENCES

ALLEN, J.R.L. (1970) *Physical Processes of Sedimentation.* Unwin University Books, London, 248 pp.

AMOS, C.L. (1987) The Atlantic Geoscience Centre sediment transport numeric models. *Geol. Surv. Can. Open File Rept* **1705** (10 sections).

AMOS, C.L. (1989) Submersible observations of Quaternary sediments and bedforms on the Scotian Shelf. *Geol. Surv. Can.* **88**, 9–26.

AMOS, C.L. & JUDGE, J.T. (1991) Sediment transport on the Canadian eastern continental shelf. *Continent. Shelf Res.* **11,** 1037–1068.

AMOS, C.L., BOWEN, A.J., HUNTLEY, D.A. & LEWIS, C.F.M. (1988) Ripple generation under the combined influences of waves and currents on the Canadian continental shelf. *Continent. Shelf Res.* **8(10)**, 1129–1153.

BARRIE, J.V. & TUCKER, K. (1992) Nearshore surficial geology of the Queen Charlotte islands—northwestern Graham Island. *Geol. Surv. Can. Open File Rept* **2523** (12 map sheets).

BARRIE, J.V. LUTERNAUER, J.L. & CONWAY, K.W. (1990) Surficial geology of the Queen Charlotte Basin; Dixon Entrance–Hecate Strait. *Geol. Surv. Can. Open File Rept* **2193** (7 map sheets).

BARRIE, J.V., BORNHOLD, B.D., CONWAY, K.W. & LUTERNAUER, J.L. (1991) Surficial geology of the northwestern Canadian continental shelf. *Continent. Shelf Res.* **11**, 701–715.

BORNHOLD, B.D. & BARRIE, J.V. (1991) Surficial sediments on the western Canadian continental shelf. *Continent. Shelf Res.* **11(8–10)**, 685–699.

BROWN, R.D., BALAKAS, N., MORTSCH, L.D., SAULESLEJA, A. & SWAIL, V.R. (1986) Marine climatological atlas—Canadian west coast. *Canadian Climate Centre, Downsview, Rept* **86–10**, 372 pp.

CHEVRON CANADA RESOURCES LIMITED (1982) Initial environmental evaluation for renewed petroleum exploration in Hecate Strait and Queen Charlotte Sound. *Chevron Canada Limited, Calgary, Rept* **1(1–3)**, 125 pp.

CLAGUE, J.J. & BORNHOLD, B.D. (1980) Morphology and littoral processes of the Pacific coast of Canada. In: *Coastline of Canada* (Ed. McCann, S.B.), *Geol. Surv. Can.* **80–10**, 339–380.

CRAWFORD, W.R. & GREISMAN, P. (1987) Investigation of permanent eddies in Dixon Entrance, British Columbia. *Continent. Shelf Res.* **7(8)**, 851–870.

CRAWFORD, W.R. & THOMSON, R.E. (1991) Physical oceanography of the western Canadian continental shelf. *Continent. Shelf Res.* **11(8–10)**, 683–699.

CRAWFORD, W.R., HUGGETT, W.S. & WOODWARD, M.J. (1988) Water transport through Hecate Strait, British Columbia. *Atmosphere-Ocean* **26(3)**, 301–320.

DODIMEAD, A.J. (1980) A general review of the oceanography of the Queen Charlotte Sound-Hecate Strait-Dixon Entrance region. *Can. Fish. aquat. Science, Nanaimo,* **1574**, 248 pp.

DUANE, D.B., FIELD, M.E., MEISBURGER, E.P., SWIFT, D.J.P. & WILLIAMS, S.J. (1972) Linear shoals on the Atlantic inner continental shelf, Florida to Long Island. In: *Shelf Sediment Transport: Process and Pattern* (Eds Swift, D.J.P., Duane, D.B. & Pilkey, O.H.), pp. 447–498. Dowden, Hutchinson & Ross, Stroudsburg.

ENGELUND, F. & HANSEN, E. (1967) *A Monograph on Sediment Transport in Alluvial Streams.* Teknisk Vorlag, Copenhagen, 62 pp.

FLATHER, R.A. (1987) A tidal model of the northeast Pacific. *Atmosphere-Ocean* **25(1)**, 22–45.

FOREMAN, M.G.G. & WALTERS, R.A. (1990) A finite-element tidal model for the southwest coast of Vancouver Island. *Atmosphere-Ocean* **28(3)**, 261–287.

GOLDEN SOFTWARE INC. (1990) *Surfer Version 4 Reference Manual.* Publ. Golden Software, Colorado.

GRANT, W.D. & GLENN, S.M. (1983) Continental Shelf Bottom Boundary Layer Model, Vol. I–III. Report to Pipeline Research Committee. *Am. Gas Ass. Proj.* **1**, Pr-153-126.

GRANT, W.D. & MADSEN, O.S. (1979) Combined wave and current interaction with a rough bottom. *J. geophys. Res.* **84(4)**, 1797–1808.

HANNAH, C.G., LEBLOND, P.H., CRAWFORD, W.R. & BUDGELL, W.P. (1991) Wind-driven depth-averaged circulation in Queen Charlotte Sound and Hecate Strait. *Atmosphere-Ocean* **29(4)**, 712–736.

HARPER, J.R. (1980) Coastal processes on northeastern Graham Island, Queen Charlotte Islands, B.C. *Geol. Surv. Can.* **80-1A**, 13–18.

HARPER, J.R. (1987) Non-carbonate sediment budgets. In: *Workshop on Coastal Processes in the South Pacific Island Nations. SOPAC Tech. Bull.* **7**, 55–58.

HODGINS, D.O. & NIKLEVA, S. (1986) On the impact of new observing sites on severe sea state warnings for the B.C. coast. Consult. Rept Dept Fish. Oceans, Sidney, B.C.

LANGFORD, R.W. (1983) *A Preliminary Environmental Assessment of Offshore Hydrocarbon Exploration and Development.* British Columbia Ministry Environment Publ., 334 pp.

LEWIS, C.J. & MORAN, M.D. (1985) Severe storms off Canada's west coast: a catalogue summary for the period 1957 to 1983. *Canadian Climate Centre, Downsview, Rept* **85–7.**

LODER, J.W. & WRIGHT, D.G. (1985) Tidal rectification and frontal circulation on the sides of Georges Bank. *J. mar. Res.* **43**, 581–604.

MADSEN, O.S. & GRANT, W.D. (1976) Sediment transport in the coastal environment. *M.I.T. Dept Civil Eng. Rept* **209**, 105 pp.

MOSLOW, T.F., LUTERNAUER, J.L. & ROHR, K. (1991) Origin and late Quaternary tectonism of a western Canadian continental shelf trough. *Continent. Shelf Res.,* **11**, 755–769.

NIELSEN, P. (1992) *Coastal Bottom Boundary Layers and Sediment Transport.* World Scientific, 324 pp.

PETRO-CANADA LIMITED (1983) *Offshore Queen Charlotte Islands Initial Environmental Evaluation.* Petro-Canada Limited, Calgary.

SMITH, S.D. (1988) Coefficients for sea surface wind stress, heat flux, and wind profiles as a function of wind speed and temperature. *J. geophys. Res.* **15**, 467–472.

STERNBERG, R.W. (1972) Predicting initial motion and bedload transport of sediment particles in the shallow marine environment. In: *Shelf Sediment Transport, Process and Pattern* (Eds Swift, D.J.P., Duane, D.B. & Pilkey, O.H.), pp. 61–83. Dowden, Hutchinson & Ross, Stroudsburg.

SWIFT, D.J.P. & THORNE, J.A. (1991) Sedimentation on continental margins, I: A general model for shelf sedimentation. In: *Shelf Sand and Sandstone Bodies* (Eds Swift, D.J.P., Oertel, G.F., Tillman, R.W. & Thorne, J.A.). *Spec. Publs int. Ass. Sediment.* **14**, 3–31.

SWIFT, D.J.P., YOUNG, R.A., CLARKE, T.L., VINCENT, C.E., NIDORODA, A. & LESHT, B. (1981) Sediment transport in the Middle Atlantic Bight of North America: synopsis of recent observations. In: *Holocene Marine Sedimentation of the North Sea* (Eds Nio, S.D., Shüttenhelm, R.T.E. & Van Weering, T.C.E.). *Spec. Publs int. Ass. Sediment.* **5**, 361–383.

THOMSON, R.E. (1981) An analysis of wind and current observations collected within Queen Charlotte Sound–Hecate Strait–Dixon Entrance region during 1954 and 1955. *Pac. mar. Science Rept* **81–10**.

Modern Tide-dominated Environments and Facies

Spec. Publs int. Ass. Sediment. (1995) **24**, 71–84

Occupation of a relict distributary system by a new tidal inlet, Quatre Bayou Pass, Louisiana

D.R. LEVIN

Department of Marine Sciences, Louisiana State University, Baton Rouge, La 70803, USA
(new affiliation: Department of Science, Bryant College Smithfield, RI 02917-1284, USA)

ABSTRACT

It has long been suspected that some Louisiana tidal inlets occupy old distributary channels because they do not migrate significant distances alongshore. Historical information coupled with an intensive coring programme supports this contention for Quatre Bayou Pass (QBP), a tide-dominated inlet located in the microtidal Barataria Bight, Louisiana.

Cat Bayou, a levee-bound distributary channel, was mapped on the east side of the QBP opening in 1842. Coincident with levee breeches prior to 1886, the backbarrier marsh became channelized producing more efficient conduits for Barataria Bay tidal flux. By 1934, the Cat Bayou channel was largely abandoned and a narrow, meandering, 8-m deep channel was established on the west side of the inlet opening. Vibracores reveal that the thalweg of this inlet is anchored in place by erosion-resistant channel walls of a relict distributary. There was no core evidence that the channel had migrated to its new position.

Apparent coastwise migration of the inlet thalweg is related to the degree of meander in the evacuated channel system. During transgression the inlet throat is resituated within the antecedent meander position. Subsurface recognition of this phenomenon requires identification of marine sediments filling a non-migratory channel that incises delta plain facies.

INTRODUCTION

The coastwise migration of tidal inlets has been the subject of numerous detailed field and map investigations (e.g. Bruun & Gerritson, 1960; Hayes, 1975; FitzGerald & Levin, 1981; FitzGerald, 1984; Boothroyd, 1985). Thalwegs are forced in the direction of longshore sediment transport as barrier sands spill into the updrift side of inlet channels. Equilibrium cross-sections are maintained by scouring the downdrift channel boundary (Bruun & Gerritsen, 1960).

The physical description of depositional units formed during this process has contributed to subsurface recognition of tidal inlet stratigraphy (Kumar & Sanders, 1975; Tye, 1985; Siringan & Anderson, 1993). In outcrop, a preserved migratory inlet channel will have a lenticular geometry oriented at right angles to the channel axis (Hoyt & Henry, 1967; Barwis & Makurath, 1978), similar to a point bar accumulation (Reinson, 1984).

Inlet stability in Louisiana is enigmatic when compared to inlets located along the east coast of the United States because they do not migrate appreciable distances alongshore (Price & Parker, 1979; Shamban, 1985; Suter & Penland, 1987). This minimal migration suggests that some Louisiana tidal inlet thalwegs are encased in erosion resistant lithologies (Dealteris & Byrne, 1975; FitzGerald, 1976). Russell (1939) and Fisk (1944) were the first to suggest that stable, non-structured Louisiana tidal inlets occupied old distributary channels.

The primary objective of this paper is to show that Quatre Bayou Pass and adjacent tidal inlets along the Louisiana coast (Fig. 1) are located in old distributary channels. A mechanism for the reoccupation of filled distributary channels is proposed. The facies architecture encasing the thalweg explains why the inlets are widening while the deep channel remains stationary.

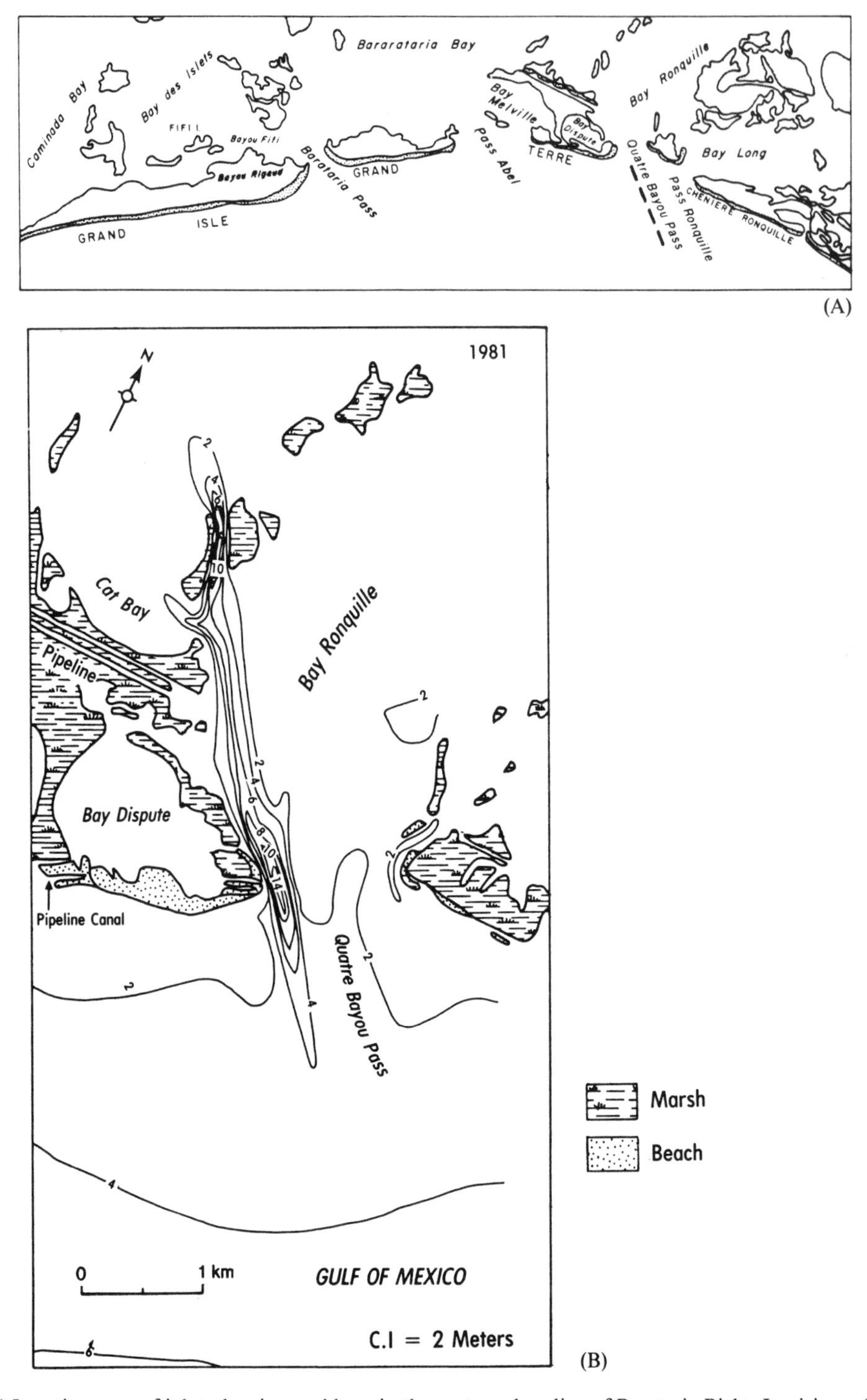

Fig. 1. (A) Location map of inlets, barriers and bays in the eastern shoreline of Barataria Bight, Louisiana. (B) Quatre Bayou Pass, 1981. A 14-m deep thalweg hugs the western side of the Pass opening. Note levee remnants bounding a shallow channel on the east side of the Pass opening.

EVOLUTION AND PHYSICAL SETTING

The Barataria Bight chronology begins with Bayou des Famille (4600–3600 yr BP) and Bayou Blue (2600–1900 yr BP) prograding into what is now Barataria Bay (Fig. 2; Levin, 1990). With abandonment, the respective delta lobes subsided and were transgressed creating accommodation space for subsequent progradations. Two sublobes of the Mississippi delta (Lobe 13) prograded into the eastern side of ancestral Barataria Bay (1000–200 yr BP; Levin, 1990). Beach ridges that grew westward from the abandoned Mississippi delta headlands are preserved in Grand Terre barrier island (500–400 yr BP). During the waning stages of the Mississippi progradation, the Late Lafourche delta (Lobe 15; 500–200 yr BP) built against the Early Lafourche delta (Lobe 14; 700–400 yr BP) along the west side of Barataria Bay (Fig. 2). When this delta passed the seaward limit of the Early Lafourche delta, its sediments were transported to the west creating the Caminada beach ridge plain (400–300 yr BP; Gerdes, 1982). Once abandoned, waves eroded the Late Lafourche headland building the Timbalier Islands to the west and Grand Isle to the east (Harper, 1977). Sand that bypassed Barataria Pass renourished Grand Terre barrier island (Shamban, 1985). The proximity of Grand Isle to Grand Terre and the building of subdeltas to the west of the modern delta (i.e. Bayou Robinson) contributed to the formation of the Barataria Bight (Fig. 2; Welder, 1959).

Tides in the Mississippi River delta plain are largely diurnal (Marmer, 1954) with a mean range of 0.35 m (US Department of Commerce, 1988). Meteorological tides of up to 1.2 m frequently account for 50% or more of the daily water level fluctuations (Wax, 1978; Boyd & Penland, 1981). Hurricanes and tropical storms can create surges of several metres (Boyd & Penland, 1981; Brower *et al*, 1972). Average wave heights along the Grand Terre coast are 0.5 m (US Army Corps of Engineers, 1972) and the dominant direction of longshore sediment transport is eastwards (Nakashima, 1988).

Grand Terre barrier is part of a transgressive shoreline fronting the eastern side of the Barataria Bay complex. The island is flanked and breached by four tidal inlets. From west to east the inlets are Barataria Press, Pass Abel, Quatre Bayou Pass (QBP) and Pass Ronquille (Fig. 1). All but Pass Ronquille are tide dominated. The tide-dominated inlets have not migrated since the first detailed maps were published in the early 1800s.

QBP is a tide-dominated inlet that connects Bay Ronquille and the Gulf of Mexico Bay. Ronquille is part of a larger bay complex dominated by Barataria Bay. A subtidal flood-tidal delta is present to the east of the main ebb channel and a large ebb-tidal delta platform extends 3 km seaward (Fig. 1; Howard, 1983). The narrow (150 m) and deep (14 m) thalweg is located at the western extreme of the pass opening. The thalweg shallows abruptly to a 1 m deep platform that extends nearly a kilometre to the eastern shore. The thalweg accounts for only 12% of the total pass width, yet contains over 60% of the inlet's throat cross-section. On the east side of the pass, levees of Cat Bayou (Fig. 1) appear as vestiges of the last distributary channel to flow through the area.

DATA ACQUISITION

Historical analysis

Maps and charts dating back to the 1700s were compared to trace the shoreline history of the study area. Scales on the early maps (pre-1841) were too small to facilitate detailed analyses of coastal change. However, helpful information regarding the location of relict distributaries, barrier islands and tidal inlets could be ascertained. Other historical evidence was gathered from work on shoreline changes carried out along specific reaches of this coastline by Gerdes (1982) and Howard (1983).

Field work and core analysis

Sixty-five vibracores averaging 6 m in depth were taken along the coast between the east end of Grand Terre and Cheniere Ronquille during the summer of 1983 as part of a larger, regional stratigraphic study (Levin, 1990). The cores were obtained using a modified concrete vibrator (Lanesky *et al.*, 1979; Smith, 1984). A subset of this data set (eight cores) representing two transects was utilized for this study.

Cores were described using parameters keyed to specific depositional environments (Levin, 1990). Nearly 200 discrete sand samples taken from the cores were analysed by sieving to detect intracore grading and to facilitate intercore correlation. X-ray

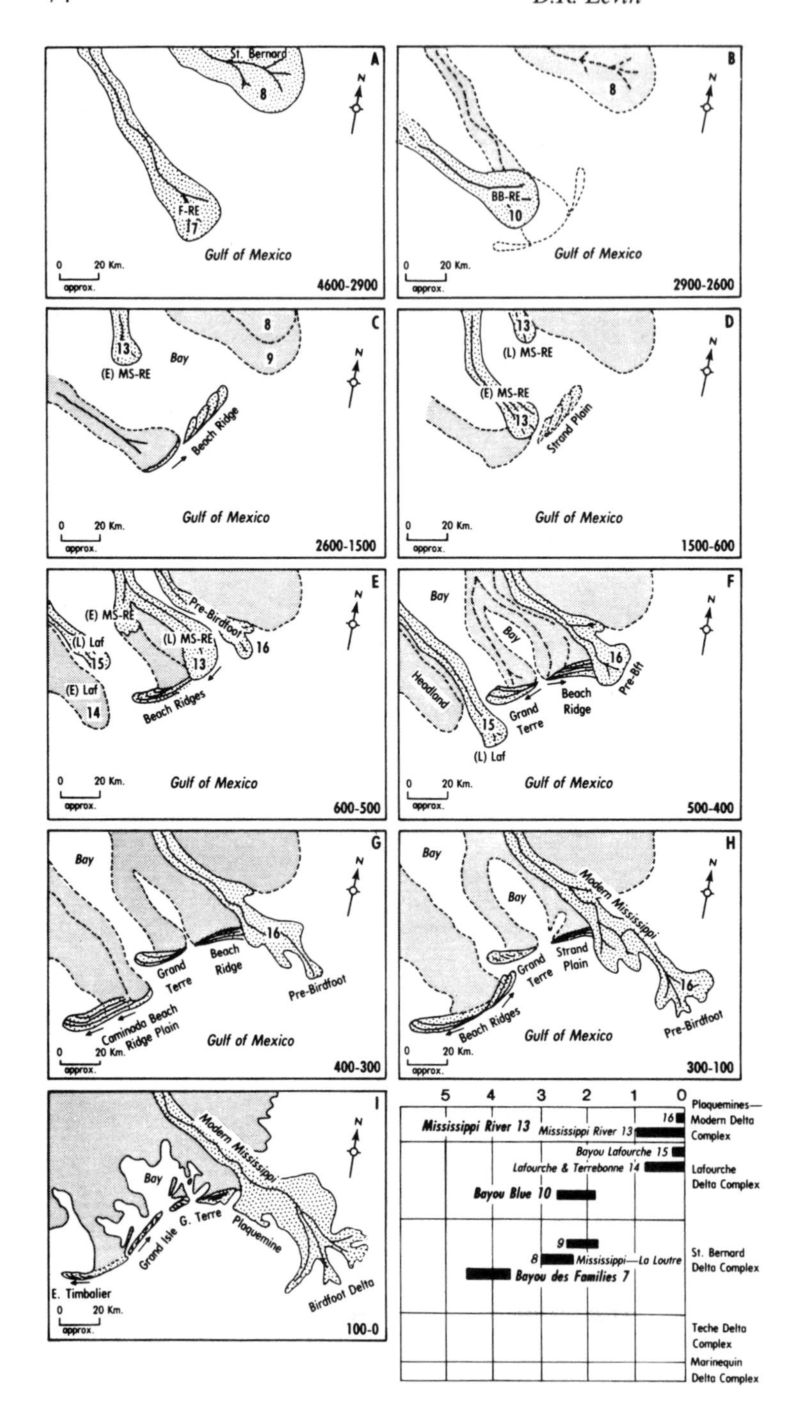

Fig. 2. Summary diagram depicting locations and periods of regression and transgression in the Barataria area of the Mississippi River delta plain, 4600 yr BP to present.

radiographs of 420 fine-grained units (Hamblin, 1962) revealed stratification, burrows and root structures in strata that appeared homogeneous to the unaided eye (Roberts *et al.*, 1976). Figure 3 provides a key of symbols and abbreviations used in core illustrations and figures depicting the areal distributions of various depositional environments.

CORE DESCRIPTION SYMBOLS

mean grain size	trough cross-bedded
fining upward	climbing ripples
coarsening upward	rip-up clasts
C-14 date	shell disart/frag
burrowing	shell articulated
erosional contact	roots
horizontal laminae	wood debris
wavy laminae	organics
flaser	distortion
lenticular	sand
clay drapes	$CaCO_3$nodule
ripple bedded	micro-fault

FACIES ABBREVIATIONS

BF — Bayfill	BK — Backshore
DF — Delta Front	PP — Pass Platform
C — Channel	PM — Pass Margin
CA — Abandoned Channel	S — Spit
OP — Proximal Overbank	BY — Bay
OD — Distal Overbank	FTD — Flood-Tidal Delta
M — Marsh	ETD — Ebb-Tidal Delta
B — Barrier	RV — Ravinement

Fig. 3. Core description symbols and facies abbreviations used in the text and in labels of Figs 4 & 5.

LITHOFACIES ANALYSIS

The stratigraphic record of the Barataria area (+1.0 m to −21.0 m MSL) contains a wide range of lithofacies. The surficial morphology consists of shallow bays, delta plain marsh, and barrier/tidal inlet complexes. The subsurface geology is dominated by regressive components of past Holocene delta lobes, including channel, overbank (levee), bayfill (prodelta) and delta front deposits. Facies lithology in the Barataria region has been fully described in relation to depositional processes elsewhere by Levin (1990). Processes inferred from lithofacies descriptions assist in the interpretation of the facies architecture of the Barataria region (Table 1).

Subsurface lithofacies characterization

Bayfills comprise laminated clays and silts with rare sand seams (Table 1). X-ray radiography reveals distinct laminations sometimes interbedded with burrowed, non-shelly strata. Thin (<2 cm) graded beds contain woody detritus, shell fragments, ripples, wavy lenticular bedding and soft sediment deformation structures, root structures being uncommon. Bayfills are commonly part of an overall coarsening-upward sequence. The thickness of this facies averages 60 cm, ranging from 50 to 100 cm.

Delta front facies contain dm-thick upward fining subunits within an overall coarsening sequence. The subunits fine from trough cross-bedded and ripple-bedded sands to laminated clays with lenticular bedding. Bedded-then-burrowed strata frequently alternate with non-burrowed intervals. Towards the top of this facies trough cross-bedded channel sands become prevalent, except where it lies beneath delta plain (marsh) deposits. Seams of wood and fine plant debris (<1 cm) are rare to common and shells are rare. Average delta front thickness is 90 cm, ranging from 25 to 250 cm. Without exception this lithofacies overlies bayfill deposits (Table 1). Channel (CH) facies have been partitioned into basal (CH-B), middle (CH-M) and upper (CA) subfacies, representing abandoned channels. CH-B usually overlies delta front facies, occasionally cutting into or through it. Where the CH-B cuts into bayfill or marine facies, mud rip-up clasts may be present. Immature grey, well-sorted sands (150 μm) commonly exhibit ripples, climbing ripples and small-scale trough cross-stratification. The grey sands are rich in ilmenite and magnetite. Marine faunal remains are absent, except when ripped from underlying strata.

Middle channel subfacies deposits (CH-M) are characterized by interbedded fine-grained sands (90–150 μm) and silts that are sometimes separated by clay drapes. Ripple-bedded units may be truncated. Organic detritus is common but burrowing rare. Abandoned channel subfacies deposits (CA) are comprised of bioturbated silts and clays with little or no sand. Soft-sediment deformation, roots, plant remains, leaf impressions and woody debris are common. Primary structures or shells are rare. Overall thicknesses of the entire channel facies ranged from 50–350 cm, averaging at 130 cm. Overbank facies contain stacked graded units in an overall fining-upward sequence averaging 110 cm in thickness, ranging from 25 to 350 cm. Two lithologically distinct OB subfacies have been identified: overbank proximal (OP) and overbank distal (OD). Generally, OP sequences exhibit better preserved primary structures and coarser lithologies than OD

R RARE Nm Normal
C COMMON Rv Reverse
Ø NOT OBSERVED

FACIES / CHARACTERISTICS	THICKNESS (Avg) cm	THICKNESS RANGE Mx/Mn	OVERALL GRADING Nm Rv	INTERNAL GRADING Nm Rv	SAND % / MUD % approx	SAND LAMINAE 1 cm	CLAY LAMINAE 1 cm	ORGANIC DETRITUS	ROOTING	SHELLS In-Situ	SHELL FRAGMENTS (F) VALVES (V)	BURROWING	BEDDED THEN BURROWED	ALTERNATING BEDDED & BURROWED	REGULAR LAYERED NO BURROWS	RIPPLE BEDDED	CLIMBING RIPPLES	X TROUGH BEDDING	PARALLEL LAMINATIONS	WAVY LAMINATIONS	FOSSIL TYPES RANGIA CUNEATA	MULINIA SP	CRASSOSTREA SP	OTHERS	UNDERLYING	OVERLYING
1 BAY FILL	60	100/50	Rv	Nm	10/90	R	C	R	R	R-C	R-C	R	R	C	R	R	Ø	Ø	C	C	R-C	R-C	R	R	9 8 14	2 3 8 13
2 DELTA FRONT	90	250/25	Rv	Nm	60/40	C	C	C	Ø	Ø	R	R	R	R	C	C	C	R	C	C	Ø	Ø	Ø	Ø	1	3 4 6 7 8
3 ACTIVE CHANNEL	130	350/50	Nm	Rv	100/0	C	Ø	R	Ø	Ø	R	Ø	Ø	Ø	C	C	C	C	R	R	Ø	Ø	Ø	Ø	1 2 5 9	4 5
4 MID AB CHANNEL	″	350/50	Nm	Nm	60/40	C	C	C	R	Ø	R	R	Ø	R	C	C	R	R	C	C	R	R	Ø	Ø	3	5 8 9
5 UPR AB CHANNEL	″	350/50	Nm	Nm	10/90	R	C	C	C	C	C	C	R	Ø	Ø	Ø	Ø	Ø	R	R	R-C	R-C	Ø	Ø	4	8 9 1
6 PROXIMAL OVERBANK	110	350/25	Nm	Nm	90/10	C	C	C	R	Ø	R	C	C	C	R	C	R	Ø	C	C	R	R	R	Ø	9 8	7 8 9
7 DISTAL OVERBANK	″	350/25	Nm	Nm	40/60	R	C	C	C	R	R	C	C	R	R	R	Ø	Ø	C	C	R	R	R	Ø	6 8 9	6 9
8 MARSH	90	200/25	Ø	Ø	10/90	R	Ø	C	C	R	C	C	Ø	Ø	Ø	Ø	Ø	Ø	Ø	Ø	Ø	Ø	R	R	1 5 11	6 7 9
9 BAY	100	300/25	Ø	Ø	30/70	R	C	R	Ø	C	C	C	R	Ø	Ø	R	Ø	Ø	R	R	C	C	C	C	8 5	1 3 13
10 BARRIER	40	175/25	Rv	Nm	100/0	C	Ø	R	R	R	C	C	C	Ø	Ø	Ø	Ø	Ø	C	Ø	C	C	C	C	8 9	1 11
11 BACKBARRIER	100	125/25	Nm	Rv	80/20	C	R	C	C	R	C	C	C	C	Ø	R	Ø	Ø	Ø	Ø	C	C	C	C	8 9	10
12 SPITS	90	350/70	Rv	Ø	100/0	Ø	Ø	R	R	Ø	R-C	C	C	R	Ø	C	Ø	Ø	C	Ø	C	C	C	C	8 9	1 9
13 FLOOD TIDAL DELTA	60	100/30	Rv	Ø	100/0	Ø	Ø	Ø	Ø	Ø	C	C	Ø	Ø	Ø	C	Ø	Ø	Ø	Ø	C	C	C	C	9 14	1
14 PASS PLATFORM	70	200/40	Rv	Ø	100/0	C	Ø	Ø	Ø	R	C	C	C	R	Ø	C	Ø	Ø	C	Ø	R	R	C	C	9	1 13
15 PASS MARGIN	110	150/30	Rv	Nm	90/10	C	C	C	C	R	C	C	C	Ø	Ø	R	R	Ø	Ø	Ø	C	C	C	C	1 8	1 6

Table 1. Facies characteristics in the Barataria shoreline region

units. OP contains fine-grained sandy silts (90–120 μm) interbedded with macerated organics and wood chips. The basal contact of OP is commonly sharp. Climbing ripples and small-scale cross-bedding are common, clay drapes and soft sediment deformation are not. Rootlets will sometimes grow into OP from above. Facies OP frequently grades to more heterolithic bedding, e.g. parallel and wavy laminations and lenticular bedding.

Overbank distal facies consist of interbedded silts and clays. Sands are rare. Without the aid of X-ray radiography it can appear homogeneous and structureless. Subfacies OD can contain organic detritus, rooting, and can be heavily burrowed. OD commonly grades into marsh.

Bay (BY) facies range from clays to sandy clays. Silt and sand laminae are uncommon. Burrows and brackish water faunal remains (disarticulated shells and fragments) are common. In the non-burrowed intervals, horizontal laminations and small, clay-draped oscillatory ripples are seen. Organic content is common to rare. Bay facies average 95 cm in thickness, ranging from 25 to 300 cm. Bayfills or components of barrer/inlet systems commonly overlie bay facies.

Backbarrier (BKBR) facies comprise stacked subunits of bedded shelly sands that fine into burrowed, organic and sometimes rooted muds. The average facies thickness is 100 cm, ranging between 30 and 175 cm. Sands at the subunit base may be interbedded with laminated muds and small bits of wood. Burrows in the upper subunit are attributed to the fiddler crab *Uca pugilator* (Neese, 1984). Disarticulated shells of brackish water molluscs are commonly found in this unit because it forms part of a transgressed delta plain.

Subsurface characterization of exposed environments

Tops of vibracores taken from the Barataria region were assumed representative of the environment exposed at the surface (Fig. 4). These facies are described in terms that allow their subsurface recognition.

Marsh (M) lithofacies are thick-bedded (>20 cm), mottled to structureless muds with at least 20% rooting (Levin, 1990). Remnants of deteriorating marsh grass (*Spartina alterniflora*) comprise the bulk of the detrital organics. Shells are rare. X-ray

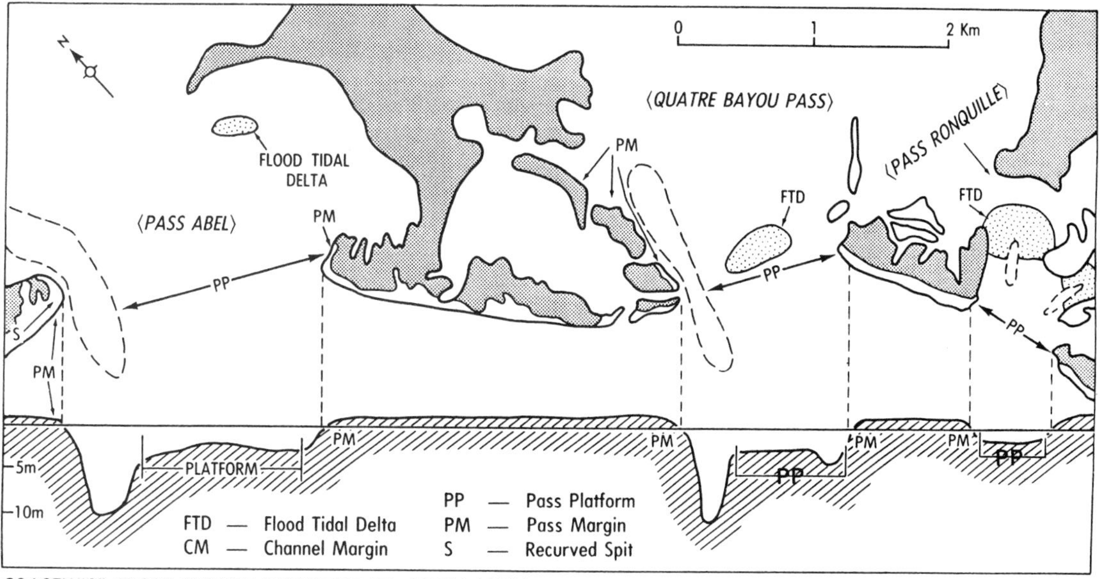

Fig. 4. Areal distribution of depositional environments and bathymetric profiles of Quatre Bayou Pass and Pass Abel. Both pass profiles show deep and narrow thalwegs to the west and adjacent shallow, wide pass platforms. Both inlet thalwegs are being filled from the west by eastward directed longshore sediment transport.

radiography reveals early diagenesis around roots, including pyrite and siderite replacement. The sediment consistency ranges from a watery organic ooze to firm silty clays. Sandy laminations are rare. Average marsh thickness is 90 cm, ranging from 25 to 200 cm. Marsh facies lie conformably above bay, overbank and channel fill deposits. Buried upper marsh contacts (marine flooding surface) always appear sharp. Recurved spit (S) facies consist of upward-coarsening shelly sand beds with infrequent *Ophiomorpha*-type burrows. Spit thickness averages at less than 100 cm. However, Grand Terre spit is 450 cm thick where it fills the west side of Pass Abel channel. Tabular bedding is observed at the subaerial spit margin. This facies is commonly underlain by bay, marsh or channel fill deposits. Spit facies appear identical to the lower portions of barrier, pass platform and flood deltas.

Pass platforms (PP) are wide, shelf-like shallow subtidal features (<1 m deep) that lie adjacent to the inlet thalweg. It is a sandy coarsening-upward facies, ranging in thickness from 40 to 200 cm. Contacts between the sands and underlying bay muds are usually sharp. The lower part of the sandy (120 μm) unit is burrowed and contains shells. Burrowing decreases while shell content and thin beds of horizontally bedded, rippled sands increase higher in the unit. Pass platforms are the foundation for flood-tidal deltas.

Flood-tidal deltas (FTD) commonly cap burrowed shelly sands. The facies coarsen to well-sorted sands (130 μm) and shell content increases towards the delta crest. The average FTD thickness is 63 cm.

Pass margins (PM) are the intertidal boundaries of tidal inlet/pass openings. They contain well-sorted quartzose sands and a moderate shell content (Table 1). Burrowing is common and *Spartina* sp. roots are found in supratidal areas. Channel margins average 110 cm in thickness. Basal contacts are commonly sharp and erosional, often overlying bay or marsh deposits. Pass platform facies usually bury channel margins as the pass opening widens.

Barriers (B) lie above backbarrier/bay units or the rooted marsh environments of eroding headlands along this transgressive coastline. Roots and/or humic staining left in the sand by roots are rare to common. The sands contain low-angle to horizontal bedding, frequently interbedded with seams of disarticulated, imbricated shells. *Ophiomorpha*-type burrows are common in the berm portion of the beach. Barrier thickness averages at 70 cm, ranging up to 170 cm.

STRATIGRAPHICAL ANALYSIS

Quatre Bayou Pass stratigraphy

Two transects containing eight cores were used to characterize the facies architecture in the vicinity of the inlet thalweg. Transect QBP-B was positioned adjacent to the channel axis and QBP-S across the pass opening (Fig. 5).

Transect QBP-B: description and interpretation

Parasequence BB-Re (penetration limit at − 5.5 m) contains proximal overbank subunits overlain by marsh. In cores B-1 through B-4 (− 5.5 m), BB-Re is truncated by a bay ravinement surface identified by sands containing large fragments of *Crassostrea virginica* and *Thais hemastoma* (salinity tolerant marine fauna). This debris was ripped from the shoreface. The marine flooding surface disappears landwards of core B-4. In core B-5 (− 6.5 to − 3.0 m) bayfill muds grade into delta-front and active channel deposits. No ravinement surface is present.

Parasequence BB-Re represents the regressive phase of the Bayou Blue delta lobe. Marsh and overbank facies in this transect indicate that the main channel was close by, perhaps even in the present Quatre Bayou thalweg location. In cores B-1 through B-4 the ravinement has been buried by a progradational sequence. The limit of this transgression lies landwards of core B-4.

Laminated bayfill muds (− 4.8 m) coarsen to delta-front sands at the base of Ms-Re in cores B-1 through B-4 (− 2.8 m). A thin stratum of grey immature sands, correlated between cores B-1 through B-5 (− 2.5 m and − 1.5 m), contained coarse *in situ* roots. The radiocarbon-dated roots demonstrate that a nearby channel overflowed its bank some 710 yr BP. Delta-plain marsh is ubiquitous in this parasequence above − 1.5 m. Parasequence Ms-Re corresponds to the progradation of the Mississippi Delta lobe system.

There is a sharp contact (core B-1, − 0.2 m) between the upper marsh of Ms-Re and the shelly, quartzose sands of the modern transgression (Md-Tr).

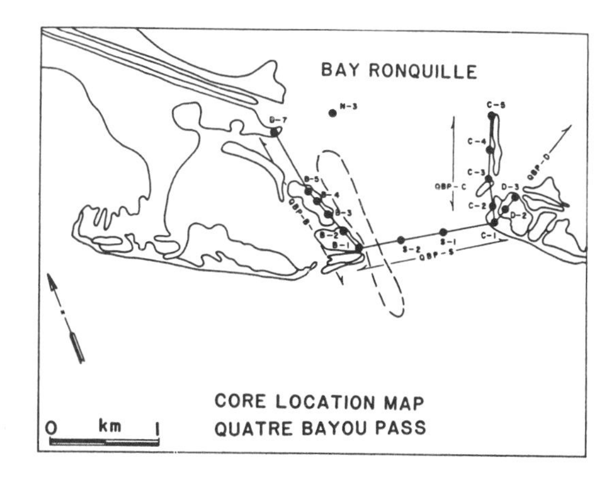

CROSS SECTION QBP-S

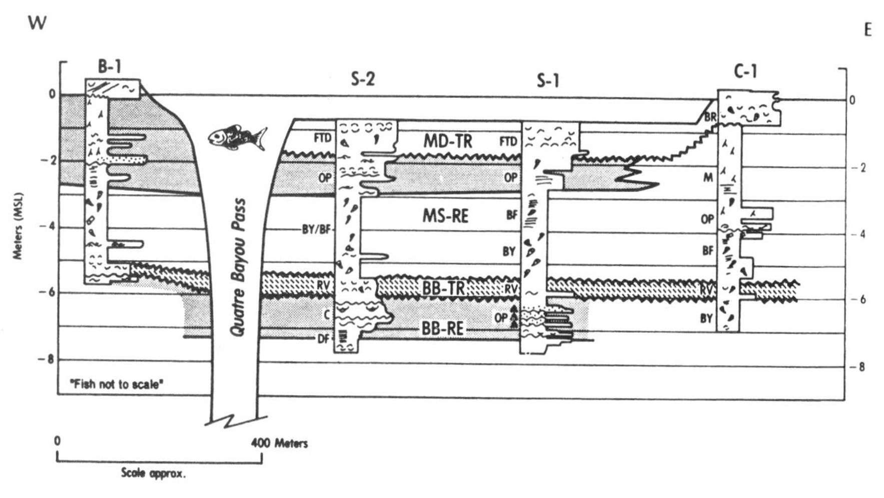

CROSS SECTION QBP-B

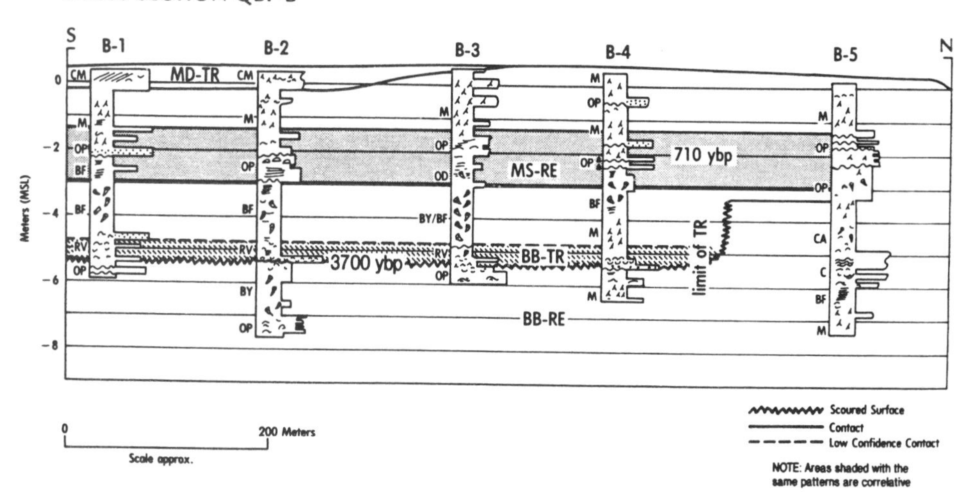

Fig. 5. Vibracore transects QBP-B and QBP-S. In QBP-B two periods of overbank facies are separated by a transgressive shell lag (– 5.5 to – 5.0 m). Distributaries that occupied the present QBP thalweg were responsible for the overbank deposition. In QBP-S lithofacies change from proximal to distal overbank (cores S-2 and S-1, – 7 to – 6 m), supporting the hypothesis that a distributary flowed through the present thalweg position.

Transect QBP-S: description and interpretation

Regressive facies dominate the lower portion of transect QBP-S. In core S-2 (– 7.5 to – 6.0 m), a sequence of laminated bayfill clays grade into delta-front and then active channel sands. Various hues of heavy minerals within the channel sands (– 6.5 to – 6.0 m) accentuate trough cross-bedding in the core. Core S-1 (0.4 km to the east of core S-2) contained three graded units (– 7 to – 6 m). In core C-1 (to the east) this interval contained laminated and burrowed bayfill muds. A sharp erosion contact above the sands in core S-1 can be traced eastwards across the transect.

BB-Re describes the Bayou Blue progradation. The facies change between cores S-2 and S-1 suggests that a distributary channel was flowing in the present Quatre Bayou Pass thalweg location. A 0.5-m thick unit of mature shelly sands (core S-2, – 6.0 to – 5.5 m) has a sharp basal contact that scours into the underlying channel. A correlative stratum of shelly sands (core C-1, – 5.5 m) grades eastwards into bedded sands in core S-1. This is the transgressive component of the Bayou Blue progradation.

Parasequence Ms-Re is thickest in core C-1. Burrowed, shelly bay muds (core C1, – 6.0 m) are buried by bayfill, overbank sands (– 4.0 to – 3.0 m) and marsh (– 3.0 to – 0.75 m). To the east cores S-1, S-2 and B-1 contain interdistributary bay muds and bayfill from – 5 to – 3 m. Overbank deposits of Ms-Re flank the present channel location from – 3.0 to – 1.8 m. In core B-1 they occur between – 2.5 and – 1.0 m.

Parasequence Ms-Re describes a phased delta lobe progradation. The overbank facies located in core C-1 preceded those flanking the western channel. It is hypothesized that levee failure upstream of

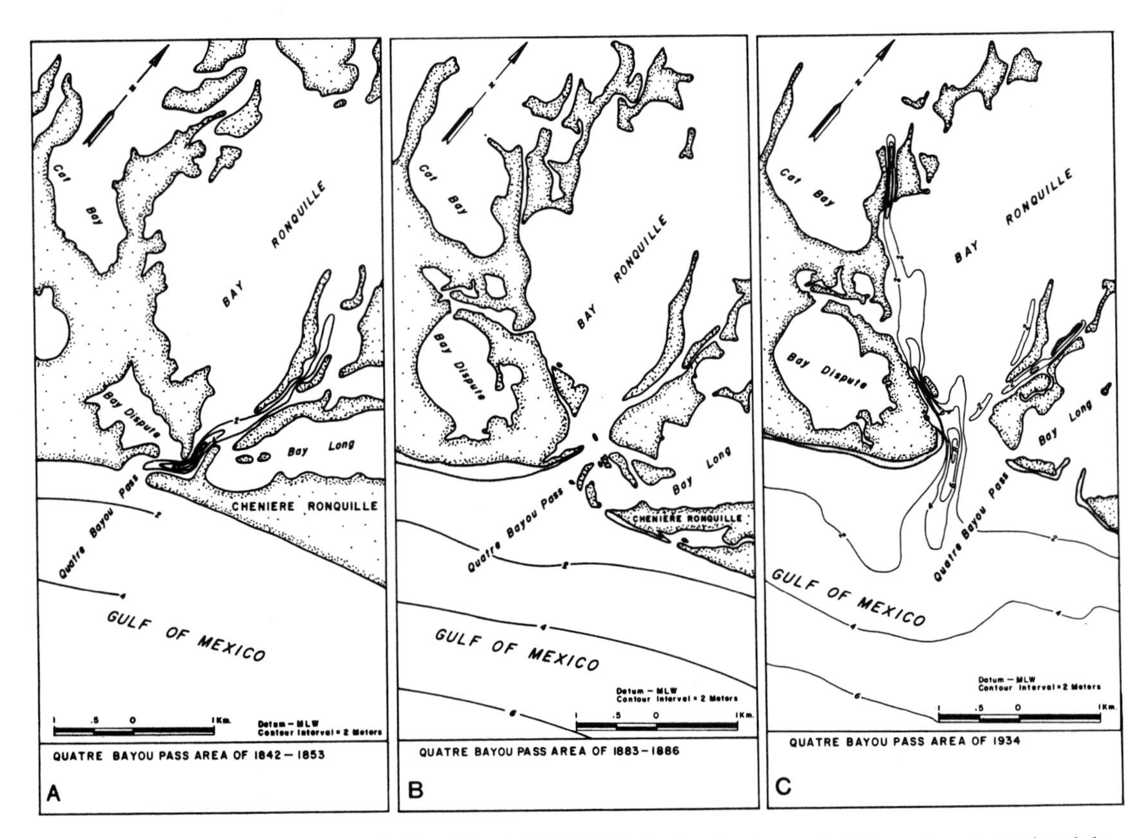

Fig. 6. Quatre Bayou Pass in (A) 1842, (B) 1883 and (C) 1934. Note the abandonment of the eastern channel and the establishment of a new channel on the west side of the Pass between 1842 and 1934. Also note the temporal expansion of the ebb-tidal delta.

the established eastern channel caused the switch to a more westerly channel location (Levin, 1990).

Parasequence Md-Tr caps the entire QBP-S transect between - 2.0 m and the present surface. It is significantly coarser than underlying units, containing moderate to heavy concentrations of shell fragments. Parasequence Md-Tr represents the influx of marine sediments into the Quatre Bayou Pass area over the past hundred years.

ANALYSIS OF QUATRE BAYOU PASS MAP DATA

Map history

In an 1816 map, an opening is depicted on the east end of Grand Terre, roughly where QBP is presently located. No bathymetry or shoaling is marked on this map that would permit comment on inlet morphology.

The earliest reliable chart (1842; Fig. 5) shows a levee-bound channel running several kilometres inland from the Quatre Bayou Pass opening. Splay deposits spill into Bay Long through a breach in the east levee. Contoured isobaths do not reveal an ebb-tidal delta.

A compilation of hydrographic and planimetric surveys by Howard (1983) allows a reasonably accurate depiction of QBP around 1886 (Fig. 6). Compared to 1841, the levees of Cat Bayou have deteriorated and are riddled with breaches. Quatre Bayou Pass has a mixed-energy morphology (Hubbard *et al.*, 1979) with shoals located in the inlet throat. A small ebb-tidal delta is present. Coincident with the changes in pass morphology the marsh has fragmented, producing new conduits for bay tidal flux.

A narrow, meandering 8-m deep channel was mapped on the west side of the inlet opening where none had previously been located. Maximum depths in the abandoned eastern channel were barely half that of the new channel. The levees of the eastern channel continued to erode while Quatre Bayou Pass widened. The terminal lobe of the ebb delta (terminology of Hayes, 1975) extended nearly a kilometre further seawards than noted in 1883.

By 1981, the QBP thalweg had scoured to depths exceeding 14 m and the shallow pass platform had widened (Fig. 1). The old channel filled and small vestiges of the levees remained. The terminal lobe of the ebb-tidal delta lay further offshore than in 1934 (Howard, 1983). The seaward end of the main ebb channel had filled and the new thalweg axis had a south–south-east orientation.

Temporal changes in channel dimensions

Changes in the width and cross-sectional area of QBP were plotted by Howard (1983) over 130-year period between 1860 and 1920. In this 60-year span, the channel thalweg width increased by less than 20 m while a 200 m pass platform formed to the east. By 1920 the thalweg width had stabilized, pass widening continuing at an average rate of 3.4 m yr^{-1} (Howard, 1983). By 1982 the width of QBP had expanded to just over 1 km. Between February and May of 1983 wave-induced erosion caused QBP to widen an additional 15 m (Levin, 1990). For comparison, the rate of shoreface retreat in this coastal reach approaches 20 m yr^{-1} (Nakashima, 1988).

DISCUSSION

Inlets occupying antecedent distributary channels

Stratigraphical and historical data support the hypothesis that the new Quatre Bayou Pass thalweg occupies a relict distributary location. The following mechanism for inlet reoccupation of abandoned distributary thalwegs is proposed: as Barataria and Bay Ronquille increased in size the tidal prism exchanged by inlets of Grand Terre also increased (Levin, 1993). Bay area increase was accomplished, in part, by fragmentation of the marsh system. Abandoned distributaries were well masked in the marsh environment, occupying imperceptible depressions that followed old channel axes. With delta plain subsidence the slightly lower channel elevation flooded and became a preferred path for tidal exchange.

With bay expansion the tidal prism grew, requiring larger inlet cross-sections to transfer tidal waters between the Gulf and the inland bays (O'Brien, 1969; Jarrett, 1976). Excavation of relict distributary channels required that the fill be less resistant to erosion than the surrounding facies. Seismic and vibracore data collected by the Louisiana Geological Survey from the offshore region of Barataria Bight revealed several sandy channel fills interpreted as offshore extensions of abandoned distributaries

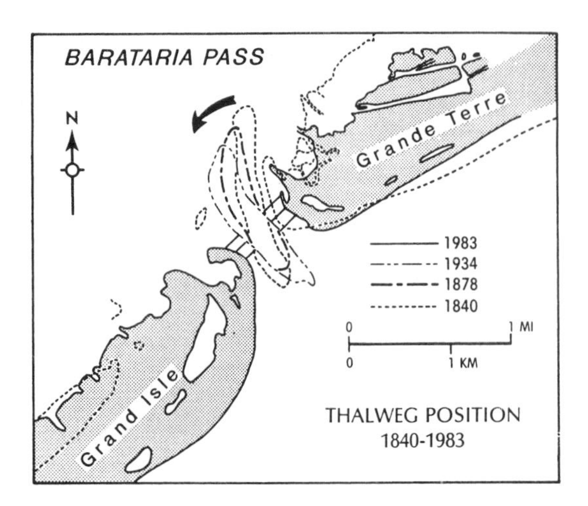

Fig. 7. Counter-clockwise pivot of the Barataria Pass inlet throat between 1840 and 1983 (from Shamban, 1985). As the shoreline recedes, the inlet throat is being resituated in a meander bend of an old distributary system.

(Levin, 1990). Downward excavation of the sandy, abandoned channel fill halted when the erosion resistant muddy side-walls (Hjulstrom, 1955) of the levee, bayfill and bay facies were exposed. As a result, the thalweg could no longer be modified. However, this did not prevent the inlet from modifying its cross-section by lateral erosion of adjacent barrier sands and expansion of the pass platform.

Inlet migration in the Barataria Bight shoreline

Thalwegs of inlets in the eastern Barataria Bay shoreline do not migrate. They are anchored in place by the cohesive, fine-grained lithology of the channel side-walls. Barataria Pass has 'pivoted' from a northerly to north-westerly orientation between 1840 and 1983 without moving downdrift (Fig. 7; Shamban, 1985). At Pass Abel the eastern spit of Grand Terre is presently filling the west bank of the Pass thalweg (Fig. 4). However, the eastern thalweg bank has not eroded to compensate for the concomitant decrease in inlet cross-section. Instead, the channel cross-section is increased by eroding the veneer of unconsolidated barrier island sands off the fine-grained platform (Fig. 8).

The stability of Barataria Pass, Quatre Bayou Pass and Pass Abel is explained by facies architecture. At Quatre Bayou Pass and Pass Abel, a deep and narrow thalweg is cradled by delta plain facies on the western extreme of a wide and shallow pass platform (Fig. 4; Levin, 1990). Any apparent coastwise migration by these inlets is related to the meander-fills being reoccupied. As the shoreline surrounding the pass recedes, the inlet throat will be resituated in an antecedent meander position. In fact, if adjacent passes (e.g. Pass Abel and Quatre Bayou Pass) are reoccupying genetically related channel bifurcations, then shoreline erosion would cause the upstream distance between these to decrease. Without knowledge of the regional stratigraphy, longshore sediment transport directions would appear opposite within the same coastal cell.

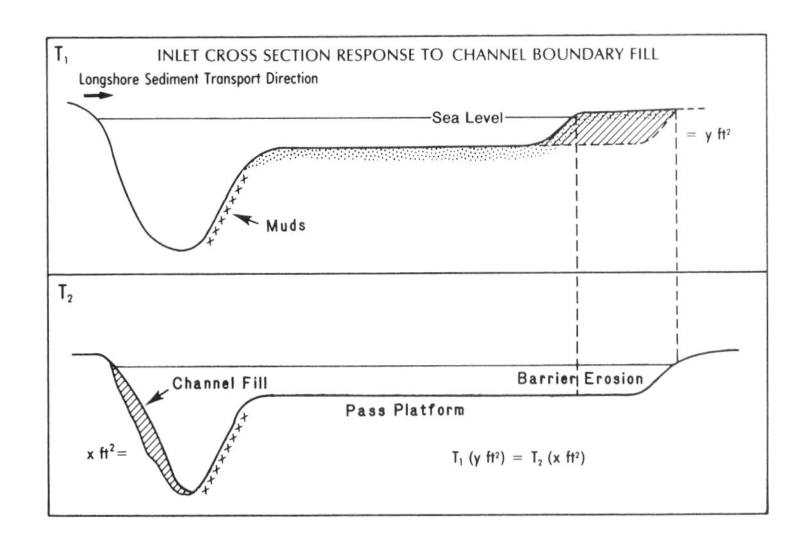

Fig. 8. Diagram of inlet cross-section modification by lateral planation of barrier islands. As sands fill the updrift channel boundary, the fine-grained channel side-walls downdrift will not erode. To keep cross-sectional equilibrium with tidal prism, a commensurate area of barrier sands is eroded, whereby barrier erosion (y ft^2) equals the volume of thalweg fill (x ft^2).

Subsurface identification of inlet-occupied distributaries

As the coast is transgressed, barrier island sands will either be pushed landwards (Moslow, 1980), stranded and drowned in the nearshore area (Howard, 1983) or redistributed onto the nearshore shelf (Snedden *et al.*, 1988). Facies with the highest likelihood of surviving transgressions are inlet channels that cut beneath the erosional wave base (Hubbard & Barwis, 1974; Dalrymple *et al.*, 1992). Non-migratory inlet channel systems will not be preserved as coastwise, lenticular sand bodies in the rock record. In the rock record a tidal inlet deposit occurring along a transgressed Barataria Bight-type coast will show a distributary channel filled disconformably with sands derived from adjacent derelict barrier islands and ebb-tidal deltas.

CONCLUSIONS

1 QBP and other tidal inlets in eastern Barataria Bay are located in old distributary channels. Reoccupation of previously abandoned distributaries is caused by temporally increasing tidal prisms that cause a commensurate boost in tidal current velocities. The swift currents cause the excavation of unconsolidated channel fills. Once the cohesive channel side-walls of the surrounding facies are encountered, further modification and lateral migration of the thalweg is inhibited.
2 Inlet cross-sections of the Barataria shoreline are expanded by sweeping the thin veneer of unconsolidated barrier sands off the platform of fine-grained bay clays. Rapid lateral planation of the barrier is the result of increasing tidal prisms, decreasing sediment supply and the location of tidal inlets in old distributary channels. The rate of inlet widening is nearly equal to the rate of the overall shoreline transgression.
3 Minimal lateral migration of tidal inlet channels is tied to the cradling of the pass thalweg within old distributary channel reaches. As the shoreline retreats, the pass is resituated in the path of the antecedent channel.

ACKNOWLEDGEMENTS

This work was funded, in part, by Sohio, Arco, GCAGS, Rockefeller Scholarships, and the Department of Marine Sciences, Louisiana State University. Pete Lackey, Eliza McClennen, Clarence Dupe and Mary Lee Eggert drafted the figures. Darryl Keith, USEPA/ERL Naragansett, reviewed the preliminary draft of this manuscript. Anonymous reviewers contributed significantly to the quality of the manuscript.

REFERENCES

Barwis, H.H. & Makurath, J.H. (1978) Recognition of ancient tidal inlet sequences: an example from the Upper Silurian Keyser Limestone in Virginia. *Sedimentology* **25**, 61–82.

Boothroyd, J.C. (1985) Mesotidal inlets and estuaries. In: *Coastal Sedimentary Environments* (Ed. Davis, R.A. Jr), pp. 287–360. Springer-Verlag, New York, 650 pp.

Boyd, R. & Penland, S. (1981) Washover of deltaic barriers on the Louisiana Coast. *Trans. Gulf Coast Assoc. Geol. Soc.* **31**, 243–248.

Brower, W.A., Meserve, J.M. & Quayle, R.G. (1972) *Environmental Guide for the U.S. Gulf Coast.* NOAA, Environmental Data Service, National Climatic Center, Asheville, North Carolina.

Bruun, P. & Gerritsen, F. (1960) *Stability of Coastal Inlets.* North Holland, Amsterdam, 123 pp.

Dalrymple, R.W., Zaitlan, B.A. & Boyd, R. (1992) Estuarine facies models: conceptual basis and stratigraphic implications. *J. sediment. Petrol.* **62**, 1130–1146.

Dealteris, J.T. & Byrne, R.J. (1975) The recent history of Wachapreague Inlet, Virgina. In: *Estuarine Research* (Ed. Cronin, L.E.), pp. 167–181. Academic Press, New York.

Fisk, H.N. (1944) Geological investigation of the alluvial valley of the lower Mississippi River. *U.S. Army Corps Eng., Mississippi River Comm.*, Vicksburg, Mississippi, 78 pp.

FitzGerald, D.M. (1976) Ebb-tidal delta of Price Inlet, South Carolina: geomorphology, physical processes, and associated inlet shoreline changes, In: *Terrigenous Clastic Depositional Environments* (Eds Hayes, M.O. & Kana, T.), pp. 158–171. Coastal Res. Div., Univ. South Carolina, Columbia.

FitzGerald, D.M. (1984) Interactions between the ebb tidal delta and landward shoreline: Price Inlet, South Carolina. *J. sediment. Petrol.* **54**, 1303–1313.

FitzGerald, D.M. & Levin, D.R. (1981) Hydraulics, morphology and sediment transport patterns at Pamet River Inlet: Truro, Massachusetts. *J. northeastern Geol.* **3**, 216–224.

Gerdes, R. (1982) *Stratigraphy and history of development of the Caminda-Moreau beach ridge plain, Southeast Louisiana.* MSc thesis, Dept Geol., Louisiana State Univ., Baton Rouge, La.

Hamblin, W.H. (1962) X-ray radiography in the study of structures in homogeneous sediments *J. sediment. Petrol.* **32**, 201–210.

Harper, J.R. (1977) Sediment dispersal trends of the

Caminada-Moreau beach ridge system. *Trans. Gulf Coast Assoc. Geol. Soc.* **27**, 283–289.

HAYES, M.O. (1975) Morphology of sand accumulations in estuaries. In: *Estuarine Research* **2**. *Geology and Engineering* (Ed. Cronin, L.E.), pp. 3–22. Academic Press, New York.

HJULSTROM, F. (1955) Transportation of detritus by moving water. In: *Recent Marine Sediments* (Ed. Trask, P.D.), pp. 5–31. S.E.P.M. **4**, Tulsa, Oklahoma.

HOWARD, P.C. (1983) *Quatre Bayou Pass, Louisiana: Analysis of currents, sediments and history.* MSc thesis, Dept Geol., Louisiana State Univ., Baton Rouge, La.

HOYT, J.H. & HENRY, V.J. JR. (1967) Influence of inlet migration on barrier island sedimentation. *Geol. Soc. Am. Bull.* **82**, 2131–2158.

HUBBARD, D.K. & BARWIS, J.N. (1974) Discussion of tidal inlet sand deposits: Example from the S. Carolina coast. In: *Terrigenous Clastic Depositional Environments* (Eds Hayes, M.O. & Kana, T.W.), pp. 158–171. Tech. Rept **11-CRDa**, University of South Carolina, Columbia.

HUBBARD, D.K., OERTEL, G. & NUMMEDAL, D. (1979) The role of waves and tidal currents in the development of tidal inlet sedimentary structures and sand body geometry: examples from North Carolina, South Carolina, and Georgia. *J. sediment. Petrol.* **49**, 1073–1092.

JARRET, J.T. (1976) Tidal prism-inlet area relationships. General investigation of tidal inlets. *U.S. Army Corps Eng., Coastal Eng. Res. Center,* Ft Belvoir, Va., Rept **3**, 45 pp.

KUMAR, N. & SANDERS, J.E. (1975) Inlet sequence: a vertical succession of sedimentary structures and textures created by the lateral migration of tidal inlets. *Sedimentology* **21**, 491–532.

LANESKY, D.E., LOGAN, B.W., BROWN, R.G. & HINE, A.C. (1979) A new approach to portable vibracoring underwater and on land. *J. sediment. Petrol.* **49**, 654–657.

LEVIN, D.R. (1990) *Transgressions and regressions in the Barataria Bight region of coastal Louisiana.* PhD thesis, Dept Mar. Sci., Louisiana State Univ., Baton Rouge, La.

LEVIN, D.R. (1993) Tidal inlet evolution in the Mississippi River delta plain. *J. coast. Res.* **9**, 462–480.

MARMER, H.A. (1954) The currents in Barataria Bay: *Texas. A & M Research Foundation,* Proj. **9**, 30 pp.

MOSLOW, T.F. (1980) *Stratigraphy of mesotidal barrier islands.* PhD thesis, Univ. South Carolina, Columbia, SC.

NAKASHIMA, L.D. (1988) Short-term changes in beach morphology on the Louisiana Coast. *Trans. Gulf Coast Assoc. Geol. Soc.* **38**, 323–327.

NEESE, K.J. (1984) *Stratigraphy and geologic evolution of Isles Dernieres, Terrebone Parish, Louisiana.* MSc thesis, Dept Geol., Louisiana State Univ., Baton Rouge, La.

N.O.A.A. (1988) *Tide Tables, east coast of North and South America.* US Dept Commerce.

O'BRIEN, M.P. (1969) Equilibrium flow area of inlets on sandy coasts. *A.S.C.E., J. Waterw. Harb. Div.* **95**, 43–52.

PRICE, W.A. & PARKER, R.H. (1979) Origins of permanent inlets separating barrier islands and influence of drowned valleys on tidal records along the Gulf coast of Texas. *Trans. Gulf Coast Assoc. Geol. Soc.* **29**, 371–385:

REINSON, G.E. (1984) Barrier-island and associated strandplain systems. In: *Facies Models*, 2nd Edn (Ed. Walker, R.G.), pp. 119–140. Geosci. Canada Repr. Ser. **1**, Geol. Ass. Canada Publ., Toronto.

ROBERTS, H.H., CRATSLEY, D.W., WHELAN, T. & COLEMAN, J.M. (1976) Stability of Mississippi delta sediments as evaluated by analysis of structural features in sediment borings. *8th Ann. Offshore Tech. Conf.*, 9–28.

RUSSELL, R.J. (1939) Louisiana stream patterns. *Am. Ass. Petrol. Geol. Bull.* **23**, 1199–1227.

SHAMBAN, A. (1985) *Historical evolution, morphology and processes of Barataria Pass, Louisiana.* MSc thesis, Dept Geol., Louisiana State Univ. Baton Rouge, La.

SIRINGAN, F.P. & ANDERSON, J.B. (1993) Seismic facies and architecture of a tidal inlet/delta complex. *J. sediment. Petrol.* **63**, 749–808.

SMITH, D.G. (1984) Vibracoring fluvial and deltaic sediments: Tips on improving penetration and recovery. *J. sediment. Petrol.* **54**, 660–663.

SNEDDEN, J.W., NUMMEDAL, D. & AMOS, A. (1988) Storm- and fair-weather combined flow on the central Texas continental shelf. *J. sediment. Petrol.* **58**, 580–595.

SUTER, J.R. & PENLAND, S. (1987) Evolution of Cat Island Pass, Louisiana. *Coastal Sediments '87, Am. Soc. Civil Eng.*, 16 pp.

TYE, R.S. (1985) Geomorphic evolution and stratigraphic framework of Price and Capers Inlets, South Carolina. *Sedimentology* **31**, 655–674.

US ARMY CORPS OF ENGINEERS (1972) Grand Isle and vicinity. Louisiana review report: beach erosion and hurricane protection. New Orleans District, New Orleans, La.

US DEPARTMENT OF COMMERCE, National Oceanic Atmospheric Administration (1988) Tide Tables, East Coast of North and South America, Washington, D.C.

WAX, C.L. (1978) *Barataria Basin: Synoptic weather types and environmental responses.* Coastal and Wetland Resources, Louisiana State Univ. **CWR-LSU**, 60 pp.

WELDER, F.A. (1959) Processes of deltaic sedimentation in the lower Mississippi River. *Coastal Stud. Inst., Louisiana State Univ., Baton Rouge, Tech. Rept* **12**, 174 pp.

Spec. Publs int. Ass. Sediment. (1995) **24**, 85–99

Morphological response characteristics of the Zoutkamperlaag, Frisian inlet (The Netherlands), to a sudden reduction in basin area

E. BIEGEL* *and* P. HOEKSTRA†

**Institute for Coastal and Marine Management, Ministry of Transport and Public Works, PO Box 20907, 2500 EX The Hague, The Netherlands; and*
†Institute for Marine and Atmospheric Research, Utrecht University, PO Box 80.115, 3508 TC Utrecht, The Netherlands

ABSTRACT

The Frisian inlet which separates the barrier islands of Ameland and Schiermonnikoog bifurcates into two smaller inlets that are separated by a supratidal shoal. The two smaller inlets are the Pinke and the Zoutkamperlaag inlets. In 1969 the closure of the Lauwers Sea reduced the flood tidal basin of the Zoutkamperlaag system. The change in tidal dynamics modified the equilibrium between the inlets and the basin tidal prism. In this study, morphological changes occurring since 1969 in the Zoutkamperlaag and its ebb-tidal delta as well as along the coast of the adjacent island of Schiermonnikoog and in the flood-tidal basin are investigated. The ebb-tidal delta has gradually reduced in size while depositional processes were dominant in the flood-tidal basin, reducing channel cross-sections. Empirical relationships between ebb and flood volumes and the cross-sectional areas of the channels show that the channels are adjusting towards a new equilibrium.

This study suggests that exponential models classically used to describe the morphological adaptation during disequilibrium are not generally applicable. There is strong spatial variability with respect to the cross-sectional behaviour, for example due to autonomous processes in the flood-tidal basin. In the inlet throat itself, the exponential model appears to be adequate.

INTRODUCTION

The Frisian inlet is located within a chain of barrier islands that fringes the North Sea from the northern part of The Netherlands to Denmark. The inlet separates the two barrier islands of Ameland and Schiermonnikoog (Figs 1A, 1B). The island of Ameland has a length of 25 km and a width of 3 km, whereas the island of Schiermonnikoog is 15 km long and 3 km wide. The inlet between the two islands has a width of about 11 km at the throat. However, due to the presence of a large supratidal shoal (Engelsmanplaat) halfway between the islands, the Frisian inlet is actually divided into two smaller inlets. These inlets are the Pinke inlet to the west and the larger Zoutkamperlaag to the east (Fig. 1B). The supratidal shoal of the Engelsmanplaat is part of the tidal backbarrier drainage divide which separates the flood-tidal basins in the backbarrier region. There are no flood-tidal deltas in the tidal basins.

Within the framework of the Delta Act of 1953 it was decided that the coastal defence works of The Netherlands should be enhanced by a number of measures: (i) by building storm-surge barriers; (ii) by improving the quality and height of the dikes; and (iii) by a partial reduction of the total length of the dike system along the Dutch coast. The latter was partly achieved by closing off a number of tidal inlets and estuaries in both the SW (Rhine Delta region) and the NE of the country (Wadden Sea). As part of these measures the Lauwers Sea was closed

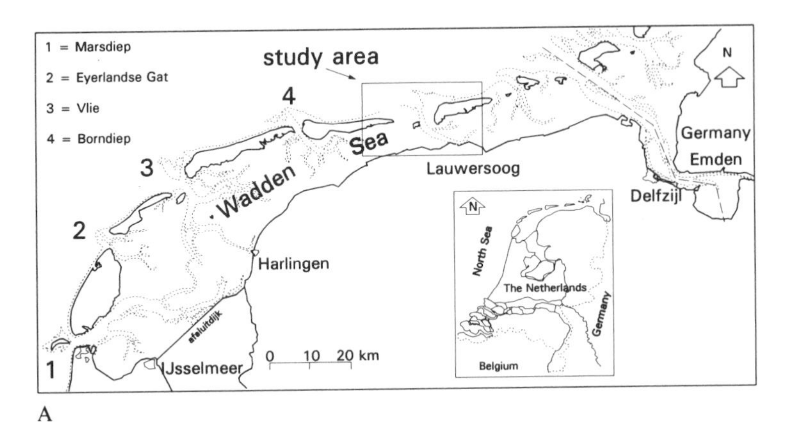

A

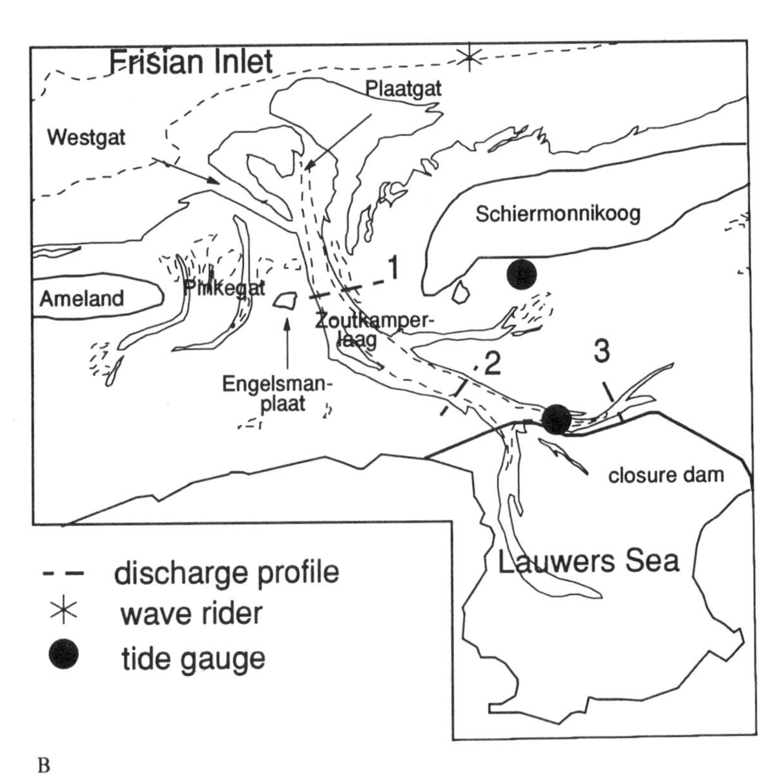

B

Fig. 1. (A) Location map of the Wadden Sea tidal inlets, The Netherlands. The Frisian inlet system is located in the box 'study area'. Numbers on the map refer to other tidal inlets in the western Wadden Sea. (B) The research area of the Frisian inlet system (the Lauwers Sea was closed off in 1969). Locations of the discharge profiles (1–3), tide gauges and offshore wave buoy are indicated on the map.

in 1969, thereby reducing the size of the tidal basin of the Zoutkamperlaag system and turning the Lauwers Sea into a tideless lake with wetlands (Fig. 1B). The resulting change in the hydrodynamics triggered a morphodynamic response of the total inlet system.

The main purpose of this study is to illustrate and explain the hydrodynamic and morphodynamic changes in the Frisian inlet region, in its associated outer ebb-tidal delta, on the coast of the adjacent island of Schiermonnikoog and in the flood-tidal basin since 1969. This was accomplished by collecting field data consisting of hydrodynamic measurements (e.g. wave parameters, tides and tidal prism), meteorological observations and regular soundings to describe the morphodynamic changes both qualitatively and quantitatively.

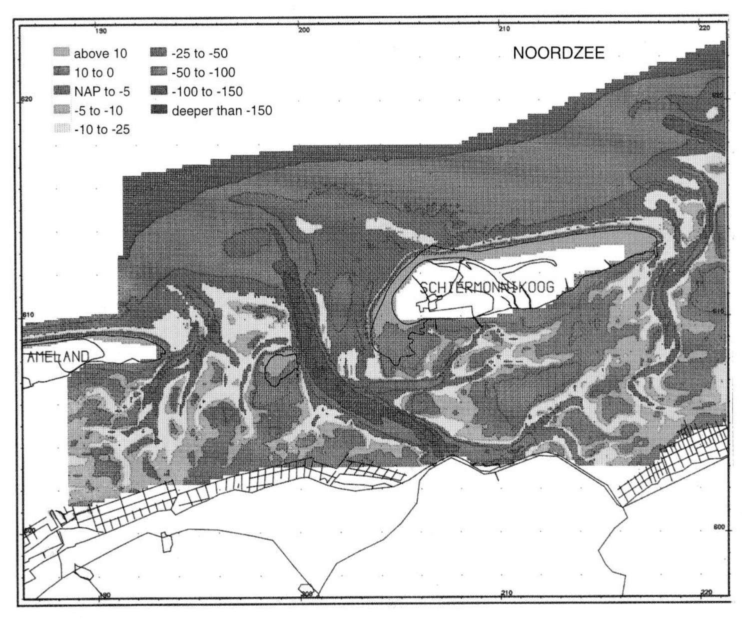

(a)

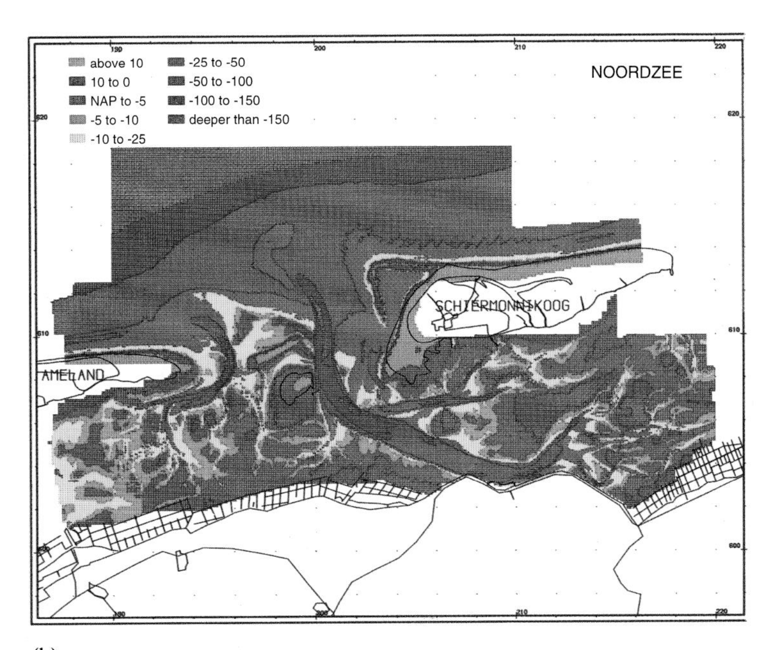

(b)

Plate 1 Bathymetry of the Frisian inlet system in (a) 1970 and (b) 1987. Depths are given in dm below or above Dutch Ordnance Datum (NAP). The NAP-level approximates mean sea-level.

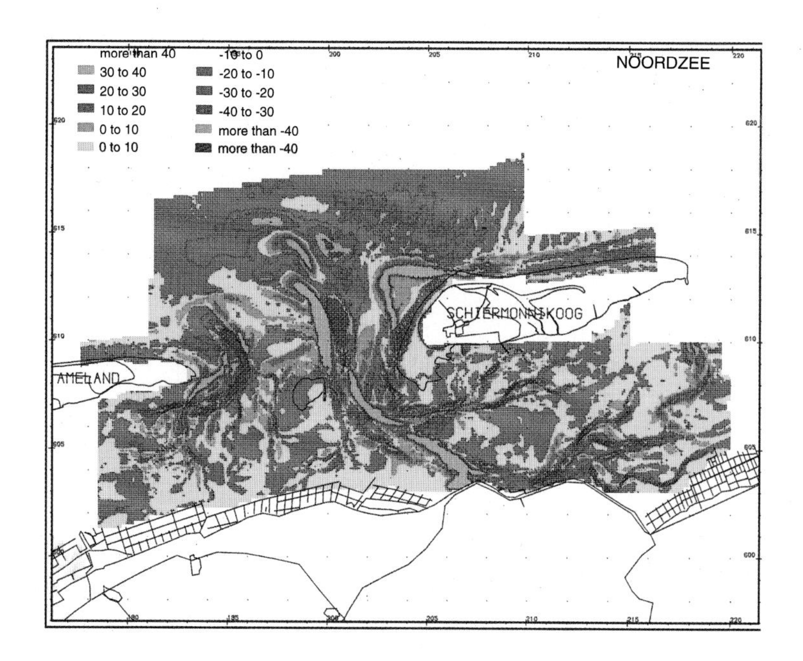

Plate 2 Morphological changes in the Frisian inlet/Zoutkamperlaag system over the period 1970–1987. The ebb-tidal delta is partly eroded, whereas the west coast of Schirmonnikoog and the Zoutkamperlaag show accretion.

STABILITY THEORY

In the recent past considerable efforts have been made to determine empirical relationships between geometric and hydrodynamic parameters for ebb-tidal deltas, tidal inlets and flood-tidal basins. Such relationships form the basis of equilibrium or stability concepts. In the Dutch Wadden Sea, for example, empirical relationships of this type have been established by Gerritsen & de Jong (1985), Misdorp *et al.* (1990), Sha (1990), Eysink (1990) and Eysink & Biegel (1992). In other parts of the world similar observations along sandy coasts were made by, amongst others, Bruun & Gerritsen (1960), Jarret (1976), Walton & Adams (1976) and Dieckman *et al.* (1988).

Hydrodynamic and morphodynamic stability relationships are commonly used in compound empirical models by means of which the morphodynamic behaviour of coastal systems is simulated. When a system is out of equilibrium such models are used to predict the conditions required for the restoration of equilibrium. In one respect these models were found to be unsatisfactory. Thus, in the case of larger morphological units, e.g. tidal channels, the time-scale of response was usually not taken into account (see FitzGerald & Nummedal, 1983).

Such an equilibrium concept is applied in this paper by analysing the hydrodynamic and morphodynamic responses to a sudden decrease of basin area and the associated tidal prism. In particular, the morphological time-scale factor is discussed for a number of selected cases of the Zoutkamperlaag channel system.

PHYSICAL SETTING

The Zoutkamperlaag is by far the largest tidal system of the Frisian inlet. Its outer delta extends seawards by over 5 km. The Zoutkamperlaag has a throat width of 4 km and a depth of over 18 m below NAP (Dutch Ordinance Level). In the northern part of the flood-tidal basin the main channel is oriented almost north–south. Within the ebb-delta region the Zoutkamperlaag bifurcates into a main ebb channel oriented N–S and a marginal flood channel oriented E–W (Fig. 1; Hayes, 1979).

The tidal basin encloses an area of about 210 km^2. Two main channels may be distinguished, i.e. the Zoutkamperlaag which extends far into the basin, and the Gat van Schiermonnikoog which meets the Zoutkamperlaag south of the inlet. The channels are separated by subtidal and intertidal shoals and flats. In the tidal basin the Zoutkamperlaag bends in a SE direction. In the past, the southernmost extension of the Zoutkamperlaag reached into the former Lauwers Sea. The tidal basin is separated from adjacent basins, i.e. the Pinkegat in the west and the Eilanderbalg in the east, by two backbarrier drainage divides which are located below mean high water. It is believed that transport of water and sediment across these divides is significant only during storms (de Boer, 1979).

According to Winkelmolen & Veenstra (1974) and Veenstra & Winkelmolen (1976), the grain-size distribution pattern in the outer delta is rather complicated. A lag deposit on the frontal part of the delta can be distinguished from finer sand which enters and continuously re-enters the inlet until it finally nourishes the beach on the downdrift island. The grain size at the cross-section in the inlet throat is below 500 μm with a d_{50} of 160 μm. Towards the south, the grain size of the channel sediments drops to <250 μm with a median size of 120 μm. According to de Glopper (1967), the upper 25 cm of the tidal flats consist of sand with a clay content of 1.5–3%. Median grain sizes range from 80 to 110 μm (Postma, 1957).

The tide along this part of the coast is semi-diurnal (Fig. 2) with a mean tidal range of about 2.2 m. The coast can therefore be classified as mesotidal (Davies, 1964; Hayes, 1979). Tidal ranges show a periodic fluctuation on an 18.6-year scale, reflecting the influence of the lunar cycle. Accordingly, the largest recent tidal range occurred in 1982, showing an increase of 8% with respect to that of 1970 (Fig. 2).

The dominant wave mode seawards of the inlet (Fig. 1) at a water depth of 19 m lies between 0.5 and 1.0 m, being dominantly incident from the west. According to Niemeyer (1985), wave penetration through the inlets of the Wadden Sea is small. Calculations using wave propagation models for the Marsdiep tidal basin show that wave refraction concentrates the major part of the wave energy along the island heads (Andorka Gal & Voortman, 1992; Delft Hydraulics, 1993), while a significant reduction in wave energy is achieved by bottom friction and wave breaking. The radial spreading of wave energy also reduces the total amount of wave energy per unit length along the wave front and

Table 1. Ebb (EV) and flood (FV) volumes (in 10^6 m^3) and cross-sectional areas Ac (in m^2) for different sections of the Zoutkamperlaag, measured in various years. The location of the cross-sections is given in Fig. 1

	Frisian Inlet (1)			Rode Hoofd (2)		Oort (3)	
Year	FV	EV	Ac	FV	EV	EV	EV
1958				204	191		
1961	312	313		225	211	60	47
1963				210	166	63	47
1963				208	178		
1968	320	321		233	220	59	51
1968						59	44
Closure							
1969						54	45
1971	196	210	25753	99	84		
1975	199	196	23387	112	74		
1981	239	239	20957	122	94	80	54
1987	218	194	19578			71	57

therefore limits the penetration through the inlet (see also Niemeyer, 1986). The mean wave height in the tidal basin is considered to be less than 0.5 m, waves being mainly locally generated.

HYDRAULIC HISTORY

In 1969 the original tidal basin was reduced in size by the building of a 10-km long closure dam (Fig. 1) which reduced the basin area by *c.* 100 km^2. Measurements taken in the inlet throat between the Engelsmanplaat and Schiermonnikoog show a tidal volume (i.e. ebb + flood volume) of 641 million m^3 in 1968. By 1971 this had reduced to 406 million m^3 or 70% of the original volume. Although the differences between flood and ebb dominance were minor, it was nevertheless clear that the ebb dominance prior to 1969 changed into a flood dominance after the closure (Table 1).

Discharge measurements along a profile located just seawards of the closure dam in the Zoutkamperlaag channel (section 2) show that the tidal volume was reduced from 453 million m^3 to 183 million m^3, which amounts to a 60% reduction. By 1981, this volume had again increased to 216 million m^3. These measurements suggest that the system was flood-dominated both before and after the closure.

Measurements carried out in 1987 in a cross-sectional profile east of the closure dam (section 3) indicate an increase in tidal volume by 18% with respect to 1968 and by 29% with respect to 1969. In addition, the local flood dominance had increased to *c.* 14 million m^3 by 1982 (Table 1). All computed volumes were corrected for tidal differences prevailing at the time of the measurements, thus making direct comparisons possible.

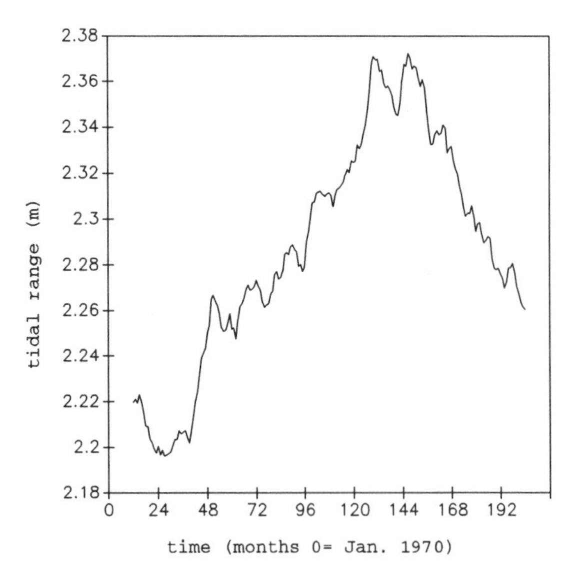

Fig. 2. Changes in the tidal range calculated from measurements at the gauge of Schiermonnikoog. The long-term variation in mean tidal range is caused by the 18.6-year lunar cycle.

STUDY METHODS

Accurate and detailed soundings taken in the Frisian inlet system in 1970, 1975, 1979, 1982 and 1987 made it possible to reconstruct the morphological changes that took place over this 18-year period. The sounding data were digitized and processed by means of a GIS. After interpolation, digital elevation models were computed on the basis of which bathymetric maps were generated and hypsometric and volumetric curves were constructed. Examples of the bathymetric maps are shown in Plate 1a (1970) and Plate 1b (facing p.86) (1987).

Sounding errors have been defined in extensive studies carried out by the Rijkswaterstaat whereby several sources of error were identified (Nanninga, 1985). Of these, the so-called 'ship effect', i.e. systematic deviations between the results of different survey vessels, has the largest influence. For the Frisian inlet, the standard deviation between successive soundings performed in the tidal basin and the ebb-tidal delta amounts to 4.1 and 6.5 cm respectively (Hartman & Pastoor, 1985). Apart from the sounding error, an additional error is introduced through the interpolation procedure. This is related to the size of the grid cells, the topography of the bottom, the orientation of the sounding lines and the interpolation method itself (Oost & de Haas, 1992). The work of de Looff (1975) suggests that the maximum standard error is about 1 cm for sounding lines spaced at a maximum of 200 m. For the interpolation method itself, no standard error can be determined. However, the method applied in this study was specially developed for this type of structured data and is considered by the authors to be the most accurate procedure currently available by means of which interpolation errors can be reduced.

MORPHOLOGICAL CHANGES SINCE 1970

Ebb-tidal delta

In 1970 the ebb-tidal delta was characterized by the presence of two major channels separated by a sandy shoal (Plate 1a). East of the channels a triangular swash platform (26 km^2) was located at average depths of 2.5–5 m below NAP. This swash platform was directly connected to the west coast of the barrier island of Schiermonnikoog. The highest parts of the platform were situated along its northern and western edges (Plate 1a). The northern edge consisted of a shallow, linear swash bar situated at a depth of 1.0–2.5 m below NAP. The swash bar was oriented WNW–ESE, forming an angle of 20–25° with the coastline of Schiermonnikoog. On a smaller morphological scale, an undulating sea-bed topography existed on parts of the swash platform and on the delta lobe to the north at water depths of 2.5–10 m below NAP (i.e. relative to mean sea-level, Plate 1a). The regular variations in height (maximum 1.5 m), the spacings (600–700 m) and the orientation of the topographic features (mainly N–S) suggest the presence of a field of large-scale bedforms such as sandwaves (Biegel, 1991).

Over the period 1900–1975 a clockwise migration (or rotation) of the channels is evident (Oost & de Haas, 1992). Simultaneously, the swash platform reduced in size and its NW-edge has migrated landwards (to the SE) over a distance of 2 km. The linear swash bar that formed the northern part of the platform can be seen to have amalgamated with the coast, while still being oriented obliquely to the shoreline. After 1975 similar trends are observed, although the throat of the inlet has remained stable in shape and position over the whole period. From 1975 to 1979 the swash platform further reduced in size and the swash bar grew in height. By 1979 the top of the bar had reached a height of 1 m above NAP. In addition its crestline had rotated in an anti-clockwise direction, the bar now lying almost parallel to the shoreline and a small spit having formed on its western side.

Finally, by 1982 the former swash bar had completely welded to the island beach, to form an extensive terrace exposed at low tide. The western edge subsequently began to accrete in a southward direction and the spit continued to grow (from NNW to SSE; Plate 1b). Moreover, the northern part of the spit moved 300 m to the south, now being completely in line with the local shoreline. By 1987 the northern section of the recurved bar had a total length (W–E) of *c.* 3 km, the N–S oriented part of the spit being approx. 2.5 km in length (Plates 1b, 2). Over the same period (1982–1987) the flood-dominated channel, located in the inlet throat, migrated 300 m to the east (compare Plate 2).

According to consecutive soundings between 1970 and 1987 the small-scale bedforms migrated in an easterly direction. North of the swash platform both the crests and troughs of the bedforms were initially eroded by currents and waves as the ebb-tidal delta adjusted to the new hydraulic conditions. At a later stage, the field of bedforms to the north of the recurved bar extended in a southerly direction and the lengths of the crestlines increased.

From 1970 to 1987 erosion and sedimentation rates related to depth were calculated for both the outer delta and the tidal basin (Fig. 3). From the data it was evident that the outer delta experienced erosion below - 4 m NAP, while above - 4 m NAP sedimentation dominated. The magnitudes of sedimentation and erosion for each time interval are presented in columns 2 and 3 of Table 2. Sediment budget calculations indicate a net loss of 21 million m^3 in the outer delta region.

Table 2. Sediment balance of the Frisian Inlet System (ebb tidal delta and flood tidal basin (in 10^6 m^3)

Period	Tidal basin	Outer delta	Sum
1970–1975	13.7	- 12.8	+ 0.9
1975–1979	8.9	6.1	15.0
1979–1982	3.6	- 5.4	- 1.8
1982–1987	4.6	- 8.5	- 3.9
1970–1987	30.8	- 20.7	10.2

Flood-tidal basin

Between 1970 and 1978 the decrease in tidal prism caused a general reduction of all main channel cross-sectional areas in the flood-tidal basin. Major depositional areas were located adjacent to the closure dam. After 1979 the zone of significant deposition gradually shifted towards the northern part of the channel, closer to the inlet throat. In the tidal inlet itself a major adjustment is seen up to 1975, the inlet throat becoming shallower. No major modifications are observed thereafter and the average depth and cross-sectional areas remain more or less the same. In the final period (1981–1987) the shallower part of the inlet moved somewhat further to the east, i.e. towards the island of Schiermonnikoog. As a consequence, the island shore began to erode locally.

The positions of the backbarrier drainage divides turned out to be rather mobile. The main divide south of Schiermonnikoog migrated towards the east. Since this migration already began in the early 1960s, i.e. before the closure of the Lauwers Sea, it can obviously not be entirely attributed to this closure. From 1963/1964 to 1981 the northern part of the divide gradually migrated eastwards by a distance of about 1 km. In the period 1981–1987 the position of this part of the divide remained more or less fixed. By contrast, the central and

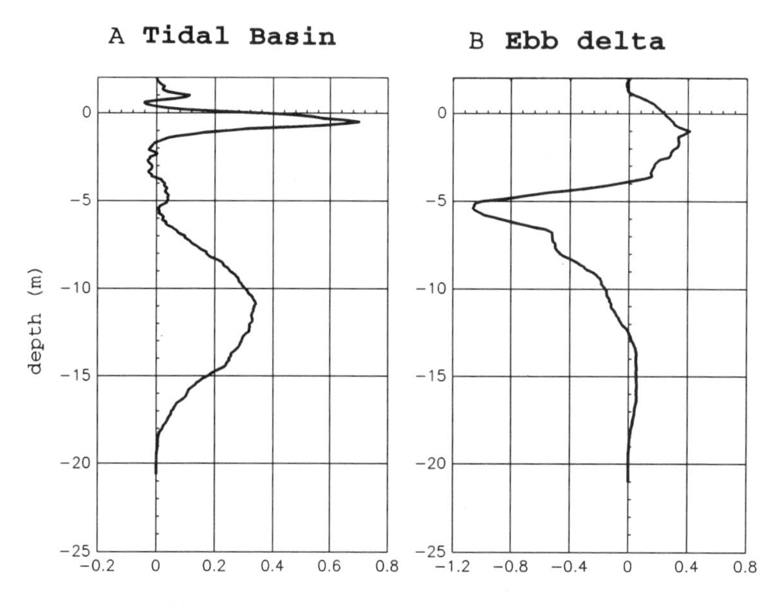

Fig. 3. Sedimentation (+) and erosion (-) with depth in the period 1970–1987 for the tidal basin (A) and the outer delta (B) of the Zoutkamperlaag system.

southern parts of the divide shifted by 1.5 and approximately 3.0 km respectively (period 1960–1987). This watershed migration had a rather spasmodic character. Between 1963 and 1970 the average migration amounted to about 800–900 m. Remarkable is the fact that in the period 1970–1981 there was hardly any change. In other words, in the period characterized by strong morphological responses of the tidal basin to the new hydraulic conditions, the backbarrier drainage divide along the main tidal channel remained in position. This suggests that the observed migration of the divide was a natural, autonomous process which was temporarily disrupted by the closure of the Lauwers Sea.

With the migration of the divide a synchronous process of erosion is observed. Thus, the smaller channels in the east of the basin began to erode, showing an increase in depths and a further extension eastwards. Meanwhile the backbarrier drainage divide south of the shoal of the Engelsmansplaat also shifted towards the east. In spite of this watershed migration, the total surface area of the flood-tidal basin remained about the same.

Over the period 1970–1987 the total sediment input into the tidal basin amounted to 31 million m^3 (Fig. 3), part of which was used to fill the channel system (sedimentation between −20 m and −5 m NAP). Compared with the net loss in the ebb-tidal delta, one has to conclude that the sedimentary system of the Zoutkamperlaag, including the ebb-tidal delta, is not a closed system. At least 10 million m^3 of sediment were imported from adjacent coastal areas.

Discussion

Summarizing the morphological trends outlined above, it is clear that by reducing the tidal prism in 1970 the equilibrium of the tidal system was disturbed. The reduction of the tidal prism by 35% and the commensurate reduction in flow velocities reduced the transport capacity of the ebb current to such an extent that it was no longer capable of maintaining the ebb-tidal delta against the corresponding proportional increase in wave influence (compare Kohsiek, 1988). As a result, the seaward parts of the tidal delta were reworked, the sediments being transported onshore due to wave asymmetry. The loss of sediment below the −4 m NAP and the gain higher up in the profile is a clear indication of this process (Fig. 3). Wave-driven longshore currents, alternately enhanced by tidal currents, further contributed to the longshore distribution of sediment.

The onshore and longshore transport of sediment by waves breaking on the swash platform gave rise to a concentration of sediment in (sub)tidal shoals which subsequently developed into a linear sandbar. On the western side of this bar a spit started to grow in a southerly direction. The growth and orientation of the spit was possible because of an abundance of sand in the large linear bar and the prevailing sediment transport directions in the inlet throat, the latter resulting from a combination of wave-driven currents and tides. The incident wave spectrum is dominated by the W–NW component, but waves from other directions will also refract around the bar. In both cases a southward flowing, wave-driven longshore current is generated. The tidal currents in the inlet will alternately enhance or reduce the effect of these wave-driven currents.

Around 1982 the linear bar was remodelled into a recurved bar or sand hook. Although the entire bar system has the appearance of a recurved spit, only the N–S oriented part is actually generated by typical spit-forming processes, i.e. longshore drift. Radiometric observations of heavy minerals also confirm this interpretation (de Meyer *et al.*, 1989). If a net longshore drift were responsible for the development of the W–E part of the bar, this should have been reflected by a significant increase or decrease in heavy mineral concentrations in a direction parallel to the bar due to longshore sediment sorting. Field observations by de Meyer *et al.* (1989) demonstrate that such sorting processes are absent along the W–E portion of the bar. The N–S portion, on the other hand, does show grain-size sorting and a decrease of heavy minerals in a southerly direction.

Substantial deposition has taken place in the tidal inlet and the flood-tidal basin. The sediment input is most obvious in the filling of the channel system, as revealed by the reduction of cross-sectional areas (see Table 1). The dynamics of the drainage divides, on the other hand, appears to be driven by an autonomous process. A similar phenomenon is evident in the divide south of Ameland. In the period 1831–1980 this divide shifted eastwards by a distance of 5 km, *c.* 800 m of this displacement having occurred between 1970 and 1980 (de Boer *et al.*, 1991). In this latter tidal basin anthropogenic influences are not self-evident.

The migration of tidal watersheds under 'natural' conditions, as observed in the period 1963–1970, may have been induced by the frequent occurrence of westerly winds and storms. During storms, both water and sediment are transported across the drainage divide, not only in the channels but especially on the shoals. Owing to the eastward displacement of a divide, the affected channel systems have to drain greater surface areas and hence respond by headward erosion. Another potential mechanism is the presence of a tide-induced slope of water levels across a divide caused by a phase lag of the tide. In the present case such a phase lag could have resulted from: (i) deposition along the east coast of Schiermonnikoog; (ii) a change in propagation velocity of the tidal wave; or (iii) a change in resonance characteristics of the tidal basin.

EMPIRICAL STABILITY

Since the beginning of this century several attempts have been made to deduce empirical relationships between geometrical characteristics of channels and hydraulic parameters. For the Wadden Sea some recent studies to this effect were made by Gerritsen & de Jong (1985), Stive & Eysink (1989), Gerritsen (1990) and Misdorp *et al.* (1990).

All the relationships established in the course of these studies apply to systems that were in hydraulic equilibrium at the time of study. Furthermore, all the data were collected from channels which were of similar size to that of the Zoutkamperlaag channel. Since derivatives from geometric and hydraulic parameters appear to only enlarge the amount of variance (Eysink & Biegel, 1992), only basic relationships were chosen to describe the state of equilibrium of the Zoutkamperlaag system after the closure. In this paper only a small selection of the large number of relationships will be discussed. Using data of Gerritsen (1990), the following equations appear to be particularly relevant (see also Figs 4A, 4B):

$$DV = 17519A_c - 75.4 * 10^6 \quad (1)$$

and, since this involves a totally different hydraulic parameter:

$$Q_{max} = 1.27A_c' - 6937 \quad (2)$$

where A_c = flow area below NAP (mean sea-level) (m^2), A_c' = flow area below water level during maximum flow (m^2), Q_{max} = maximum discharge volume during ebb or flood (m^3), DV = dominant volume (i.e. the maximum integrated discharge over either flood or ebb phase) (m^3).

In earlier equations derived by Gerritsen & de Jong (1985) for the Frisian inlet system, the tidal volume (TV) for spring tide conditions was used as the independent variable instead of the selected dominant volume (DV) applied here (see equation 1).

Equations 1 and 2 have been plotted in Figs 4A & 4B together with the Frisian inlet data. The other data, derived from several other inlets of the Wadden Sea, are discussed by Gerritsen (1990). Figures 4A & 4B illustrate that equilibrium is gradually restored after the hydraulic environment of a system has been disturbed. They also demonstrate the simplicity of determining a new equilibrium value using basic empirical relationships. Since the flow area or cross-section turns out to be a reliable parameter for empirical research, it has been used as the main parameter to study the morphological behaviour of the Zoutkamperlaag through time.

EQUILIBRIUM AND TIME-SCALE OF MORPHOLOGICAL RESPONSE

Human interferences can have long-term effects on a coastal system and hence can determine large-scale coastal behaviour (LSCB; e.g. de Vriend, 1991). Such LSCB refers to coastal evolution at spatial scales larger than tens of kilometres and time-scales of decades. In many cases it is not only the equilibrium parameters that are of interest (see above), but especially also the hydrodynamic and morphological transitions from one state of equilibrium into another. Morphological changes in the Zoutkamperlaag provided a good opportunity to study both the time-scale of such a response and the spatial adjustment of the system.

It is generally accepted that the process of morphological response to new hydrodynamic conditions is initially rather quick (FitzGerald & Nummedal, 1983). The process then decelerates progressively as the new equilibrium is approached. Morphological adjustments thus usually behave as an exponential function of time (Stive & Eysink, 1989; Gerritsen, 1990; de Vriend, 1991):

$$\frac{\delta x}{\delta t} = -\frac{1}{\tau}\Delta(x - x_e) \quad (3)$$

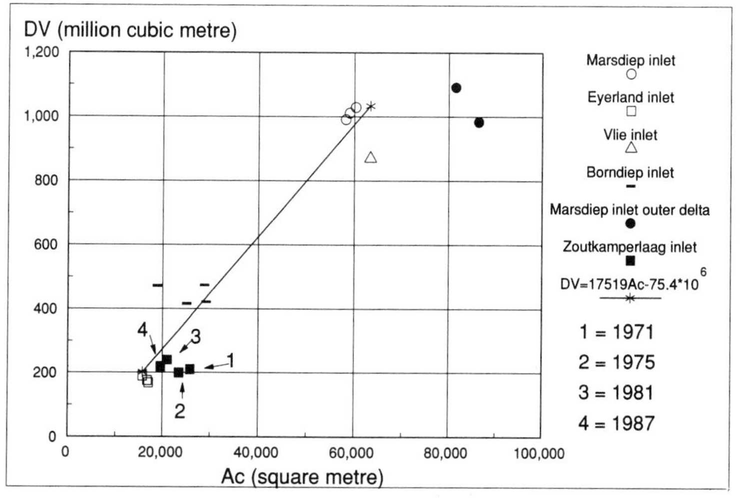

A

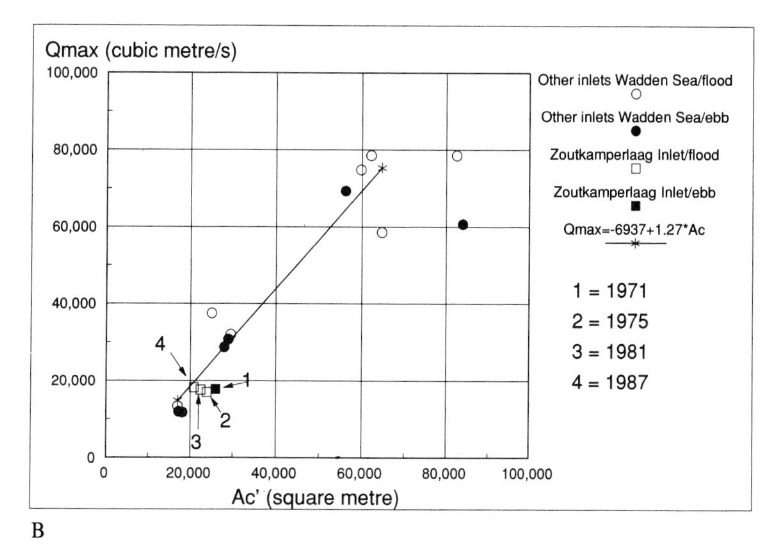

B

Fig. 4. (A) Empirical relationship between the dominant volume ($\boldsymbol{DV}$ = ebb + flood volume) and the flow area $\boldsymbol{A_c}$ (data for the Zoutkamperlaag are based on discharge profile 1 (see Fig. 1); partly based on Gerritsen, 1990). (B) Empirical relationship between the maximum discharge $\boldsymbol{Q_{\max}}$ and the flow area $\boldsymbol{A_c'}$ (discharge data are based on measurements in profile 1, see Fig. 1).

where x = morphological parameter in question, x_e = new equilibrium value of x following the change, τ = time-scale (year) and t = time (year).

In this study the cross-sectional area is considered to be a representative morphological parameter and integration thus yields:

$$\frac{(A - A_e)}{(A_i - A_e)} = e^{-t/\tau} \tag{4}$$

where A = flow area below NAP (m^2), A_i = initial flow area following the change (m^2), and A_e = equilibrium flow area following the change (m^2)

Time-scale of adjustment of the Zoutkamperlaag

Thirty-one channel cross-sections spaced at approximately 500 m were selected (Fig. 5). Profiles 1–4 were located in the outer delta, profiles 5–7 in the inlet throat and profiles 23–28 near the closure dam, having one fixed boundary. Temporal changes were measured using six gridded bathymetric maps for the years 1967 (data from de Oost & Haas, 1993), 1970, 1975, 1979, 1982 and 1987. Boundaries between tidal channels and flats were chosen individually, based on information of the same maps and the shape of the profiles. The flow areas

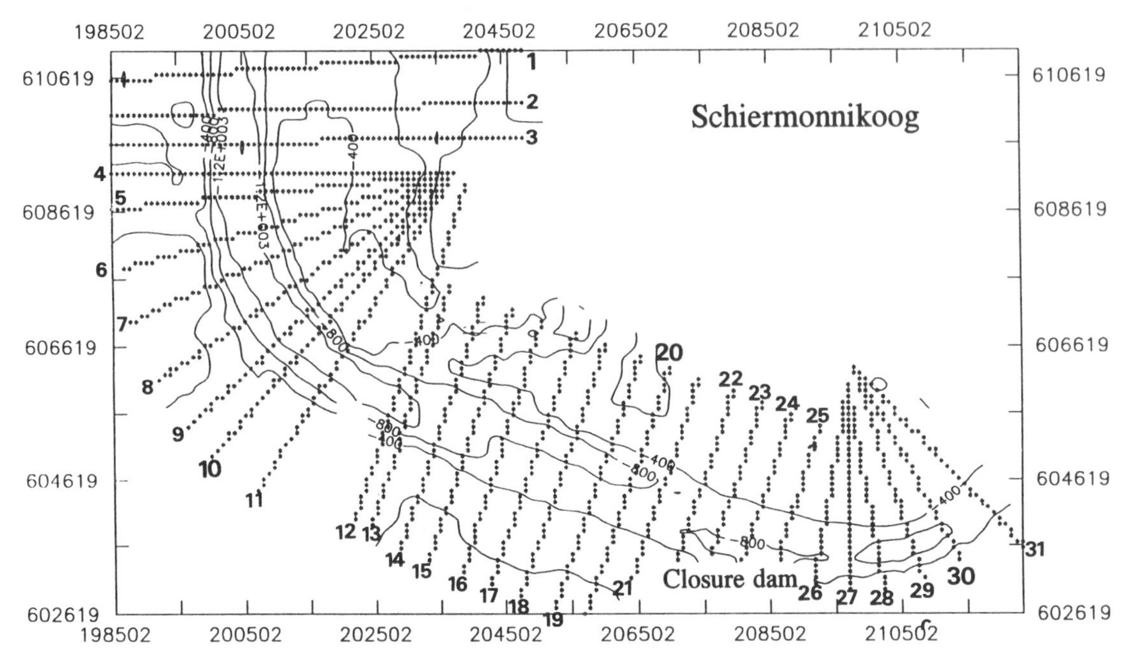

Fig. 5. Location of the 31 cross-sectional profiles in the main channel of the Zoutkamperlaag (depths are in cm).

below NAP ($A_{c,NAP}$) were calculated using triangulation. It is believed that no changes occurred from 1967 to 1968. In order to calculate the time-scale accurately, flow areas were normalized with respect to the year 1968. They have been plotted against time for all 6 years in Fig. 6. Two major groups may be distinguished: (i) flow areas decreasing with time, located west of the closure dam; and (ii) flow areas increasing with time, located east of the closure dam.

For the outer delta (profiles 1–4) the parameter $A_{c,NAP}$ has its fastest rate of decrease in the period 1970–1975. After 1975, the value of $A_{c,NAP}$ appears to be stable. Although the scatter is high, the behaviour in the inlet throat (sections 5–7) seems to follow more or less an exponential function. The sections up to cross-section 13 show their largest decrease of A_c immediately before the closure and, with some delay, also after 1975. The cross-sectional areas of profiles 14–18 are characterized by a rapid decrease around the time of closure and an almost linear decrease in time since 1975. Near the closure dam, the morphological adjustment in some cases starts with an increase of A_c between 1968 and 1970 (profiles 19, 20 and 21). In general though, the flow area decreases exponentially since the closure (profiles 19–22). Profile 23 is remarkably stable in shape and size and marks the transition to a totally different environment near the backbarrier drainage divide. By contrast, flow areas actually increase by up to 31% eastwards of the closure dam and in the direction of the divide (the data are summarized in Fig. 7).

Discussion of response time-scales

The results clearly show that there is a large spatial variability in the degree of morphological change with time. It is obvious, however, that the exponential relationship for morphological adjustment is not as generally applicable as one might have expected. Quite often an exponential curve is not the best fit and has to be replaced by a linear relationship (e.g. profiles 15 and 23) or by a combination of an exponential and a linear relationship (profiles 4 and 18). For those profiles in which the adjustment initially resulted in an increase of the flow area (A), the time-scale may be calculated by using the model of equation 3, but this time-scale is then without any physical meaning. The time-scale factor included in equation 3 is basically related to the intial rate of adaptation (Stive & Eysink, 1989).

The morphological developments in the eastern part of the tidal basin, on the other hand, are in total

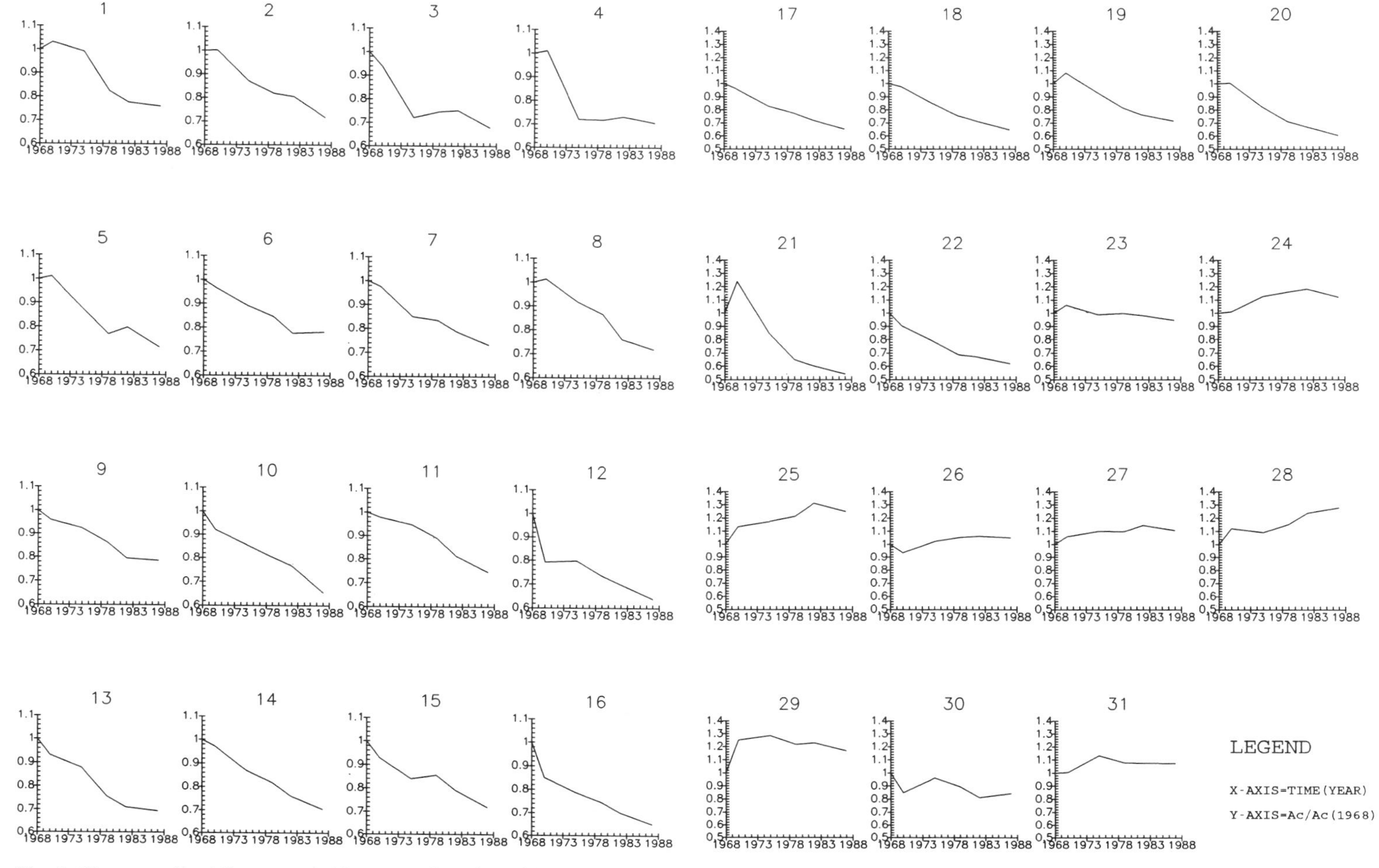

Fig. 6. The normalized flow area A_c/A_{c1968} as a function of time t for the 31 profiles indicated in Fig. 5.

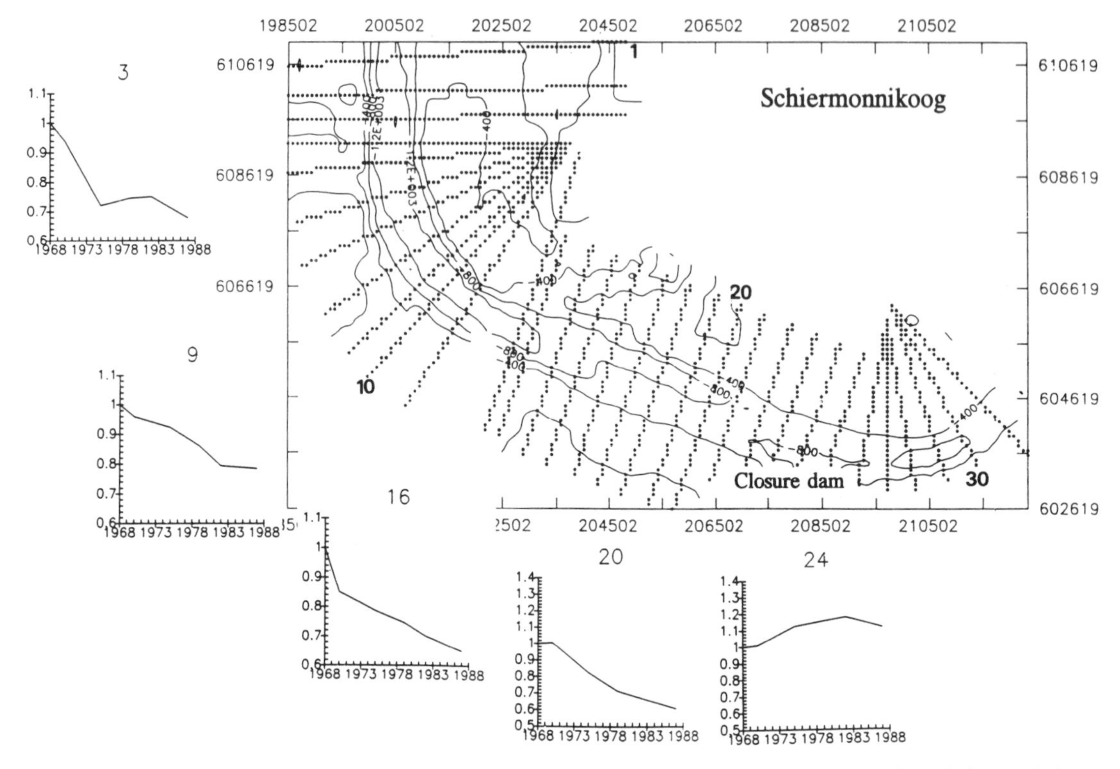

Fig. 7. Five characteristic responses of the normalized flow areas to the closure of the Lauwers Sea relative to their positions in the Zoutkamperlaag system.

contradiction to the theoretical principles of equation 3, in that A increases with time instead of decreasing with time (Figs 6 & 7).

For a number of selected cases, however, the application of the exponential relationship is definitely justified by the field data. These profiles are located in the throat (profiles 6 and 7), to the west of the closure dam (profiles 12, 14, 16 and 17) and just north of the dam (profile 22). By using equation 4, the morphological time-scale factor is calculated to vary from almost 6 years to 12 years. Profile 14 deviates from the general pattern by having an adaptation time-scale of *c.* 26 years (Table 3).

It would appear that the time-scale is small for profiles in the ebb-tidal delta as well as near the dam. Between these, time-scales are either larger or cannot be calculated because the morphological evolution departs substantially from the exponential model.

The large geographical variability in morphological behaviour through time and the deviation from the exponential model can be explained in several ways. Every morphological unit, be it an ebb-tidal delta or a channel in the flood-tidal basin, is always characterized by a certain degree of inertia. It takes time before sediment is mobilized, redistributed and redeposited again. The time lag of response or the relaxation time will depend on the spatial characteristics of the morphological element involved, the amount of sediment available and the energy present to redistribute sediment. There is a

Table 3. Equilibrium values of A_e and related time-scales of response (τ)

Profile	$A_i - A_e$ (m^2)	A_e (m^2)	τ (years)
6	4647	12190	10.0
7	5781	12478	10.1
12	5310	11136	7.3
14	10684	5772	26.1
16	4592	9962	5.8
17	6523	8032	12.1
22	4871	7743	6.2

considerable difference in this respect between the ebb-tidal delta and the flood-tidal basin. In the first 10 years some 22.6 million m^3 of sediment (i.e. 70% of the total amount over the period 1970–1987) was deposited in the flood-tidal basin, whereas the ebb-tidal delta lost 6.7 million m^3 (i.e. 33% of the total loss) over the same period. This proves that the sediment demand of the channels (i.e. the amount of sediment required to restore equilibrium; see Eysink, 1990) is only partly satisfied by the supply from the ebb-tidal delta. This is also visible in greater detail in the Zoutkamperlaag tidal channel. In the inlet throat sediment from the ebb-tidal delta is readily available, thus leading to a relatively rapid adjustment of the throat cross-section (Fig. 6; sections 3 and 4). Since a significant amount of sand from the ebb-tidal delta is also caught in the spit, it is not immediately available for the morphological adjustment of the channels. In addition, sediment transport rates into the tidal basin—and therefore morphological adjustment—will depend on the net tidal excursion during each tidal phase relative to the actual distance between source and sink of the sediment.

Further into the basin the system is less well adapted (time-scales being larger), often showing a time lag in the character of morphological response, as can be seen near the closure dam. Here the closure works themselves are expected to be responsible for the initial increase in cross-sectional area (Fig. 6; profiles 19–21). The construction of the dam resulted in a gradual reduction of the flow section previously draining the Lauwers Sea. This inevitably led to an increase in flow velocities, significant scour in the channels and an initial increase in flow area at the time of closure. Soon after the closure, rapid adjustment is evident (profiles 19–22). This suggests that the sediment filling this part of the inlet is not derived from distant (e.g. seaward) sources. Instead, the sand must have been derived either from the tidal flats, the island headlands or the adjacent tidal basins by crossing the backbarrier drainage divides.

A significant amount of the sediment is expected to be fine-grained, consisting of silts and muds. This material is transported in suspension and is predominantly imported from the North Sea—there being no freshwater or riverine source—to be deposited under the specific physicochemical and biological conditions in the Wadden Sea (Postma, 1982).

The rapid local adjustment of the channel system close to the dam is quite logical because of the drastic physical changes produced by the construction works. The southern channel, for example, lost its drainage basin in the course of the closure, the sudden flow reduction resulting in rapid deposition.

Developments further to the east are mainly determined by the dynamics of the drainage divide south of Schiermonnikoog. Since these were only partly affected by the closure works, the expected gradual reduction in channel capacity did not occur (sections 24–31). Effects of the closure only manifested themselves in the stable position of the central and southern parts of the drainage divide over the period 1970–1980. The eastward migration of the divide under natural conditions would have required a local increase in channel capacity. However, due to the reduction in overall tidal prisms this capacity was already available. As a result, no further migration of this part of the drainage divide was observed in the first decade after the closure. In the north, on the other hand, the natural displacement of the drainage divide continued. After 1980 the entire drainage divide resumed its natural dynamic behaviour. In response, the local increase in tidal discharge resulted in channel scour, thereby enlarging the cross-sectional area (sections 24–31).

GENERAL CONCLUSIONS

In 1969 the basin area of the Zoutkamperlaag channel system, which forms part of the Frisian inlet, was reduced by *c.* 100 km^2 with the construction of a closure dam. This resulted in a reduction of the total tidal volume (i.e. ebb + flood volume) by about 30%. As a consequence, the equilibrium between the morphology of the inlet system and the tidal prism of the catchment basin changed. The ebb-tidal delta gradually reduced in size due to erosion by currents and waves, whereas depositional processes dominated in the flood-tidal basin, reducing channel cross-sections. Within the first decade some 23 million m^3 of sediment (70% of the total over the period 1970–1987) was deposited in the flood-tidal basin. At the same time the ebb-tidal delta lost approximately 7 million m^3 of sediment, which amounts to 33% of the total loss over the period 1970–1987. In combination with quantitative data concerning the time-scales of adjustment (of the order of 10–12 years), this would seem to suggest that in the flood-tidal basin the main morphological adaptation took place in the period

1970–1980/1982. Thereafter natural processes were once again dominant, especially in the eastern parts of the basin.

Morphological adjustment varied throughout the tidal basin. In the inlet throat, the behaviour of the cross-sectional area can be described as an exponential function of time. Sediment from the ebb-tidal delta is rapidly released, leading to a relatively quick adjustment. In other parts of the tidal system the morphological response and the relaxation time depend on: (i) the spatial characteristics of the morphological system; (ii) the amount of sediment available; (iii) the energy present to redistribute the sediment; and (iv) the effects of ongoing natural processes.

As indicated by the sediment budgets for the ebb-tidal delta, a more gradual response to changing hydrodynamic conditions caused by the basin closure is observed.

ACKNOWLEDGEMENTS

This project was funded by the Tidal Waters Division of the Ministry of Transport and Public Works (Rijkswaterstaat) as part of the project ISOS*2 (Impact Sealevel Rise on Society) and the Coastal Genesis Programme. Albert Oost and Henk de Haas (Comparative Sedimentology Division, Faculty of Earth Sciences, Utrecht University) kindly provided us with the digitized bathymetric map of the year 1967.

REFERENCES

Andorka Gal, J.H. & Voortman, H.G. (1992) Onderzoek naar de golfdoordringing in het Zeegat van Texel met numerieke modellen. *Rijkswaterstaat Rept* **GWAO-92.212x**.

Biegel, E.J. (1991) De ontwikkelingen van de ebgetijde delta en het kombergingsgebied van het Friesche Zeegat in relatie tot de sluiting van de Lauwerszee. *Fac. Geograph. Sci., Univ. Utrecht, Rept* **GEOPRO 1991.07**, 79 pp.

Boer, M. de (1979) Morfologisch Onderzoek Ameland. Verslag van het onderzoek op het Amelander Wantij in 1973. *Rijkswaterstaat Rept* **WWKZ-79.H005**.

Boer, M. de, Kool, G & Lieshout, M.F. (1991) Erosie en sedimentatie in de binnendelta van het Zeegat van Ameland 1926-1984: deelonderzoek no. 4. *Rijkswaterstaat Rept* **ANVX-91.H202**, 42 pp.

Bruun, P. & Gerritsen, F. (1960) *Stability of Coastal Inlets*. North Holland, Amsterdam, 123 pp.

Davies, J.L. (1964) A morphogenetic approach to world shorelines. *Zeits. f. Geomorph.* **8**, 127–142.

Delft Hydraulics (1993) Wave conditions along the Dutch Coast. *Delft Hydraulics Rept* **H1355**.

Dieckman, R., Osterthun, M. & Partenscky, H.W. (1988) A comparison between German and North American tidal inlets. *Proc. 21st Coastal Eng. Conf.* **3**, 2681–2691.

Eysink, W.D. (1990) Morphological response of tidal basins to changes. *Proc. 22nd Coastal Eng. Conf.* **2**, 1948–1961.

Eysink, W.D. & Biegel, E.J. (1992) Impact of sea level rise on the morphology of the Wadden Sea in the scope of its ecological function: Investigations on empirical morphological relations. *Delft Hydraulics Rept* **H1300**, 73 pp.

FitzGerald, M.D. & Nummedal, D. (1983) Response characteristics of an ebb-dominated tidal inlet channel. *J. sediment. Petrol.* **53**, 833–845.

Gerritsen, F. (1990) Morphological stability of inlets and channels of the Western Wadden Sea. *Rijkswaterstaat Rept* **GWAO-90.019**, 139 pp.

Gerritsen, F. & Jong, H. de (1985) Stabiliteit van doorstroomprofielen in het Waddengebied. *Rijkswaterstaat Rept* **WWKZ 84.v016**.

Glopper, R.J. de (1967) *Over de bodemgesteldheid van het Waddengebied.* Tjeenk Willink Zwolle, 67 pp.

Hartman, J. & Pastoor, K. (1985) Onderzoek naar de verandering van de bodemligging voor de Oostelijke Waddenzee en buitendelta's voor de perioden 1975–1979 en 1970–1979. *Rijkswaterstaat Rept* **85-24**.

Hayes, M.O. (1979) Barrier island morphology as a function of tidal and wave regime. In: *Barrier Islands* (Ed. Leatherman, S.), pp. 1–28. Academic Press, New York.

Jarret, J.T. (1976) Tidal prism-inlet area relationship. *Coastal Eng. Res. Center, Ft Belvoir, Virginina, Rept* **3**, 32 pp.

Kohsiek, L.H.M. (1988) Reworking of former ebb-tidal deltas into large longshore bars following the artificial closure of tidal inlets in the south west of the Netherlands. In: *Tide-influenced Sedimentary Environments and Facies* (Eds Boer, P.L. de, Gelder, S. van & Nio, S.D.), pp. 113–132. Reidel, Dordrecht.

Looff, D. de (1975) Enkele opmerkingen over de nauwkeurigheid van de bij de inhoudsberekeningen voor de Westerschelde toegepaste methode. *Rijkswaterstaat Rept* **75.1**.

Meyer, R.J. de, Prakken, A. & Wiersma, J. (1989) Radiometry of the west spit of Schiermonnikoog. In: *KVI Annual Rept 1989.* Kernfysisch Versneller Instituut, Groningen.

Misdorp, R.F., Steyeart, F., Hallie, F. & Ronde, J. (1990) Climate change, sea level rise and morphological developments in the Dutch Wadden Sea, a marine wetland. In: *Expected Effects of Climate Change on Marine Coastal Ecosystems* (Eds Beukema, J.J., Wolff, W.J. & Brouns, J.J.W.M.), pp 123–131. Kluwer, Dordrecht.

Nanninga, M. (1985) The accuracy of echo sounding: description in a mathematical model. *Rijkswaterstaat Rept* **WWKZ 85.H016**.

Niemeyer, H.D. (1985) Ausbreitung und Dämpfung des Seegangs im See- und Wattengebiet von Norderney. *Jber. 1985, Forsch. Stelle Küste* **37**, 49–97.

OOST, A.P. & HAAS, H. DE (1992) Het Friesche Zeegat: Morfologisch-sedimentologische veranderingen in de periode 1970–1987. *Inst. Aardwetenschappen, Afd. Sediment., Univ. Utrecht*, 93 pp.

OOST, A.P. & HAAS, H. DE (1993) Het Friesche Zeegat. Morfologische–sedimentologische veranderingen in de periode 1927–1970. Cyclische veranderingen in een tidal inlet systeem. *Inst. Aardwetenschappen, Afd. Sediment., Univ. Utrecht, Rept Kustgenese* **1(1-2)**, 94 pp.

POSTMA, H. (1957) Size frequency distribution of sands in the Dutch Wadden Sea. *Archs. Neerl. Zool.* **12**, 319–349.

POSTMA, H. (1982) *Hydrography of the Wadden Sea: Movements and Properties of Water and Particulate Matter.* Wadden Sea Working Group, Balkema, Rotterdam, 75 pp.

SHA, L.P. (1989) Variation in ebb-delta morphologies along the West and East Frisian Islands, The Netherlands and Germany. *Mar. Geol.* **89**, 11–28.

SHA, L.P. (1990) *Sedimentological studies of the ebb-tidal deltas along the West Frisian Islands, The Netherlands.* Thesis Rijksuniv. Utrecht, *Geologica Ultrajectina* **64**, 160 pp. (unpubl.).

STIVE, M.J.F. & EYSINK, W.D. (1989) Voorspelling ontwikkeling kustlijn 1990–2090. 3.1: Dynamisch model van het Nederlandse Kustsysteem. *Delft Hydraulics*, 57 pp.

VEENSTRA, H.J. & WINKELMOLEN, A.M. (1976) Size, shape and density sorting around two barrier islands along the north coast of Holland. *Geol. Mijnb.* **55**, 87–104.

VRIEND, H.J. DE (1991) Mathematical modelling and large-scale coastal behaviour. 1: Physical processes. *J. hydraul. Res.* **29**, 727–740.

WALTON, R.L. & ADAMS, W.D. (1976) Capacity of inlet outer bars to store sand. *Proc. 15th Coastal Eng. Conf., ASCE* **112**, 1919–1937.

WINKELMOLEN, A.M. & VEENSTRA, H.J. (1974) Size and shape sorting in a Dutch tidal inlet. *Sedimentology* **21**, 107–126.

Spec. Publs int. Ass. Sediment. (1995) **24**, 101–119

Sedimentological implications of morphodynamic changes in the ebb-tidal delta, the inlet and the drainage basin of the Zoutkamperlaag tidal inlet (Dutch Wadden Sea), induced by a sudden decrease in the tidal prism

A.P. OOST

Comparative Sedimentology Division, Institute of Earth Sciences, University of Utrecht, PO Box 80.021, NL 3508 TA Utrecht, The Netherlands

ABSTRACT

In 1969 the Lauwerszee, an embayment of the Wadden Sea, was diked and the tidal volume of the Zoutkamperlaag tidal inlet decreased from $305 * 10^6$ m^3 to $200 * 10^6$ m^3. The sudden decrease in tidal prism caused significant sedimentary changes. In order to document and investigate these changes, high-resolution computer studies were conducted and sedimentary cores were analysed.

Reduction of the tidal prism led to strong erosion of the ebb-tidal delta, a downdrift shift and partial fill of the inlet gorge, the formation of a large bar in the ebb-tidal delta, the partial fill of the main backbarrier channel, a shift of the eastern watershed and channel fill east of that watershed. The abandoned channel fills in particular have a significant preservation potential. Such fills reflect the sudden decrease in tidal prism and consists of clays, sands or alternations of both, deposited under relatively quiet conditions. The alternations reflect seasonal depositional cycles.

Sudden reductions of tidal prisms also occur under natural conditions, both in tectonically active and inactive lagoonal/estuarine settings. Apart from tectonics, several other factors can lead to a reduction of tidal prism and to the formation of abandoned channel fills. Several cases of abandoned channel fills from the fossil record can thus be explained in terms of local reduction of tidal prisms.

INTRODUCTION

Ebb-tidal deltas and backbarrier drainage basins have been investigated extensively both in recent environments (Bruun, 1978; FitzGerald & Nummedal, 1983; FitzGerald *et al.*, 1984a; FitzGerald & Penland, 1987; Aubrey & Weishar, 1988; Sha, 1990a) and in the fossil records (Beets *et al.*, 1979; Hamberg, 1991; Sloan & Williams, 1991; van der Spek & Beets, 1992). In recent systems, study periods generally span relatively short time intervals (decades). Consequently, apart from short-term cyclic and random variations in the behaviour of channels and shoals, these systems are commonly considered to be in dynamic equilibrium with prevailing hydrodynamic regimes. Long-term temporal variability is often overlooked also in the case of fossil ebb-tidal delta deposits. Indeed, these are commonly interpreted as representing a specific set of conditions that differs from that of prior and subsequent depositional events. For recent as well as fossil environments, therefore, the response time of barrier systems to changes in external conditions has to date generally not been considered.

In 1969 the tidal prism of the Zoutkamperlaag tidal inlet system was decreased by one-third (van Sijp, 1989) due to the diking of the Lauwerszee embayment (Fig. 1). Subsequent sedimentological changes in the ebb-tidal delta, the inlet and the backbarrier drainage basin were analysed with high precision. The purpose of this paper is to discuss the geological relevance of these observations.

SETTING

The Zoutkamperlaag tidal inlet is situated in the barrier island chain along the North Sea coast of

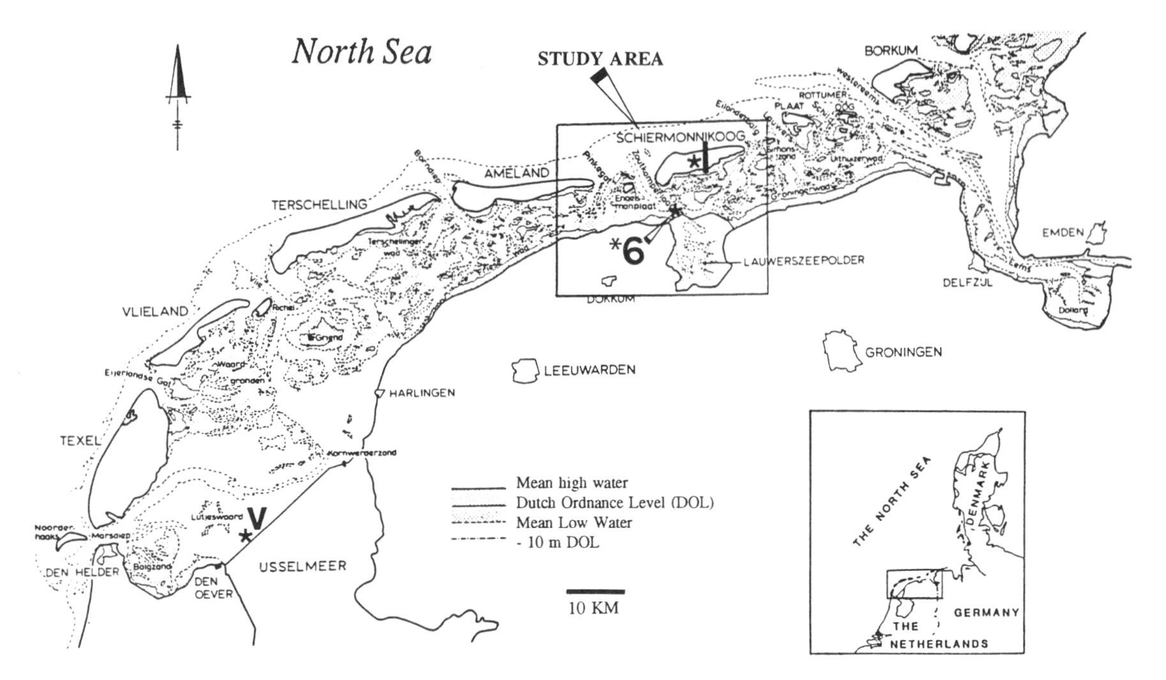

Fig. 1. Overview of the area. Asterisk 6 (***6**) refers to the location of the core in Fig. 6. Asterisk I (***I**) indicates the former position of the Lauwers inlet in AD 1300. Asterisk V (***V**) indicates the position of the former channel Vlieter (for details see text; after Steijn, 1991). The inset gives an overview of the North Sea.

The Netherlands, Germany and part of Denmark (Fig. 1). The North Sea is a broad and shallow shelf sea on a passive continental margin. The tidal wave propagates from west to east, with a residual current towards the east. The tidal regime is semi-diurnal. In the area under discussion the tidal amplitude is 2.3 m and the tidal regime is therefore mesotidal (Hayes, 1979; Postma, 1982). The wind strength varies seasonally, average wind velocities reaching 15 m s^{-1} in winter and 7 m s^{-1} in summer. Waves are highest in winter, reaching 2 m at 20 m water depth. The coast can be classified as a mixed-energy shoreline, influenced by both tides and waves (Hayes, 1975). The island has a typical 'drumstick' form characteristic for mesotidal coasts (Hayes, 1979).

The Zoutkamperlaag and the Pinkegat inlet systems are situated between the barrier islands of Schiermonnikoog and Ameland (Fig. 1). The inlets and their largely intertidal backbarrier flood-basins are separated by a supratidal shoal and its adjacent watershed, situated on top of a massive, relatively stable clay body at − 5 m with reference to Dutch Ordinance Level (DOL ≈ mean sea-level) (Sha, 1992; Oost & de Haas, 1992). In this paper attention is focused on the larger of the two inlet systems, the Zoutkamperlaag inlet (Fig. 1)

METHODS

Depth measurements and soundings, densely spaced in the channels and on the intertidal and supratidal flats, were made by the Ministry of Waterworks during the years 1967, 1970, 1975, 1979, 1982, 1987 and 1991. The data were condensed into a grid of 90 ∗ 90 m areas. The CONLOD Program, especially developed to handle sounding data (van den Boogert & Noordstra, 1988; van den Boogert, 1991), was used to fill empty cells with weighted interpolated values. Accuracy of both the original soundings and the interpolation technique is high. Thus, for the difference between two successive years, the standard deviation (1σ) is 0.06 m for the interpolated data over larger areas (Oost & de Haas, 1992, 1993). From the filled grid cells, depth maps were generated which show the development of the area between Ameland and Schiermonnikoog over the last 20 years (Plates 1–7 (facing p. 110); Oost & de Haas, 1992, 1993).

Net sedimentation maps were produced by subtracting grid values of consecutive years. Plate 8 shows the erosion and sedimentation for 1970–1987, i.e. the period after the closure of the Lauwerszee. In addition, quantitative data of net erosion and sedimentation were calculated for several areas (Figs 3–4, Oost & de Haas, 1992, 1993). Owing to the slightly different sizes of the study areas, the net values presented here differ from those of Biegel & Hoekstra (this volume, pp. 85–99); the general picture, however, is the same.

CHANGES IN THE ZOUTKAMPERLAAG SYSTEM

Owing to the closure of the Lauwerszee in 1969 (Fig. 1; Plates 1–2), the tidal prism of the Zoutkamperlaag was reduced from $305 * 10^6 \, m^3$ to $200 * 10^6 \, m^3$ (van Sijp, 1989). As a result, the system was no longer in equilibrium with existing hydrodynamic conditions and the ebb-tidal delta and the backbarrier drainage basin started to shift towards a new morphodynamic equilibrium (Plates 1–8; Biegel, 1991a, b; Oost & de Haas, 1992, 1993; Biegel & Hoekstra, pp. 85–99). The following changes were observed (for full description, see de Oost & de Haas, 1992, 1993).

Ebb-tidal delta

The ebb-tidal delta of the Zoutkamperlaag measures about 94 km^2. Total net erosion measured $26 * 10^3$ in the ebb-tidal delta for the time-span 1970–1987. Net erosion dominated during the periods 1970–1975, 1979–1982, and 1982–1987. Net erosion was mainly confined to the zones −4 to −12 m DOL (1970–1975, 1982–1987) or −4 to −8.7 m DOL (1975–1979) (Fig. 2). From 1970 to 1991, the −10 m DOL contour on the North Sea side of the ebb-tidal delta retreated shorewards, whereas the −15 m DOL contour remained in place (Plates 1–7).

Drainage basin

The backbarrier drainage basin of the Zoutkamperlaag tidal system measures about 126 km^2. In the period 1970–1987 sedimentation amounted to $30 * 10^6 \, m^3$. Sedimentation rates were highest soon after the reduction of the tidal prism (Fig. 3).

Bar development

In 1970 a triangular swash platform of 26 km^2 was present north-west of Schiermonnikoog (Plate 2). The top of this platform was largely above −5 m DOL, consisting of fine to medium sand, with a median size of 100–180 μm (Winkelmolen & Veenstra, 1974). Especially from 1970 to 1982, part of this swash platform eroded strongly (Plates 2–5). Southeastward sediment transport by waves and probably tidal currents is inferred from the migration of shoals and the refraction of waves (Veenstra & Winkelmolen, 1976).

A large recurved bar developed in the ebb-tidal delta at the western side of Schiermonnikoog. More than $6 * 10^6 \, m^3$ sand was stored in this bar above −2 m DOL (Noordstra, 1989). The E–W oriented part of the bar evolved partly as a swash bar from a subtidal shoal (Plate 1) that migrated towards the south-east in the period 1970–1975 (Plates 2 & 3). Aerial photographs show that structures observed in the beach and shallow foreshore region of Schiermonnikoog are largely wave-built. Judging from these structures, the observed development in the period 1970–1982, and the shape characteristics of the sediment, sand was transported partly parallel and partly perpendicular to the coast (Winkelmolen, 1969, 1982; Winkelmolen & Veenstra, 1980). During the period 1975–1982 this part of the bar rotated slightly counter-clockwise (Plates 3–5). After 1982 it migrated southwards (Plates 5–7).

The N–S oriented part of the bar developed after 1979 under the influence of waves and, to a lesser extent, probably also of tidal currents (Plates 4–7). Swash bars observed on aerial photographs indicate that the formation of this part of the bar was dominated by fairweather wave action, with waves approaching mainly from the WNW and refracting around the developing bar (Oost & de Haas, 1992). Quiet conditions prevailed in the embayment behind the bar (Plate 6), allowing sedimentation of muds alternating with sands. The southward migration of the bar over these deposits resulted in a coarsening-upward sequence. By 1987 the bar had become partly supratidal (Plate 6). Since 1989 the bar has become subject to strong erosion by tides and storms (J. Wiersma, personal communication). In 1991 the bar was again intertidal and was breached in several places by small tidal channels visible down to −5 m DOL (Plate 7).

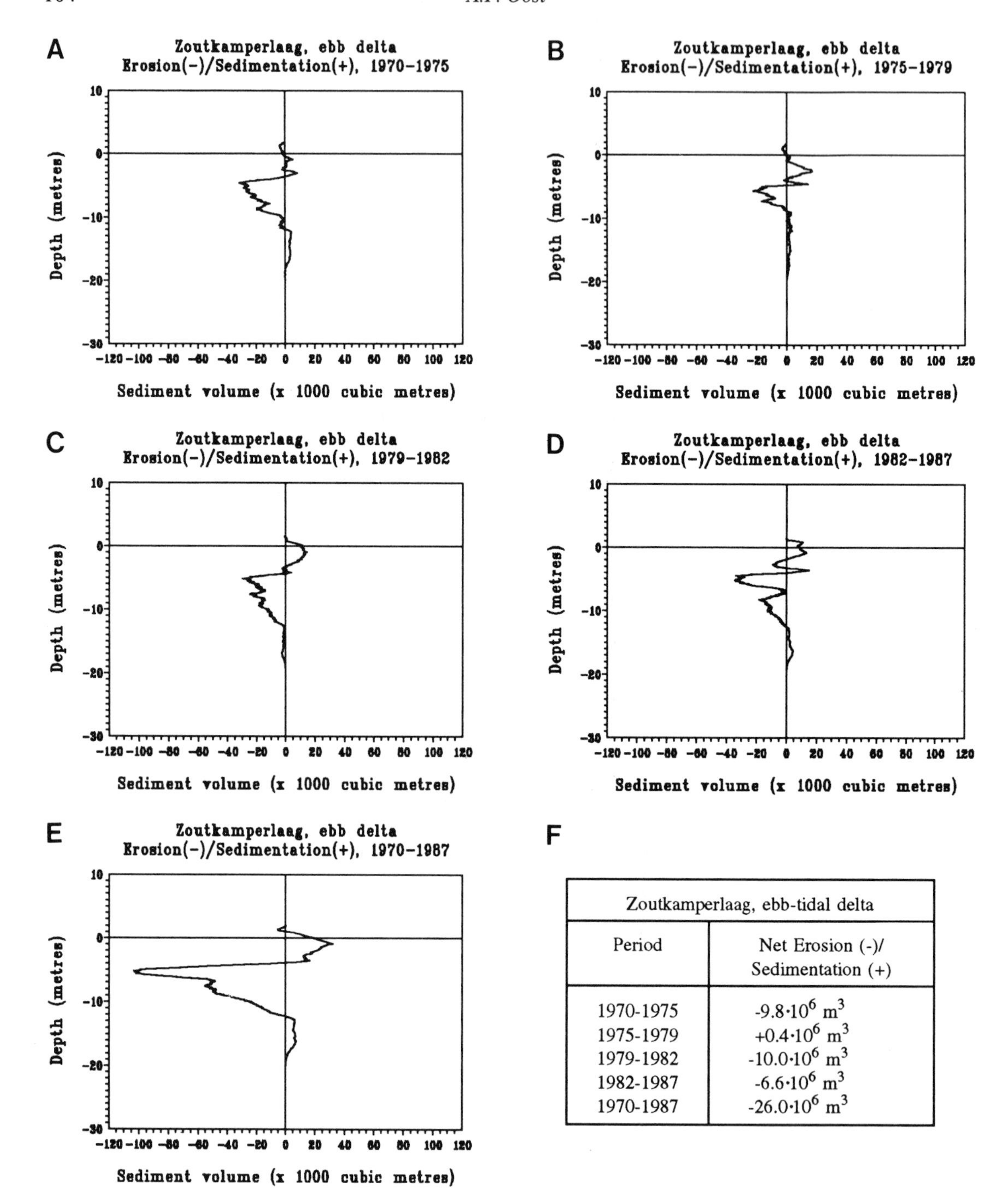

Zoutkamperlaag, ebb-tidal delta	
Period	Net Erosion (-)/ Sedimentation (+)
1970-1975	$-9.8 \cdot 10^6$ m^3
1975-1979	$+0.4 \cdot 10^6$ m^3
1979-1982	$-10.0 \cdot 10^6$ m^3
1982-1987	$-6.6 \cdot 10^6$ m^3
1970-1987	$-26.0 \cdot 10^6$ m^3

Fig. 2. Sedimentation (positive) and erosion (negative) as a function of depth in 0.01-m increments in the ebb-tidal delta of the Zoutkamperlaag inlet over various time intervals. A = 1970–1975, B = 1975–1979, C = 1979–1982, D = 1982–1987, E = 1970–1987, F = net erosion or sedimentation over these periods. Note the similarity of trends in all periods. Strong erosion was mainly confined to the zone between −4 m to −12 m DOL, whereas sedimentation dominated above and below this zone.

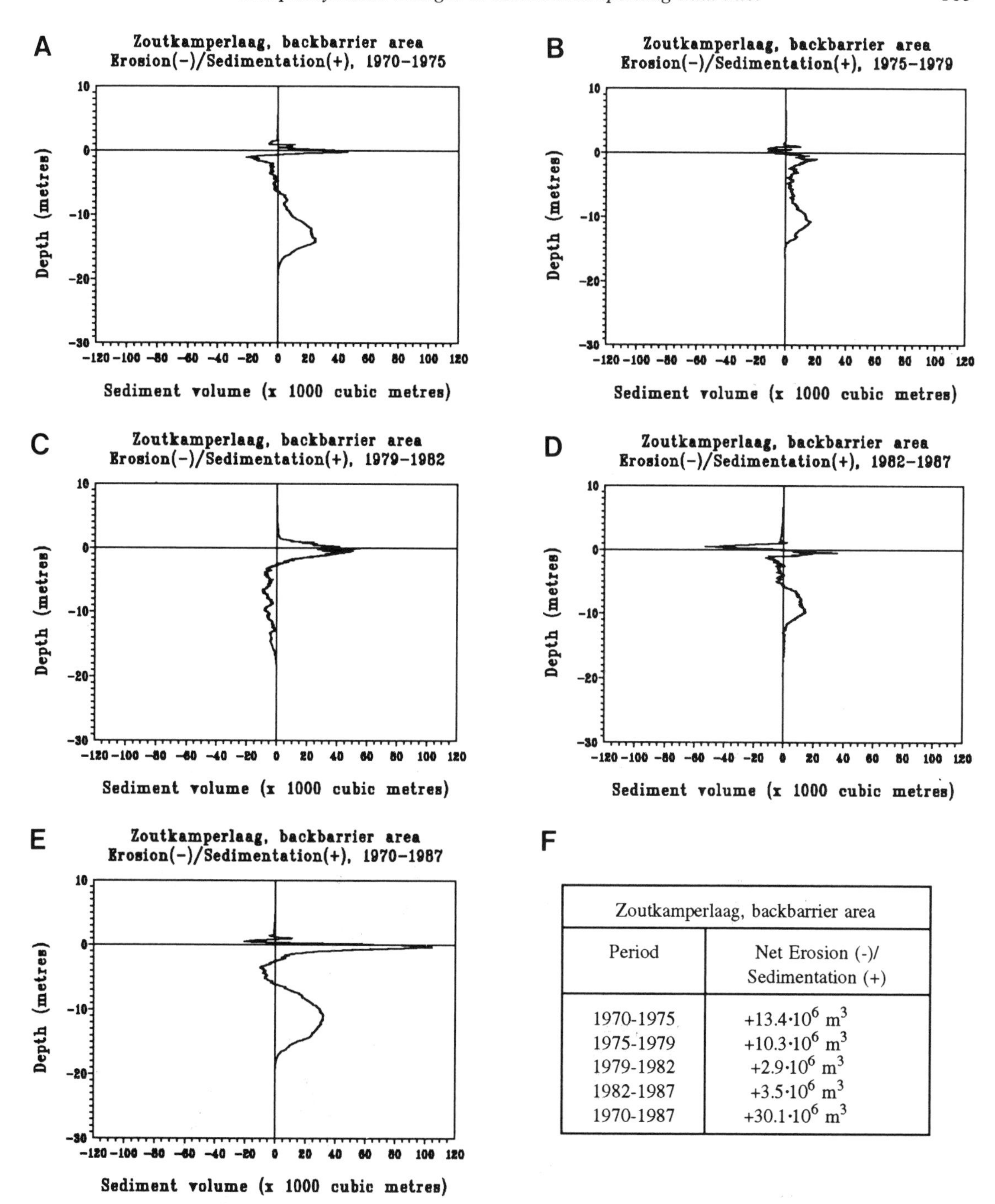

Zoutkamperlaag, backbarrier area	
Period	Net Erosion (-)/ Sedimentation (+)
1970-1975	$+13.4 \cdot 10^6$ m^3
1975-1979	$+10.3 \cdot 10^6$ m^3
1979-1982	$+2.9 \cdot 10^6$ m^3
1982-1987	$+3.5 \cdot 10^6$ m^3
1970-1987	$+30.1 \cdot 10^6$ m^3

Fig. 3. Sedimentation (positive) and erosion (negative) as a function of depth in 0.01-m increments in the rear of the Zoutkamperlaag inlet over various time intervals. A = 1970–1975, B = 1975–1979, C = 1979–1982, D = 1982–1987, E = 1970–1987, F = net erosion or sedimentation over these periods. Note the comparable patterns for most periods. Pronounced net sedimentation occurred in the deeper parts due to the fill of the main channel, except in the period 1979–1982 when the formation of drainage channels west of the watershed dominated the sedimentation pattern. Deposition at shallower depths was mainly due to migration of the watershed and accretion near Schiermonnikoog and the mainland.

Channel rotation

In 1970 the Zoutkamperlaag inlet had two outer channels, a flood-dominated channel to the west and an ebb-dominated channel to the north-west (Plate 2). These channels were separated by a subtidal shoal (- 2.5 to - 5 m DOL) of about 2 km^2.

Both channels rotated during the period 1970–1982 in a clockwise mode (Plate 2–5). Furthermore, they shifted laterally over distances of up to 300 m and 1000 m respectively. Especially during the period 1970–1975, the axes of the flood-dominated and ebb-dominated channels experienced strong lateral shifts and rotation of 10° and 20° respectively (Plates 2 & 3). Channel depths decreased by several metres, especially that of the ebb-dominated channel. At the same time the width of this channel increased, mainly due to strong erosion of the higher parts of the channel wall at the eastern side. From 1975 to 1982 the ebb-dominated channel split into two north-facing channels, the eastern (original) one being abandoned in the period 1982–1987 (Plates 3–6). Concurrently, the flood-dominated channel reverted to a more E–W orientation.

Fill and shift of inlet gorge

Below - 12 m DOL, net deposition of $3.3 * 10^6$ m^3 of fine sand took place (Sha, 1992), mainly in the gorge of the inlet (Fig. 4). In 1970 the total length of that part of the gorge lying at - 15 m DOL was 5.5 km, the maximum depth reaching - 17 m DOL (Plate 2). During the period 1970–1991 the seaward part of the main gorge shifted downdrift (to the east) and the length of the backbarrier channel region below - 15 m DOL decreased to 3.5 km, mainly between 1970 and 1975 (Plates 2–6).

Fill of main channel

The sediments were deposited mainly in the main (1 km wide) backbarrier channel which originally connected the Lauwerszee with the North Sea (Plates 1–8). As a result the cross-sectional area and the depth of the channel were reduced (Fig. 5). Sedimentation rates were high, reaching 1 m yr^{-1} (Fig. 6). The sediments of the fill consist of fine sand or clay, or alternations of both. The alternations have a rhythmic appearance; climbing ripple structures, lenticular bedding, loadcasts and bioturbation are common (Fig. 6). Net sedimentation dominated in the deeper parts of the channel at first but shifted progressively upwards as the channel became shallower (Figs 3, 5 & 6; Oost *et al.*, 1993). The deeper parts were dominated by erosion only in the period 1979–1982, when the zone above - 3 m DOL experienced strong sedimentation (Fig. 3).

On the tidal flats adjacent to the main channel, erosion occurred in the period 1970–1975 and, to a lesser extent, also from 1975 to 1979 (Fig. 3). Winkelmolen & Veenstra (1974) showed convincingly that the sediments filling up the main channel

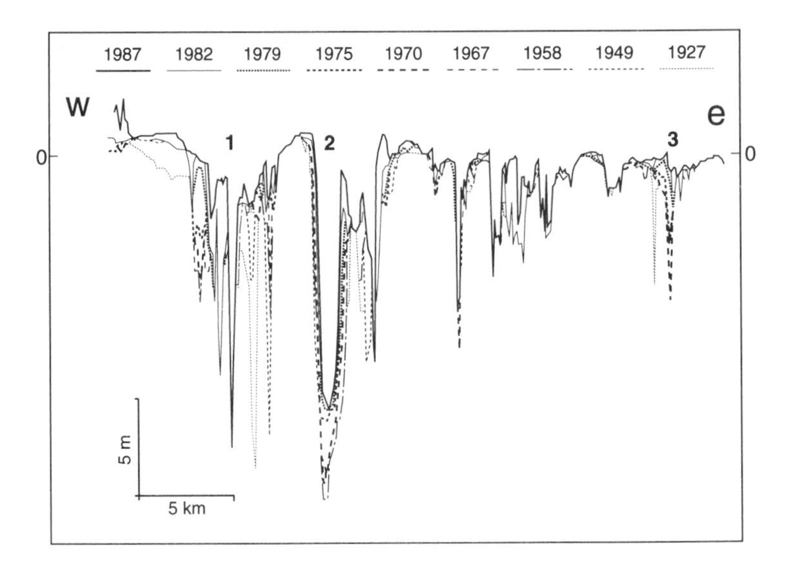

Fig. 4. W–E bathymetric transect along the line w–e (see Plate 1), showing the profiles between 1927 and 1987 which have not been eroded by subsequent morphological changes. At the left 0 indicates the level of DOL (~mean sea-level). The superimposed profiles illustrate the marked changes in the channel positions in the Pinkegat inlet (1), the vertical fill of the inlet gorge of the Zoutkamperlaag inlet (2) and the vertical infill of the inlet system east of Schiermonnikoog (3).

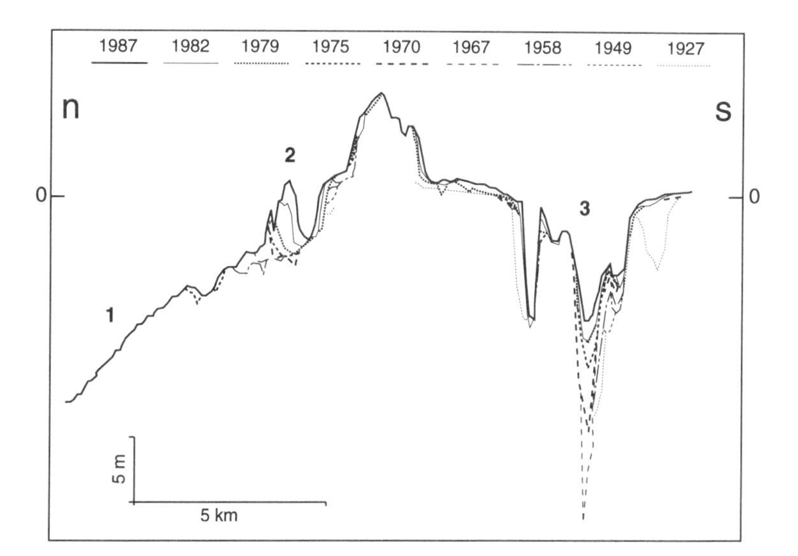

Fig. 5. N–S bathymetric transect along the line n–s (see Plate 1), showing the profiles between 1927 and 1987 which have not been eroded by subsequent morphological changes. At the left, 0 indicates the level of DOL. Visible is the strong erosion in the ebb-tidal delta, leaving only the 1987 profile line (1), the vertical growth and coastward shift of the recurved bar (2) and the fill of the main channel of the Zoutkamperlaag tidal system (3).

in the drainage basin shortly after closure of the Lauwerszee were mainly derived from adjacent tidal flats. Erosion of these tidal flats did not continue after 1979 (Plates 4–6).

Shift of watershed

Historical maps show that the position of the morphological watershed of Schiermonnikoog (south) was quite stable between 1891 (when the inlet east of Schiermonnikoog had developed at the cost of the more eastern Lauwers inlet) and 1927. In the period 1927–1965 the watershed shifted 1 km to the west.

In the period 1970–1979 the southern part of the watershed of Schiermonnikoog slowly shifted towards the east (Plates 2–4) (van Parreeren, 1980). Soon after the closure of the Lauwerszee in 1969, and lasting at least up to 1983, the backbarrier channels near the watershed became deeper and wider (Postma & van Parreeren, 1982; Postma, 1983). The main channel of the inlet system east of Schiermonnikoog retreated by rapid infilling in the period 1970–1979 (Plates 2–4). After 1979 the eastward migration rate of the southern part of the watershed accelerated, total migration over the period 1970–1987 amounting to 3–4 km. West of the watershed new small channels were formed (Plates 4–6). The northern part of the watershed migrated only *c.* 1 km over the period 1970–1987.

INTERPRETATION OF CHANGES

Ebb-tidal delta

Ebb-tidal deltas are controlled by tidal currents, wave action and the tidal wave along the coast (Dean & Walton, 1975; Oertel, 1975; Hayes, 1975, 1979, 1980; Walton & Adams, 1976; Hubbard *et al.*, 1977; Nummedal *et al.*, 1977; Nummedal & Fischer, 1978; FitzGerald *et al.*, 1984a; Sha, 1989a–d, 1990a, b; Sha & de Boer, 1991; Steijn, 1991). Before the closure of the Lauwerszee in 1969, the system was in a state of dynamic equilibrium (Oost & de Haas, 1993).

The above data show that the reduction of the tidal prism of the Zoutkamperlaag tidal inlet system resulted in a reduction in size of the ebb-tidal delta. As the ebb currents slowed down with the decrease in tidal prism, the relative importance of wave action increased, leading to net erosion of sand from the ebb-tidal delta (cf. Carter *et al.*, 1987). Indeed, net erosion took place down to the storm erosion base of − 12 to − 13 m DOL (Plate 1; Oost & de Haas, 1992).

Transport of sediment within the ebb-tidal delta occurs through the channel system, by migration of subtidal shoals on the swash platform and by migration of marginal breaker bars off Schiermonnikoog Island (cf. Hayes, 1979). The fine sand is transported both by saltation (Veenstra & Winkel-

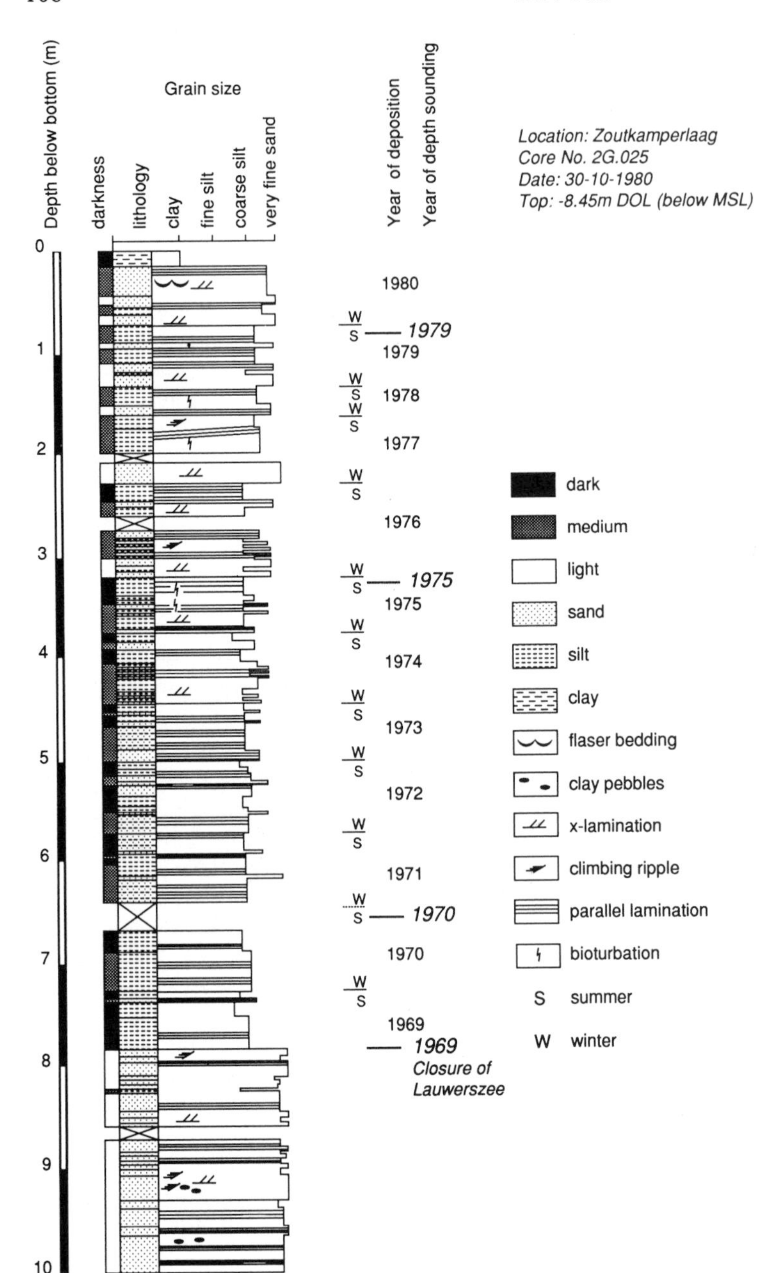

Fig. 6. Sedimentary log of the sequence formed in the main backbarrier channel of the Zoutkamperlaag system after the closure of the Lauwerszee. For location see Fig. 1. The inferred year of deposition given to the right of the log correlates with the available depth sounding data of 1969, 1970, 1975 and 1979 (after Oost *et al.*, 1993).

molen, 1976) and by bedload as is evident from the abundance of dunes and current ripples observed in box-cores, on sonar records and on aerial photographs (Oost & de Haas, 1992).

The ebb-tidal delta lost $26 * 10^6$ m^3 of sediment in the period 1970–1987. Sediment budgets for adjacent areas to the west show that no sediment was transported in that direction (Oost & de Haas, 1992). The area east of the ebb-tidal delta, i.e. the coast of Schiermonnikoog, also showed a small sediment surplus over that period. A net sediment export must therefore have taken place mainly towards the north and/or south.

Reineck & Singh (1972) and Aigner & Reineck

(1982) have demonstrated that seaward sediment transport to below erosional wave base occurs in the German Bight during storms (see also Aigner, 1985). In this way sediment could also have been transferred from the ebb-tidal delta to the offshore. However, in the present case net offshore transport cannot have been substantial as indicated by the stable position of the - 15 m DOL line at the outer rim of the delta (Plates 1–7). Moreover, cores taken at the seaward side of the ebb-tidal delta show that a thin veneer of recent North Sea sands, only a few dm thick, overlies older Holocene–Pleistocene deposits (Sha, 1992). It is therefore concluded that the larger part of the eroded sediment was transported through the inlet into the drainage basin of the Zoutkamperlaag by the flood current and wave action, resulting in the deposition of $30 * 10^6$ m^3 (Plate 8; Oost & de Haas, 1992).

Bar development

Within the ebb-tidal delta, part of the sediment that was eroded from the swash platform contributed to the development of the large recurved bar northwest of Schiermonnikoog (Plate 6, Fig. 5). The formation of such bars is a common phenomenon along the Frisian barrier island coast, having been documented, e.g. off Texel in 1908 (Sha, 1989a, 1990a), Schiermonnikoog in 1927, and Ameland in 1974 (data from nautical maps). Such bars form when marginal flood channels are displaced towards the islands where they are abandoned, so that the upstream shoals (with reference to the residual tidal current) merge with the islands (cf. Moslow & Tye, 1985). Because of the decrease of tidal energy in such marginal channels, the relative influence of waves increases locally, resulting in coastward sediment transport and the formation of linear swash bars parallel to the coast. In many cases there is subsequent development of bars oriented parallel to the inlets. The nautical map of Schiermonnikoog of 1967 (Plate 1), i.e. before the closure of the Lauwerszee in 1969, shows that the present bar started to develop in precisely this manner, i.e. after abandonment of a marginal flood channel. Such bars are commonly only a few km long. The unusually large size of the bar off Schiermonnikoog is explained by additional sediment supply from the eroding ebb-tidal delta (Plates 6–8; Oost & de Haas, 1992). After 1987 the supply of sand from the ebb-tidal delta largely ceased, resulting in rapid erosion of the bar by storms and tides since 1989 (J. Wiersma, personal communication; Plate 7).

Channel rotation

After closure of the Lauwerszee the outer tidal channels in the ebb-tidal delta of the Zoutkamperlaag inlet rotated rapidly (Plates 1–8). Clockwise rotation and translation of outer channels in ebb-tidal deltas of the Wadden Sea are common phenomena (Joustra, 1971). Longshore sediment supply and interactions of hydrodynamics and morphology force channels to migrate (Johnson, 1919; FitzGerald, 1988; Sha, 1989a). Shoals shift towards the adjacent downstream island, largely driven by wave action and tidal transport. Adjacent channels are forced to rotate and migrate until they become abandoned and are filled up (Sha, 1989a). The abandonment of channels in the downdrift part of the ebb-tidal delta stimulates the migration and rotation of other channels within the system, as well as the creation of new channels (Joustra, 1971; Sha, 1989a; Sha & de Boer, 1991). The rotations and translations are partly the result of gradual lateral accretion, enhanced by the above processes. Moreover, massive sediment transport during storms (both subtidal and supratidal) probably plays an important role in this respect, especially in triggering the ending of a rotational channel movement (abandonment) and in the formation of new channels (Sha, 1989a; Sha & de Boer, 1991).

Historical maps of the area show that the channels of the West Frisian Islands commonly rotate at a rate of about $1° yr^{-1}$ (Joustra, 1971; Sha, 1990a; Oost & de Haas, 1992). In the period 1970–1975, both the flood-dominated and the ebb-dominated channels rotated on average by 2° and $4° yr^{-1}$ respectively. This rapid rotation is ascribed to the sudden reduction in tidal prism and the resulting increase of relative wave influence, which tend to give the ebb-tidal delta a more downdrift orientation (with reference to the residual current) (Sha & de Boer, 1991). Generally, the outer channels maintain their cross-sectional areas because the strength of the ebb current removes sand delivered by littoral drift at the western side of the channels. Moreover, any net deposition along the upstream side of the channels is compensated by erosion along the eastern side, as a result of which channel migration and rotation occur. In the present case, due to the decrease in the tidal prism, wave-generated sediment supply at the updrift side of the channels (Biegel, 1991a), could not be compensated by the weaker tidal currents (Oost & de Haas, 1992). As a result, deposition towards the southeast increased, thereby filling up the deeper parts of the channel. This process is particularly well recorded in the ebb-dominated channel which was originally oriented NW–SE.

Along the eastern sides (downstream sides) of the channels, erosion of the upper channel slopes was enhanced by the relative increase of wave energy. After the closure of the Lauwerszee the weaker ebb-current could no longer compensate the wave-induced loss of sand, especially above - 12 m DOL. Sedimentation along the western side, coupled with erosion of the eastern side of the outer channels, resulted in faster than normal rotation (Oost & de Haas, 1992).

Fill and shift of inlet gorge

From Figs 2, 4 & 5 it is evident that net deposition within the ebb-tidal delta below - 12 m DOL was largely confined to the main gorge, where *c.* $3 * 10^6$ m^3 sand was deposited. The cross-sectional area, depth and width of an inlet are correlated with the tidal volume passing through it (O'Brien, 1969; Walther, 1972; Jarret, 1976; Dieckmann *et al.*, 1988; Gerritsen, 1990; Hume & Herdendorf, 1990; Niemeyer, 1990; Sha, 1990a; Steijn, 1991; Biegel, 1991b; Flemming, 1991; Eysink & Biegel, 1992). Therefore, sedimentation in the main gorge documented by the decrease in length of the - 15 m DOL region (Plates 2–6 & 8) is logically explained by the decrease in tidal prism and the related strong reduction of the ebb-tidal flow velocities in the main gorge (Biegel & Hoekstra, this volume, pp. 85–99; Fig. 4). The downdrift shift of the inlet gorge is attributed to the relative increase of wave forces (Sha & de Boer, 1991).

Backbarrier area

Before the closure of the Lauwerszee a dynamic equilibrium existed in the drainage basin, with periods of net erosion alternating with periods of net sedimentation (Oost & de Haas, 1993).

Owing to the reduction in tidal prism, net deposition of $30 * 10^6$ m^3 occurred in the drainage basin in the period 1970–1987 (Fig. 3). As outlined above, the larger part of this sediment was derived from the ebb-tidal delta, from which $26 * 10^6$ m^3 were eroded. Erosion in the drainage basin of the Pinkegat inlet and in other, more westerly tidal flats was not substantial (de Boer, 1979; de Boer *et al.*, 1991; Oost & de Haas, 1992). These areas can therefore be disregarded as possible sediment sources. The remaining $4 * 10^6$ m^3 of sediment probably comprise muds that were carried into the drainage basin in suspension (Oost & de Haas, 1992).

Fill of main channel

The rapid vertical accretion of the main channel in the drainage basin (Fig. 5) was caused by the strong decrease of tidal current velocities after the closure, comparable to the development in the main gorge (Biegel, 1991b; Biegel & Hoekstra, this volume, pp. 85–99). Initially these sediments were derived from the adjacent tidal flats (Winkelmolen & Veenstra, 1974), weaker tidal currents being unable to transport them back to the flats (Oost & de Haas, 1992). The intercalated fine-grained material in the deposits show that current velocities in the channel were at times very low. Analyses of cores from these channel fills (Fig. 6) (Oost *et al.*, 1993) suggest that the sedimentation pattern was seasonal, comparable to the example given by van den Berg (1981), with deposition of fines and bioturbation during the quiet summer season (de Haas & Eisma, 1993) and deposition of sandy sediments in winter, when tides in the northern hemisphere are stronger and storms dominate the North Sea coast. The inferred timing of deposition in the main channel correlates well with available depth sounding data (Fig. 6; Oost & de Haas, 1992). Fine sediments are more prominent in the channel fill of the drainage basin than in the fill of the inlet gorge. This is in accordance with the decrease in current energy and the increasing proportion of fines towards the mainland coast (van Straaten, 1954; van Straaten & Kuenen, 1957, 1958; Postma, 1961, 1967; Flemming & Nyandwi, 1994; Oost & de Boer, 1994). Moreover, the formation of pellets of faeces and pseudofaeces by filter-feeders (especially *Mytilus edulis* and *Cerastoderma edule*) and deposit-feeders (for instance, *Macoma balthica*) enhances sedimentation of clay and silt in the drainage basin, especially during the summer (e.g. Flemming & Delafontaine, 1994). It should be noted that the abrupt upward decrease in grain size of sediments in the abandoned channel fill is indicative of a sudden, rather than gradual drop in current velocity. In channels such grain-size 'jumps'

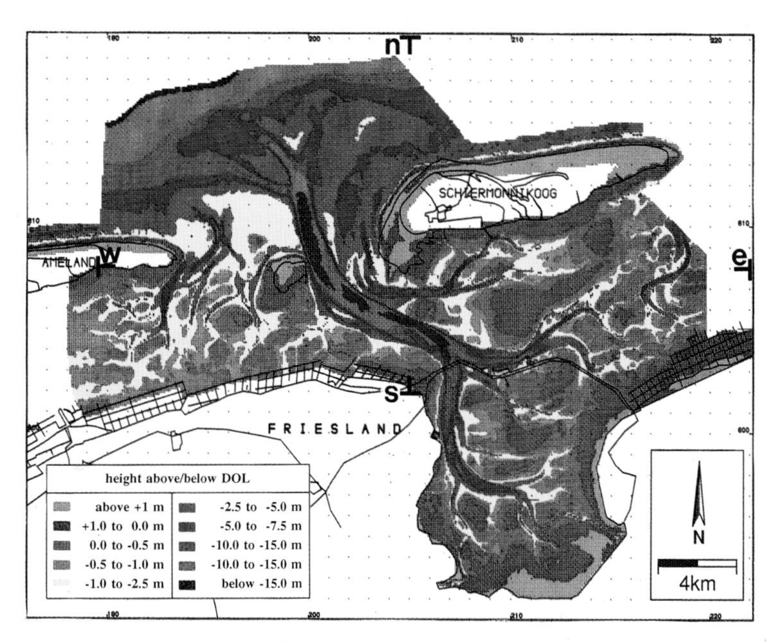

Plate 1 Bathymetry of the study area in 1967. The coastlines represent the situation in 1978. The W–E profile line (Fig. 4) is indicated by w and e, the N–S profile line (Fig. 5) by n and s. The map shows the area 2 years before the closure of the Lauwerszee embayment (lower embayment). The Zoutkamperlaag system (west of barrier island Schiermonnikoog) has a large ebb-tidal delta (N) and a deep main channel. Note the shallow (above –7.5 m) remnants of an originally flood-dominated channel cutting through the triangular swash platform directly northwest of Schiermonnikoog and the small shoals (1–2.5 m) in the northern part of the platform (after Oost & de Haas, 1993).

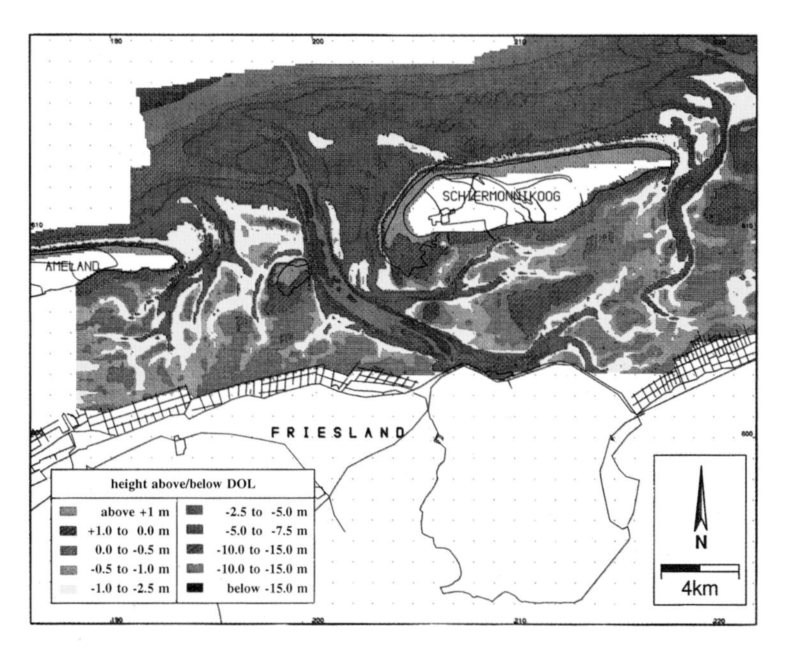

Plate 2 Bathymetry of the study area in 1970, 1 year after closure of the Lauwerszee. Note the formation of the E–W oriented part of the bar on the ebb-tidal delta (swash bar northwest of Schiermonnikoog) from the small shoals of 1967. In the ebb-tidal delta an E–W oriented flood-dominated outer channel and a NW–SE ebb-dominated outer channel are visible. Also note the shallowing of the main channels in the backbarrier area of the Zoutkamperlaag inlet and of the inlet system east of the watershed of Schiermonnikoog (after Oost and de Haas, 1992).

(*Facing page 110*)

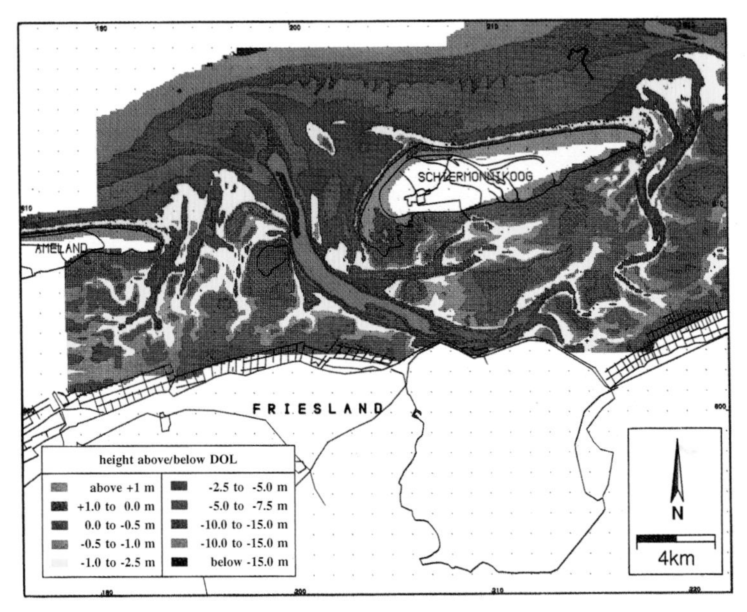

Plate 3 Bathymetry of the study area in 1975. The triangular swash platform retreated coastwards in the period 1970–1975. The shoal in the northwest and the E–W oriented swash bar have become higher. Note the rotation of the outer channels. The southern part of the inlet gorge has filled up. The shallowing of the main channels in the backbarrier area of the Zoutkamperlaag tidal inlet system and of the system east of Schiermonnikoog has continued. The shift of the interjacent watershed has commenced. The Pinkegat system west of Ameland has developed into a double inlet system (after Oost and de Haas, 1992).

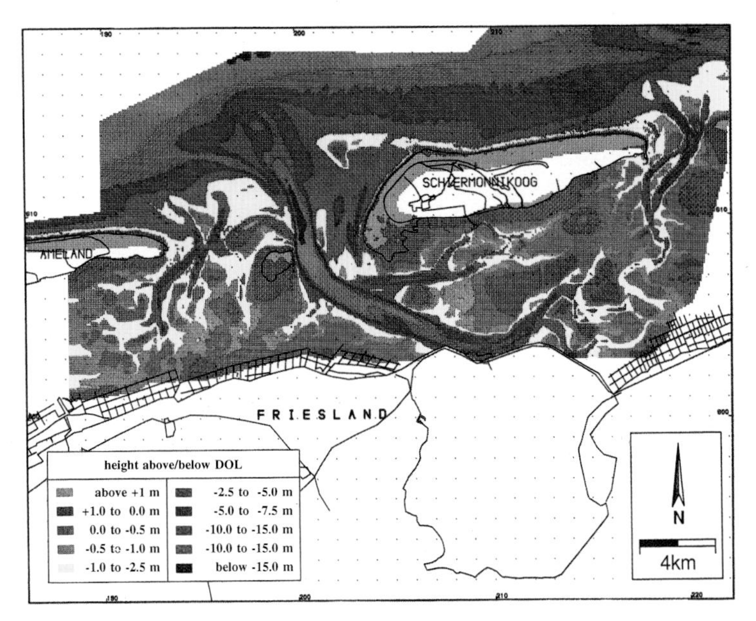

Plate 4 Bathymetry of the study area in 1979. The erosion and retreat of the ebb-tidal delta of the Zoutkamperlaag inlet has continued and the formation of the N–S oriented part of the bar has begun. Rotation of the outer channels and shallowing of the main channel in the backbarrier area of the Zoutkamperlaag inlet have continued. Pronounced retreat and shallowing of the system east of Schiermonnikoog is evident. Note the abandonment of smaller channels south of Ameland due to the marked changes in the configuration of the Pinkegat system (after Oost & de Haas, 1992).

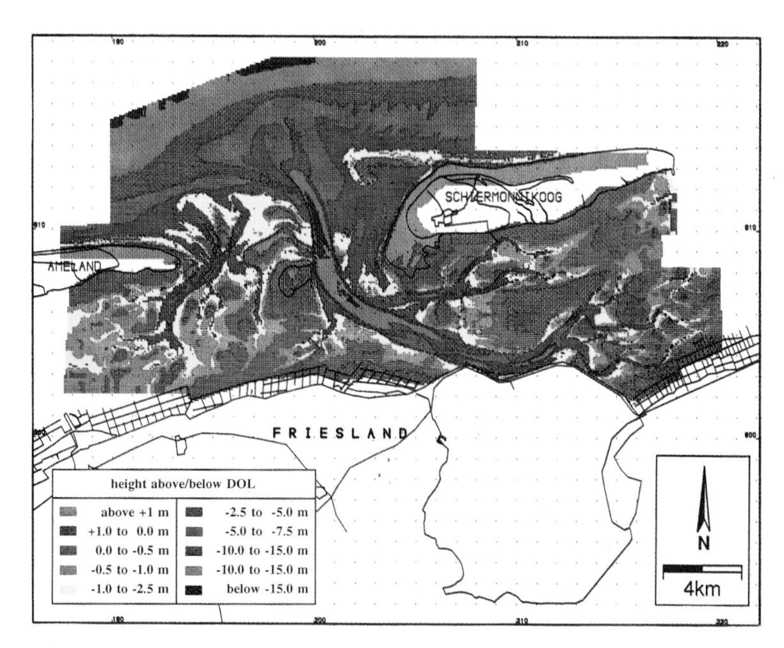

Plate 5 Bathymetry of the study area in 1982. Erosion and retreat of the ebb-tidal delta of the Zoutkamperlaag inlet and bar formation off Schiermonnikoog have continued. The shallowing of the main channel in the backbarrier area of the Zoutkamperlaag system has decelerated slightly. A pronounced shift of the watershed south of Schiermonnikoog has occurred in combination with the erosion of channels at the western side (after Oost & de Haas, 1992).

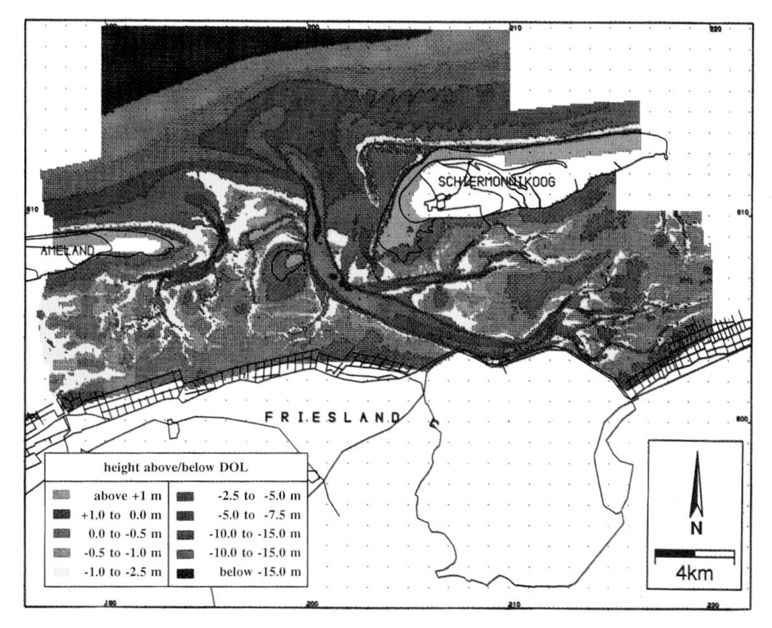

Plate 6 Bathymetry of the study area in 1987. Erosion and retreat of the ebb-tidal delta of the Zoutkamperlaag inlet and formation of the bar, which reached full maturity and became supratidal in 1987, have continued. The eastern branch of the outer ebb-dominated channel has been abandoned. The shallowing of the main channel in the backbarrier area of the Zoutkamperlaag inlet and erosion and channel formation at the western side of the watershed south of Schiermonnikoog have continued. Note that the Pinkegat inlet has once more become a single inlet system. In this cyclic process several larger outer channels (compare 1982 and 1987) and smaller backbarrier channels have been abandoned (after Oost & de Haas, 1992).

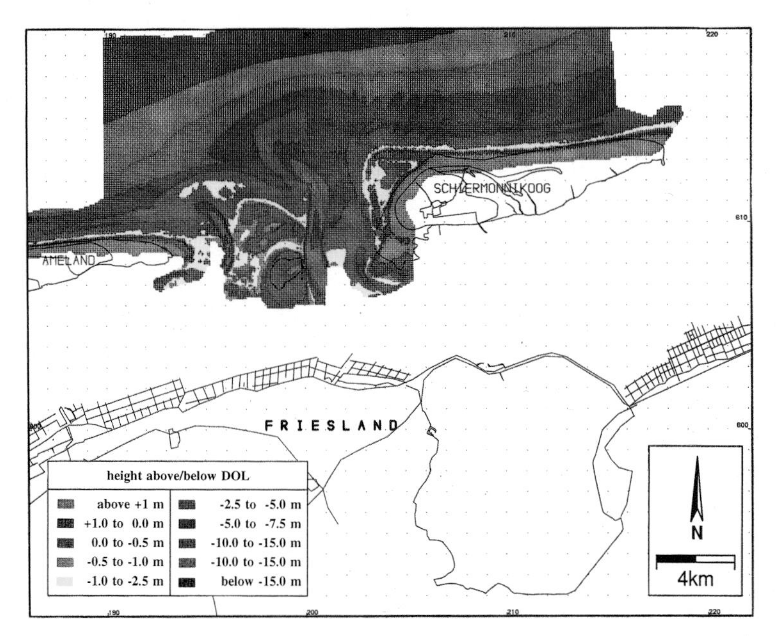

Plate 7 Bathymetry of the study area in 1991. Sounding data only of the ebb-tidal deltas were available. The ebb-tidal delta of the Zoutkamperlaag inlet has retreated further, but at a reduced rate. The bar encloses a small embayment and is in decay, having become intertidal and breached by several channels (up to 5 m deep) (after Oost & de Haas, 1993).

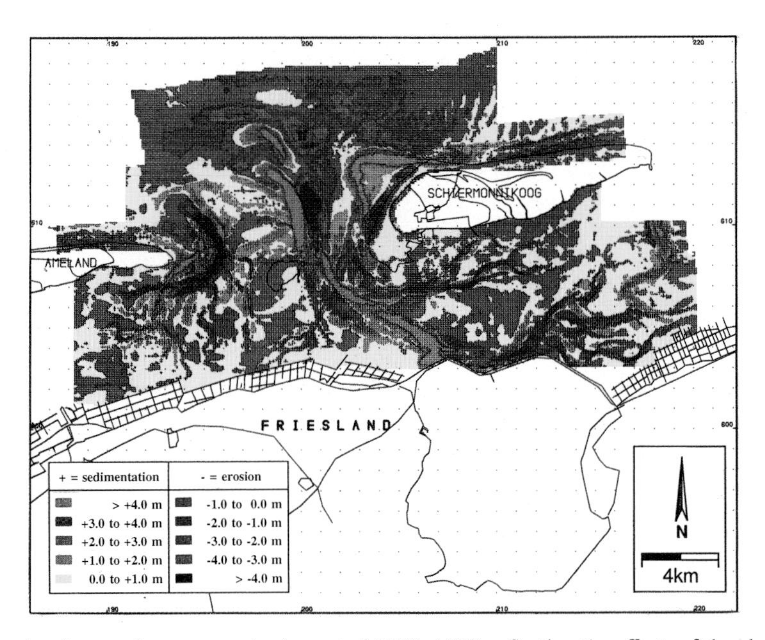

Plate 8 Erosion and sedimentation patterns in the period 1970–1987, reflecting the effects of the closure of the Lauwerszee embayment. Visible are the erosion of the ebb-tidal delta of the Zoutkamperlaag inlet, the formation of the bar, sedimentation and erosion due to outer channel shift, the vertical fill of the main backbarrier channel in the backbarrier area of the Zoutkamperlaag inlet, the formation of new channels at the western side of the watershed and the fill of the main channel southeast of Schiermonnikoog. The erosion and sedimentation pattern in the Pinkegat system is a poor reflection of the rapid changes that have occurred in that system (compare Plates 1–7; after Oost & de Haas, 1992).

can be considered to be indicative of sudden decreases in tidal flow strength.

The observation that the filling of the partly abandoned channel continued for more than a decade, i.e. at least from 1969 until 1987, indicates that the ability of the system to supply sediment largely determines the rate of deposition in such channels. Such constraints on sediment supply in tide-influenced sedimentary systems are also demonstrated by the sedimentary fill of the channel Vlieter. This channel was completely blocked at the landward side by a dike enclosing the IJsselmeer embayment in 1932 (Fig. 1). Since it almost totally lost its sediment transport capacity after the closure, sedimentation rates in the channel have only been 6–7 cm yr^{-1} (Eisma *et al.*, 1987). This is in contrast to the high sedimentation rates of several dm yr^{-1} in the Zoutkamperlaag main channel (Oost *et al.*, 1993) and in the partly abandoned channel studied by van den Berg (1981).

Shift of watershed

Before the closure, the phase difference of the tidal wave between the inlet gorges east and west of Schiermonnikoog was 20 min. Also, the resistance of the western channel system was less than that of the eastern channel system. These two effects resulted in an asymmetrical position of the tidal and morphological watersheds south of the eastern part of Schiermonnikoog (van Parreeren, 1980, personal communication) which was relatively stable.

The eastward migration of especially the southern part of the morphological watershed south of Schiermonnikoog, starting after the closure of the Lauwerszee in 1969 (Postma, 1983), implies an increase of phase difference between the tidal waves at either side of the morphological watershed. Indeed, although some eastward migration had already occurred, direct measurements in 1980 showed that the tidal wave at the western side of the southern part of the morphological watershed was still up to several tens of minutes earlier than the wave at the eastern side (van Parreeren, 1980).

The main channel in the drainage basin of the Zoutkamperlaag system drained the Lauwerszee area before its closure (Plate 1). After closure, the large width of the channel facilitated a faster propagation of both the vertical and horizontal tidal components towards the southern part of the morphological watershed (van Parreeren, 1980; Postma & van Parreeren, 1982). This was further enhanced by the formation of new channels and erosion west of the watershed (Plates 5, 6, van Parreeren, 1980; Postma, 1983; Oost & de Haas, 1992). Thus, the tidal (hydrological) watershed and, as a result, also the morphological watershed shifted towards the east. A comparable, but less pronounced, displacement can be observed in the northern part of the watershed, where the wider inlet gorge probably enhanced the propagation of the tide.

The fact that the shift of the southern part of the watershed was rather slow in the period 1970–1979 (Plates 2–4) can be explained by analysing the development of the tidal inlet system southeast of the barrier island Schiermonnikoog. A rapid shallowing and retreat of the main channel of this system occurred in the period 1970–1979 (Plates 2–4, Fig. 4), because part of the drainage was taken over by the Zoutkamperlaag inlet (van Parreeren, 1980). The time required for morphological adjustment of the partly abandoned eastern inlet system to the new hydrodynamic conditions is the main reason for the initially slow shift of the watershed. At that time flood waters entering the system through the Zoutkamperlaag inlet drained through the eastern inlet (van Parreeren, 1980). After filling up of the eastern inlet system, new channels started to erode at the western side (Zoutkamperlaag) and the shift of the watershed accelerated (Plates 4–6). This suggests that, with some delay, the lateral shift of the watershed was indeed brought about by the closure of the Lauwerszee. Wind-forced migration of the watershed as proposed by FitzGerald & Penland (1987) and Biegel & Hoekstra (this volume, pp. 85–99) may have played an additional but certainly minor role, as is apparent from the relatively stable position of the watershed before the closure of the Lauwerszee.

Historical evidence shows that eastward shifts of watersheds and barrier islands are associated with land-reclamation schemes and related reductions in tidal prisms at the updrift side, e.g. Schiermonnikoog, as from 1550; Memmert, as from 1750; Juist, as from 1600; Norderney, as from 1650; Baltrum, as from 1650; Langeoog, as from 1650; and Wangerooge as from 1650 (Homeier & Luck, 1969; FitzGerald & Penland, 1987; FitzGerald, 1988; Ligtendag, 1990; Oost, unpublished data). It seems logical to conclude that the shifts of the watersheds were also brought about by the decreases in tidal prisms, comparable to the shift in the backbarrier system of the Zoutkamperlaag inlet after the closure of the Lauwerszee, and possibly also wind forcing (FitzGerald & Penland, 1987).

To summarize, a dynamic equilibrium was maintained in the ebb-tidal delta and the drainage basin of the Zoutkamperlaag inlet before the closure of the Lauwerszee. After the closure in 1969 the ebb-tidal delta of the Zoutkamperlaag inlet was affected by erosion, $26 * 10^6$ m^3 of sand being transferred to the drainage basin in the period 1970–1987. Sediment was also concentrated by wave action in a large intertidal recurved bar situated north-west of Schiermonnikoog. Offshore transport by storms was subordinate or absent.

After the closure a rapid rotation of the outer channels brought about by an increase in sedimentation at the western side and an increase in erosion at the eastern side of the channels, induced by the decrease in tidal prism and the relative increase in wave influence. A substantial deposition of sand in, and a reorientation of the main gorge occurred due to the decrease of the tidal prism.

In the drainage basin the change in tidal prism caused partial abandonment of the main channel resulting in rapid vertical accretion of sands and clays. Furthermore, the rapid propagation of the tides through the wide main channel in the drainage basin of the westerly tidal system (Zoutkamperlaag) was the main cause of an eastward shift of the tidal watershed. Thus the inlet system east of Schiermonnikoog became partially abandoned and, as a result, it was partly filled. Especially after the fill, the southern part of the morphological watershed between the two systems moved rapidly towards the east.

DISCUSSION

Preservation potential of deposits

The deepest parts of a barrier system, i.e. the channels, the inlet gorge and the lower part of the ebb-tidal delta, have the highest preservation potential, as they suffer least from subsequent erosion by tidal currents and waves (Moslow & Tye, 1985; Sha, 1990a).

Ebb-tidal delta

The strong erosion of the ebb-tidal delta of the Zoutkamperlaag inlet (Fig. 5) shows that such deposits, if located above the storm erosion base, have little preservation potential in the wake of a sudden decrease of the tidal volume along a mixed-energy coast. The only sediments that have some preservation potential are the deposits formed by the migrating outer channels and the fine sandy deposits in the inlet gorge (Fig. 4).

An accelerated rotation of the outer channels will result in a series of laterally accreting deposits inclined to the east. Unless these are documented on a short-term (annual) basis, they cannot be distinguished from the deposits commonly formed by eastward shifting inlet channels as described by Sha (1992). Their preservation depends on the extent of erosion of the ebb-tidal delta. A comparison of maps from 1967 to 1991 indicates that part of this lateral accretion surface is currently preserved in the ebb-tidal delta of the Zoutkamperlaag.

The sediments in the inlet gorge form nested channel deposits and a vertical channel fill in the non-migrating part and the shallowing lateral accretion surfaces of the downdrift migrating channel section (Fig. 4), features which are commonly observed in the fossil record. Preservation potential of the deposits which fill the original inlet gorge is high because the newly established inlet channel will be shallower than the preceeding one.

The intertidal to supratidal recurved bar southwest of Schiermonnikoog (Fig. 5) has, notwithstanding its size, no significant preservation potential. Ever since sand supply from the ebb-tidal delta has declined, erosion has prevailed and the bar deposits are likely to disappear within the near future, quite comparable to other bars in the Wadden Sea.

A general empirical relationship between sand volume (above the normal island shore profile) and tidal prism has been proposed for ebb-tidal deltas (Dean & Walton, 1975; Walton & Adams, 1976; Steijn, 1991):

$$V = 65.6 * 10^{-4} P^{1.23}$$

where V is the sand volume of the ebb-tidal delta and P is the mean tidal prism. At equilibrium, ebb-tidal delta volume decreases with increasing relative wave influence (Dean, 1988). A linear empirical relationship has been proposed by Gerritsen & de Jong (1985) and Gerritsen (1990) between the tidal prism and the cross-sectional area of the main channels of the Dutch Wadden Sea. It follows that, depending on the length and geometry of the channels in the drainage basin and the size of the ebb-tidal delta, a decrease in tidal prism can generate a surplus of sediment, In general, the larger part of such a surplus would probably be reworked, to be either deposited in the offshore as storm deposits (Reineck & Singh, 1972; Aigner & Reineck, 1982; Aigner, 1985) or driven onshore and added to existing barrier islands (FitzGerald *et al.*, 1984b; Sha, 1989a, 1990a; Sha & de Boer, 1991; Flemming & Davis, 1994).

Backbarrier drainage basin

The considerable deposition of sands and fines in the main backbarrier channel of the Zoutkamperlaag tidal system following upon the decrease in tidal prism is comparable in scale to the infill of the gorge. The sediments form nested channel deposits (Fig. 5) reflected in the vertical channel fill sequences (Fig. 6). Here too, preservation potential is high because the new, main backbarrier channel is shallower than its predecessor.

The observed migration of the watersheds in the drainage basin, caused by changes in tidal prism, will produce significant features in the geological record. The channel fill of the inlet system east of the barrier island Schiermonnikoog is now partly covered by intertidal watershed deposits (Plates 2–6, Fig. 4). Although on a somewhat smaller scale, the infilling of the channels will be comparable in sedimentary sequence to that observed in the Zoutkamperlaag main channel, since here too the tidal prism has decreased. Obviously these infilled channel deposits, being situated below the new watershed, have a high preservation potential. By contrast, the watershed deposits which were present in 1969 became incised and eroded by new channels (Plates 3–8).

In conclusion, only sediments deposited in the deeper parts of the inlet system have a high preservation potential. These comprise the sedimentary fills of the partially abandoned inlet, the main channel in the backbarrier drainage basin and the main channel of the adjacent (east) inlet system. The fills are characterized by sand in the inlet and sand, mud or an alternation of both in the backbarrier channels.

Natural equivalents

The decrease of the tidal prism in the Zoutkamperlaag tidal inlet by one-third has resulted in rapid and substantial changes in the morphology of both the outer delta and the drainage basin. The change was artificially induced and one might wonder whether equivalent processes occur under natural conditions. In this respect two different settings must be considered: tectonically active areas and tectonically inactive areas. The changes considered here are so rapid (i.e. they occur within years) that absolute sea-level fluctuations can be ignored.

Tectonically active areas

About 22% of all barrier coasts are located along collision zones (Pilkey *et al.*, 1988). Studies show that average uplift rates in such areas range from several mm yr^{-1} (Hails, 1983; Bishop, 1991; Carter *et al.*, 1991; Fortuin & de Smet, 1991; Flint *et al.*, 1991) to >0.1 m yr^{-1} (Prince *et al.*, 1974). Between 1964 and 1982 a rapid uplift of 0.2–0.3 m yr^{-1} occurred in Portage, Alaska (Atwater *et al.*, 1991). During the 1964 Alaska earthquake, sudden localized uplift of the shoreline amounted to 10 m and, during the Yakutat Bay earthquake of 1899, the shoreline rose by 15 m (Bolt, 1978). If such uplifts were to occur in backbarrier settings, they would result in a reduction in tidal prism and hence in morphological changes comparable to those reported here.

Tectonically related upwarping of barrier sequences has been documented in the Silurian of SW Ireland (Sloan & Williams, 1991), in the Pleistocence of the Naracoorte region in SE South Australia (Hails, 1983), in Holocene deposits of Hainan Island (15–20 m) (Wang, 1986) (all related to volcanic activity), in Holocene marine deposits of the Spencer Gulf, South Australia (2.5 m, seismotectonics), in the Penholoway shoreline deposits (6 m) and in the Wicomico barrier deposits (20 m), the last two in northern Florida (Hails, 1983). Glacio-eustatic uplift of coastal lagoon deposits has been reported from, amongst other places, Finland (Eronen, 1983), Norway (Hafsten, 1983) and Scotland (Sissons, 1983). The exact morphodynamic responses of depositional environments affected by such changes strongly depend on the orientation of the tectonic lines with reference to barrier orientation and the exact nature of the tectonic regime. The above examples, however, demonstrate that preservation of coastal deposits, that have experienced rapid changes in their relative vertical positions, has occurred in the past.

In the case of an extensive tectonic regime one might expect that the tidal volume will increase by subsidence of a backbarrier region or estuary

(Nelson, 1988; Sloan & Williams, 1991). This would result in an increase of the tidal prism and hence in the deepeening of channels and enlargement of the ebb-tidal delta.

Tectonically inactive areas

About 78% of all barrier island chains are located along passive margins (Pilkey *et al.*, 1988), mostly in an inactive tectonic setting. In such areas a sudden decrease of the tidal volume passing through a channel can be produced by several mechanisms.

1 A washover channel changes into a tidal inlet. This can occur when a washover channel is excavated during a hurricane or severe storm (e.g. Hayes, 1967; Mehta & Brooks, 1973; Penland & Suter, 1986). Another possibility is that after a shift of the watershed (e.g. by lateral migration of the barrier island), a washover that was originally situated adjacent to a watershed begins to drain a part of the backbarrier area, thereby developing into an active tidal inlet.

2 Sudden infilling of part of a backbarrier drainage basin and/or blocking of the channel occurs. This can happen when massive sedimentation occurs by ice rafted deposits (cf. Dionne, 1987), wind transport (Schoorl, 1973), shore-parallel transport of sediment (for instance, spit growth) (cf. Moslow & Tye, 1985; Jenings & Smyth, 1988; Massa & Sanders, 1989), storm-related sedimentation (cf. Penland & Suter, 1986), mass transport deposition (cf. McKnight, 1969) or volcanic deposition (Sloan & Williams, 1991). Total infilling, however, must be considered a rare event because large amounts of sediment have to be deposited in a relatively short time.

3 An inlet system takes over part of the drainage of an adjacent inlet system (Bruun, 1978; Moslow & Tye, 1985). This depends on the relative phase difference of the tidal wave between two competitive inlet systems. Historical data suggest that between *c.* 1350 and 1550 such an inlet take-over occurred in the Lauwerszee embayment (Reitsma, 1991; Oost & Dijkema, 1993). Before about 1350 the precursor of the Zoutkamperlaag inlet (west of Schiermonnikoog) was probably small and had no connection with the Lauwerszee embayment, the latter being drained by the Lauwers inlet east of Schiermonnikoog (cf. Bosch & Vos, 1992). The watershed of Schiermonnikoog was probably breached in the period 1350–1450 (Sha, 1992; Oost & Dijkema, 1993). Around 1500 the Lauwerszee was drained by both the Lauwers inlet and the Zoutkamperlaag inlet. Thereafter the drainage was rapidly taken over by the Zoutkamperlaag. As a result of the increase in tidal prism the cross-sectional area of the channels increased. Judicial information shows that in 1556 the Lauwers inlet had completely lost its connection with the Lauwerszee (Formsma, 1954, 1958). The Lauwers inlet was largely abandoned, its channels and inlet decreased in cross-sectional area and the whole system shifted eastwards.

4 On a smaller scale, channels of the same inlet system may switch drainage areas, locally leading to partial or total abandonment of channels. This can happen both in the ebb-tidal delta (Moslow & Tye, 1985; Sha, 1989a; Sha & de Boer, 1991) and in the backbarrier area (Moslow & Tye, 1985). The filling-up can be rather gradual, depending on the rate of take-over (Moslow & Tye, 1985). Several examples of channel abandonment can be observed in the Pinkegat inlet, which developed cyclically from an initially single-inlet system into a multiple-inlet system, reverting back to the former within a period of 20–55 years (Plates 2–7, Fig. 4) (Oost & de Haas, 1993). Owing to the considerable changes in the channel patterns of this system, many of the inner and outer channels became abandoned in the course of each cycle (e.g. Plates 3–6).

5 Sudden changes in tidal amplitude generated by changes elsewhere in the system (mostly changes in resonance and interference of tidal waves). The closure of the IJsselmeer (Fig. 1), for example, is known to have influenced the tidal patterns up to the inlet west of Ameland, i.e. over a distance of 45 km (Sha, 1990a; de Boer *et al.*, 1991; van Parreeren, personal communication). Before the closure, the size of the IJsselmeer basin was such that the reflected outgoing tidal wave interfered with the incoming tidal wave, resulting in a standing wave of low tidal amplitude (Klok & Schalkers, 1980). After closure, the tidal amplitude increased considerably (by approximately 20% near Texel inlet) and the catchment area of Texel inlet within the remaining part of the Wadden Sea became larger at the cost of other inlets (Sha, 1990a).

Fossil examples

From the rock record several sequences have been described that are comparable to the sedimentary sequences formed in the Zoutkamperlaag.

Hardeberga Fm; Lower Cambrian of southern Sweden

Above a basal fluviodeltaic deposit, the Lower Cambrian Hardeberga Formation in Scania (Hamberg, 1991) consists of three 30–50 m thick, vertically stacked sequences. Within the backbarrier deposits of these sequences, abandoned channel fills, 1–4 m thick, are present on top of sediments deposited in an active channel setting. The abandoned channel fills comprise between 2 and 10 stacked annual microsequences of 0.2–0.8 m thick, large-scale, cross-bedded sandstone which pass upward into bioturbated and/or tidally-bedded sandstones. The channels were thought to have been abandoned by continued migration of the watersheds downwind and downdrift (FitzGerald & Penland, 1987; Hamberg, 1991). It is difficult to understand how such processes can lead to thick abandoned channel fill deposits, indicative of sedimentation events over periods of up to 10 years, because abandoned channels near the watershed are mostly shallow and fill up within a year. Also, the observed bimodality and the occurrence of the large-scale cross-bedding is at variance with sedimentation near watersheds. A more likely explanation is that larger channels of the same inlet system or another inlet system switched drainage areas which locally resulted in partial or total abandonment of channels (cf. Moslow & Tye, 1985).

Rocky Mountains molasse, Mesozoic of North America

Partly or completely mud-filled channels are quite common in Mesozoic sequences of the Rocky Mountains molasse (Rahmani, 1986). Brownridge & Moslow (1991) describe Lower Cretaceous estuarine channel fills of the Glauconitic Member. These tidal channel fills consist of a single fining upward succession or multiple successions separated by erosive contacts. Part of these successions comprise massive mudstones (with thin sandstone beds), typically 1–5 m thick and occasionally up to 15 m thick, which abruptly or gradationally overlie either heterolithic point-bar deposits or cross-bedded sandstones. This mudstone facies is interpreted to represent abandoned channel and channel margin deposits (Brownridge & Moslow, 1991). Especially the mudstones displaying sharp basal contacts and the thicker sequences of massive mudstone may have formed by abrupt channel and main channel abandonment, respectively.

Tegelen Formation, Pleistocene of Holland/Belgium

A part of the Tegelen Formation has been interpreted as micro- to mesotidal. Within the sediments (Turnhout Member) several fine-grained channel deposits up to 2 m thick have been observed. These have been interpreted as abandoned channel fills (Kasse, 1988). In addition, thicker channel fills reaching 10 m in thickness and consisting of muds alternating with sands were observed which probably also represent abandoned channel fills. In all cases the change from sands to clays is abrupt, a feature associated with a sudden drop in tidal amplitude (Kasse, 1988) which resulted in a reduction of the tidal prism and allowed (partial) abandonment of the channels as observed after the artificial reduction of the Zoutkamperlaag tidal prism.

CONCLUSIONS

The artificial closure of a part of the backbarrier drainage basin of the Zoutkamperlaag tidal inlet caused a sudden decrease in the tidal prism. As a result, the morphology of the inlet system was no longer in equilibrium with the new hydraulic conditions. After the closure the system adapted towards a new equilibrium. This was mainly achieved by erosion of sediment in the ebb-tidal delta and deposition in the inlet gorge and in the main channel of the drainage basin. Also, a redistribution of sediments within the ebb-tidal delta occurred, thereby enhancing the formation of a large recurved bar. Moreover, the observed retreat and fill of the main channel of the inlet system east of Schiermonnikoog and the subsequent shift of the watershed to the east were brought about by the reduction in the tidal prism.

Of all the sedimentary features that changed, only the vertical inlet and channel fills consisting of sands, clays or an alternation of both (abandoned channel fill) have a significant preservation potential.

Rapid changes in tidal prisms can also occur in natural settings, both in tectonically active and tectonically passive regions. In the former, the size and depth of the backbarrier area can change by sudden vertical movements, the exact nature of which depends on the tectonic regime and the orientation of the coastline with reference to the tectonic lines. In the latter, a sudden decrease in tidal prism can be generated by the formation of new channels which take over the drainage of others by sudden changes in tidal amplitude or by sudden massive sedimentation in the backbarrier area.

Abandoned channel deposits in the rock record suggest that sudden decreases of tidal prisms flowing through tidal channels are not unusual. The fine-grained deposits of channel fills can thus be used as indicators of sudden changes in the hydrodynamic regime.

ACKNOWLEDGEMENTS

First I thank Henk de Haas, with whom I did several preparatory studies, for his enthusiastic support. Computer studies of the area were made possible by the Ministry of Traffic and Public Works (R.W.S.), Direction Tidal Waters (R.I.K.Z.) & Direction Friesland, The Netherlands as part of the Coastal Genesis Studies. The author thanks Albert Prakken, Jaap van den Boogert, Dirk Reitsma and Pieter Noordstra (R.W.S. Direction Friesland) for their extensive support in these studies. Hans Wiersma (R.W.S. Direction North Sea), Dirk Beets (Geological Survey, Haarlem) and Henk van Parreeren (R.W.S. Direction Groningen) all kindly provided data. Edwin Biegel (R.W.S., R.I.K.Z.), with whom I did part of a preparatory study, and Piet Hoekstra (Utrecht University) are thanked for their co-operation. The technical support of Brigitta Benders, Wil den Hartog, Jaco Bergenhenegouwen and Fred Trappenburg is gratefully acknowledged. Thanks go to Poppe de Boer (Utrecht University), Jürgen Ehlers (Geological Survey, Hamburg) and Burg Flemming (Senckenberg Institute, Wilhelmshaven) for critically commenting on the manuscript.

REFERENCES

AIGNER, T. (1985) *Storm Depositional Systems*. Lect. Notes Earth Sci. **3**. Springer-Verlag, New York, 174 pp.

AIGNER, T. & REINECK, H.E. (1982) Proximality trends in modern storm sands from the Helgoländer Bight (North Sea) and their implications for basin analysis. *Senckenbergiana marit.* **14**, 183–215.

ATWATER, B.F., OBERMEIER, S.F., TABACZYNSKI, D.A., POND, E.C. & MARTIN, J.R. (1991) Holocene shaking in southern coastal Washington. *EOS, Trans. Am. Geophys. Un.* **72**, 44, 313.

AUBREY, D.G. & WEISHAR, L. (Eds) (1988) *Hydrodynamics and Sediment Dynamics of Tidal Inlets*, Preface. Lect. Notes Coastal Estuar. Stud. **29**. Springer-Verlag, New York, 456 pp.

BEETS, D.J., VALK, L. VAN DER & STIVE, M.J.F. (1979) Holocene evolution of the coast of Holland. *Mar. Geol.* **103**, 423–443.

BERG, J.H. VAN DEN (1981) Rhythmic seasonal layering in a mesotidal channel fill sequence, Oosterschelde Mouth, the Netherlands. In: *Holocene Marine Sedimentation in the North Sea Basin* (Eds Nio, S.D., Schüttenhelm, R.T.E. & Weering, Tj., C.E. van. *Spec. Publs int. Ass. Sediment.* **5**, 147–159.

BIEGEL, E.J. (1991a) De ontwikkelingen van de ebgetijde delta en het kombergingsgebied van het Friese Zeegat in relatie tot de sluiting van de Lauwerszee. *GEOPRO* **1991.07**. Univ. Utrecht, Utrecht, 57 pp.

BIEGEL, E.J. (1991b) Equilibrium relations in the ebb tidal delta, inlet and backbarrier area of the Frisian inlet system. *GEOPRO* **1991.028/GWAAO-91.016**. Univ. Utrecht, Utrecht, 79 pp.

BISHOP, D.G. (1991) High-level marine terraces in western and southern New Zealand: indicators of the tectonic tempo of an active continental margin. In: *Sedimentation, Tectonics and Eustasy, Sea-level Changes at Active Margins* (Ed. Macdonald, D.I.M.), p. 69–78. *Spec. Publs int. Ass. Sediment.* **12**.

BOER, M. DE (1979) Morfologisch onderzoek Ameland. Verslag van het onderzoekop het Amelander wantij in 1973. *Rijkswaterstaat Rept* **WWKZ-79.H005**, 67 pp.

BOER, M. DE, KOOL, G. & LIESHOUT, M.F. (1991) Erosie en sedimentatie in de binnendelta van het Zeegat van Ameland 1926–1984, deelondezoek no. 4. *Directie Noord-Holland Rept* **ANVX-91.H202**, 42 pp.

BOLT, B.A. (1979) Earthquake Hazards. *EOS* **59**, 11, 946–962.

BOOGERT, J.M. VAN DEN (1991) Beschrijving van de Conlod programmatuur. *Rijkswaterstaat Rept* **ANW.91.05**, 107 pp.

BOOGERT, J.M. VAN DEN & NOORDSTRA, P. (1988) Beschrijving van de Conlod programmatuur. *Rijkswaterstaat Rept* **ANW88.30**, 65 pp.

BOSCH, A. & VOS, P.C. (1992) Paleogeografische reconstructie van het Lauwersmeergebied, concept. *State Geol. Surv. Proj.* **40009**, 22 pp.

BROWNRIDGE, S. & MOSLOW, T.F. (1991) Tidal estuary and marine facies of the Glauconitic Member, Drayton Valley, central Alberta. In: *Clastic Tidal Sedimentology* (Eds Smith, D.G., Reinson, G.E., Zaitlin, B.A. & Rahmani, R.A.), *Mem. Soc. petr. Geol.* **16**, 107–122.

BRUUN, P. (1978) *Stability of Tidal Inlets, Theory and Engineering*. Develop. Geotech. Eng. **23** Elsevier, Amsterdam, 510 pp.

CARTER, R.W.G., JOHNSTON, T.W., MCKENNA, J. & ORFORD, J.D. (1987) Sea-level, sediment supply and coastal changes: Examples from the coast of Ireland. In: *The Hydrodynamic and Sedimentary Consequences of Sea-level Change* (Eds Carter, R.W.G. & Devoy, R.J.N.). *Prog. Oceanogr.* **18**, 79–101.

CARTER, R.M., ABBOTT, S.T., FULTHORPE, C.S., HAYWICK, D.W. & HENDERSON, R.A. (1991) Application of global sea-level and sequence-stratigraphic models in Southern Hemisphere Neogene strata from New Zealand. In:

Sedimentation, Tectonics and Eustasy, Sea-level Changes at Active Margins (Ed Macdonald, D.I.M.). *Spec. Publs int. Ass. Sediment.* **12**, 41–65.

DEAN, R.G. (1988) Sediment interaction at modified coastal inlets: processes and policies. In: *Hydrodynamics and Sediment Dynamics of Tidal Inlets* (Eds Aubrey, D.G. & Weishar, L.), pp. 412–439. Lect. Notes Coastal Estuar. Stud. **29**. Springer-Verlag, New York.

DEAN, R.G. & WALTON, T.L. (1975) Sediment transport processes in the vicinity of inlets with special reference to sand trapping. In: *Estuarine Research* **2**, *Geology and Engineering* (Ed. Cronin, L.E.), pp. 129–150. Academic Press, New York.

DIECKMANN, R., OSTERTHUN, M. & PARTENSCKY, H.W. (1988) A comparison between German and North American tidal inlets. *Proc. 21st Conf. Coastal Eng., ASCE* **3**, 199, 2681–2691.

DIONNE, J.C. (1987) Characteristic features of modern tidal flats in cold regions. In: *Tide-influenced Sedimentary Environments and Facies* (Eds Boer, P.L. De, Gelder, A. van & Nio, S.D.), pp. 301–332. Reidel, Dordrecht.

EISMA, D., BERGER, G.W., CHEN, W.Y. & SHEN, J. (1987) Pb-210 as a tracer for sediment transport and deposition in the Dutch-German Wadden Sea. In: *Proc. KNGMG Symp. Coastal Lowlands, Geology and Geotechnology* (Eds Linden, W.J.M. van der, *et al.*), pp. 237–253. Kluwer, Dordrecht.

ERONEN, M. (1983) Late Weichselian and Holocene shore displacement in Finland. In: *Shorelines and Isotasy* (Eds Smith, D.E. & Dawson, A.G.). *Spec. Publs inst. Brit. Geogr.* **16**, 183–207.

EYSINK, W.D. & BIEGEL, E.J. (1992) Impact of sea-level rise on the morphology of the Wadden Sea in the scope of its ecological function. Investigations on empirical morphological relations. *Rijkswaterstaat ISOS* **2**, 73 pp.

FITZGERALD, D.M. (1988) Shoreline erosional-depositional processes associated with tidal inlets. In: *Hydrodynamics and Sediment Dynamics of Tidal Inlets* (Eds Aubrey, D.G. & Weishar, L.), pp. 186–225. Lect. Notes Coastal Estuar. Stud. **29**. Springer-Verlag, New York.

FITZGERALD, D.M. & NUMMEDAL, D. (1983) Response characteristics of an ebb-dominated tidal inlet channel. *J. sediment. Petrol.* **53**, 833–845.

FITZGERALD, D.M., & PENLAND, S. (1987) Backbarrier dynamics of the east Friesian Islands. *J. sediment. Petrol.* **57**, 746–754.

FITZGERALD, D.M., PENLAND, S. & NUMMEDAL, D. (1984a) Control of barrier island shape by inlet sediment bypassing: East Frisian Islands, West Germany. *Mar. Geol.* **60**, 355–376

FITZGERALD, D.M., PENLAND, S. & NUMMEDAL, D. (1984b) Changes in tidal inlet geometry due to the backbarrier filling: East Frisian Islands, West Germany. *Shore & Beach* **52**, 3–8.

FLEMMING, B.W. (1991) Holozäne Entwicklung, Morphologie und fazielle Gliederung der Osfriesischen Insel Spiekeroog (südliche Nordsee). *Senckenberg-am-Meer* **91/3**, 51 pp.

FLEMMING, B.W. & DAVIS, R.A. (1994) Holocence evolution, morphodynamics and sedimentology of the Spiekeroog barrier island system (southern North Sea). *Senckenbergiana marit.* **24**, 117–156.

FLEMMING, B.W. & DELAFONTAINE, M.T. (1994) Biodeposition in a juvenile mussel bed of the East Frisian Wadden Sea (southern North Sea). *Neth. J. aquat. Ecol.* **28**, 289–297.

FLEMMING, B.W. & NYANDWI, N. (1994) Land reclamation as a cause of fine-grained sediment depletion in backbarrier tidal flats (southern North Sea), *Neth. J. aquat. Ecol.* **28**, 299–307.

FLINT, S., TURNER, P & JOLLY, E.J. (1991) Depositional architecture of Quaternary fan-delta deposits of the Andean fore-arc: relative sea-level changes as a response to a seismic ridge subduction. In: *Sedimentation, Tectonics and Eustasy, Sea-level Changes at Active Margins* (Ed. Macdonald, D.I.M.). *Spec. Publs int. Ass. Sediment.* **12**, 91–104.

FORMSMA, W.J. (1954) Een reis naar Schiermonnikoog in 1556. *Groningse Volksalmanak 1954*, 65–71.

FORMSMA, W.J. (1958) De grens tussen de provinces Groningen en Friesland in de Wadden. *Groningse Volksalmanak 1958*, 27–42.

FORTUIN, A.R. & SMET, M.E.M. DE (1991) Rates and magnitudes of late Cenozoic vertical movements in the Indonesian Banda Arc and the distinction of eustatic effects. In: *Sedimentation, Tectonics and Eustasy, Sea-level Changes at Active Margins* (Ed. Macdonald, D.I.M.). *Spec. Publs int. Ass. Sediment.* **12**, 79–90.

GERRITSEN, F. (1990) Morphological stability of inlets and channels of the western Wadden Sea. *Rijkswaterstaat Rept* **GWAO-90.019**, 86 pp.

GERRITSEN, F. & JONG, H. DE (1985) Stabiliteit van doorstroomprofielen in het Waddengebied. *Rijkswaterstaat Rept* **WWKZ-84.V016**, 53 pp.

HAAS, H. DE & EISMA, D. (1993) Suspended sediment transport in the Dollard Estuary. *Neth. J. Sea Res.* **31**, 37–42.

HARSTEN, U. (1983) Shore-level changes in South Norway during the last 13000 years, traced by biostratigraphical methods and radiometric datings. *Norsk Geogr. Tidsskrift* **37**, 63–79.

HAILS, J.R. (1983) Coastal processes, relict shorelines and changes in sea level on selected mid- and low-latitude coasts. In: *Shorelines and Isostasy* (Eds Smith, D.E. & Dawson, A.G.). *Spec. Publs inst. Brit. Geogr.* **16**, 29–51.

HAMBERG, L. (1991) Tidal and seasonal cycles in a Lower Cambrian shallow marine sandstone (Hardeberga Fm.), Scania, Southern Sweden. In: *Clastic Tidal Sedimentology* (Eds Smith, D.G., Reinson, G.E., Zaitlin, B.A. & Rahmani, R.A.). *Mem. Can. Soc. Petr. Geol.* **16**, 255–273.

HAYES, M.O. (1967) Hurricanes as geological agents, South Texas coast. *Am. Ass. Petr. Geol. Bull.* **51**, 937–942.

HAYES, M.O. (1975) Morphology of sand accumulations in estuaries. In: *Estuarine Research* **2**, *Geology and Engineering* (Ed. Cronin, L.E.), pp. 3–22. Academic Press, New York.

HAYES, M.O. (1979) Barrier island morphology as a function of tidal and wave regime. In: *Proc. Coastal Symp. Barrier Islands* (Ed. Leatherman, S.), pp. 1–28. Academic Press, New York.

HAYES, M.O. (1980) General morphology and sediment patterns in tidal inlets. *Sediment. Geol.* **26**, 139–156.

HOMEIER, H. & LUCK, G. (1969) Das historische

Kartenwerk 1:50,000 der Niedersächsischen Wasserwirtschaftsverwaltung als Ergebnis historisch-topographischer Untersuchungen und Grundlage zur kausalen Deutung der Hydrovorgänge im Küstengebiet. Göttingen.

HUBBARD, D.K., BARWIS, J.H. & NUMMEDAL, D. (1977) Sediment transport in four South Carolina inlets. *Proc. Coastal Sediments '77, ASCE,* 582–601.

HUME, T.M. & HERDENDORF, C.E. (1990) Morphological and hydraulic characteristics of tidal inlets on a headland dominated low littoral drift coast, Northeastern New Zealand. *Proc. Skagen Symp., Spec. Iss. J. coast. Res.*, 527–563.

JARRET, J.T. (1976) Tidal prism-inlet area relationships. *GITI Rept* **3**, US Army Eng. Waterw. Exp. Stat., Vicksburg.

JENINGS, S. & SMITH, C. (1988) The application of coastal and sea-level research to problems of coastal management in southern England. *Abstr. Int. Symp. Theoretical and Applied Aspects of Coastal and Shelf Evolution, Past and Future, Amsterdam, The Netherlands, 19–24 Sept. 1988*, 51–52.

JOHNSON, J.W. (1919) *Shore Processes and Shoreline Development*. John Wiley, New York, 584 pp.

JOUSTRA, D.SJ. (1971) Geulbeweging in de buitendeltas van de Waddenzee. *Rijkswaterstaat Rept* **WWK 71-14**, 27 pp.

KASSE, C. (1988) *Early-Pleistocene tidal and fluviatile environments in the southern Netherlands and northern Belgium*. Thesis V.U. Amsterdam, Free Univ. Press.

KLOK, B. & SCHALKERS, K.M. (1980) De veranderingen in de Waddenzee ten gevolge van de afsluiting van de Zuiderzee. *Rijkswaterstaat Rept* **78.H238**, 13 pp.

LIGTENDAG, W.A. (1990) Van IJzer tot Jade, een reconstructie van de zuidelijke Noordzeekust in de jaren 1600 en 1750. *Rijkswaterstaat*, 55 pp.

MASSA, A.A. & SANDERS, J.E. (1989) Quantitative determinations of unusually rapid rates of spit growth and longshore currents on a moderate-wave-energy microtidal coast (Fire Island NY). *Abstr. Geol. Soc. Am.* **21 (6)**, 174.

MCKNIGHT, D.G. (1969) A recent, possibly catastrophic burial in a marine molluscan community. *New Zeal. J. mar. freshw. Res.* **3**, 177–179.

MEHTA, A.J. & BROOKS, H.K. (1973) Mosquito Lagoon barrier beach study. *Shore and Beach* **41**, 27–34.

MOSLOW, T.F. & TYE, R.S. (1985) Recognition and characterization of Holocene tidal inlet sequences. *Mar. Geol.* **63**, 129–151.

NELSON, A.R. (1988) Prospects for identification of small coseismic rises in relative sea level during the Holocene along tectonically active coasts: preliminary studies in South-Central Oregon, U.S.A. *Abstr. Int. Symp. Theoretical and Applied Aspects of Coastal and Shelf Evolution, Past and Future, Amsterdam, The Netherlands, 19–24 Sept. 1988*, 84–87.

NIEMEYER, H.D. (199) CM5 Morphodynamics of tidal inlets. In: *Coastal Morphology*. Stichting Postacad. Ond. Civ. Techn. en Bouwtechn., Delft.

NOORDSTRA, P. (1989) Zandhaak Schiermonnikoog, volumebepaling. *Rijkswaterstaat Rept* **ANW89.51**, 4 pp.

NUMMEDAL, D. & FISCHER, I.A. (1978) Process-response models for depositional shorelines: The German and Georgia Bights. *Proc. 16th coastal Eng. Conf., ASCE,* 1215–1231.

NUMMEDAL, D., OERTEL, G.F., HUBBARD, D.K. & HINE, A.C. (1977) Tidal inlet variability: Cape Hatteras to Cape Canaveral. *Proc. coastal Sediments '77, ASCE,* 543–562.

O'BRIEN, M.P. (1969) Equilibrium flow areas of inlets on sandy coasts. *J. Waterw. Harb. Div., ASCE* **WWI**, 43–52.

OERTEL, G.F. (1975) Ebb-tidal deltas of Georgia estuaries. In: *Estuarine Research* **2**, *Geology and Engineering* (Ed. Cronin, L.E.), pp. 267–276. Academic Press, New York.

OOST, A.P. & BOER, P.L. DE (1994) Sedimentology and development of barrier islands, ebb-tidal deltas, inlets and backbarrier areas of the Dutch Wadden Sea. *Senckenbergiana marit.* **24**, 65–116.

OOST, A.P. & DIJKEMA, K.S. (1993) Effecten van bodemdaling door gaswinning in de Waddenzee. *IBN Rept* **025**, 133 pp.

OOST, A.P. & HAAS, H. DE (1992) Het Friesche Zeegat, morfologisch-sedimentologische veranderingen in de periode 1970–1987, een getijde inlet systeem uit evenwicht. *Kustgenese* **1(1-2)**, 68 pp.

OOST, A.P. & HAAS, H. DE (1993) Het Friesche Zeegat morfologisch-sedimentologische veranderingen in de periode 1927–1970. *Kustgenese* **1(1-2)**, 94 pp.

OOST, A.P., HAAS, H. DE, IJNSEN, F., BOOGERT, J.M. VAN DEN & BOER, P.L. DE (1993) The 18.6 year nodal cycle and its impact on tidal sedimentation. *Sedimen. Geol.* **87**, 1–11.

PARREEREN, D.V. VAN (1980) Waterloopkundige aspecten van de doorbaggering wantij 'Schiermonnikoog'. *Rijkswaterstaat Rept* **80–29**, 40 pp.

PENLAND, S. & SUTER, J.R. (1986) Mapping the morphodynamic signature of hurricane impacts in the Northern Gulf of Mexico. *Abstr. 12th Int. Sediment. Congr., Int. Ass. Sediment., Canberra, 24–30 Aug. 1986*, 239.

PILKEY, O.R., HENDERSON, V. & KEYSWORTH, A. (1988) Controls of barrier island chain morphology and distribution: a global view. *Abstr. Int. Symp. Theoretical and Applied Aspects of Coastal and Shelf Evolution, Past and Future, Amsterdam, The Netherlands, 19–24 Sept. 1988*, 103–106.

POSTMA, H. (1961) Transport and accumulation of suspended matter in the Dutch Wadden Sea. *Neth. J. Sea Res.* **1**, 148–190.

POSTMA, H. (1967) Sediment transport and sedimentation in the estuarine environment. In: *Estuaries* (Ed. Lauff, G.A.), pp. 158–179. Am. Ass. Adv. Sci., Washington DC.

POSTMA, H. (Ed.) (1982) Hydrography of the Wadden Sea: Movements and properties of water and particulate matter. *Wadden Sea Working Group Rept* **2**, 75 pp.

POSTMA, J.T. (1983) De Groningerbalg als verbinding tussen het Oort en de Eilanderbalg respectievelijk het Hornhuizerwad. *Rijkswaterstaat Rept* **83–28**, 17 pp.

POSTMA, J.T. & PARREEREN, D.V. VAN (1982) Onderzoek naar de bevaarbaarheid van de wantijen Engelsmanplaat, Lutjewad en Hornhuizerwad. *Rijkswaterstaat Rept* **82–7**, 26 pp.

PRINCE, R.A., RESIG, J.M., KULM, L.D. & MOORE, T.C., JR. (1972) Uplifted turbidite basins on the seaward wall of the Peru Trench. *Geology* **2**, 607–611.

RAHMANI, R.A. (1986) Mesozoic mud-filled channels; abandoned fluvial channels or active river estuaries? *Reservoir* **13**, 1–2.

REINECK, H.E. & SINGH, I.B. (1972) Genesis of laminated sand and graded rhythmites in storm-sand layers of shelf mud. *Sedimentology* **18**, 123–128.

REITSMA, D.T. (1991) Historie oostelijk Waddengebied. *Waddenbulletin* **26**, 176–178.

SCHOORL, H. (1973) *Zeshonderd jaar water en land, Bijdrage tot de historische Geo-en Hydrografie van de Kop van Noord-Holland in de periode ± 1150–1750.* Aardrijksk. Genoots. **2**, Wolters-Noordhoff nv, Groningen, 534 pp.

SHA, L.P. (1989a) Cyclic morphological changes of the ebb-tidal, Texel Inlet, The Netherlands. *Geol. Mijnb.* **68**, 35–49.

SHA, L.P. (1989b) Sand transport patterns in the ebb-tidal delta off Texel Inlet, Wadden Sea, The Netherlands. *Mar. Geol.* **86**, 137–154.

SHA, L.P. (1989c) Variation in ebb-delta morphologies along the West and East Frisian Islands, The Netherlands and Germany. *Mar. Geol.* **89**, 11–28

SHA, L.P. (1989d) Holocene–Pleistocene interface and three-dimensional geometry of the ebb-delta complex, Texel Inlet, The Netherlands. *Mar. Geol.* **89**, 207–228.

SHA, L.P. (1990a) *Sedimentological studies of the ebb-tidal deltas along the West Frisian Islands, the Netherlands.* Ph.D. thesis, University of Utrecht, Geologica Ultraiectina **64**.

SHA, L.P. (1990b) Surface sediment characteristics in the ebb-tidal delta of Texel Inlet, The Netherlands. *Sediment. Geol.* **68**, 125–141.

SHA, L.P. (1992) Geological Research in the Ebb-Tidal Delta of 'Het Friesche Zeegat', The Netherlands. *R.G.D. Proj.* **40010**, 20 pp.

SHA, L.P. & BOER, P.L. DE (1991) Ebb-tidal delta deposits along the West Frisian Islands (The Netherlands): processes, facies architecture and preservation. In: *Clastic Tidal Sedimentology* (Eds Smith, D.G., Reinson, G.E., Zaitlin, B.A. & Rahmani, R.A.). *Can. Soc. Petrol. Geol. Mem.* **16**, 199–218.

SIJP, D. VAN (1989) Correcties op gemeten eb- en vloedvolume bij de omrekening naar gemiddeld getij, in het Friese Zeegat. *Rijkswaterstaat Rept* **ANW 89–02**, 4 pp.

SISSONS, J.B. (1983) Shorelines and isostasy in Scotland. In: *Shorelines and Isotasy* (Eds Smith, D.E. & Dawson, A.G.). *Spec. Publs Inst. brit. Geogr.* **16**, 209–225.

SLOAN, R.J. & WILLIAMS, B.P.J. (1991) Volcano-tectonic control of offshore to tidal-flat regressive cycles from the Dunquin Group (Silurian) of southwest Ireland. In: *Sedimentation, Tectonics and Eustasy, Sea-level Changes at Active Margins* (Ed. Macdonald, D.I.M.) *Spec. Publs int. Ass. Sediment.* **12**, 105–119.

SPEK, A.J.F. VAN DER & BEETS, D.J. (1992) Mid-Holocene evolution of a tidal basin in the western Netherlands: a model for future changes in the northern Netherlands under conditions of accelerated sea-level rise? *Sediment. Geol.* **80**, 185–197.

STEIJN, R.C. (1991) Some considerations on tidal inlets. A literature survey on hydrodynamic and morphodynamic characteristics of tidal inlets with special attention to 'Het Friesche Zeegat' Delft Hydraulics. *Coastal Genesis Rept* **H840.45**, 109 pp.

STRAATEN, L.M.J.U. VAN (1954) Composition and structure of recent marine sediments in the Netherlands. *Leidse Geol. Meded.* **19**, 1–110.

STRAATEN, L.M.J.U. VAN & KUENEN, PH.H (1957) Accumulation of fine grained sediments in the Dutch Wadden Sea. *Geol. Mijnb.* **19**, 329–354.

STRAATEN, L.M.J.U. VAN & KUENEN, PH.H (1958) Tidal action as cause of clay accumulation. *J. sediment. Petrol.* **28**, 406–413.

VEENSTRA, H.J. & WINKELMOLEN, A.M. (1976) Size, shape and density sorting around two barrier islands along the north coast of Holland. *Geol. Mijnb.* **55**, 87–104.

WALTHER, F. (1972) Zusammenhänge zwischen der Größe der Ostfriesischen Seegaten mit ihrem Wattgebieten sowie den Gezeiten und Strömungen. *Jber. Forschungsstelle Küste* **23**, 7–32.

WALTON, T.L. & ADAMS, W.D. (1976) Capacity of inlet bars to store sand. *Proc. 15th Conf. Coastal Eng., A.S.C.E.*, 1919–1938.

WANG, Y. (1986) Sedimentary characteristics of the coast of Hainan Island: effects of a tropical climate and active tectonism. *Abstr. 12th Int. Sediment. Congr., Int. Ass. Sediment., Canberra, 24–30 Aug. 1986*, 318.

WINKELMOLEN, A.M. (1969) *Experimental Rollability and Natural Shape Sorting of Sand* (2nd Edn). Holland, 141 pp.

WINKELMOLEN, A.M. (1982) Critical remarks on grain parameters with special emphasis on shape. *Sedimentology* **29**, 225–265.

WINKELMOLEN, A.M. & VEENSTRA, H.J. (1974) Size and shape sorting in a Dutch tidal inlet. *Sedimentology* **21**, 107–126.

WINKELMOLEN, A.M. VEENSTRA, H.J. (1980) The effect of a storm surge on near-shore sediments in the Ameland–Schiermonnikoog area, N Netherlands. *Geol. Mijnb.* **59**, 97–111.

Spec. Publs int. Ass. Sediment. (1995) **24**, 121–132

Stratigraphy of a combined wave- and tide-dominated intertidal sand body: Martens Plate, East Frisian Wadden Sea, Germany

R.A. DAVIS, JR* *and* B.W. FLEMMING†

**Coastal Research Laboratory, Department of Geology, University of South Florida, Tampa, Florida 33620, USA; and*

†Senckenberg Institute, Schleusenstrasse 39a, 26382 Wilhelmshaven, Germany

ABSTRACT

Intertidal sand bodies are generally assumed to be tide-dominated, especially those in meso- to macrotidal settings. Vibracores ranging in length from 1.4 to 4.6 m recovered from Martens Plate, an intertidal sand body situated landward of the mesotidal Harle inlet in the East Frisian Wadden Sea of Germany, show a near equal mixture of wave- and tide-dominated sedimentary structures. The sediments of the cores are dominated by well-sorted fine sand, typically containing less than 2% mud. Bioturbation is not conspicuous throughout the cores, although the tidal flats are in places abundantly populated by the tube-forming polychaetes *Arenicola marina* and *Lanice conchilega.* Physical sedimentary structures include parallel lamination, small-scale ripple cross-stratification and mesoscale cross-bedding. Tidal bedding and tidal bundles can be recognized within several of the cores. The core data are consistent with physical surface structures observed on the present tidal flat/channel margin complex where the intertidal flat surface is dominated by wave-generated structures, whereas the channels and channel margins are dominated by current-generated bedforms.

It is concluded that it would be difficult to associate depositional sequences observed in the rock record with depositional environments corresponding to the specific backbarrier settings of sand bodies such as Martens Plate.

INTRODUCTION

The German Bight, forming the south-eastern portion of the North Sea basin, is bounded in the west by the north coast of The Netherlands, in the south-east and east by the North Sea coast of Germany and in the north-east by the southern Jutland coast of Denmark. This funnel-shaped bay, with its western and eastern coastlines converging at almost right angles towards the Elbe river estuary, displays a well-defined trend in tidal gradient (cf. Nummedal *et al.*, 1977; Ehlers, 1988), ranging from microtidal at Texel (The Netherlands) in the west, to lower macrotidal at the apex of the inner German Bight (Jade Bay, Weser and Elbe river estuaries), and back to microtidal at Skallingen (Denmark) to the north-east (Fig. 1).

The East Frisian Islands along the north coast of Germany (Fig. 2) are situated in the western mesotidal section of this tidal gradient, being characterized by short drumstick barrier islands and wide, deep inlets (FitzGerald *et al.*, 1984). According to the morphogenetic barrier island classification of Hayes (1979), the tidal range and wave climate conditions define the coast as a mixed-energy environment. The mean annual significant wave height is 1.6 m at Norderney, storm waves reaching maximum heights of 4.8 m along the open coast (Dette, 1977). The mean tidal range in the study area amounts to 2.8 m, reaching >3.0 m spring tide. The tidal prism draining the Harle catchment between the islands of Spiekeroog and Wangerooge amounts to about 60 million m^3.

Intertidal sand bodies in close proximity to barrier island inlets, such as the Martens Plate, are notoriously difficult to access, let alone be subjected

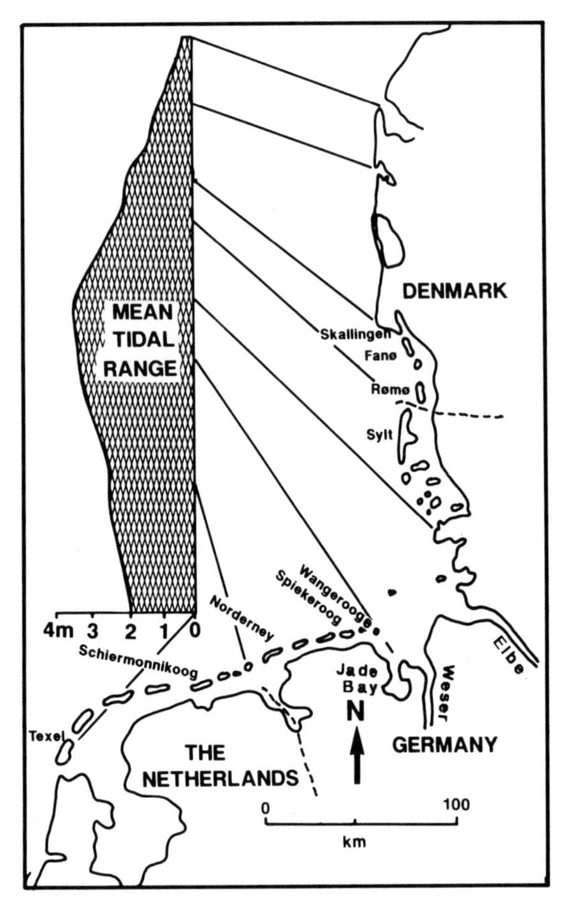

Fig. 1. Locality map of the German Bight and the regional pattern of tidal ranges.

to an in-depth study. As a consequence, stratigraphical information about such deposits are scarce and incomplete (Hanisch, 1981). The objectives of this study, therefore, were to document the internal sedimentary structures of such a sand body and to develop a generalized facies model that could be of use in the interpretation of similar deposits in the rock record.

PHYSIOGRAPHICAL SETTING

The study area is situated along the western margin of the large Harle tidal channel to the south-east of the eastern head of Spiekeroog island (Fig. 2). It comprises two adjacent intertidal sand bodies, the Martens Plate and the Muschelbank, which are separated by the smaller Muschelbalje tidal channel complex (e.g. Hempel, 1985). The surface of the Martens Plate sand body slopes gently toward the south-west, its elevation ranging from + 0.4 m to − 0.5 m relative to the German topographic chart datum (NN). The Muschelbank slopes southward and is somewhat lower on average. Since the chart datum (0 m NN) is locally elevated by about 0.3 m above the mean tide-level, large areas of both sand bodies emerge above mean sea-level at low tide. The adjacent tidal channels are 2–4 m deep at low tide (Davis & Flemming, 1991).

Large surfaces of the intertidal sand bodies are covered by a combination of small ripples and the burrows of the polychaete *Arenicola marina*

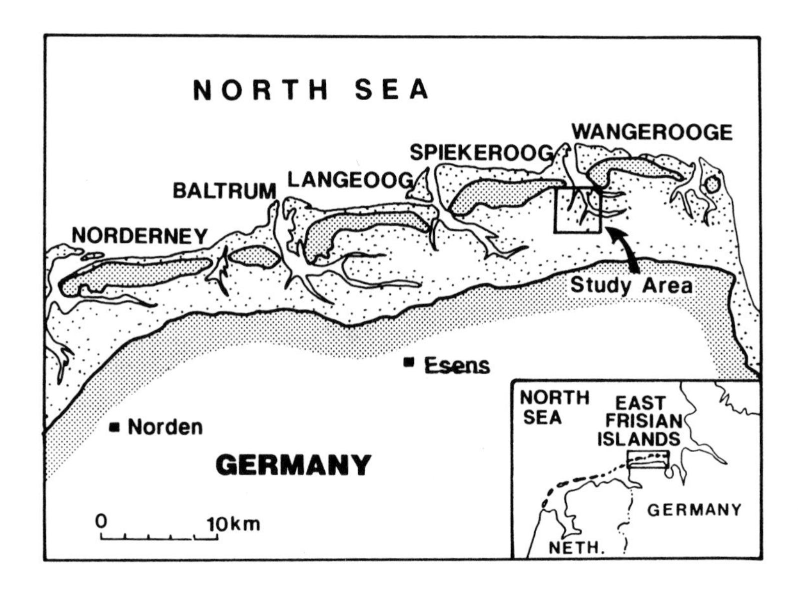

Fig. 2. The East Frisian barrier islands showing the location of the study area between Spiekeroog and Wangerooge.

(Fig. 3A). Locally, especially toward the south, the polychaete *Lanice conchilega* may be more prominent (Knecht, 1991). By contrast, most channel beds and the northern margins of the intertidal sand bodies are dominated by small dunes that are typically ebb-dominated (Fig. 3B).

The sediment is uniformly graded throughout the survey area, comprising very well-sorted fine sands with mean diameters ranging from 2.3 to 2.7 phi (0.2–0.15 mm) (Brosinsky, 1991; Knecht, 1991). Only along the more wave-exposed northern and eastern margins of Martens Plate can slightly coarser grain sizes be encountered. The mud content shows a strong north–south gradient, reaching levels of only <0.2% in the north, 0.2–0.5% in the centre and 0.5–5% just north of the mussel beds in the south (Knecht, 1991). Mud values of >30% occur locally where biogenically deposited muds at and near the sediment surface are associated with extensive beds of the black mussel *Mytilus edulis* (Flemming & Delafontaine, 1994). The carbonate content of the sediment is mostly low, except in the vicinity of mussel beds where shells and shell hash of *M. edulis* and

A

B

Fig. 3. (A) Wave ripples and burrows of *Arenicola marina* on Martens Plate. (B) Small, ebb-oriented dunes along the north-western margin of Martens Plate.

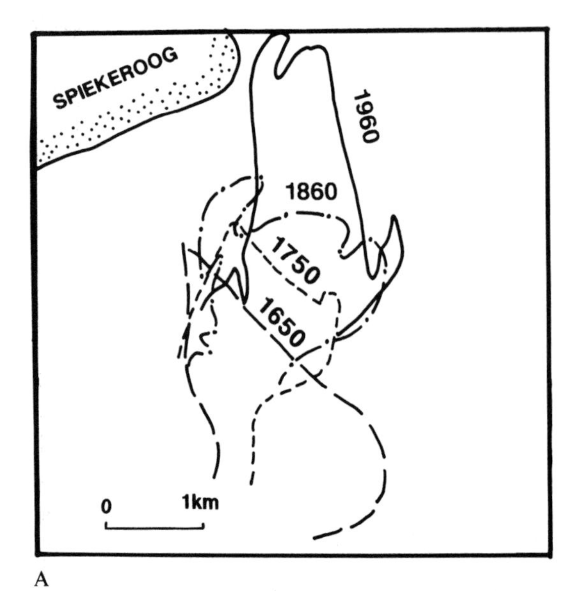

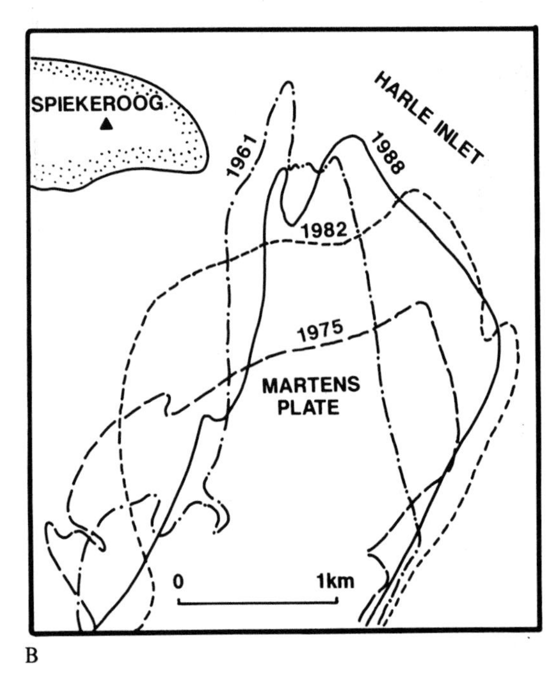

Fig. 4. (A) Morphological evolution of Martens Plate in historical times. (B) Morphodynamic behaviour of Martens Plate over the past few decades.

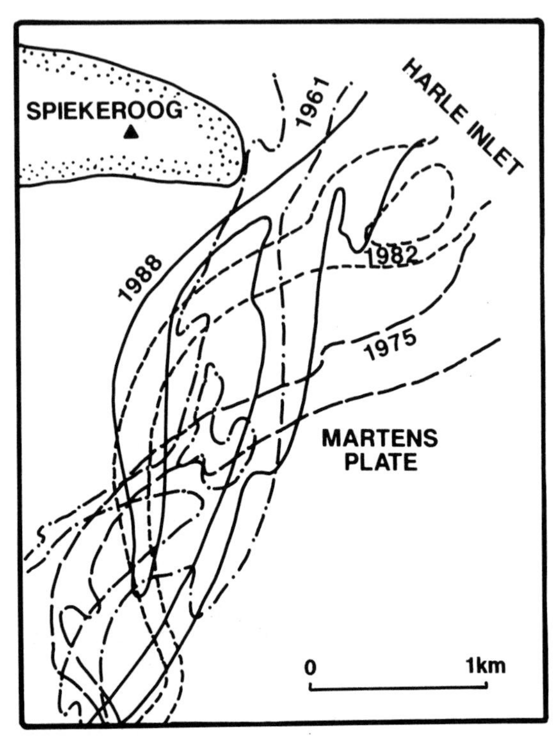

Fig. 5. Morphodynamic behaviour of the Muschelbalje channel over the past few decades.

the cockle *Cerastoderma edule* can form extensive pavements.

The East Frisian Islands have been inhabited by man for many centuries and the main physiographical features of the region have thus been regularly documented on fairly accurate charts dating back to AD 1650. These historical records reveal that Martens Plate has been steadily prograding northwards since earliest documentation (Homeier, 1979), reaching its most northerly, i.e. most seaward, extension in 1960 (Fig. 4A). Since then, the position and shape of Martens Plate have experienced frequent changes (Fig. 4B), mostly reflecting landward erosion due to storm action, the most severe one in modern times having occurred in 1962. Since 1975 Martens Plate is again actively prograding, now both toward the north and the east (Fig. 4B).

Corresponding to the morphodynamic adjustments of Martens Plate in recent times, the adjacent tidal channel (Muschelbalje) also displayed considerable mobility (Fig. 5). As demonstrated by Flemming & Davis (1994), the dynamics of the channel system can be described by an oscillating meandering motion that is fixed at the tidal watershed and increases in stroke toward its confluence with the larger Harle channel. In doing so, parts of the adjacent tidal sand bodies are reworked and rebuilt at regular intervals. It was shown that the relationship between the length of the channel axis and its catchment area corresponded closely to the general rule governing the same parameters in fluvial systems (e.g. Hack, 1957; Leopold *et al.*, 1964; Leeder, 1991). This phenomenon is evidently controlled by hydraulic factors such as discharge volume and slope, rather than tectonic factors as hitherto assumed (e.g. Leeder, 1991).

CORING METHOD AND STRATEGY

Coring positions on the intertidal flats and in the adjacent channels (Fig. 6) were located on the basis of the available historical maps (Figs 4 & 5). In addition, a number of coring sites were chosen along the modern channel margins and in the area of significant mud accumulation in the south. This approach was chosen in an attempt to obtain information on the internal sedimentary structures generated in the process of the historical progradation of Martens Plate, as well as from all the modern facies observed in the course of concurrent studies (Brosinsky, 1991; Knecht, 1991). In all, 31 vibracores were collected in the summers of 1988 and 1990.

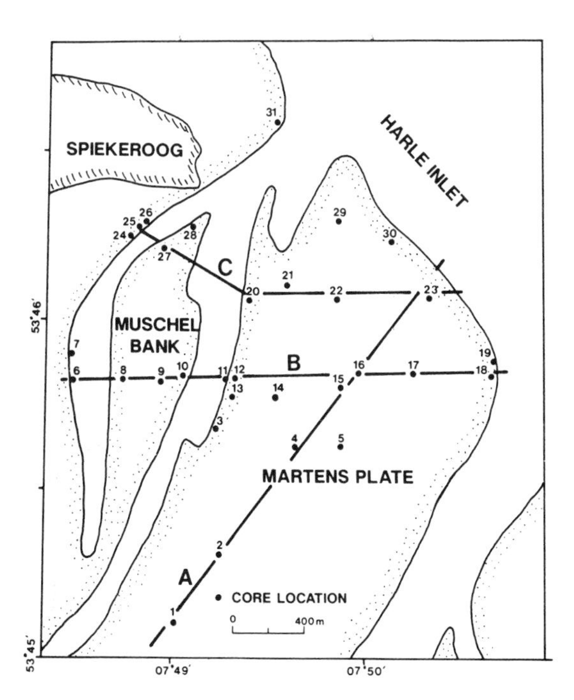

Fig. 6. Study area showing the location and numbering of vibracores and the position of stratigraphical cross-sections.

Two methods of vibracoring were applied, both making use of an electrically driven vibrator (concrete vibrator and hammer vibrator). In the case of the concrete vibrator, the coring tubes consisted of aluminium with an internal diameter of 8 cm. In the case of the hammer vibrator, larger square-box barrels (15 * 15 cm) consisting of stainless steel were used. The cores were in both cases extracted by chain pulley suspended from a tripod. Penetration was not always satisfactory due to the extremely compact packing of the well-sorted fine sands, recovered core lengths ranging from 4.6 m to <3 m. Although other coring techniques achieve greater penetration, they do not preserve the sedimentary structures and are hence unsuitable for detailed facies analyses (e.g. Streif, 1989)

FACIES ANALYSIS

Four facies were recognized in the course of core examination. Two of these, i.e. a wave-dominated and a tide-dominated sand facies, occurred widely. The other two, a bioturbated sand facies and a channel-lag facies, were less common and of small thickness.

The *channel-lag facies* is characterized by an abundance of coarse, shelly particles on a sharply defined erosional base, being typically 20–30 cm thick. Along the modern channel bed of the Muschelbalje this facies was less conspicuous, being restricted to isolated patches, perhaps because in most places the bed was dominated by dune bedforms which covered the shell lags with up to 2-m thick sand layers. Indeed, the existence of extensive shell lags in the deeper channels of the Wadden Sea has been demonstrated by Flemming *et al.* (1992). In addition, relatively thick shell lags were encountered in the vicinity of the mussel beds, where they amass in shallow, intertidal creeks.

The *bioturbated sand facies* is characterized by intense bioturbation produced by the burrowing activity of bivalves and polychaetes, especially the lug worm *A. marina*. Considering the widespread occurrence of this species, the facies is surprisingly limited in extent. It is more frequently encountered only toward the southern margin of

Martens Plate, where physical energy levels are relatively low.

The *wave-dominated sand facies* is primarily characterized by ripple cross-stratification and, to a lesser extent, also by plane beds. The sediment consists of well-sorted fine sands with small amounts of mud and organic debris (coffee grounds) which are locally incorporated in form of flasers. Rare, thin and discontinuous shell beds may also occur.

The *tide-dominated sand facies* occurs in two distinct subfacies. One is dominated by horizontal tidal bedding (Fig. 7A), the other by cross-bedding displaying tidal bundles (Fig. 7B). The latter is more common and in most cases is characterized by mesoscale cross-stratification. The sediment consists of well-sorted fine sands containing scattered shell debris. Although well sorted, the sands are not as well sorted as those of the wave-dominated sand facies. Mud is common, occurring in the interbeds of the horizontal tidal bedding units (Fig. 7A), or in form of drapes on the mesoscale cross-beds displaying tidal bundles (Fig. 7B). Ripple cross-stratification and small flasers are not commonly observed in this facies.

Fig. 7. Tide-generated sedimentary structures in vibracores from Martens Plate. A: Horizontal tidal bedding. B: Cross-bedding with tidal bundles and mud drapes.

STRATIGRAPHICAL ANALYSIS

The individual cores serve as good examples of the various facies associations observed on Martens Plate and the Muschelbank. One of the shortcomings of the present stratigraphical data set is the limited penetration that was achieved with the coring equipment. Cores rarely extended beyond the beds of interbank channels. Nevertheless, most cores did at least penetrate significantly below the spring low-tide level (SLT).

Grain-size patterns

Grain-size analyses are shown for three typical cores (cores 7, 12 and 23) in Fig. 8. Sample depths have in all cases been normalized to the topographic chart datum (0 m NN). All the samples, with the exception of one, fall into the fine sand range. Core 7, located in the westernmost position of the coring grid (Fig. 6), comprises very fine sands (up to 3.2 phi) at the surface. It then coarsens down to about − 1.4 m NN, attaining 2.7 phi before fining gradually down to a depth of − 3.75 m NN (2.88 phi). Thereafter, it again coarsens slightly to reach 2.8 phi at the core base at − 4.5 m NN.

Core 12, located along the western margin of Martens Plate, does not show any conspicuous downcore trend, except that the grain sizes fluctuate between 2.65 phi and 2.71 phi.

Core 23, located adjacent to the large Harle tidal channel to the northeast, at first fines from 2.48 phi to 2.71 phi down to a core depth of − 1.4 m NN. Thereafter, the mean grain size remains almost constant at 2.69 phi.

The greatest variation in mean grain size occurs above − 1.4 m NN in all cases, a depth that closely corresponds to the mean low-tide level. This would suggest that deposition in the subtidal environment is physically more stable than in the intertidal environment.

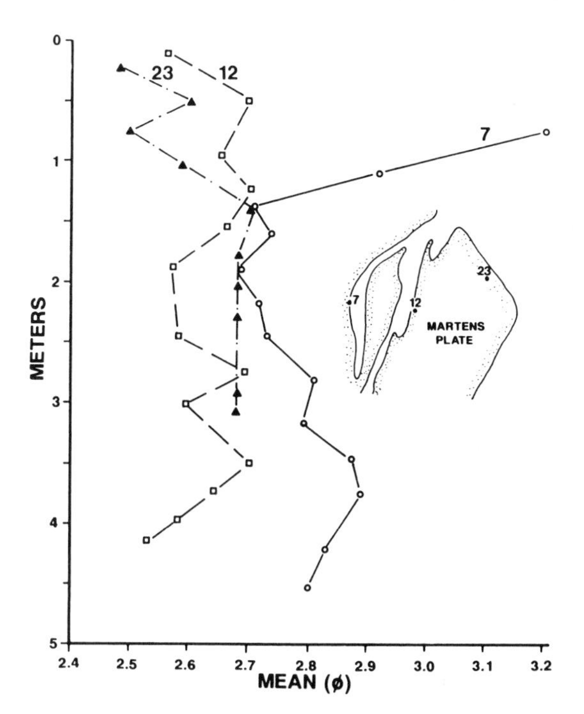

Fig. 8. Typical grain-size trends observed in selected vibracores.

Core descriptions

The stratigraphical variability within the Martens Plate and Muschelbank area is also illustrated by three selected cores (Fig. 9). In this case the cores 15, 23 and 26 have been chosen. Core 15 (Fig. 9 left) is located in the central part of Martens Plate (cf. Fig. 6). It penetrates to a depth of nearly 3 m below mean sea-level. The entire core is characterized by well-sorted fine sand with a few mud clasts near the top. It also displays a combination of flaser bedding and ripple cross-stratification throughout the core length. No channel deposits or bioturbated sections are present. This association of internal sedimentary structures suggests that the entire core was deposited under the dominating influence of wave action.

Core 26 (Fig. 9 centre) was recovered from the western bank of a muddy side channel, located in the north-western part of the study area opposite the northern tip of the Muschelbank (cf. Fig. 6). It contrasts markedly with all other cores. The core begins at −0.4 m NN and penetrates to just over −3.0 m NN. The upper metre contains some shell layers and flasers, indicating the influence of both wave action and sluggish, channelized flow. Most of the remaining core displays horizontally interlayered tidal and flaser bedding. Coffee grounds are generally mixed with mud and are often incorporated within the tidal bedding and the flasers. These features indicate that tidal currents dominated deposition in the lower 2 m of the core. The upper part probably formed very recently in the shelter of the prograding eastern head of Spiekeroog Island since 1962.

The clearest evidence for both wave- and current-domination within a single core is presented in core 23 (Fig. 9 right), located along the northeastern margin of Martens Plate adjacent to the large Harle channel (Fig. 6). The core is 3.2 m long, beginning at about 0 m NN. The downcore grain size characteristics are shown in Fig. 8. They consist of well-sorted fine sand with a marked change in trend at about −1.4 m NN. The upper half contains a few scattered mud drapes and clasts but no discernible primary sedimentary structures. In spite of the absence of distinct ripple cross-stratification and flasers, this section is interpreted as wave-dominated on the basis of its location and the sharp transition to the lower, clearly tide-dominated section of the core. This lower section is dominated by ebb-oriented, mesoscale cross-bedding, most of which display mud drapes. The existence of such dune-bedding can also be postulated from the results of a detailed study in the Muschelbalje tidal channel presented by Davis & Flemming (1991).

STRATIGRAPHICAL CROSS-SECTIONS

Using the data logs from all 31 cores, three stratigraphical cross-sections were constructed (see Fig. 6 for location) in order to show the distribution and arrangement of the various facies resulting from the cut and fill processes in the course of channel migration over time.

Cross-section A (Fig. 10) extends from the south-west to the north-east along the central axis of the Martens Plate sand body. It crosses six core locations and was chosen to illustrate the stratigraphy of the deep channel that has been filled as Martens Plate prograded northward and eastward in historical times. None of the evidently deeper-lying tidal-channel deposits have been reached in the cores. However, all the cores penetrated to at least the spring low-tide level. All four facies, identified in various cores, are represented in this cross-

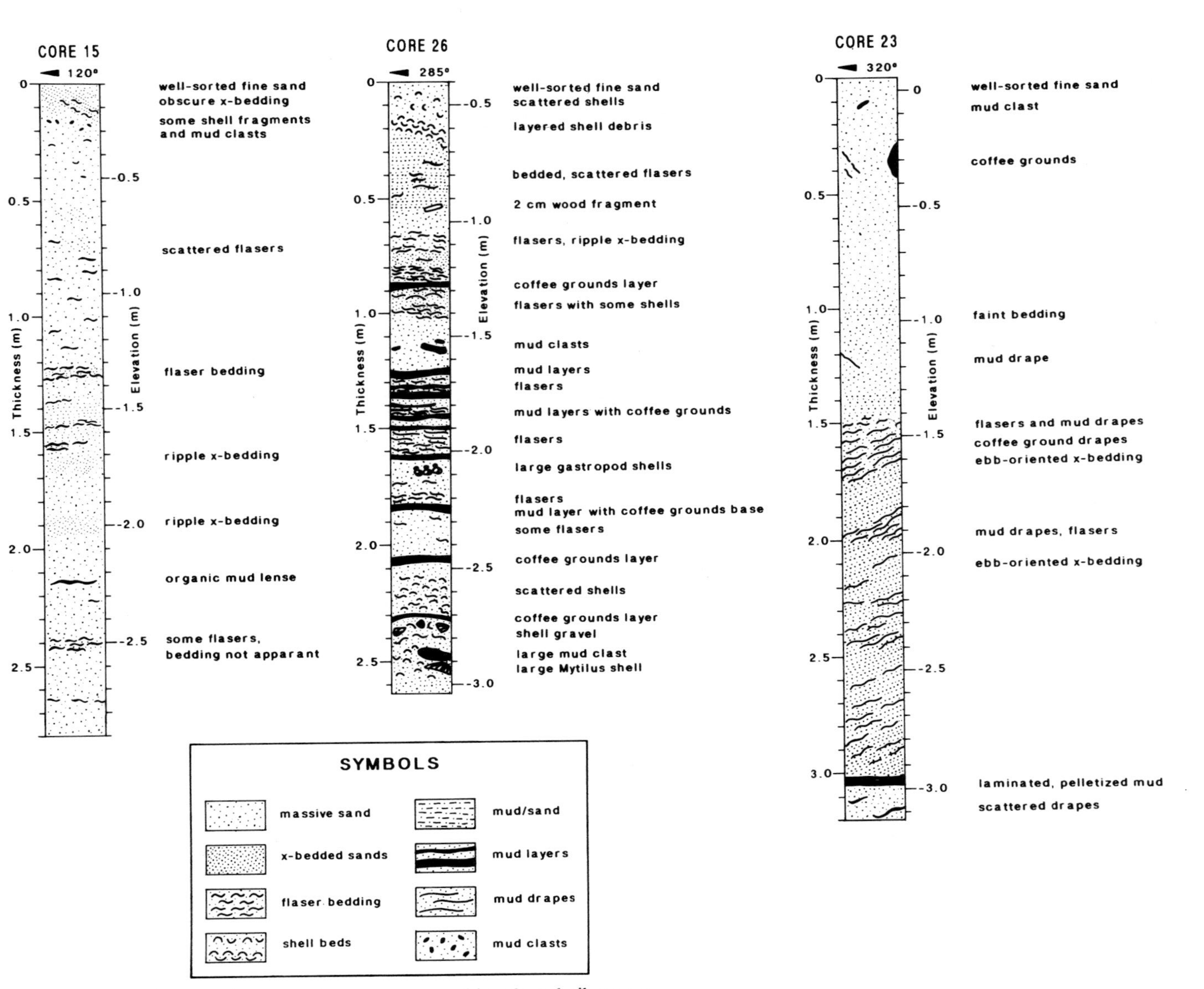

Fig. 9. Typical primary sedimentary structures observed in selected vibracores.

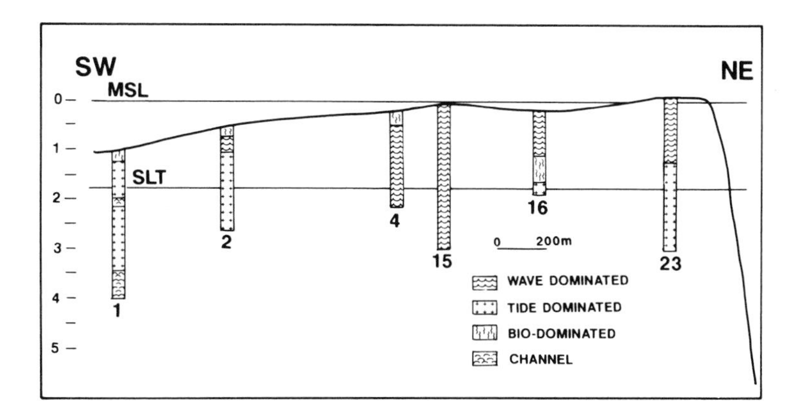

Fig. 10. Stratigraphy of cores along cross-section A.

section. Although the seaward displacement of the northern margin of Martens Plate since the 1970s must have buried sandy tidal channel and channel-margin environments along its way, a close examination of this cross-section provides little evidence of stratigraphic signatures relating to a tidal-channel environment. Only core 1 contains channel lag deposits at its base and in a thin layer within its upper half, the latter revealing the passage of a shallow tidal creek in the vicinity of a mussel bed. Significantly, the deeper channel deposits of core 1 are situated at a depth greater than that penetrated by any of the other cores along this cross-section. Conversely, only cores 1, 2, 16 and 23 reveal the tide-dominated sand facies in their central or lower parts, the remaining cores being characterized by wave domination over most of their lengths, with the exception of short core sections showing evidence of the strongly bioturbated sand facies. The near-surface presence of this facies in some cores is explained by the relative absence of wave action in the more remote, low-lying parts of the tidal flat sand body.

Cross-section B (Fig. 11) extends from west to east across the Muschelbank and the Martens Plate sand body complex, including two persistent migrating tidal channels. The stratigraphy of this cross-section is based on 10 cores which display a consistent pattern. The bioturbated sand facies and the channel-lag facies are both rare and scattered. The tide-dominated sand facies typically occurs at the channel margins or below the spring low-tide level. Exceptions can be seen in cores 16 and 17, where some mesoscale cross-beds can be seen just above SLT. The wave-dominated facies is pervasive above SLT and extends down to or close to the bottom of several cores, in one case even at a channel margin (core 7).

Cross-section C (Fig. 12) also runs from west to east but is situated more northerly than cross-section B. It displays a similar stratigraphical pattern as the more southerly one. The tide-dominated

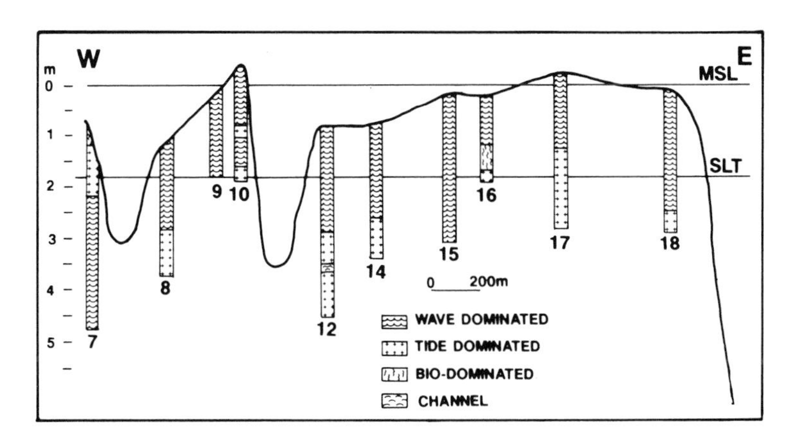

Fig. 11. Stratigraphy of cores along cross-section B.

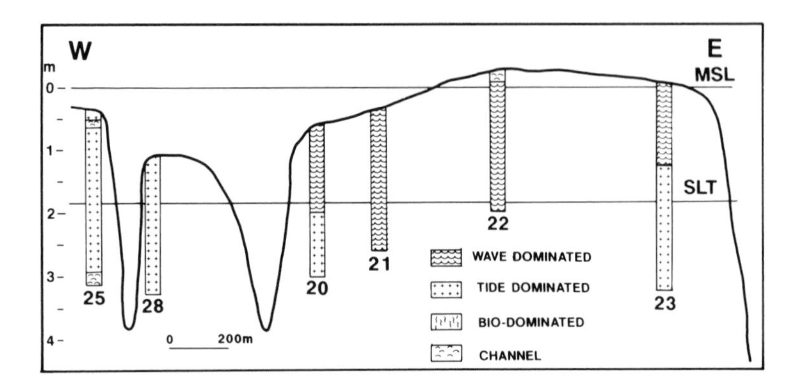

Fig. 12. Stratigraphy of cores along cross-section C.

sand facies only occurs near the margins of tidal channels, or below SLT in the one case where it is overlain by the wave-dominated sand facies (core 20). Three cores situated close to the cross-section, but which are not shown here (cores 27, 24 and 26), are quite different from the other cores in that the tide-dominated signature primarily relates to horizontal tidal bedding incorporating significant amounts of mud (cf. core 26 on Fig. 9). Furthermore, two thin channel-lag deposits can be seen in core 25. The one near the surface relates to a shallow intertidal creek commonly observed in this area, whereas the one at the base of the core most probably reflects the lateral migration of the deeper adjacent channel.

DISCUSSION AND CONCLUSIONS

The sedimentary and biological features that characterize the modern depositional environment of the Martens Plate and Muschelbank can be used, although with some caution in the latter case, to reconstruct the stratigraphy of this tidal sand body complex. The abundance and distribution of the infauna, especially bivalves and polychaetes, produce a range of bioturbation structures in the near-surface sediments which are most intense in muddy and low-lying tidal flat areas, where physical energy levels are low. Since the sand body is situated in an area dominated by physical processes, the stratigraphical preservation potential of bioturbation structures is low in most places.

Channel-lag deposits are limited to shallow, ephemeral intertidal creeks which, under present conditions, accumulate shell material and some mud clasts. In addition, these shallow channels are the only environment in which current ripples occur, larger bedforms being absent. Such relatively thin, shelly lag pavements occur widely scattered in both the modern intertidal environment and the stratigraphical column, reconstructed on the basis of vibracore data.

Tidal deposits characterized by mesoscale cross-bedding with tidal bundles are restricted to subtidal beds of bigger channels and to intertidal channel margins. They are invariably ebb-dominated. By contrast, horizontally-bedded tidal deposits can only be found in areas of low current velocity. The presence of such structures in the stratigraphical record can therefore be associated with similar depositional environments. The locations of tide-dominated facies observed in vibracores from Martens Plate are thus indicative of the presence or proximity of deep tidal channels.

The evaluation of historical maps and modern charts has revealed that almost the whole area covered by the coring programme has been reworked by migrating channels at some time during the past few hundred years. Most of the cores have sections displaying the tide-dominated sand facies. These are typically located below the spring low-tide level. This suggests that the tide-dominated facies is effectively reworked at higher locations on Marten Plate to produce wave-ripple cross-bedding and flasers. In some cores this type of bedding extends down to -4 m NN, suggesting that substantial reworking of mesoscale cross-bedded deposits must also occur in some places as the tidal flat sand body progrades over former channel margins and beds.

The continuity and consistency of the core data made it possible to reconstruct the larger-scale facies architecture within the sand body complex of

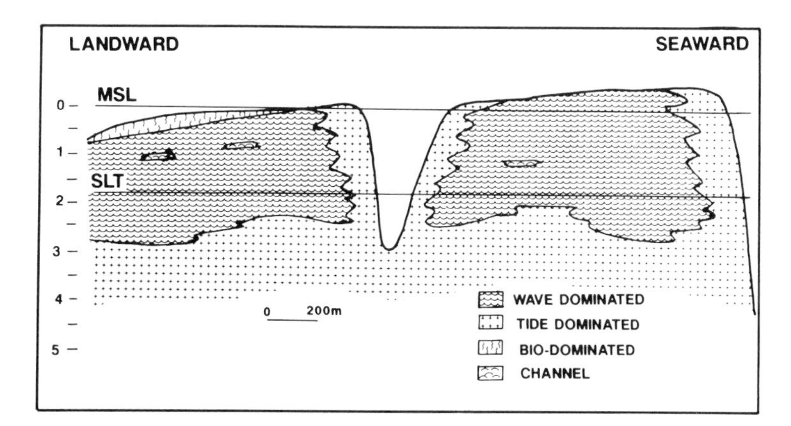

Fig. 13. Generalized facies architecture of Martens Plate based on vibracore data.

the Martens Plate and the Muschelbank (Fig. 13). The primary architectural elements include the four facies associations identified and discussed above. It must be emphasized, however, that the model presented here does not include the deeper parts of the sand bodies which must also exist, but were not penetrated by the coring equipment.

The wave-dominated sand facies, the bioturbated sand facies and the channel-lag facies are typically restricted to the upper, intertidal parts of the tidal flat complex. The latter two are mostly thin and occur widely scattered, the bioturbated facies tending to occur near the present sediment surface in places of low physical energy. By contrast, the tide-dominated facies dominates the lower, subtidal parts and the intertidal margins of the sand body complex.

This type of facies distribution and arrangement is probably typical of sand bodies situated in proximity to wide and deep tidal inlets as found, for example, along the southern North Sea coast. In such locations wave energy is highest, thus explaining the predominance of the wave-dominated sand facies. By implication, mud deposits and bioturbation structures are rare by comparison. These are expected to become increasingly more dominant both landward and toward the tidal watersheds, where tidal channels are shallower and less mobile.

It is important to note that funnel-shaped, meso- to macrotidal estuaries, as discussed by Dalrymple *et al.* (1992) and Allen & Posamentier (1993), do not show this general pattern of apparent wave domination in the sediment record. Only from the Oosterschelde estuary have wave-dominated facies been reported (Berg, 1981). Furthermore, the mesotidal backbarrier tidal flats of the Georgia Bight, USA, are also devoid of wave-induced sedimentary structures, being dominated instead by mud-flats and salt marsh deposits (Frey & Howard, 1986).

The stratigraphical architecture of the Martens Plate sand body complex, and probably of many similar deposits in the Wadden Sea region, presents a picture that would usually not have been associated with a mesotidal backbarrier setting. This would suggest that ancient examples displaying similar features may have to be reinterpreted.

ACKNOWLEDGEMENTS

Much of this work was carried out while the first author was a DAAD visiting fellow at the Senckenberg Institute in Wilhelmshaven in 1988. The logistical backup and equipment was supplied by the institute. The authors wish to thank the captain and crew of the research vessel *Senckenberg* for their assistance under not always favourable weather conditions. Special thanks are also due to the students and technicians who assisted with the field work, in particular Ute Mann, Dorothee Brosinky, Sylke Knecht and Karl-Heinz Naujokat.

REFERENCES

Allen, G.P. & Posamentier, H.W. (1993) Sequence stratigraphy and facies model of an incised valley fill: the Gironde estuary, France. *J. sediment. Petrol.* **63**, 378–391.

Berg, J.H. van den (1981) Rhythmic seasonal layering in a mesotidal channel fill sequence, Oosterschelde Mouth, The Netherlands. In: *Holocene Marine Sedimentation in the North Sea Basin* (Eds Nio, S.D., Schüttenhelm, R.T.E. & Weering, Tj.C.E. van). *Spec. Publs Int. Ass. Sediment.* **5**, 147–159.

BRONSINSKY, D. (1991) *Morphodynamische Prozesse und Sedimentverteilung im Bereich der Muschelbalje (östliches Spiekerooger Watt).* MSc thesis, Univ. Giessen (Germany).

DALRYMPLE, R.W., ZAITLIN, B.A. & BOYD, R. (1992) Estuarine facies models: conceptual basis and stratigraphic implications. *J. sediment. Petrol.* **62**, 1130–1146.

DAVIS, R.A., JR & FLEMMING, B.W. (1991) Time-series study of mesotidal bedforms, Martens Plate, Wadden Sea, Germany. In: *Clastic Tidal Sedimentology* (Eds Smith, D.G., Reinson, G.E., Zaitlin, B.A. & Rahmani, R.A.). *Can. Soc. Petrol. Geol. Mem.* **16**, 275–282.

DETTE, H. (1977) Ein Vorschlag zur Analyse eines Wellenklimas. *Die Küste* **31**, 166–180.

EHLERS, J. (1988) *Morphodynamics of the Wadden Sea.* Balkema, Rotterdam, 397 pp.

FITZGERALD, D.M., PENLAND, S. & NUMMEDAL, D. (1984) Control of barrier island shape by inlet sediment bypassing: East Frisian Islands, West Germany. *Mar. Geol.* **60**, 355–376.

FLEMMING, B.W. & DAVIS, R.A., JR. (1994) Holocene evolution, morphodynamics and sedimentology of the Spiekeroog barrier island system (southern North Sea). In: *Tidal Flats and Barrier Systems of Continental Europe: A Selective Overview* (Eds FLEMMING, B.W. & HERTWECK, G.). *Senckenbergiana marit.* **24**, 117–155.

FLEMMING, B.W. & DELAFONTAINE, M.T. (1994) Biodeposition in a juvenile mussel bed of the East Frisian Wadden Sea (southern North Sea). *Neth. J. aquat. Ecol.* **28**, 289–297.

FLEMMING, B.W., SCHUBERT, H., HERTWECK, G. & MÜLLER, K. (1992) Bioclastic tidal-channel lag deposits: a genetic model. *Senckenbergiana marit.* **22**, 109–129.

FREY, R.W. & HOWARD, J.D. (1986) Mesotidal estuarine sequences: a perspective from the Georgia Bight. *J. sediment Petrol.* **56**, 911–924.

HACK, J.T. (1957) Studies of longitudinal stream profiles in Virginia and Maryland. *U.S. geol. Surv. Prof. Paper* **219B**, 97 pp.

HANISCH, J. (1981) Sand transport in the tidal inlet between Wangerooge and Spiekeroog (W. Germany). In: *Holocene Marine Sedimentation in the North Sea Basin* (Eds Nio, S.D., Schüttenhelm, R.T.E. & Weering, Tj. C.E. van). *Spec. Publs int. Ass. Sediment.* **5**, 175–185.

HAYES, M.O. (1979) Barrier island morphology as a function of tidal and wave regime. In: *Barrier Islands from the Gulf of St Lawrence to the Gulf of Mexico* (Ed. Leatherman, S.P.), pp. 1–27. Academic Press, New York.

HEMPEL, L. (1985) *Erläuterungen zur Geomorphologischen Karte 1:100,000 der Bundesrepublik Deutschland, GMK 100 Blatt 4 C 2310/C 2314 Esens/Langen.* Berlin, 87 pp.

HOMEIER, H. (1979) Die Verlandung der Harlebucht bis 1600 auf der Grundlage neuer Befunde. *Forschungsstelle Norderney, Jber.* **30**, 106–115.

KNECHT, S. (1991) *Morphologische Entwicklung. Sedimentverteilung und Faziesdifferenzierung der Martens Plate (östliches Spiekerooger Watt).* MSc thesis, Univ. Giessen (Germany).

LEEDER, M.R. (1991) Denudation, vertical crustal movement and sedimentary infill. *Geol. Rundschau* **80**, 441–458.

LEOPOLD, L.B., WOLMAN, M.G. & MILLER, J.P. (1964) *Fluvial Processes in Geomorphology.* Freeman, San Francisco, 525 pp.

NUMMEDAL, D., OERTEL, G.F., HUBBARD, D.K. & HINE, A.C. (1977) Tidal inlet variability—Cape Hatteras to Cape Canaveral. *A.S.C.E., Coastal Sediments '77*, 543–562.

STREIF, H.J. (1989) Barrier islands, tidal flats, and coastal marshes resulting from a relative rise of sea level in East Frisia and the German North Sea coast. *Proc. KNGMG Symp. Coastal Lowlands, Geology and Geotechnology*, 213–223.

Spec. Publs int. Ass. Sediment. (1995) **24**, 133–149

Sedimentation in the mesotidal Rías Bajas of Galicia (north-western Spain): Ensenada de San Simón, Inner Ría de Vigo

M.A. NOMBELA*, F. VILAS* *and* G. EVANS†‡

**Departamento Recursos Naturales y Medio Ambiente, Universidad de Vigo, Spain; and*
†Department of Geology, Royal School of Mines, Imperial College,
South Kensington, London SW7

ABSTRACT

The mesotidal (2–4 m) Rías Bajas of Galicia are fault-bounded depressions drowned by the sea during the Flandrian transgression to form elongate coastal embayments, which are being infilled by land-derived siliciclastic detritus and locally produced biogenic carbonate.

Small fluvio-tidal bay-head deltas have developed at the landward extremities where the rivers debouch into the rías. Muds and sands accumulate in narrow marshes, on *Zostera nana* covered mud-flats and on sand-flats, which are crossed by estuarine channels floored with sands fashioned into a wide variety of bedforms.

Seawards of the sand-flats associated with the fluvio-tidal deltas, the inner subtidal parts of the rías are covered with a dense carpet of *Zostera marina* which traps and prevents the seaward dispersal of sands. These are stratified and rich in *Cerastoderma edule* (Facies III), floor the inner rías, and pass seawards into diffusely stratified muds which are poor in skeletal debris and rich in organic matter (5–6%) (Facies II). Structures produced by organically generated gas are common in these sediments. The surficial sediments are underlain (<2 m) by better stratified muds with higher amounts of skeletal debris (Facies I), indicating recent changes in depositional conditions in the inner rías.

Generally, if the subtidal sediments were preserved in the geologic record, the tidal character of the deposits would not be recognized unless the adjacent intertidal deposits were also preserved.

Man has influenced the environment by dumping sand to produce beds for *Cerastoderma edule*, thereby artificially extending the areas of natural sand deposition seawards of the fluvio-tidal deltas. In addition, the deposition of fine-grained sediment has been increased by the introduction of mussel rafts for *Mytilus galloprovincialis*.

INTRODUCTION

The Ensenada de San Simón is the inner part of the Ría de Vigo (Fig. 1), one of the classical rías originally described by Richthofen (1886) when he introduced the term into the geological literature (the Spanish term 'ensenada' commonly denotes a bay or marine inlet). These rías have developed on the passive margin of Galicia, NW Spain (Boillot & Malod, 1988). It has been suggested that their orientation is controlled by two Alpine fault systems (NE–SW and N–S) which have guided the rivers to the coast (Boucart, 1938; Pannekoek, 1966, 1970). This coastline has been the subject of many geomorphological studies (Carle, 1947; Torre-Enciso, 1958; Mensching, 1961; Pannekoek, 1966, 1970) recently reviewed by Sala (1984).

The area has attracted considerable attention from oceanographers because of its remarkable fertility and its important commercial fisheries. This fertility has been attributed to upwelling on the adjacent shelf and intrusion of nutrient-rich waters into the rías (Margalef, 1956a; Saiz *et al.*, 1957, 1961; Fraga, 1960, 1967, 1981; Anadón *et al.*, 1961; Prego & Fraga, 1992; see Fraga, 1981 for a general review). More recent studies have investigated the influx of fresh water into the Ría de Vigo

‡*Correspondence address*: La Caumiethe, Trinity, Jersey, Channel Islands, UK.

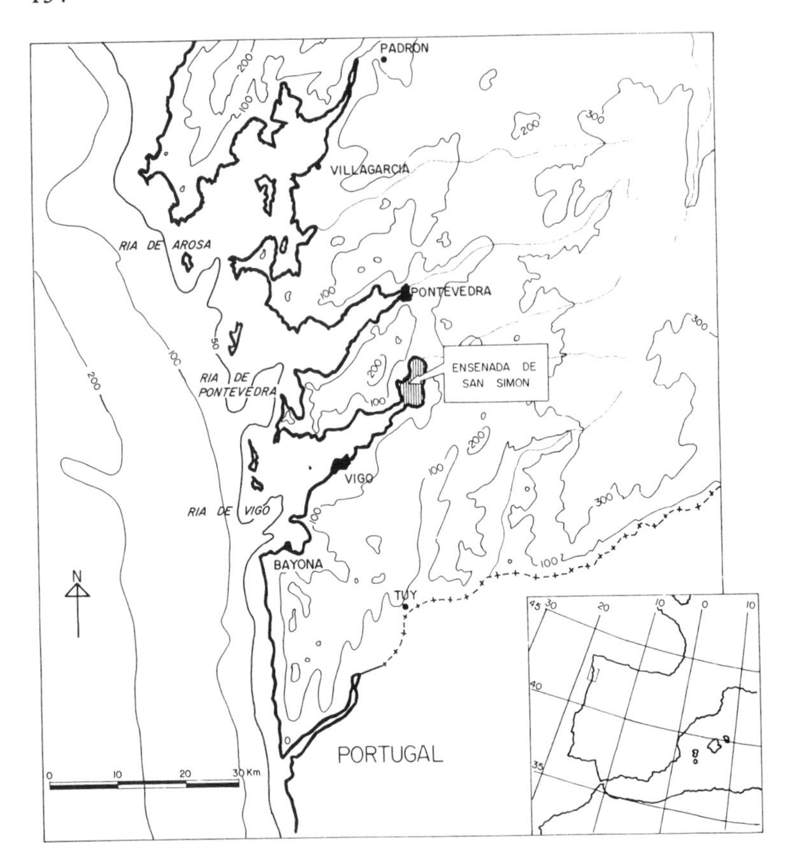

Fig. 1. Locality map of the Ensenada de San Simón (depths are in metres).

(Ríos *et al.*, 1992), water exchange between the Ría de Vigo and the adjacent shelf (Prego *et al.*, 1990), and the effects of upwelling on fish production in the neighbouring Ría de Arosa (Tenore & González, 1975; Tenore *et al.*, 1982; Varela *et al.*, 1984; Blanton *et al.*, 1987).

Detailed studies of macrofauna distribution patterns in the Ría de Arosa have been carried out by Cadée (1968) and Mora (1980). Ecological aspects of intertidal communities have been investigated by Anadón (1977, 1980) and Vieitez (1976, 1978, 1979) while substantial information exists on the taxonomy and general distribution patterns of various species (Rolán, 1983a,b; Rolán *et al.*, 1989). In contrast, similarly rigorous studies have by and large not been carried out in the Ría de Vigo. For instance, until recently the only available data on subtidal faunal composition were those of Margalef (1958). López-Jamar & Cal (1990) have in the meantime defined a series of macrobenthic communities in the main Ría de Vigo. Although this work did not extend to the Ensenada de San Simón, the latter is expected to show similar trends because it is characterized by environmental conditions similar to those of the inner part of the main Ría de Vigo.

The microfauna, particularly the foraminifers, of the Galicia rías have been described in a series of classical papers by Colom (1952, 1953), Mateu (undated) and, with particular reference to the Ría de Arosa, Voorthuysen (1973). The phytoplankton and its importance to fisheries has been investigated by, amongst others, Margalef (1956, 1958), Durán *et al.* (1956) and Vives & López-Benito (1958).

The first comprehensive analyses of the sediments of the Ría de Arosa were carried out at least 25 years ago (Pannekoek, 1966, 1970; Koldijk, 1968; de Jong & Poortman, 1970). By comparison, the neighbouring Ría de Vigo has received much less attention in this respect. Preliminary studies by Vilas (1978, 1979) have in the meantime been complemented by more detailed investigations of both the subtidal and intertidal sedimentary environments of the Ría de Vigo (Vilas, 1981a–c, 1983; Vilas & Nombela, 1985; Vilas & Rólan, 1985;

Nombela, 1989; Nombela & Vilas, 1986–87; Nombela *et al.*, 1987a,b; Alejo & Vilas, 1987; Alejo *et al.*, 1987, 1990a,b). Much of the data concerning the inner part of the Ría de Vigo—the Ensenada de San Simón—has been published only as abstracts (Nombela *et al.*, 1987a, 1990). Thus, the present work aims at providing a more comprehensive account of sedimentation in this area within a general framework defined by the regional environmental setting.

ENVIRONMENTAL SETTING

Climate

The area has a warm, humid climate (Pérez-Alberti, 1982) with average temperatures of 18.5°C in summer and 9.5°C in winter. Rainfall measures approximately 1000 mm yr^{-1} and, as in other Mediterranean countries, is heaviest in winter. The dominant winds are from the south-west with a strong northerly component during the summer months.

Oceanography

The study area is mesotidal with an average tidal range of 2.2 m. Current measurements made in the Ensenada de San Simón between spring and neap tides documented velocities of up to 35 cm s^{-1} at the surface and 18 cm s^{-1} at the bottom in the centre of the bay, decreasing to 25 cm s^{-1} and 12 cm s^{-1}, respectively, in the most landward parts seawards of the bay-head delta at the head of the Ensenada (Fig. 2). Similar velocities were found outside the bay in the main Ría de Vigo. Stronger currents, reaching 50 cm s^{-1} at the surface and 10–25 cm s^{-1} at the bottom, are found in the entrance of the main ría on either side of the Islas Cíes. Strong winds, particularly from the north during summer, can generate waves of up to 0.5 m, and the sediments on the floor of the Ensenada then appear to be vigorously reworked as the water becomes turbid over most of the bay.

The salinity of the outer part of the Ría de Vigo averages 36‰, decreasing to 31–32‰ in the entrance of the Ensenada de San Simón. More extreme values are found in the estuaries of the small rivers at the landward head of the Ensenada. During periods of high river discharge, particularly in spring, salinity gradients develop whereby a thin layer of low salinity water extends seawards of the Estrecho de Rande into the main ría.

Water temperatures vary from 11 to 12°C in

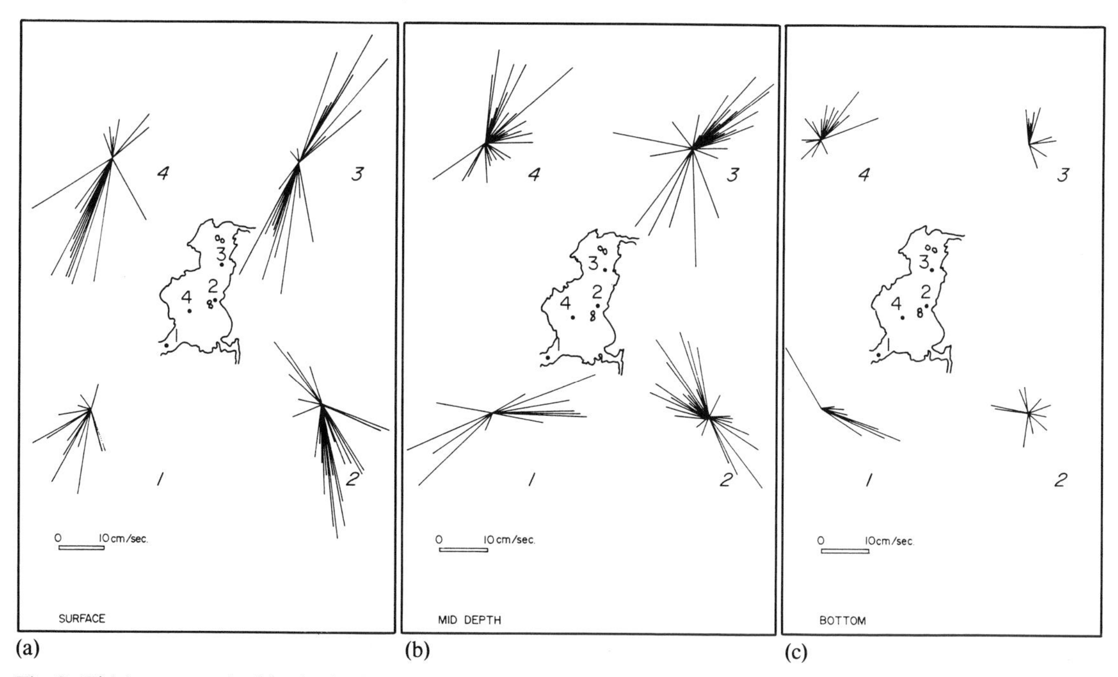

Fig. 2. Tidal current velocities in the Ensenada de San Simón. (a) Surface; (b) mid-depth; (c) bottom.

winter to 19–20°C in summer at the surface. A well-developed thermocline in the summer is enhanced in the outer ría by strong upwelling on the shelf outside the entrance of the main Ría de Vigo (see review by Fraga, 1981). This upwelled water intrudes into the ría and brings a rich supply of nutrients to the area, causing algal blooms which produce characteristic 'red tides' (Margalef, 1958; Fraga, 1981).

Geology

The geology of the area is dominated by igneous and metamorphic rocks of Precambrian–Palaeozoic age, which form a series of high hills (400–500 m) bordering the rías and the hinterland. The only younger rocks comprise Miocene–Quaternary continental sediments, which infill a series of N–S graben structures inland, and Quaternary fluvial and coastal sediments (Fig. 3). Earlier writers have claimed that there are three plantation surfaces which cut across both Precambrian and Palaeozoic rocks: a summit plain at approximately 1000–2000 m which was considered to be Oligocene in age, and two lower surfaces at 700–850 m and 600–700 m respectively. However, subsequent workers (Birot & Solé-Sabaris, 1954; Nonn, 1966) suggested that there were only two important levels. According to Sala (1984), the fundamental surface present in Galicia is between 200 and 500 m, probably being of pre-Miocene age. The higher surface at approximately 2000 m is likely to be very ancient, possibly related to Triassic and late Mesozoic planations. Several lower planation surfaces and terraces, thought to have formed at times of higher sea-levels, are present around the rías and along the adjacent coastline (Nonn, 1958, 1966). However, no detailed descriptions of any of the deposits on these features are as yet known.

Direct evidence for lower sea-levels in the area is rather sparse. On the basis of the presence of a terrestrial surface in boreholes near the harbour of

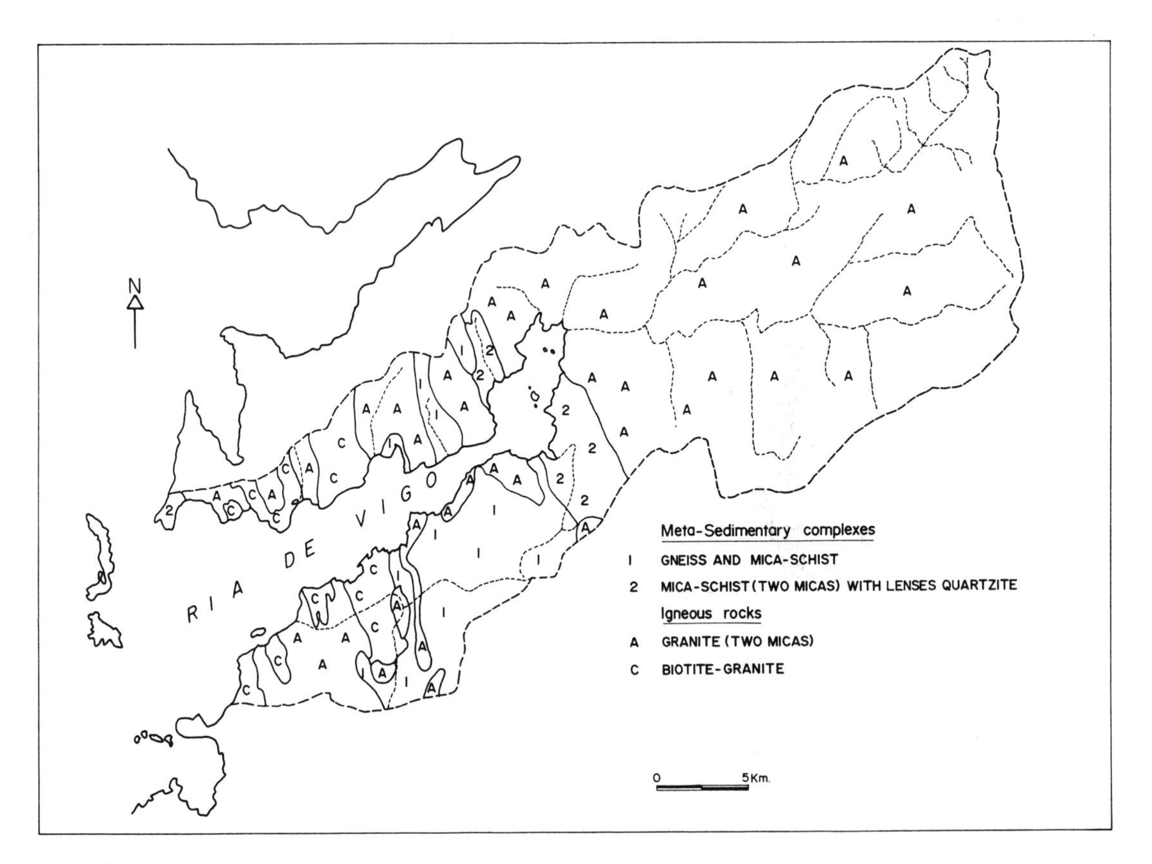

Fig. 3. General geology.

Vigo, Margalef (1956a) claimed that sea-level was 25–30 m below present-day level in the Ría de Vigo around 8500 yr BP. Also, a peaty layer with remains of marsh plants dated at 3940 ± 140 yr BP was discovered 5 m below the surface (i.e. approximately 7 m below present-day MSL) of the muddy intertidal flats near Arcade at the landward head of the Ensenada de San Simón. As has already been pointed out by several earlier writers (e.g. Boucart, 1938), the area appears to have developed its present morphology by the drowning of structurally controlled river valleys during the Flandrian transgression, followed by partial infilling with sediment. Saa-Otero & Díaz-Fierros (1988) have provided one of the few descriptions of the Holocene deposits. Recent tide-gauge records from Vigo show a mean annual rise in relative sea-level of 2.9 mm yr^{-1} between 1940 and the early 1960s (Emery & Aubrey, 1991), a process presumably continuing today.

CATCHMENT

The catchment of the entire Ría de Vigo, of which the Ensenada de San Simón is only a small part, is 578.2 km^2. Most of this area (434.8 km^2) drains into the Ensenada de San Simón and only a small part (143.4 km^2) drains directly into the main Ría de Vigo (Fig. 4).

In many places the catchment surface, situated at heights of approximately 200 m, is covered with a coarse clayey conglomerate with a red matrix which is markedly different from the underlying rocks (Nonn, 1966). This conglomerate has been considered equivalent to the 'rañas' of Castilla and, although it is less rubified and formed under conditions different from those of today (possibly more arid?), it is thought to have been associated with Mediterranean pluvials of Villefranchian age (Sala, 1984). Also, Nonn & Tricart (1960) claim evidence of periglacial activity. Some of the hilltops and

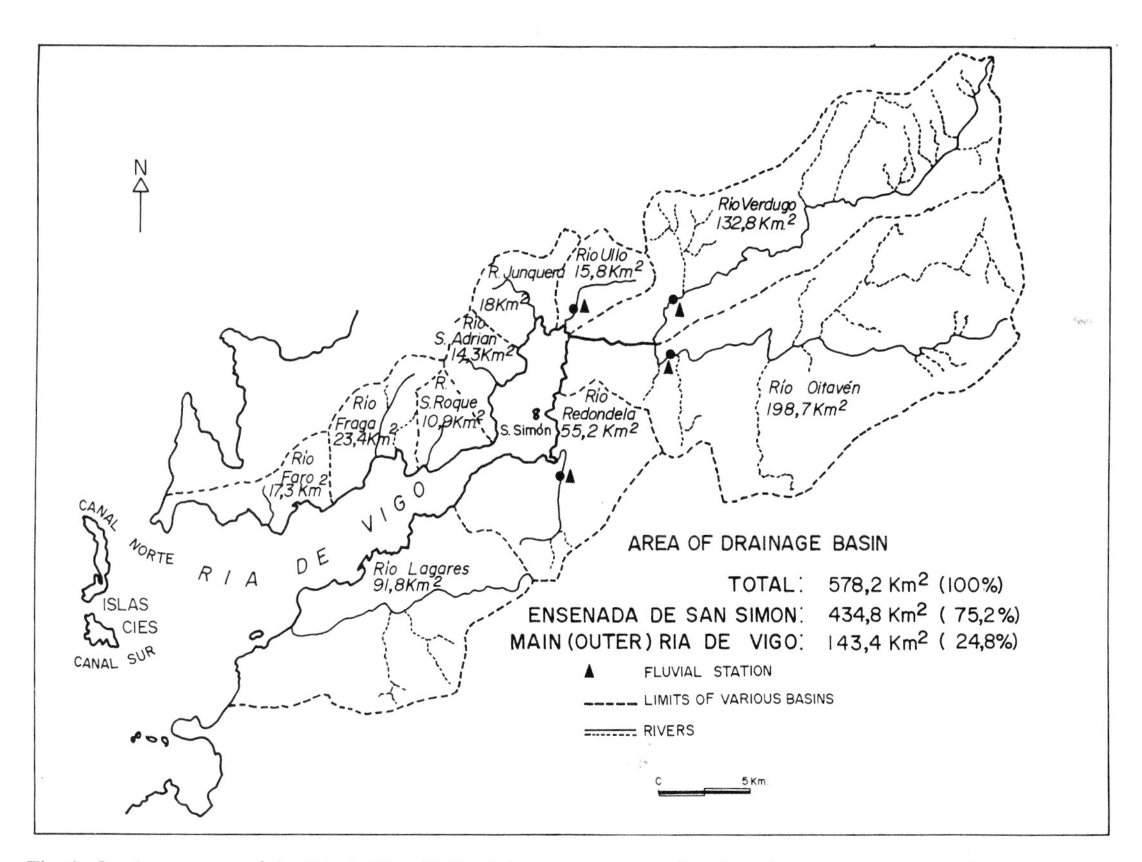

Fig. 4. Catchment area of the Ría de Vigo (N.B. all the major streams flow into the Ensenada de San Simón).

hillslopes have bare rock exposed today. However, colluvium is present elsewhere on the slopes. Small fans associated with torrential downpours are found in some places and the valleys have well-developed terraces. Roadcuts reveal that the surface rock has in places been deeply weathered. This well-developed regolith is almost certainly polygenetic and is probably due to deep Tertiary and later Quaternary weathering. There are several discontinuities in the modern soil profiles. Radiocarbon dating has shown that the latest period of soil development (Subatlantic) is interrupted by an erosional discontinuity which formed 2500–1000 yr BP. Other discontinuities were dated at 4000 yr BP and 10 000–9500 yr BP (Benito *et al.*, 1991).

Today, the hinterland is heavily vegetated, except where it has been cleared for cultivation or where forest fires have stripped the landscape. Fires are frequent in the area, especially in summer. Over the past 25 years 800–900-thousand fires have been reported to have affected 45% of the uncultivated lands of Galicia.

The dense cover of vegetation appears to prevent soil erosion effectively except where it has been disturbed by man; this presumably explains why the streams, even after heavy rains, usually contain only small amounts of suspended sediment. Estimates of erosion based on river discharge data coupled with appropriate empirical formulae (Benito *et al.*, 1991) show that erosion rates in Galicia are amongst the lowest reported worldwide (Young, 1969).

The main discharge into the Ría de Vigo is via some moderately small rivers (the Río Redondela, Río Oitabén, Río Verdugo and Río Ullo) which all enter the Ensenada de San Simón at the landward head of the Ría de Vigo. Some other small rivers flow into the main ría along its length. Stream discharge data are not abundant; however, average total discharge (based on a 1.5-year record together with some data of the Confederación Hidrográfica del Norte de España) is approximately 19.2 $m^3 s^{-1}$. Of this, the Río Oitabén supplies 10.39 $m^3 s^{-1}$, the Río Verdugo 6.92 $m^3 s^{-1}$, the Río Redondela 0.98 $m^3 s^{-1}$ and the Río Ullo 0.91 $m^3 s^{-1}$. The Río Oitabén is the only river regularly monitored, having an average discharge of 348 000 Hm^3 with a standard deviation of 134 000 Hm^3 and a range of 135 000–619 000 Hm^3 in the period 1970–1986 (i.e. a fluctuation of about 40–175% around the mean value).

Preliminary measurements of the sediment load conducted over 1.5 years during a period of fairly average discharge have shown that only small amounts of sediment are being carried in suspension today. Typical average values of suspended sediment concentration are 2.9 $mg\,l^{-1}$ for both the Río Oitabén and the Río Verdugo, 1.1 $mg\,l^{-1}$ for the Río Ullo and 225 $mg\,l^{-1}$ for the Río Redondela. Hence, in spite of being a river with one of the smallest water discharges, the Río Redondela is the main supplier of sediment to the Ensenada de San Simón today, albeit at highly variable levels. This pronounced variability in sediment discharge (120–2044 $mg\,l^{-1}$) appears to be caused by anthropogenic activities, as this river drains the most intensively farmed and industrialized area of the whole catchment. No data are available on the transport of bedload material. However, the rivers are generally clear, being floored by sand, gravel and boulders. The presence of sand on the river banks, above the average water level, clearly indictes that sand is carried during periods of high discharge. It appears that the present contribution of bedload material to the Ensenada de San Simón and the main Ría de Vigo is very small and probably episodic.

With the limited data set available at present, it is difficult to understand the apparent abundance of sediment in the Ría de Vigo and, especially, in the Ensenada de San Simón which has an extensive area of intertidal flats at its landward extremity. Possibly much of this sediment was supplied earlier when there was a different pattern of runoff and, today, this sediment is merely being redistributed by marine/estuarine processes with only minor additions of fresh sediment from the hinterland. Alternatively, sediment supply is extremely episodic so that the very limited records do not sufficiently reflect the true situation. Also, the present supply of sediment may be intercepted upstream, thus not reaching the lower parts of the rivers and the ría. Indeed, a dam was constructed on the Río Oitabén in 1977. Hence the data reported here may not be representative of the late Holocene history of the area. It is to be expected that erosion has generally increased over the last few centuries due to expanding anthropogenic activities, with perhaps a decrease in some areas due to the abandonment of fields and their return to scrub. Interestingly, Benito *et al.* (1991) claimed that there has been an increase in sediment runoff over the last 30 years.

SEDIMENTOLOGY OF THE RIA DE VIGO

The main Ría de Vigo is approximately 23 km long. It gradually narrows landwards from 6 km near the mouth to 0.6 km near the Estrecho de Rande. Its outer parts are sheltered from the open sea by the Islas Cíes, which divides the entrance into two channels (Fig. 1). Its depth in the outer parts, landwards of the Islas Cíes, decreases from 35 m to 25 m further inland. There is a current scoured trough of 30 m where the main Ría de Vigo passes through the narrows of the Estrecho de Rande into the Ensenada de San Simón.

In its entrance channels to the north and south of the Islas Cíes, the floor of the outer Ría de Vigo is covered with mixed siliciclastic and skeletal gravels (with a sandy or muddy component) which, along the axis of the ría, pass landwards into sands and clayey silts. Towards the shoreline the sediments become coarser, grading through various intermediate sediment types into clean carbonate skeletal sands or mixed carbonate siliciclastic sands. Elsewhere, particularly in the inner parts, fine-grained sediments persist up to the shoreline. Generally, the finer-grained sediments contain 4–5% organic carbon (Nombela, 1989). In some places, particularly in the outer parts of the ría, banks and blankets of the calcareous algae *Lithothamnion corralloides* and *Phymatolithon calcareum* are found, the remains of these algae being common in the sediments (Adey & McKibbin, 1970).

In the outer, more seaward parts of the Ría de Vigo, broad beaches backed by dunes infill many of the bays along the margins between low cliffs. Small estuaries are sometimes found, with sandy spits enclosing backbarrier marshes. By contrast, in the inner parts of the ría near the Estrecho de Rande, the beaches become narrower and cliffs are more common.

SEDIMENTOLOGY OF THE ENSENADA DE SAN SIMÓN

Beyond the Estrecho de Rande, the Ría widens to form the Ensenada de San Simón, a coastal embayment measuring, in N–S and E–W directions, 7 km and 4–5 km respectively. The Ensenada has an area of 19.4 km^2, of which 4 km^2 fall dry at low water (Fig. 5) (Nombela *et al.*, 1987a, 1990). It is >30 m deep in the Estrecho de Rande but rapidly shoals landwards to 10–15 m and, for the most part, is <5 m deep. The bay is fringed by a narrow muddy and sandy intertidal zone, often backed by narrow beaches and low cliffs. The intertidal zone broadens at the landward extremity of the ría where the Río Oitabén, Río Verdugo and Río Ullo enter the area. It is bordered by cliffs or rocky slopes with gentle profiles, except along the eastern shoreline where sandy beaches with low dunes and a sandy low-tide terrace backed by low cliffs have developed in response to the dominating south-westerly, and to a lesser extent, northerly winds. Several small rocky islands project above the surface of the waters of the Ensanada, the largest of which is the Isla de San Simón. In its shelter, a sandy tombolo has developed which almost links it to the shoreline. At the mouth of the Redondela river, in the south-east of the Ensenada, a sandy subtidal deltaic lobe has formed which almost falls dry at extreme low tide.

The rivers Río Oitabén, Río Verdugo, Río Ullo, and the small stream of the Río Junquera have infilled the landward part of the Ensenada de San Simón to form a tide-dominated, i.e. fluvio-deltaic bay-head delta. This feature forms the major part of the 4 km^2 intertidal area of the Ensenada as it is almost entirely exposed at low water, particularly at low water spring tides.

The intertidal complex is dominated by the estuarine channel of the Río Oitabén in the east, which is floored by medium-coarse siliciclastic sands fashioned into a series of seaward-facing bars and megaripples but with some large bedforms oriented landwards along the inner part of the estuary. At the seaward section the bedforms are modified by the flood tide but elsewhere retain their ebb orientation with the exception of the channel margins where flood-dominated bedforms prevail.

The estuarine channel is bordered by low sandy levees and muddy sand flats in the east, whereas in the west, sandy crevasse-splay deposits are fed with sediment during periods of high river discharge. Between such events the sediment is colonized by *Zostera nana* and scattered *Salicornia* sp. The remainder of the intertidal zone is formed by mud-flats covered by *Z. nana* and algae, and cut by small creeks and a sandy flat. The gastropods *Littorina obtusa* and *L. saxatilis* as well as *Hydrobia ulvae* are abundant on the muddy surface. *Scrobicularia plana* and *Nereis diversicolor* live within the sediment beneath the sea-grass and algal cover while crab burrows are frequent, particularly

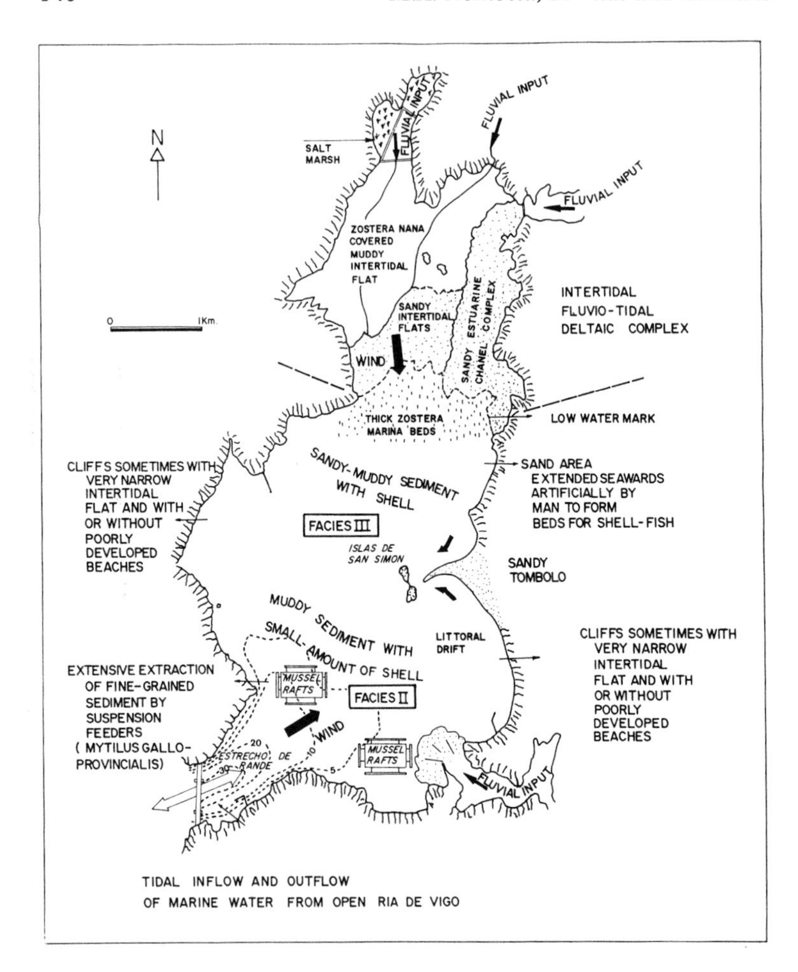

Fig. 5. The physiography, bathymetry and general pattern of sediment distribution in the Ensenada de San Simón (N.B. Facies I is not shown in the figure as it is only present in the subsurface; depths are in metres).

along the creek banks. The *Zostera*-covered surface is soft and impossible to traverse on foot. The intertidal zone is crossed by small estuarine channels of the Río Ullo and Río Junquera, both of which have sandy beds covered with dm-scale bedforms. Along the margin of the channels and creeks are small sand patches colonized by *Arenicola marina* which have been produced as spill-overs from the channels or as small point-bars at the bends. The *Zostera*-covered flats extend into the cliffed margin with the development of narrow, poorly defined beaches in some areas. In an enclosed area of the most landward parts very small patches of salt marsh are present; these have been largely embanked and, in unprotected areas, are clearly being eroded.

Seawards of the *Zostera*-covered intertidal flats are wide sand-flats which are rippled and often fashioned by the flood tide into a series of landward facing m-scale sand bars. They are relatively stable for much of the time but the smaller dm- to cm-scale bedforms show variable orientations. The sand-flats have abundant populations of *Cerastoderma edule* and *H. ulvae* while *A. marina*, *Lanice conchilega* and *Maldane glebiflex* are found on the borders of the channel. A once-dense population of *Ostrea edulis* is now reduced to a few individuals, possibly due to overexploitation (Arcade was once famous throughout Spain for its oysters).

Both the estuarine channel systems and the broad sand-flats pass seawards into a broad area of skeletal gravelly sands colonized by a dense carpet of *Z. marina* which extends landwards up the channels. This area of sand covers the whole of the inner subtidal part of the Ensenada de San Simón (Fig. 5). The inner part is exposed during equinoctial tides and there are large populations of *C. edule* and *Tellina tenuis*. The area is partly artificial as it has been extended by the dumping of sand from the estuarine channels to provide a suitable substrate for bivalve culture. Similar sands cover the subtidal delta of the Río Redondela and the area around Isla de San Simón, as well as the sandy tombolo (Fig. 5).

The subtidal sediments of the Ensenada de San Simón are composed of mud-grade sediment with admixtures of skeletal gravel and sand (Fig. 6) (Nombela *et al.*, 1987a, 1990). The coarser marginal sediment passes into skeletal gravelly-silty-sand and skeletal gravelly-sandy-silt, and ultimately into silty clay with occasional patches of clayey silt in the centre of the Ensenada. Towards the outer part, the sediments coarsen slightly to sand-silt-clays. The macrofauna is less abundant than in the sandy area of the inner Ensenada. *H. ulvae*, *Mytilus galloprovincialis*, *C. edule*, *C. minima*, *Ostrea edulis* (rare), *Hinia pygmaea*, *Alexia myosotis*, *Turboella radiata*, and *Pecten* sp. are some of the typical macrofauna present. There is a large population of vermes, prominent amongst which is the tube building form *Spiochaetopterus costarum*. Generally, the whole of the subtidal area of the Ensenada de San Simón is inhabited by the *S. costarum* community which López-Jamar & Cal (1990) have shown to dominate the inner parts of the main Ría de Vigo.

The gravel fraction of the subtidal sediment, which represents as much as 30% of the total in some places, contains less than 5% siliciclastic material. Instead, it consists mainly of weed fragments coated with sediment, articulated or single valves or fragments of *C. edule* with smaller amounts of other bivalves, whole specimens of various gastropods, notably *H. ulvae*, agglutinated

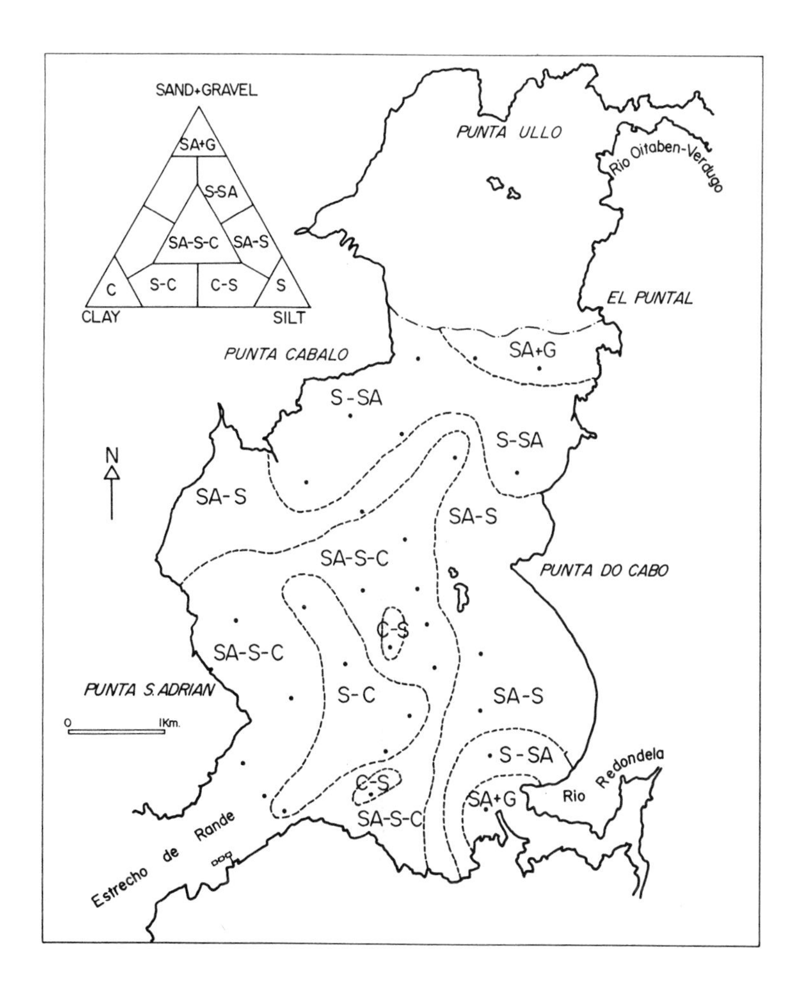

Fig. 6. Grain sizes of the surficial sediments.

worm tubes, rare echinoid species and fish bones. This skeletal debris is rarely fresh but is usually bored and in some cases only fragile lacework of shell remains. Coarse angular quartz, feldspar together with lithic fragments and mica flakes make up the siliciclastic fraction of the gravels.

The sand fraction is mainly composed of siliciclastic material, mainly quartz, feldspar and mica (usually colourless but also in brown and green varieties) together with some lithic grains and dark minerals. Mixed with these minerals are variable quantities of faecal pellets composed of silt and clay. Pellets sometimes dominate the sand fraction, particularly in the finer-grained sediments near the seaward extremity of the Ensenada de San Simón close to the Estrecho de Rande. They are almost certainly produced by the large population of vermes (see López-Jamar & Cal (1990) for a description of the various species in a similar type of environment in the inner part of the main Ría de Vigo).

There is a considerable amount of broken skeletal debris in the sand fraction, probably produced by biological breakdown of bivalve and gastropod shells. Bivalves are rarely articulated and single valves are more common; some small whole gastropods are present. Foraminifers are commonly present in all samples of this fraction, both as calcareous perforate forms and particularly as agglutinated forms with only a few calcareous porcelainous forms. Ostracods are present but are not particularly common.

A preliminary examination of the microfauna has shown it to be of low diversity (Radford & Robinson, personal communication), the results agreeing with earlier studies by Colom (1952, 1963) and Mateus (unpublished). The foraminifers are dominated by the calcareous perforate forms *Ammonia beccarii* and *Nonion boueanum* with large numbers of the agglutinated form *Eggerella scabra* and also a few calcareous porcelainous forms of Miliolidae and Lituolidae. The ostracod assemblages are also impoverished. Faunal diversity is thus noticeably poorer in the ría than on the adjacent shelf.

Echinoid species and plates are rare. In addition sponge spicules, fish bones and fragments of crustacean carapaces occur. Diatoms are present in almost all samples. As in the gravel fraction, there is a considerable amount of weed fragments, charred wood and other vegetal matter, sometimes with fragments of a brown earthy material which is probably burnt clay.

Around the margins of the Ensenada and around the margins of the intertidal flats in the north as well as near the Río Redondela mouth in the south-east, there is a noticeable increase in the amount of siliciclastic debris (quartz, feldespar and mica) and a decrease in skeletal debris and pellets.

The clay fraction of the sediment has been studied by Vázquez & Calvo de Anta (1988), and the faecal mud produced by *Mytilus galloprovincialis* by Macías *et al.* (1991). Samples collected during the present study show that the clay (<2 μm fraction) consists mainly of illite (40–60%), smectite (25–35%), kaolinite (15–25%) and small amounts of chlorite (<10%). Neither gibbsite nor vermiculite were recorded although they have been reported by Vázquez & Calvo de Anta (1988) and Macías *et al.* (1991).

The calcium carbonate contents of the subtidal sediments vary from 0.3 to 36.3%, with lowest values being found in the finer-grained sediments (Fig. 7). Generally, contents are less than 3% for most of the area, increasing in the sandy sediments of the northern Ensenada, particularly in the *Zostera*-covered sands adjacent to the bay-head delta where the highest value (36.3%) was recorded. A smaller increase in the carbonate content occurs in the proximity of the Estrecho de Rande, where values of 2.9–9.1% have been found. The carbonate is entirely of biogenic origin as the incoming fluvial sediment has very low $CaCO_3$ contents.

The sediments are relatively rich in organic matter (Nombela *et al.*, 1987a,b, 1990; Nombela, 1989), similar values having been reported from neighbouring areas (López-Benito, 1966; Koldijk, 1968; López-Jamar & Cal, 1990).

Contents of organic carbon vary from 0.4 to 6.5%, with most of the sediments except the sands containing between 4 and 5% organic carbon (Fig. 8). As commonly observed in marine sediments, contents increase with decreasing grain size, the exact opposite to the relationship between $CaCO_3$ content and grain size.

Contents of total organic nitrogen are usually ± 0.4%, higher values occurring near the Estrecho de Rande. The carbon/nitrogen ratios vary from 3.0 to 13.3, highest values occurring in the sediments near the Estrecho de Rande and also on the deltaic lobe at the mouth of the Río Redondela. The lowest ratios are present near the mouth of the estuary of the Río Oitabén at the head of the Ensenada. Organic matter is transported to the rías from the heavily vegetated hinterland and this is supple-

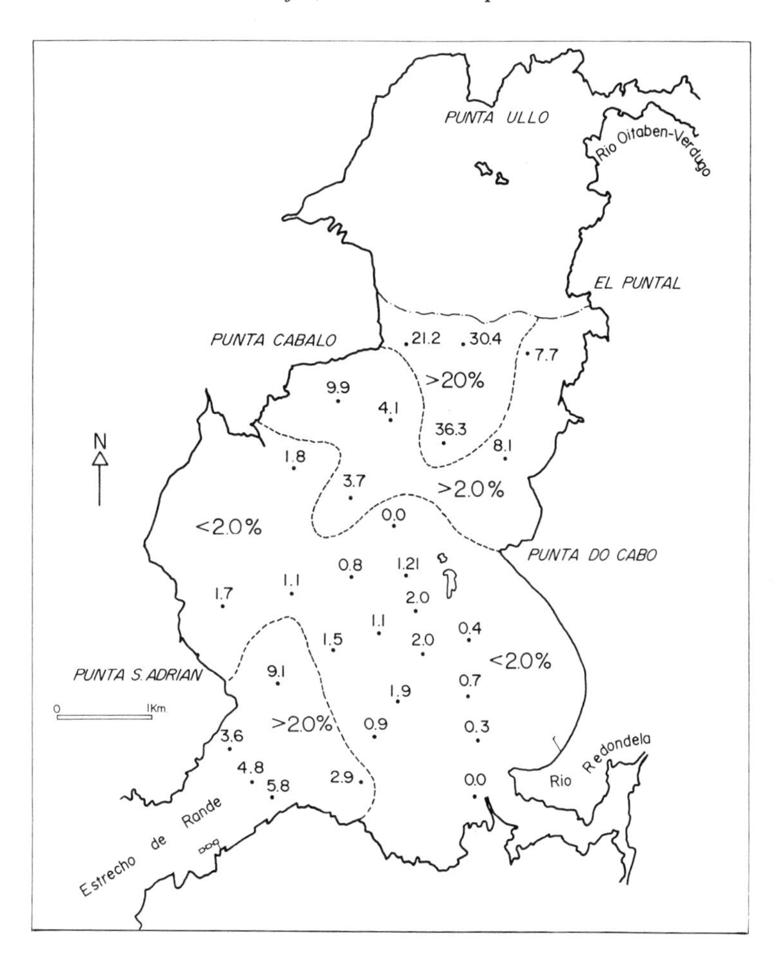

Fig. 7. Carbonate contents (percentage $CaCO_3$) of the surficial sediments.

mented by *in situ* production of algae and sea-grass as well as by additions from the open sea, notably in the form of algal blooms.

STRATIGRAPHY OF THE ENSENADA DE SAN SIMÓN

The outer part of the Ensenada de San Simón is covered with a layer of muddy sediment which is diffusely stratified (laminated) with some bioturbation and markedly low contents of skeletal debris (Fig. 9b). These sediments have been designated as Facies II. Towards the bay-head fluviotidal delta at the head of the Ensenada, these pass into coarser sediments with more sand and silt layers and a coarser stratification with some ripple cross-lamination, bioturbation and lenses of shell. These sediments contain abundant articulated, single and broken skeletal remains of various molluscs, *C. edule* being the dominant species. These sediments have been designated Facies III (Fig. 9c).

The surface sediments of Facies II and (in part) Facies III are underlain at shallow depths (i.e. <2 m) by muddy sediment with a better developed stratification (lamination) (Fig. 9a) than that of the overlying material, including some ripple cross-lamination. This sediment contains small remains of bivalves and gastropods, and has been designated Facies I. Microfaunal debris is often concentrated in shelly layers and lenses, presumably formed during storms.

The skeletal debris is noticeably finer than that found in Facies III sediments, *C. edule* being rare but occurring occasionally in the form of small individuals. Towards the landward head of the Ensenada some coarse sandy layers appear in Facies I, as in the case of the overlying coarser sediments of

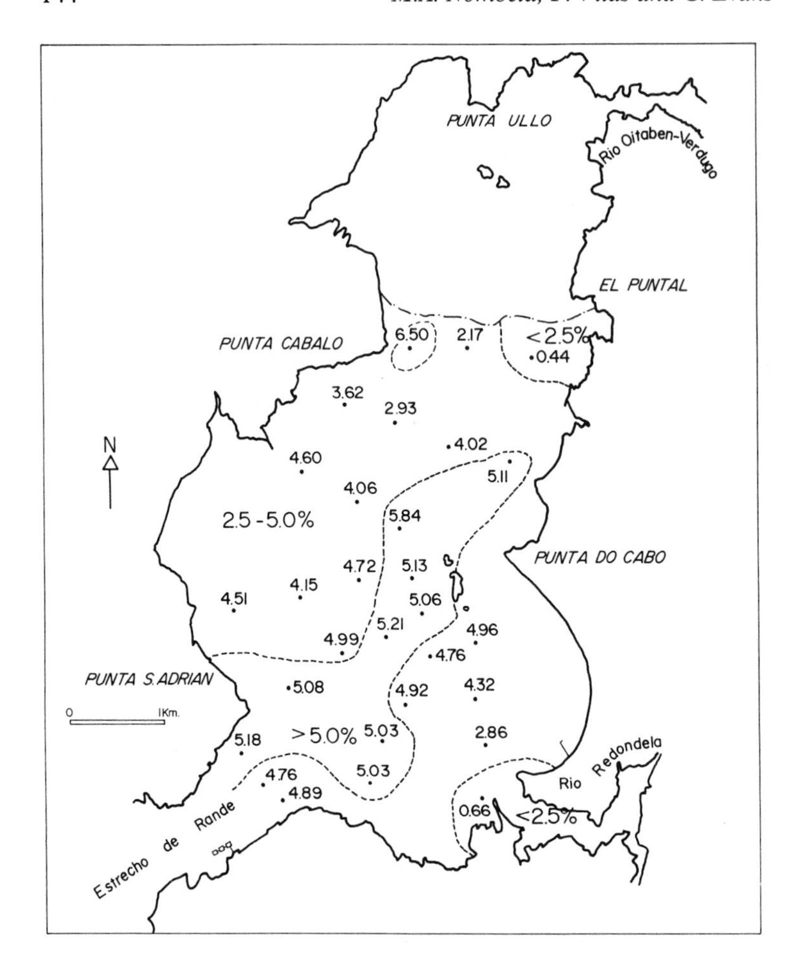

Fig. 8. Organic carbon contents (percentage) of the surficial sediments.

Facies III. The sediments appear to be transitional between Facies I and Facies III.

At all depths in the outer parts of the Ensenada de San Simón, the sediments abound in structures apparently related to gas cavities (Fig. 9b). These are probably produced by methane and/or possibly hydrogen sulphide generated by the breakdown of organic matter. Evidence for the generation of gas from organic matter in the sediments of other Galician rías has been presented by Acosta (1981).

The three-dimensional architecture of the various facies in the subsurface indicates that the finely laminated sediment containing some skeletal debris (Facies I) has been replaced with time in the seaward parts of the Ensenada de San Simón by fine-grained diffusely laminated sediments containing less skeletal remains (Facies II) and, in the more landward parts of the Ensenada, by the coarser and more crudely stratified skeletal-rich sediments of Facies III (Fig. 10).

The changes from one facies to another in the seaward part of the Ensenada de San Simón is probably due to anthropogenic interference, including the introduction of mussel rafts. These are platforms from which ropes are suspended on which *M. galloprovincialis* is commercially farmed. Such massive concentrations of *Mytilus* filter a great deal of fine, suspended sediment from the water, depositing it in the form of faecal matter on the underlying sea-floor from where it may be distributed over the Ensenada by tidal currents. It has been claimed that these organisms produce approximately 190 kg of faecal mud per raft per day (Cabanas *et al.*, 1979) Since there are 100–150 rafts installed at any one time, this means that the total amount of faecal mud deposited must be in the order of $19.0 * 10^3$ to $28.5 * 10^3$ kg per day. This is

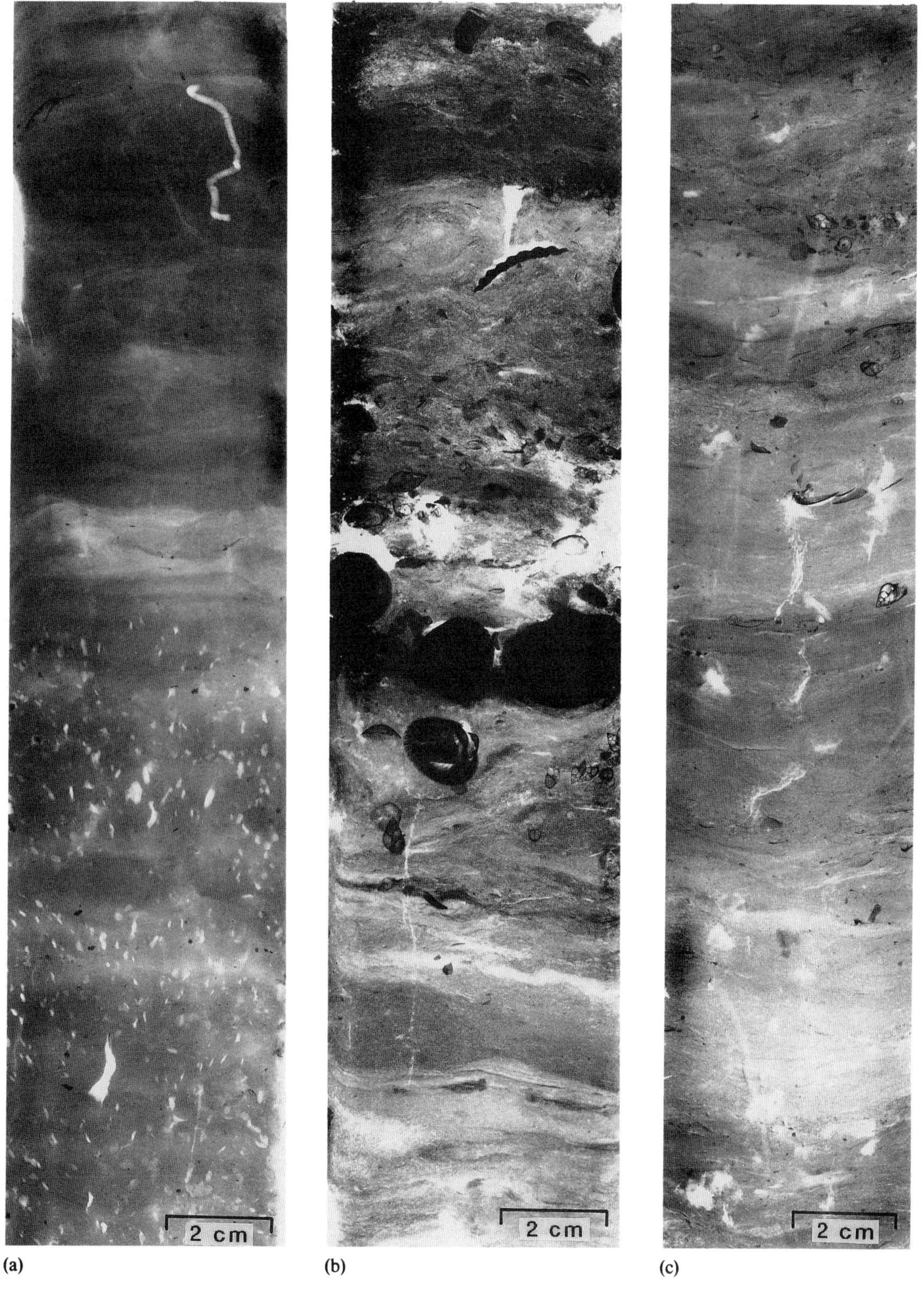

Fig. 9. Stratigraphy of the Ensenada de San Simón: (a) a typical core of Facies I; (b) a typical core of Facies II (also showing gas bubbles); (c) a typical core of Facies III (see text for a detailed description of the sediments).

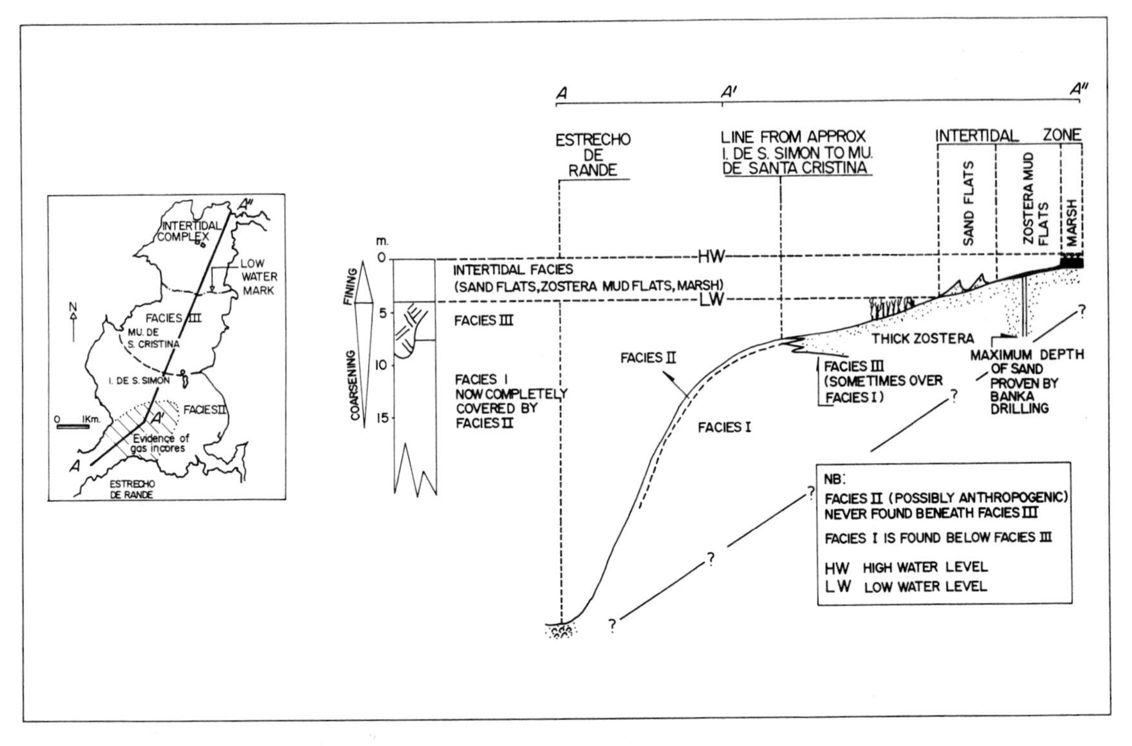

Fig. 10. Distribution of the sedimentary facies in the Ensenada de San Simón (see text for a description of the various facies) (N.B. Facies I does not appear on the surface and is therefore not shown in the inset map at the left of the figure).

comparable to the estimated average amount of suspended matter (based on 1.5 years of measurements) supplied to the Ensenada de San Simón by the local rivers (i.e. $23.5 * 10^3$ kg per day). Hence, it would appear that *M. galloprovincialis* is at present able to extract most of the suspended sediment supplied to the Ensenada de San Simón.

The introduction of similar mussel rafts in the neighbouring Ría de Arosa has also led to the deposition of a large amount of fine-grained mud over the floor of the ría (Tenore *et al.*, 1982; Blanton *et al.*, 1987). On the other hand, it cannot be excluded that man-induced changes in the hinterland and modified vegetation patterns have also contributed to the observed vertical facies changes in the Ensenada de San Simón.

Furthermore, the superposition of the coarser sandy, more crudely stratified sediments rich in skeletal remains (Facies III) on the finer laminated sediments poor in remains of *C. edule* (Facies I) is due in part to the natural progradation of the fluviotidal bay-head delta at the landward end of the Ensenada de San Simón. However, this seaward extension of sandy sediment (Facies III) has undoubtedly been promoted by artificial dumping of sand on the seaward edge of the intertidal flats bordering the bay-head delta, which provides a suitable substrate for the culture of *C. edule* and other bivalves.

CONCLUSIONS

The Ensenada de San Simón is a sheltered mesotidal coastal embayment on an otherwise exposed wave-dominated coastline. On its landward side a bay-head delta has developed which shows a similar pattern of sand-flats, mud-flats and marshes cut by estuarine channels as is commonly found in other intertidal flats of the world. However, it differs from many other intertidal flat areas, including those of northern Europe, in two aspects. Firstly, the *Zostera*-covered mud-flats are very extensive and secondly, salt marshes are very poorly developed.

The muds of the intertidal mud-flats do not appear to loose their interstitial water to become compacted, presumably because of their small grain size and the relatively small tidal range which enable them to drain completely. Furthermore, the *Z. nana*-covered mud-flats are not being invaded by the usual sequence of halophytic marsh plants. Whether this is due to some ecological factor or to the limited supply of sediment is presently unknown. The fact that there are so few salt-marshes, and that those that are present are being eroded in areas not completely protected by artificial embankments, suggests that sediment supply is indeed very limited. As a result, marsh development may not be able to keep pace with the contemporary rise in sea-level.

The subtidal areas of the Ensenada are the depositional sites of fine-grained, muddy sediment rich in organic matter, containing a low diversity of macro- and microfauna. Along the margins and in the areas adjacent to the bay-head delta, coarser skeletal-rich sediments are accumulating. These pass seawards into sediments which show poorly developed or diffuse lamination and are lacking in current-generated sedimentary structures. Furthermore, the tidal influences are poorly preserved in the subtidal sediments. If these were preserved in the rock record, it would be difficult to recognize them as having been formed in a tide-dominated environment unless the adjacent intertidal marginal and bay-head delta deposits were also preserved.

Finally, man has had an important influence on the pattern of sedimentation: firstly, by artifically extending the sand-flats at the head of the Ensenada de San Simón, a consequence of the dumping of sand for commercial shell fish cultures; secondly, by promoting the deposition of large quantities of muddy, organic-rich sediment via the large-scale installation of mussel rafts for *M. galloprovincialis* near the entrance of the Ensenada de San Simón.

On the basis of rather limited data it would appear that at present very little sediment is being supplied to the Ensenada de San Simón or to the main part of the Ría de Vigo. Furthermore, most of the suspended matter entering the system appears to be filtered by cultured mussel to produce extensive mud deposits in the Ensenada de San Simón. It is rather unlikely that a substantial amount of material can escape into the outer main ría. Indeed, any material that does escape will probably be filtered by the abundant *Mytilus* colonies in this area. As there are only a few small streams discharging into the outer main Ría de Vigo, this indicates a generally low supply of modern muds.

Furthermore, with the exception of some carbonate and also organic matter, it is difficult to envisage a seaward source for the mud. In order to explain the accummulation of at least several metres of fine sediments in the Ría de Vigo including the Ensenada de San Simón, it must be assumed that the muds were deposited under different environment conditions sometime in the past.

The presence of a large urban centre such as Vigo has obviously affected the local sediments and fauna. However, according to López-Jamar & Cal (1990), the Ría de Vigo is less affected than the Ría de Pontevedra and the Ría de Arosa and has conditions more similar to those found in Ría de Muros. Nevertheless, preliminary geochemical studies by the authors have shown a marked increase in heavy metals in the sediments of the Ensenada de San Simón over the last 30 years.

ACKNOWLEDGEMENTS

We acknowledge the support of the Comisión Interministerial de Ciencia y Tecnología (CICYT, projects: NAT 89–1075 and AMB 93–0300). Also, we wish to thank the boat crew Nucho, Basilio and Fernando. I. Alejo (Univ. de Vigo) and A. Arche (CSIC) helped with the sampling. We are grateful to M.A. Ahmed, M. Gill and Z. Ali of the Imperial College of London for the laboratory analyses. G. Evans wishes to thank the Dpto. de Estratigrafía de la Univ. Complutense. Two anonymous reviewers are thanked for their comments which greatly improved the manuscript.

REFERENCES

Acosta, J. (1981) Apantallamientos acústicos en la Ría de Muros y Noya en la margen continental. *Bol. Inst. Esp. Oceanogr.* **7**, 127–149.

Adey, W.H. & McKibbin, D.L. (1970) Studies on the maerl species *Phymatolithon calcareum* (Pallas) nov. comb. and *Lithothamnion coralloides* (Croan) in the Ría de Vigo. *Bot. Mar.* **13**, 100–106.

Alejo, I. & Vilas, F. (1987) Ambientes sedimentarios de la Ensenada de Baiona (Pontevedra). *Cuad. Lab. Xeol. de Laxe* **12**, 11–24.

Alejo, I., Ramon, M.I. de, Nombela, M.A., Reigosa, M.J. & Vilas, F. (1990a) Complejo intermareal de la Ramallosa (Bahía de Baiona, Pontevedra). 1 Ecología y evolución. *Thalassas* **8**, 45–56.

Alejo, I., Nombela, M.A., Vilas, F., Ferrero, M., Garcia-Rodeja, X. & Evans, G. (1990b) Mixed carbonate-siliciclastic sedimentation in Bahía de Baiona, Ría de Vigo, Calicia, NW Spain. *Abstr. I.A.S. Meeting, Nottingham, UK*, 14–15.

Anadón, R. (1977) *Estudio ecológico de la playa de Foz en el NW de España durante los años 1973–74.* Tesis doctoral Univ. Complutense de Madrid.

Anadón, R. (1980) Estudio ecológico de la macrofauna del estuario de Foz (NO de España). 1 Composición, estructura, variación estacional y producción de las comunidades. *Inv. Pesq.* **44**, 407–444.

Anadón, E., Saiz, F. & López-Benito, M. (1961) Estudio hidrográfico de la Ría de Vigo (3). *Inv. Pesq.* **20**, 83–130.

Benito, E., Soto, B. & Diaz-Fierros, F. (1991) Soil erosion studies in NW Spain. In: *Soil Erosion Studies in Spain* (Eds Sala, M., Rubio, J.L. & García-Ruiz, J.M.), pp. 55–74. Geoforma, Spain.

Birot, P. & Solé-Sabaris, L. (1954) Recherches morphologiques dans le Nord-Ouest de la Péninsule Ibérique. *Mén. Docums. Cent. Nat. Rech. Scinet.* **4**, 7.

Blanton, J.O., Tenore, K.R., Castillejo, F., Atkinson, L.P., Schwing, F.B. & Lavin, A. (1987) The relationship of upwelling to mussel production in the rías on the western coast of Spain. *J. mar. Res.* **45**, 497–511.

Boillot, G. & Malod, J. (1988) The north and north west Spanish continental margin: a review. *Rev. Soc. Geol. España* **1 (3–4)**, 295–316.

Boucart, J. (1938) La marge continentale. Essai sur les transgressions et régressions marines. *Bull. Géol. Soc. Fr.* **8 (5a)**, 393 pp.

Cabanas, J.M., González, J.J., Mariño, J., Pérez, R. & Román, G. (1979) Estudio del mejillón y su eficacia en los cultivos flotantes de la Ría de Arosa. III Observaciones previas sobre la retención de partículas y la bio-deposición de una batea. *Bol. Inst. Esp. Oceanogr.* **5 (268)**, 44–50.

Cadée, G.C. (1968) Molluscan biocenoses and thanatocenoses in the Ría de Arosa, Galicia, Spain. *Zöol. Verh. Rijksmus. Nat. Hist. Leiden* **95**, 1–121.

Carle, W. (1947) Las Rías Bajas gallegas. Est. Geogr. **35**, 323–330.

Colom, G. (1952) Los foraminíferos de las costas de Galicia. *Bol. Inst. Esp. Oceanogr.* **51**, 1–59.

Colom, G. (1963) Los foraminiferos de la Ría de Vigo. *Inv. Pesq.* **23**, 71–89.

Durán, M., Saiz, F., López-Benito, M. & Margalef, R.H. (1956) El fitoplancton de la Ría de Vigo, de Abril de 1954 a Junio de 1955. *Inv. Pesq.* **4**, 67–95.

Emery, K.O. & Aubrey, D.G. (1991) *Sea Levels, Land Levels and Tide Gauges.* Springer-Verlag, Berlin, 245 pp.

Fraga, F. (1960) Variación estacional de la materia orgánica suspendida y disuelta en la Ría de Vigo. Influencia de la luz y la temperatura. *Inv. Pesq.* **18**, 127–140.

Fraga, F. (1967) Hidrografia de la Ría de Vigo (1962), con especial referencia a los compuestos de nitrógeno. *Inv. Pesq.* **31 (1)**, 145–159.

Fraga, F. (1981) Upwelling off the Galician coast, NW Spain. In: *Coastal Upwelling* (Ed. Richards, F.A.), pp. 176–182. Am. Geophys. Union, Washington DC.

Jong, J.D. de & Poortman, H.H. (1970) Coastal sediments of the south eastern shores of the Ría de Arosa, Galicia, NW Spain. *Leidse. Geol. Meded.* **37**, 147–167.

Koldiijk, W.S. (1968) Bottom sediments of the Ría de Arosa, Galicia, NW Spain. *Leidse. Geol. Meded.* **37**, 77–134.

López-Benito, M. (1966) Variación estacional del contenido de materia orgánica en las arenas de la Playa de Areiño (Ría de Vigo). *Inv. Pesq.* **30**, 233–246.

López-Jamar, E. & Cal, R.M. (1990) El sistema bentónico de la zona submareal de la Ría de Vigo. Macrofauna y microbiología del sedimento. *Bol. Inst. Esp. Oceanogr.* **6 (2)**, 49–60.

Macias, F., Fernández De Landa, J.L.A. & Calvo De Anta, R. (1991) Composición química y mineralógica de biodepósitos bajo bateas de mejillón. Datos para la evaluación de su uso como fertilizante y/o enmendante de suelos de Galicia. *Thalassas* **9**, 23–29.

Margalef, R. (1956a) Paleoecología postglacial de la Ría de Vigo. *Inv. Pesq.* **5**, 89–112.

Margalef, R. (1956b) Estructura y dinámica de la purga de mar en la Ría de Vigo. *Inv. Pesq.* **5**, 113–134.

Margalef, R. (1958) La sedimentación orgánica y la vida en los fondos fangosos de la Ría de Vigo. *Inv. Pesq.* **11**, 67–100.

Mateu, G. (undated) *Unos datos y unas observaciones micropaleontológicas sobre las rías de Galicia.* 13 pp. (unpubl.).

Mensching, H. (1961) Die Rías der Galicisch-Asturischen Küste Spaniens. *Erdkunde* **15**, 210–224.

Mora, J. (1980) *Poblaciones bentónicas de la Ría de Arosa.* Tesis doctoral Univ. Santiago de Compostela.

Nombela, M.A. (1989) *Oceanografía y sedimentología de la Ría de Vigo.* Tesis doctoral Univ. Complutense de Madrid.

Nombela, M.A. & Vilas, F. (1986–87) Medios y submedios en el sector intermareal de la Ensenada de San Simón, Ría de Vigo (Pontevedra): secuencias sedimentarias características. *Acta Geol. Hisp.* **21–22**, 223–231.

Nombela, M.A. & Vilas, F. (1990) Procesos sedimentarios actuales en las playas de la Ría de Vigo. *Thalassas* **8**, 11–21.

Nombela, M.A., Evans, G. & Vilas, F. (1987a) La cuenca de San Simón: una cuenca marina interior, Galicia, NO España. *Abstr. VII Reunión EAQUA, Santander, España*, 237–238.

Nombela, M.A., Vilas, F., Rodriguez, M.D. & Ares, J.C. (1987b) Estudio sedimentológico del litoral gallego III Resultados previos sobre los sedimentos de los fondos de la Ría de Vigo. *Thalassas* **5**, 7–19.

Nombela, M.A., Vilas, F., Alejo, I., Ferrero, M., Garcia-Rodeja, X. & Evans, G. (1990) Tidally dominated 'bay-head' deltaic sedimentation. Ensenada de San Simón, Ría de Vigo, Galicia, NW Spain. *Abstr. I.A.S. Meeting, Nottingham, UK*, 166–167.

Nonn, H. (1958) Contribución al estudio de las playas antiguas de Galicia (España). Notas y Comun. *Int. Geol. Min. Esp.* **50**, 175–193.

Nonn, H. (1966) *Les régions cotières de la Galice, Espagne. Étude géomorphologique.* Thése doctorale Univ. Strasbourg.

Nonn, H. & Tricart, J. (1960) Étude d'une formation périglaciaire ancienne en Galice. *Bull. Soc. Géol. Fr.* **2**, 41–52.

PANNEKOEK, A.J. (1966) The ría problem. *Tijdschr. K. Ned. Aardrijksk. Genoot.* **83 (3)**, 289–297.

PANNEKOEK, A.J. (1970) Additional geomorphological data on the Ría de Arosa area of W Galicia (Spain). *Leidse. Geol. Meded.* **37**, 185–194.

PÉREZ-ALBERTI, A. (1982) *Xeografía de Galicia.* Ediciones Sálvora, La Coruña, España, 210 pp.

PREGO, R. & FRAGA, F. (1992) A simple model to calculate the residual flows in a Spanish ría. Hydrographic consequences in the Ría de Vigo. *Estuar. Coastal Shelf Sci.* **34**, 603–615.

PREGO, R., FRAGA, F. & RIOS, A.F. (1990) Water interchange between the Ría of Vigo and the coastal shelf. *Sci. Mar.* **54 (1)**, 95–100.

RICHTHOFEN, F. VON (1886) *Führer für Forchungsreisende.* Oppenheim, Berlin, 745 pp.

RÍOS, A.F., NOMBELA, M.A., PÉREZ, F.F., ROSÓN, G. & FRAGA, F. (1992) Calculation of run-off to an estuary Ría de Vigo. *Sci. Mar.* **56 (1)**, 29–33.

ROLÁN, E. (1983a) Moluscos de la Ría de Vigo. I Gasterópodos. *Thalassas* Anexo **1 (1)**, 383 pp.

ROLÁN, E. (1983b) *Moluscos Gasterópodos de Galicia.* Univ. Santiago de Compostela, 105 pp.

ROLÁN, E., OTERO, J. & ROLÁN, E. (1989) Moluscos de la Ría de Vigo. II Poliplacóforos, Bivalvos, Escafópodos y Cefalópodos. *Thalassas* Anexo **2**, 276 pp.

SAA-OTERO, M.P. & DÍAZ-FIERROS, F. (1988) Contribución al estudio paleobotánico mediante análisis de polen. *Estud. Geol.* **44**, 339–349.

SAIZ, F., LÓPEZ-BENITO, M. & ANADÓN, E. (1957) Estudio hidrográfico de la Ría de Vigo (1). *Inv. Pesq.* **8**, 29–87.

SAIZ, F., LÓPEZ-BENITO, M. & ANADÓN, E. (1961) Estudio hidrográfico de la Ría de Vigo (2). *Inv. Pesq.* **18**, 97–133.

SALA, M. (1984) Iberian massif. In: *Geomorphology of Europe* (Ed. Embleton, C.), pp. 294–320. Macmillan, London.

TENORE, K.R. & GONZALEZ, N. (1975) Food chain patterns in the Ría de Arosa, Spain: an area of intensive mussel aquaculture. *10th Europ. Mar. Biol. Symp.* (Eds Persoon, G. & Jasper, E.), pp. 601–619. Univers Press, Wetteren.

TENORE, K.R., BOYER, L.F., CAL, R.M. *et al.* (1982) Coastal upwelling in the Ría Bajas, NW Spain: Contrasting benthonic regimes of the Ría de Arosa and Muros. *J. mar. Res.* **40 (3)**, 701–772.

TORRE-ENCISO, E. (1958) Estado actual del conocimiento de las rías gallegas. In: *Libro en Homenaxe a Ramón Otero Pedrayo* (Editorial Galaxia, Vigo), pp. 237–250.

VARELA, M., FUENTES, J.M., PENAS, E. & CABANAS, J.M. (1984) Producción primaria de las Ría Bajas de Galicia. *Cuadernos da Area de Cieniás Mariñas, Sem. Estud. Galegos* **1**, 137–182.

VAZQUEZ, F.M. & CALVO DE ANTA, R. (1988) Arcillas y limos de sedimentos actuales de las Rías de Galicia. Consideraciones genéticas. *Geo. Sci. Aveiro* **3 (1–2)**, 179–187.

VIEITEZ, J.M. (1976) Ecología de poliquetos y moluscos de la Playa de Meira (Ría de Vigo). 1 Estudio de las Comunidades. *Inv. Pesq.* **40**, 223–248.

VIEITEZ, J.M. (1978) *Comparación ecológica de las Playas de Pontevedra y Vigo.* Tesis doctoral Univ. Complutense de Madrid.

VIEITEZ, J.M. (1979) Estudios de las comunidades bentónicas de las playas de las rías de Pontevedra y Vigo (Galicia, España). *Bol. Inst. Esp. Oceanogr.* **6 (4)**, 241–258.

VILAS, F. (1978) La sedimentación intermareal: ejemplo de la Ramallosa (Pontevedra). *Estud. Geol.* **34**, 535–541.

VILAS, F. (1979) *Características de sedimentación costera actual en las rías de Vigo y Bayona, Pontevedral.* Tesis doctoral Univ. Complutense de Madrid.

VILAS, F. (1981a) Las zonas intermareales: algunos mecanismos sobre la formación de estructuras sedimentarias intermareales. *Cuad. Lab. Xeol. de Laxe* **2**, 305–314.

VILAS, F. (1981b) Desplazamiento lateral de los canales de drenaje de las llanuras de marea: consideraciones sedimentológicas. *Cuad. Lab. Xeol. de Laxe* **2**, 315–322.

VILAS, F. (1981c) Evolución sedimentaria de la llanura intermareal de la Ramallosa (Pontevedra). *Cuad. Lab. Xeol. de Laxe* **2**, 209–217.

VILAS, F. (1983) Medios sedimentarios de transición en la Ría de Vigo: secuencias progradantes. *Thalassas* **1**, 49–55.

VILAS, F. & NOMBELA, M.A. (1985) Las zonas estuarinas de las costas de Galicia y sus merlios asociados, NW de la Península Ibérica. *Thalassas* **3 (1)**, 7–15.

VILAS, F. & ROLÁN, E. (1985) Caracterización de las lagunas costeras de Galicia, NW Península Ibérica. España. *Reunión del Cuaternario Ibérico* **1**, 253–268.

VIVES, F. & LÓPEZ-BENITO, M (1958) El fitoplancton de la Ría de Vigo y su relación con los factores térmicos y energéticos. *Inv. Pesq.* **13**, 87–125.

VOORTHUYSEN, J.H. VAN (1973) Foraminiferal ecology in the Ría de Arosa, Galicia, Spain. *Zool. Verhand.* **123**, 1–68.

YOUNG, G.A. (1969) Present rate of land erosion. *Nature* **224**, 851–852.

Spec. Publs int. Ass. Sediment. (1995) **24**, 151–170

Holocene estuarine facies along the mesotidal coast of Huelva, south-western Spain

J. BORREGO, J.A. MORALES *and* J.G. PENDON
Universidad de Huelva, Dpto de Geologia, Facultad Ciencias Experimentales, 21819 Palos Fra., Huelva, Spain

ABSTRACT

Along the wave-dominated Huelva coastline, south-west Spain, several types of mesotidal estuaries developed in the course of the Holocene transgression. Three different estuarine systems have been distinguished: (i) the drowned Guadiana fluvial valley; (ii) the Piedras estuarine lagoon; and (iii) the Odiel barrier estuary. The estuarine sequences include a number of depositional facies identified in field surveys and from sediment cores. They are grouped into an inner and an outer facies association. The inner facies association includes channels, lateral fluvial bars, active channel margins, salt marshes, sterile marshes and protected estuarine beaches. By contrast, the outer facies association incorporates sand spits, ebb-tidal deltas, open beaches and aeolian dunes. The individual facies are closely associated with critical tide levels (CTLs). The spatial distribution of the outer facies association along the three estuary mouths varies with respect to local sediment supply and process/product balance.

INTRODUCTION

General aspects

The factors controlling the physiographic evolution and depositional history of tidal environments vary from place to place. Depositional facies produced in estuarine environments are affected by, on the one hand, source-related factors and variable energy conditions such as erosional and depositional processes and, on the other hand, biological influences such as biogenic sediment production and bioturbation. For these reasons estuarine facies are often difficult to recognize in the field.

The estuaries located on the Huelva coast along the northern Gulf of Cadiz, south-western Iberian Peninsula (Fig. 1), were studied between 1988 and 1992. The results presented in this paper contribute to the refinement of facies models of mesotidal estuaries by relating the observed depositional facies to specific energy levels and associated littoral environments.

Previous work

Investigations of estuarine sedimentary environments in macro- and mesotidal settings of temperate zones include the Gironde estuary (Allen *et al.*, 1973), Willapa Bay (Clifton, 1982, 1983), the Georgia coast (Oertel, 1973, 1975; Greer, 1975; Howard & Frey, 1985; Frey & Howard, 1986), Delaware Bay (Knebel *et al.*, 1988), the Wadden Sea (Sha, 1990) and the Piedras estuary along the Huelva coast (Borrego *et al.*, 1993). Microtidal estuaries, by contrast, are less well documented, e.g. the James estuary (Nichols *et al.*, 1991).

Papers dealing with the Huelva coastal systems have focussed on coastal morphology (Vanney & Menanteau, 1979; Rodriguez, 1987a,b; Flor, 1990), beach processes (Dabrio, 1982; Dabrio & Zazo, 1987; Dabrio & Polo, 1987) and the effects of construction works along estuary shores (Borrego & Pendon, 1988, 1989a). Estuarine deposition in the Odiel and Guadiana river mouth regions has been studied by Borrego (1992) and Morales (1993), respectively, whereas the modern evolution of the Piedras estuarine lagoon has recently been investigated by Borrego *et al.* (1993).

HYDRODYNAMIC SETTING

The Huelva coast is situated in a warm temperate climate. The tidal wave propagates from east to

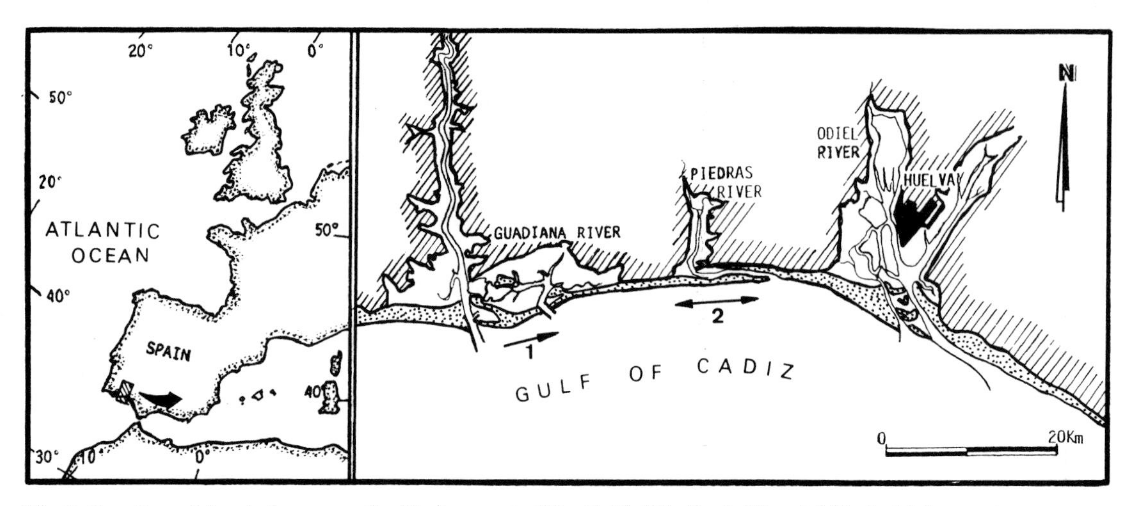

Fig. 1. Location of the study area on the Huelva coast of the Gulf of Cadiz. **1**, littoral drift; **2**, tidal currents.

west along the Gulf of Cadiz (from the Strait of Gibraltar towards the coast of Portugal), the tidal range reaching maximum values along the Huelva coast (Zazo, 1980) where a mean amplitude of 2.5 m is attained (2.8 m and 1.15 m at spring and neap tides respectively). From here it decreases gradually towards the east to reach minimum values of 1.8 m at spring tide and 0.4 m at neap tide in the Strait of Gibraltar (Ojeda, 1988).

The tidal regime is mesotidal and semi-diurnal with a slight diurnal inequality (Fig. 2). The maximum difference in tidal range between spring and neap tides is 2 m (Borrego *et al.*, 1993). The characteristics of the tidal wave along the outer littoral zones of the three estuaries allow the distinction of a variety of different critical tide levels (CTLs) *sensu* Doty (1946). These tide levels, as defined in Fig. 2, are a consequence of variations in the tidal range

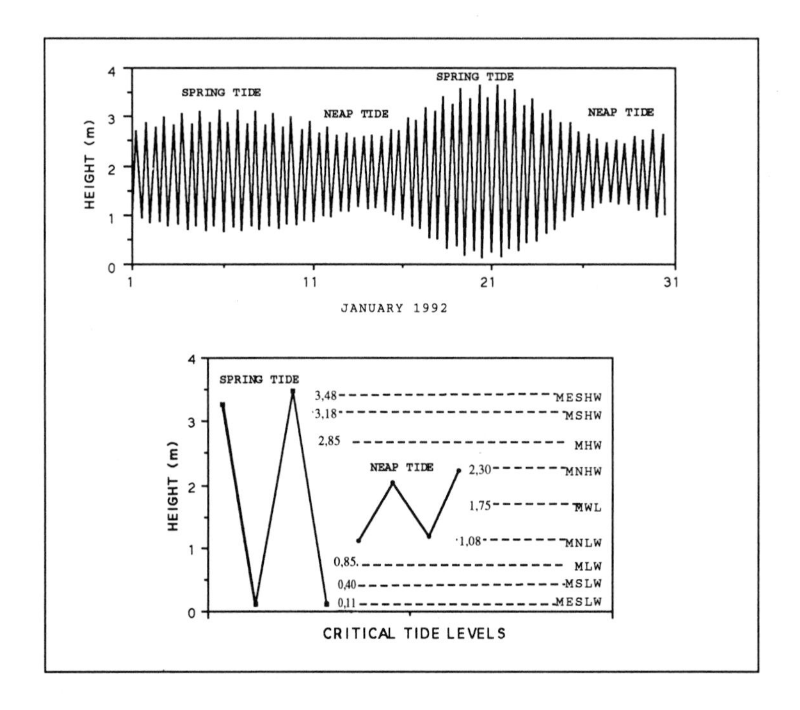

Fig. 2. Predicted tidal ranges for one month and theoretical critical tide levels (CTLs) along the Huelva littoral zone at the entrance of the Odiel estuary. MESHW, mean equinox spring high water; MSHW, mean spring high water; MHW, mean high water; MNHW, mean neap high water; MWL, mean water level; MNLW, mean neap low water; MLW, mean low water; MSLW, mean spring low water; MESLW, mean equinox spring low water. Zero refers to the lowest astronomical tide (i.e. lowest spring equinoxial tide).

produced by cycles of diverse duration and length which characterize the tidal regime along this coastal sector. As will be demonstrated in this paper, some CTLs define clear boundaries between adjacent depositional facies.

The propagation of the tidal wave is different in each of the three estuaries discussed in this paper. This is due to their individual physiographies and the resulting tidal response to both friction and convergence (Le Floch, 1961).

In the Guadiana estuary the tidal wave propagates in *synchronic* mode, the range remaining constant up to 50 km from the mouth (Morales, 1993). Even during spring tides the tidal range decreases by only 10 cm at the estuary head. In the Piedras estuary, on the other hand, the tidal range decreases landwards in *hypo-synchronic* mode, being 50 cm lower 12 km from the mouth during spring tides (Borrego *et al.*, 1993). In the Odiel estuary, by contrast, the tidal wave follows a slightly *hypso-synchronic* mode, the range progressively increasing by as much as 30 cm at a distance of 15 km from the mouth (Borrego, 1992).

This different behaviour of the tidal wave in the three estuaries is produced by the different evolution of the width/depth ratio along the main tidal channel in each case (Fig. 3). In the Piedras estuary, for example, channel depth decreases rapidly where the decrease in channel width is less pronounced. As a result of this the frictional loss of energy along the bottom is higher than the convergence effect. In the Odiel estuary, on the other hand, the main channel width decreases rapidly upstream and the channel depth thus decreases more gradually. For this reason the convergence effect dominates over bottom friction. The Guadiana estuary, by comparison, occupies an intermediate position, the width/depth ratio developing in such a manner as to compensate the effects of both friction and convergence.

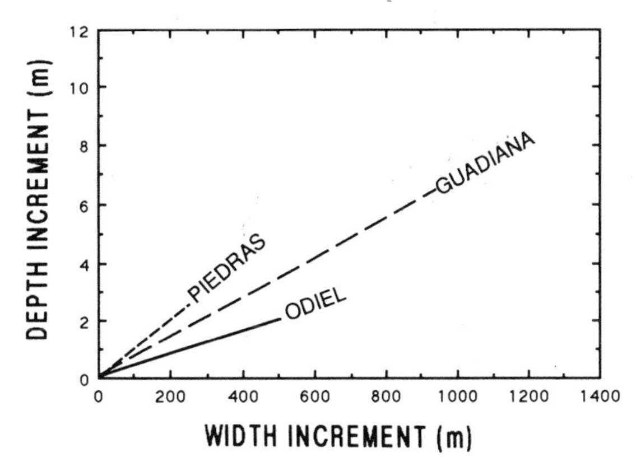

Fig. 3. Diagram illustrating variations in the height/depth ratios along the three estuaries.

The prevailing winds on the Huelva coast are from both southwest and north, whereas the wave climate is dominated by swells from the southwest with a frequency of 20% associated with the Atlantic circulation regime. Borrego *et al.* (1993) found the average significant wave height along the Huelva coast to be 0.76 m at a water depth of 22 m. Wave-induced littoral drift is eastward directed, supplying 360 000 million m^3 of sediment annually (MOPU, 1987).

The fluvial discharge of all three rivers is markedly seasonal, being high during winter and very low during summer. In addition it is very irregular on an inter-annual scale. Thus, in the case of the Odiel and Piedras rivers, the mean annual freshwater discharge is $<11.1\ m^3\ s^{-1}$. The Guadiana river, by comparison, has a mean annual discharge of $144.4\ m^3\ s^{-1}$, making it the main sediment source of the Huelva littoral zone (e.g. Ojeda, 1988).

Mixing in the Odiel estuary under mean discharge conditions is controlled by the tidal prism (Borrego, 1992). The estuary is well mixed during spring tides but, according to Simmons (1955), becomes partially stratified during neap tides (Fig. 4). The Guadiana estuary, on the other hand, is always well mixed under mean discharge conditions, irrespective of the tidal phase. These tidal conditions are responsible for the position of the fresh water/sea water interface (Morales, 1993). The Piedras River is presently being regulated by an upstream dam, the fluvial discharge being so low that mixing processes are not significant.

In terms of wave height and tidal range, the Huelva coast can be classified as a tide-dominated, mixed-energy coastline (Hayes, 1979; Davis & Hayes, 1984) with respect to absolute values of these parameters. None the less, the coastal morphology off the main river mouths is characterized by wave-dominated land forms, such as elongated sand spits.

STUDY METHODS

Facies analysis of the Huelva coast estuaries is based on hydrodynamic data supplied by local

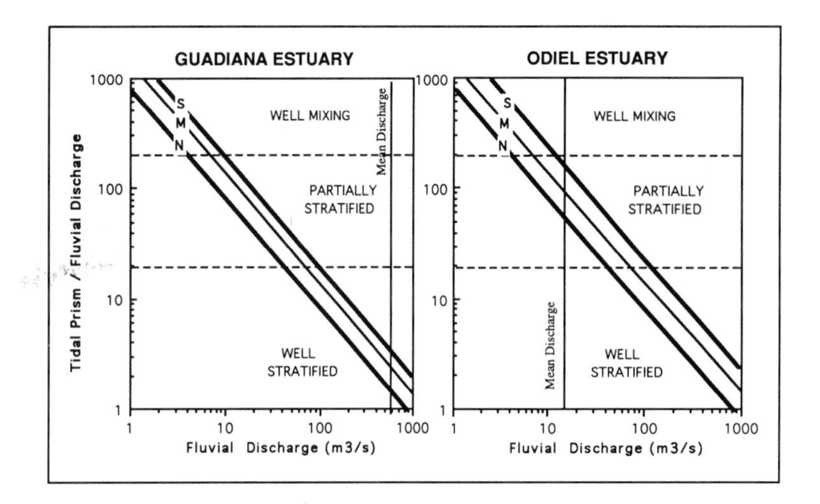

Fig. 4. Variations in the ratio between tidal volume and fluvial discharge for spring (S), mean (M) and neap (N) tides in the Guadiana and Odiel estuaries (based on data from Borrego, 1992 and Morales, 1993).

organizations, e.g. Puerto Autónomo de Huelva, and field work carried out between 1988 and 1992. Numerous bathymetric transects were run across the main tidal channels of the Guadiana, Piedras and Odiel estuaries using a fathometer. Short box cores were taken along most transects to characterize lithology and facies associations.

Sediment textures were determined by sieve analyses of the coarse fractions (>0.063 mm) and by Coulter Counter analyses (ZM model) of the fine fractions (<0.063 mm). The organic matter content was quantified by loss-on-ignition at 485°C after 4 h (Courau, 1983). Sedimentary structures were studied from X-ray radiographs of sediment cores as well as from lacquer peels.

DEPOSITIONAL FACIES

Various sedimentary subenvironments have been distinguished within the estuarine systems of the Huelva coast, each subenvironment being characterized by a different vertical and lateral lithofacies association. Each lithofacies association defines a depositional facies produced by the interaction between the available sediment and the prevailing hydrodynamic conditions. The depositional facies occur in all three estuaries, where they maintain a constant relationship with some of the CTLs and/or other factors such as wave action and fluvial effects. Nevertheless, the nature of the individual lithofacies and their lateral and vertical relationships show some differences among the three estuaries.

Lateral facies variations occur in all three estuaries, being produced by spatial changes in the sedimentary processes/products ratio. Such spatial changes are common to all estuarine systems across the world (Allen, 1971; Jouanneau & LaTouche, 1981; Woodroffe *et al.*, 1989; Dalrymple *et al.*, 1990). In the present case, the Odiel River estuary was subdivided into four domains (Borrego, 1992): (i) a fluvial domain; (ii) an upper estuarine domain; (iii) a lower estuarine domain; and (iv) a marine domain (Fig. 5). Along the Piedras River estuary three domains have been differentiated: (i) a fluvial domain; (ii) a central estuarine domain; and (iii) a marine domain (Fig. 6). The Guadiana River estuary, in turn, has also been subdivided into three domains, although these differ from those of the Piedras estuary (Morales, 1993): (i) an upper estuarine domain; (ii) a central estuarine domain; and (iii) a marine domain (Fig. 7).

On the basis of hydrodynamic criteria, depositional facies were in each case subdivided into an inner and an outer facies association. The inner facies are the result of fluvial and/or tidal processes, with little or no wave and other open marine influences (Table 1). The inner facies association comprises: (i) channels; (ii) fluvial bars; (iii) intertidal channel margins and tidal flats; (iv) salt marshes; (v) sterile marshes; and (vi) protected beaches. The outer facies association, which is dominated by open marine processes, consists of subenvironments associated with sand spits and barrier islands, e.g. ebb-tidal deltas, open beaches and coastal dunes.

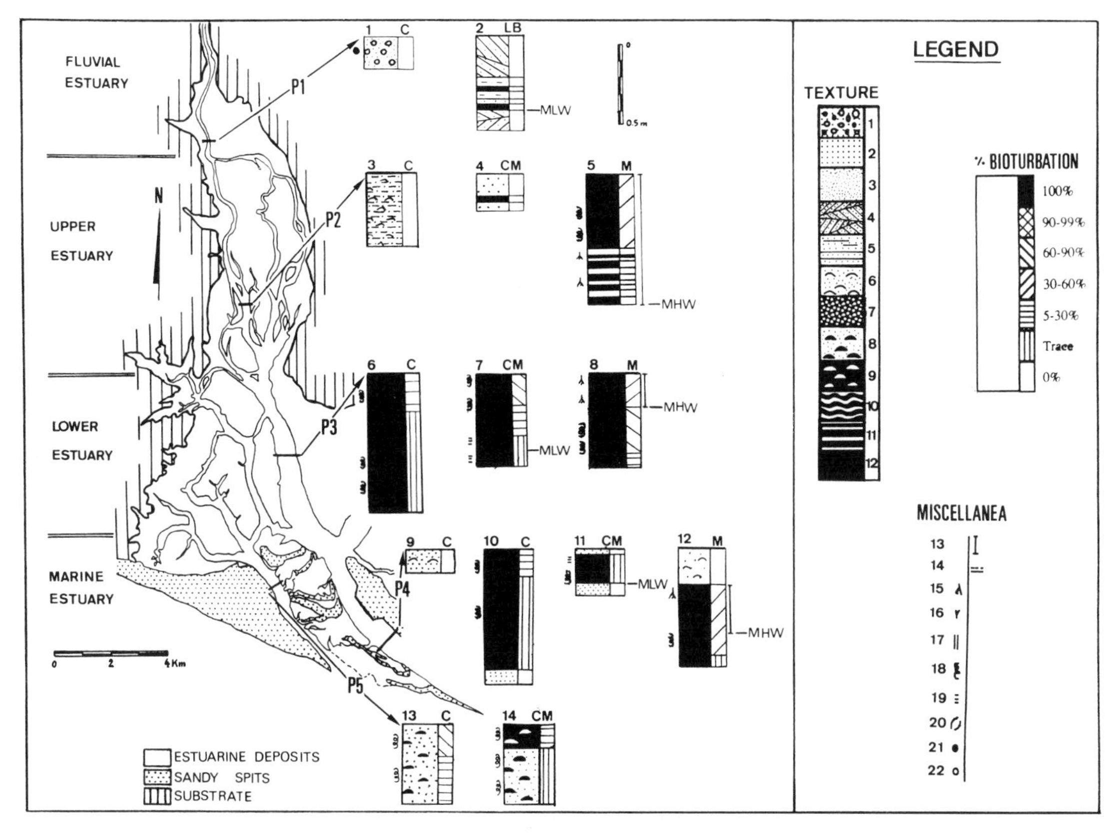

Fig. 5. Main subdivisions and coring profiles in the Odiel estuary. C, subtidal channels; CM, channel margin; M, marsh; LB, lateral fluvial bar. 1, gravel; 2, massive coarse/medium sand; 3, massive fine sand; 4, sand with ripple lamination; 5, sand with horizontal layering; 6, sand with shells; 7, muddy sand; 8, flaser bedding; 9, linsen bedding; 10, wavy bedding; 11, laminated mud; 12, massive mud; 13, oxidation level; 14, normal graded bedding; 15, root bioturbation; 16, plant remains; 17, sand-filled burrows; 18, burrows; 19, shell fragments; 20, bivalves in life position; 21, shale clasts; 22, clasts.

Inner facies association

Channel facies

Channel deposits were divided into two subfacies, depending on whether they occurred in a subtidal or an intertidal setting: (a) a subtidal channel facies, and (b) an intertidal channel facies.

(a) Subtidal channel facies These correspond to a lithofacies association occurring in non-emergent tidal channels, whose channel beds lie below mean low-tide level (Ashley & Zeff, 1988; Zeff, 1988).

In the upper estuarine domain of the Odiel River (Fig. 5), this facies is characterized by an alternation of massive or normally graded gravel and coarse sand deposits containing abundant rock fragments and red quartz. These coarse sediments show a landward textural trend typical of lag deposits (McLaren, 1981), being dominated by grain sizes <0.5 phi (i.e. >0.7 mm). Such sediments characterize the main channel along the fluvially dominated parts of the estuary. Seawards the sediment looses its lag character, becoming well-sorted with mean grain sizes in the medium-coarse sand range (Fig. 8A). The sediments are now interbedded with massive, black muddy silt layers of variable thickness (0.5–15 cm) containing abundant organic matter (>5%). There is no evidence of biological activity in these finer sediments.

In the lower estuarine domain, the channel sediments mainly consist of partly bioturbated (>25%

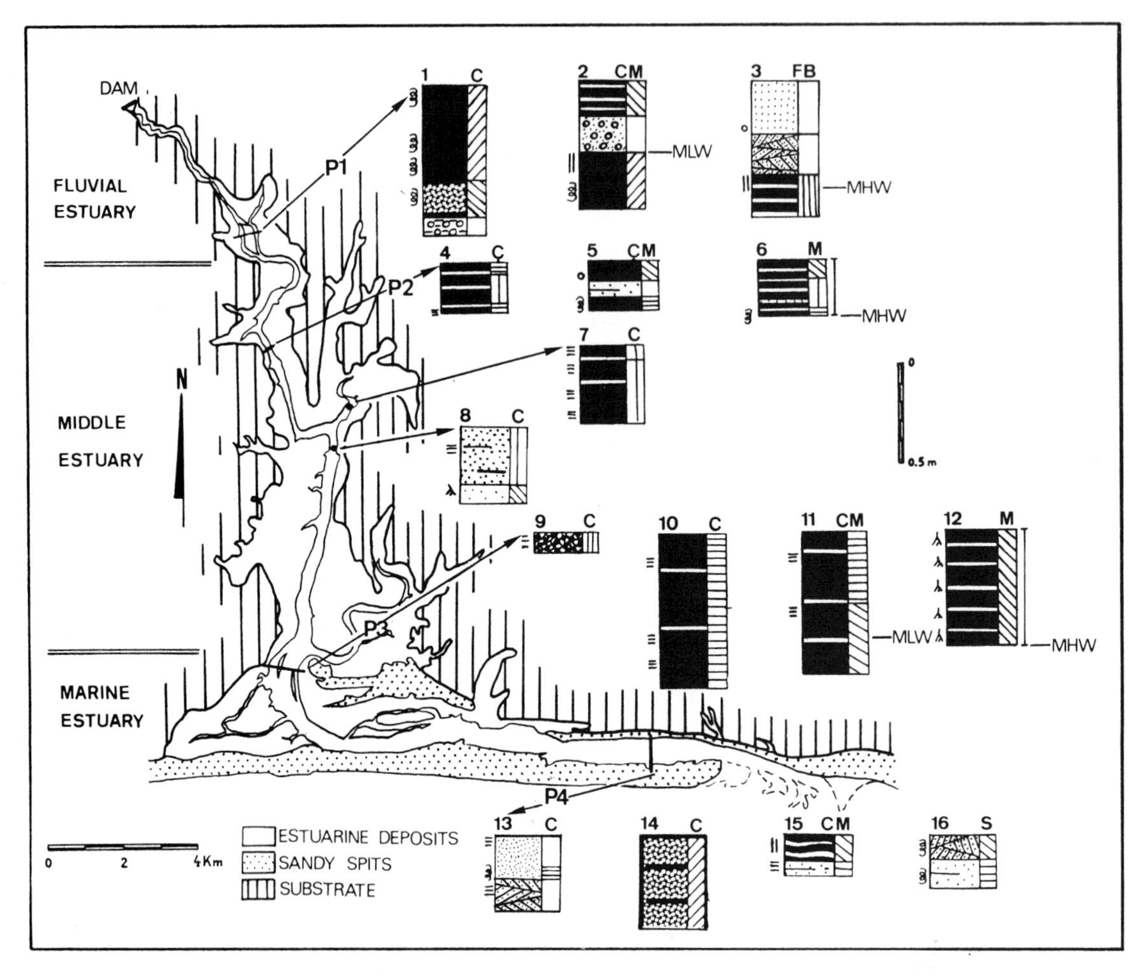

Fig. 6. Main subdivisions and coring profiles in the Piedras estuary. FB, flood basin. Other nomenclature: see Fig. 5.

burrowing), massive black muddy silts with grain sizes in the 4–8 phi range (0.063–0.004 mm). The burrows are often filled with very fine red sand (Fig. 8B). Lag deposits are very frequent in the lower Odiel estuary, two types having been distinguished: (1) A mixture of shell fragments and quartz grains without a fine matrix (Fig. 9A). The deposits are 10–30 cm thick and have a substantial lateral extent (>10 m). These lag deposits occur in channels connecting the estuary with the open sea. (2) Shell fragments in a muddy matrix, occurring on the beds of inner subtidal channels (Fig. 8C). These deposits are up to 20 cm thick, but have a smaller lateral extent than the first type. Similar deposits were found in the central estuarine domain of the Piedras River.

In the marine domain of the Odiel estuary the sediment is dominated by clean sand, passing seawards into lag deposits comprising quartz pebbles and shell fragments (Fig. 5). These lag deposits bear evidence of little or no sediment reworking.

The channel facies of the Piedras estuary (Fig. 6) has a lower textural variability than that of the Odiel estuary. In the upper estuarine domain it consists of layers (mean thickness: 10 cm) of massive, brown medium-coarse sands containing an abundance of round quartz grains. These alternate with black, organic-rich (>10%) mud layers that show a variable degree of bioturbation by annelids and bivalves and are sometimes horizontally laminated. Towards the central estuarine domain the sandy layers become less abundant and muddy

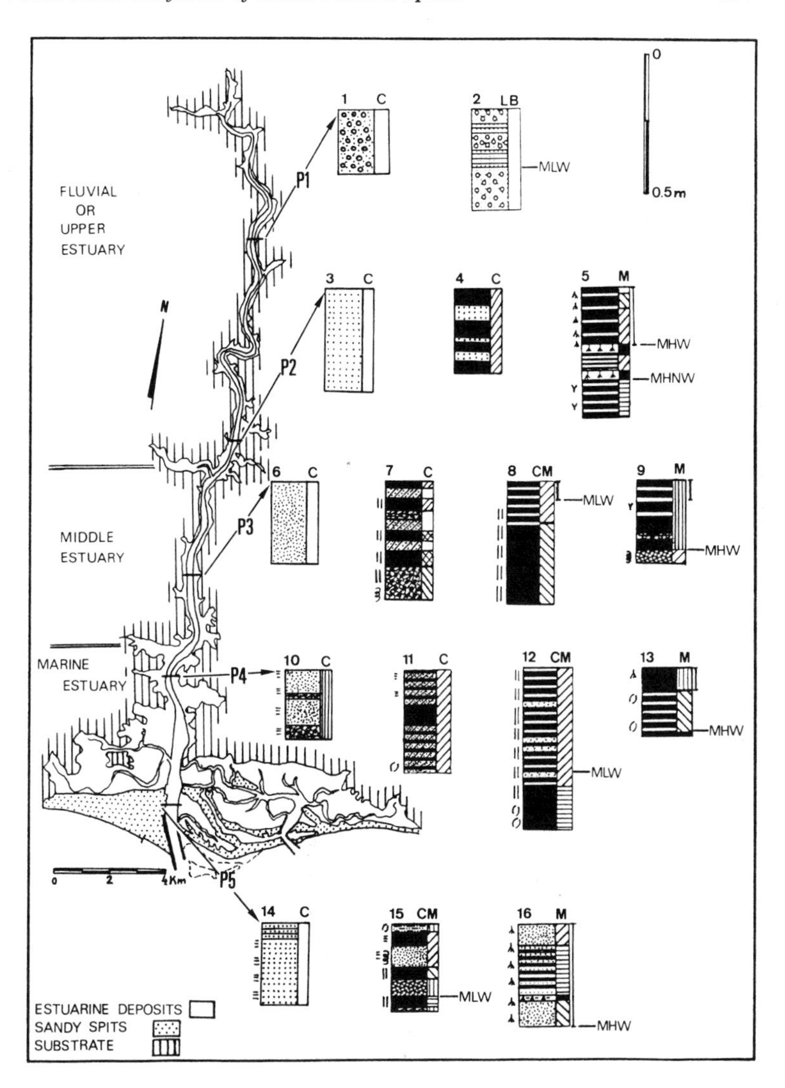

Fig. 7. Main subdivisions and coring profiles in the Guadiana estuary. Nomenclature as in Fig. 5.

sediments predominate, these being interbedded with lag deposits of quartz and shell fragments. Along the shallow zones of the channel these muddy sediments are strongly bioturbated by bivalves, and organic matter content is higher.

Within the central estuarine domain of the Piedras River the sediment is dominated by muds containing plenty of shell fragments. Bioturbation by bivalves and annelids is common but variable (Fig. 6), an exception being rare, very well-sorted layers of fine sand which appear towards the boundaries with the upper estuarine domain.

Lag deposits comprising both coarse sand and shell fragments are developed in the deeper parts of the estuarine channel of the marine domain. The shallow parts of this domain are characterized by the appearance of unsorted muddy sands, containing scattered shells as well as variably bioturbated ripple laminations (Fig. 6). In many places normally graded, parallel layers of shells and shell fragments are observed (Fig. 9B).

In the Guadiana River the subtidal channel facies shows both longitudinal and lateral variability. Poorly sorted, massive gravel beds occur in the upper estuarine domain. These gradually pass into <30 cm thick, well-sorted medium sand layers towards the marine domain. Two distinct lateral channel facies have been distinguished in the

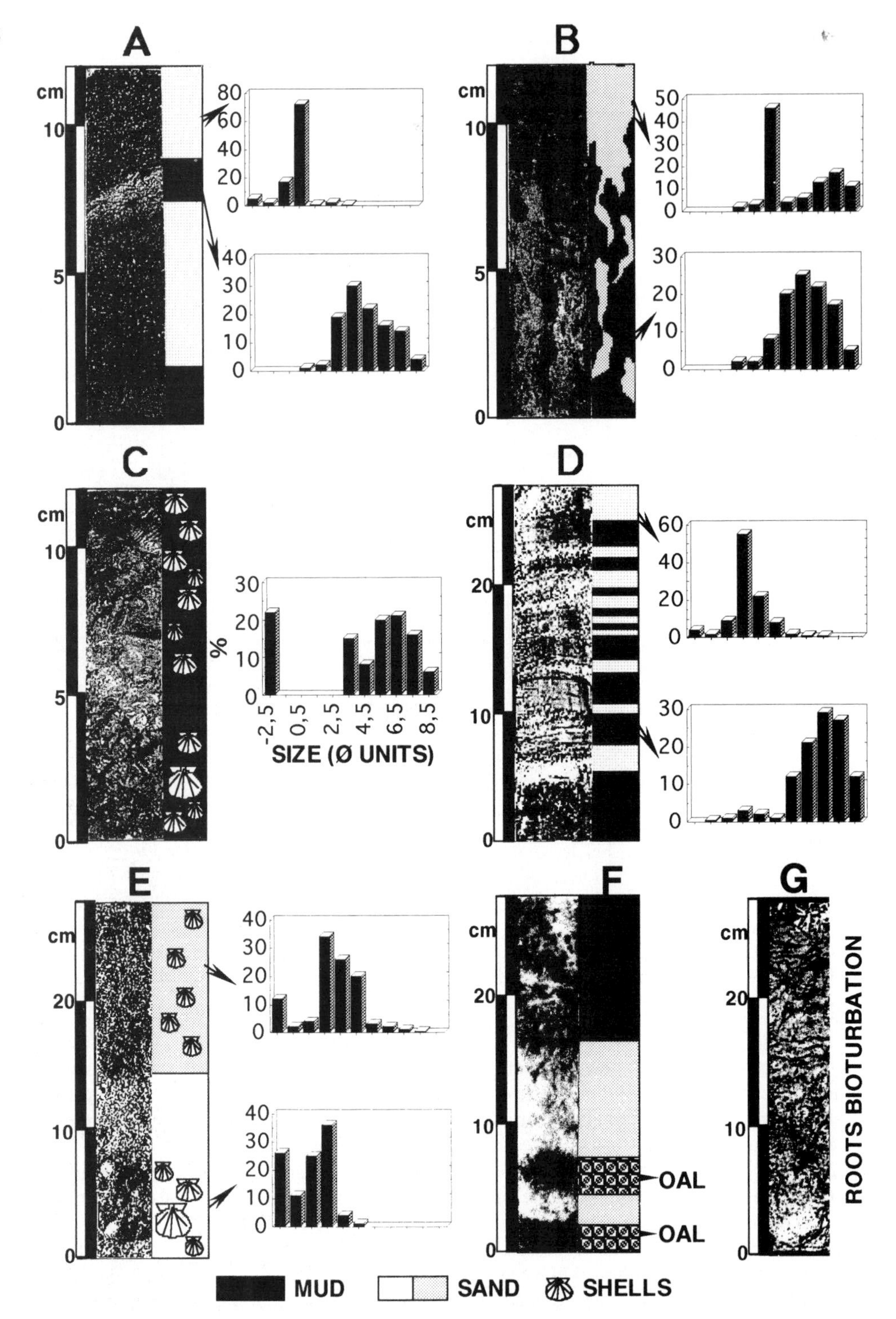

Fig. 8. Some core photographs showing the main features of the depositional facies and textural distribution of the sediment. A, alternating sandy gravels and mud layers in the Odiel fluvial estuary; B, sand-filled burrows in the upper Odiel estuary; C, lag deposits consisting of shell fragments with a muddy matrix in the central Piedras estuary; D, alternating sandy and muddy sediments in the shallow subtidal channel of the lower Guadiana estuary; E, lag deposits consisting of whole shells and shell fragments with a sandy matrix in the lower Odiel estuary; F, organic layers (OAL) along the Piedras channel margin; G, bioturbation by roots in the salt marsh facies of all three estuaries.

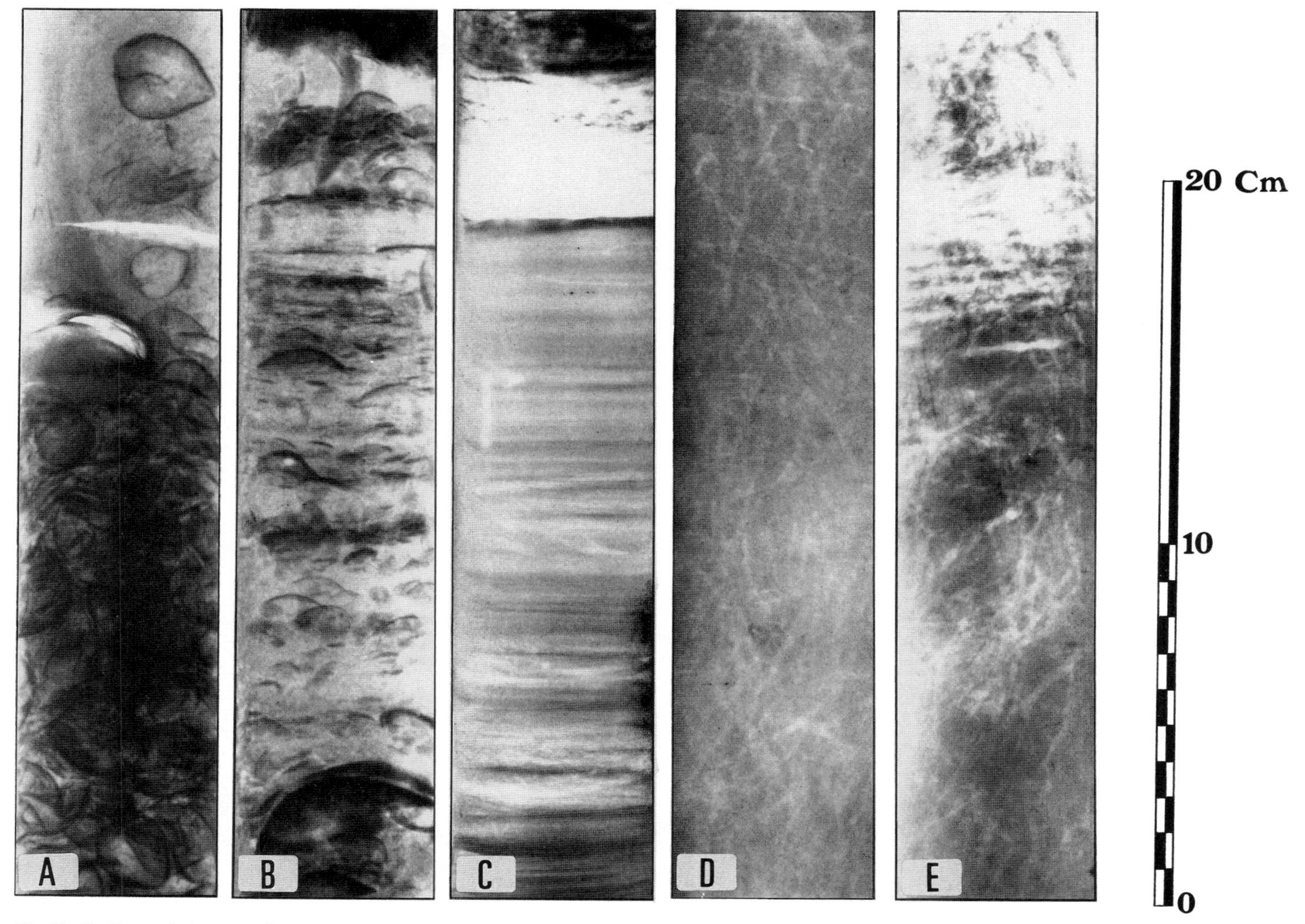

Fig. 9. Radiograph images of some cores showing the internal structures of individual depositional facies. The light colours denote sediments with low contents of organic matter and high contents of detrital grains. A, lag deposit consisting of shell fragments without matrix in the lower Odiel estuary; B, parallel layers of shells and shell fragments in an energetic subtidal channel; C, alternating silt layers without organic matter and organic-rich mud layers in the upper Odiel estuary; D, burrows of *Scrobicularia plana* in the channel margin facies; E, bioturbation by roots in the salt marsh facies.

central estuarine domain on the basis of depth and sediment texture: (i) a deep channel facies exclusively characterized by non-bioturbated coarse sands; and (ii) a shallow channel facies comprising finer sands (up to 3.5 phi) that are interbedded with muds to form flaser and wavy bedding types several dm thick (Fig. 7). The muds mainly consist of grain sizes associated with flocculation processes (i.e. 6–9 phi). Current-generated ripple lamination, frequently burrowed by annelid worms, is common in the sand layers.

The situation in the marine domain of the Guadiana River is similar in that two channel facies can again be differentiated, although both the deep and the shallow channel facies comprise sandy sediments (1–3 phi) with abundant shell fragments (Fig. 8D). A dense system of tidal channels that is not connected to the main channel is also developed along this domain. Here, the sediments dominantly consist of bioturbated muds (4–9 phi), burrowed by both annelids and bivalves (mainly *Arenicolides ecaudata, Ruditapes decussatus, Chamelea gallina* and *Cerastoderma edule*). The degree of bioturbation is greater than in the shallow zones of the connected estuarine channel. Some muddy sand layers, or even well-sorted and non-bioturbated clean fine sands occur along the head of the tidal channel network.

(b) Intertidal channel facies These are dead-end tidal channel deposits (*sensu* Ashley & Zeff, 1988; Zeff, 1988). In this case the channel bed is located above the mean low water level, the channels mostly draining the marginal salt marshes. The transverse profiles of these channels are V-shaped with steep walls. They occur in positions where the tidal cycle has a pronounced time asymmetry with a longer ebb half-cycle (Borrego, 1992).

The deposits are both muddy (>10%) and organic-rich (6–40%). Shell lags, slumped channel-margin marsh material, boulders and sand deposits are abundant, the latter displaying ripple cross-bedding reflecting the current reversals. Such sand deposits mostly occur along the spit/channel boundaries, as well as at the mouths of the feeder tidal channels (Fig. 8E) where lateral migration is common.

Lateral fluvial bar facies

This facies is located along the channel margins of the upper estuarine domain, where it is bounded by the MLW and MHW levels. Lateral fluvial bars only occur in the Guadiana and Odiel estuaries. They consist of semi-lenticular sediment bodies attached to straight channel sections. In contrast to point bars, the lateral bars migrate downstream, as observed in the Odiel estuary. Here the deposits consist of red gravel and coarse sand layers (15 cm thick) alternating with 2 cm thick, yellow silt laminae. Occasional black mud layers (5 cm thick) containing abundant organic matter (>10%) also occur. The bars are internally structured by planar cross-beds dipping downstream (Borrego, 1992). Ripples and megaripples occur, being continuously reworked by tidal and fluvial currents (Fig. 10B). Both ripple and flaser bedding structures can be found.

In contrast to the Odiel River, no bar migration was observed in the Guadiana estuary. Instead, coarse sediment (mainly gravel) accumulates by aggradation in massive layers. Thin laminae of brown silty sediment occurring on the bar surfaces have a low preservation potential (Morales, 1993). Cross-channel bar migration in both the Odiel and Guadiana estuaries is prevented because the channels are incised into Palaeozoic bedrock. The bars are sometimes vegetated by halophytes which may act as substrate for freshwater marsh.

Active channel margin facies

This geomorphic unit occurs between the MLW and MHW levels in the presence of both fluvial and intertidal processes. Some authors (e.g. Frey & Howard, 1986) do not distinguish this facies type because they consider it to be transitional between the channel and the salt marsh facies. In the present case, the active channel margins were defined as a facies in its own right because it can be recognized by a number of distinctive sedimentological features, e.g. by both the mode and intensity of bioturbation in relation to organisms that restrict their biological activity to this particular zone. The thickness of this facies depends on the tidal range or the vertical difference between MLW and MHW. This latter value can vary from one place to another along the estuary, depending on the propagation type of the tidal wave.

Morphologically, this facies is quite similar in all three estuaries discussed here. Firstly, they are all bounded by the same tide levels and secondly, the extent of the facies decreases towards the estuary heads as the slope of the exposed surface increases landwards. Because of the gentler slopes along the

Fig. 10. Panoramic views of some depositional facies. A, intertidal channel; B, megaripple sets in a lateral fluvial bar of the upper Odiel estuary; C, erosional step along the channel margin of the lower Odiel estuary; D, view of the salt marsh/channel margin transition (note the burrows of *Uca pugnax* along the upper channel margin); E, microlamination in the salt marsh facies; F, sand hummocks (wind drift) on the sterile marsh of the inner Odiel estuary.

estuary mouths (Fig. 10C), this facies has here to some extent the character of tidal flats.

On the other hand, the active channel margin facies also displays some differences between the estuaries, particularly with respect to grain size and internal structures. In the Odiel estuary (Figs 5 & 9C), for example, this facies consists of alternations of black mud and brown silts with numerous burrows in the upper estaurine domain, being progressively replaced by black silts with burrows in the lower domain. It does not occur in the fluvial sector, but reaches its broadest horizontal extent in the marine domain, where its textural differentiation progresses upwards from silty sand close to the subtidal channel into sandy silts at the contact with the salt marsh facies.

Throughout the Piedras estuary (Fig. 6) the active channel margin facies corresponds to that of the lower Odiel estuarine domain, i.e. comprising bioturbated, black silty sediments. The same applies to the Guadiana river mouth marsh (Fig. 7) and the Guadiana river main channel. In this respect it is similar to that of upper Odiel estuarine domain, although grain sizes are larger.

The boundaries between the subtidal channel and the channel margin facies are marked by the presence of the bivalve *Scrobicularia plana* (Fig. 9D), the annelid *Nereis diversicolor* (in sandy channel margins), and marine grass meadows of *Zostera noltii* and *Z. marina* which produce organic accumulation layers (OALs, Fig. 8F) by the aggradation of plant material along the central and lower estuarine domains as well as in the marine domains of all three estuaries. The channel margin facies of the central domain is characterized by intense bioturbation produced by annelids such as *Arenicolides ecaudata*, as well as scattered shell accumulations of *Crassostrea* sp., whereas the presence of some crustaceans (e.g. *Uca pugnax*) is more frequent along the boundary with the vegetated salt marsh (Fig. 10D) which is also marked by root cast bioturbation. The channel margin facies is bisected by small ebb-tidal channels which lie perpendicular to the orientation of the subtidal channels. These ebb-tidal channels have abrupt walls and thin, but not very extensive lag deposits on the channel beds. The coarse sediments are composed of shell fragments, mud balls and plant remains. Point bar deposits associated with channel migration processes are also frequent. Thin lag deposits, <10 cm in thickness, with lateral extents of a few metres occur in places. These are formed by coarse sands associated with shell fragments. They have a low preservation potential since they are easily reworked by tidal currents. The channel margin deposits have a mean thickness of 1.2 m.

Salt marsh facies

The salt marshes are tidal flats located above MHW vegetated mainly by the halophytes *Spartina* sp., *Salicornia* sp. and *Sarcocornia* sp. (Fig. 10D). The lithology of this facies varies with position in the estuary and sediment supply. The fluvial and/or upper estuarine domain of all the three estuaries is dominated by intensely bioturbated muds. Locally the sediments can consist of medium-fine sands of fluvial origin. In the lower estuarine and/or marine domain the sediments mostly consist of fine sand which can locally become muddy. These deposits are internally composed of parallel laminae which are a few mm or cm thick and can be disturbed to a variable degree by root structures (Figs 8G, 9E). The deposits incorporate some 3–5 mm thick fining- and thinning-upward microsequences that occur scattered along the outer zones of the marine domains of the estuaries (Fig. 10E). These microsequences are composed of medium-fine sand grading upwards into thinner layers of muddy sand (Table 1). The marsh deposits have a mean thickness of 0.7 m. Above spring high water a sterile marsh is developed.

Sterile marsh facies

A poorly drained environmental niche exposed over long periods of time and only covered during extreme tides or storms is developed on salt marsh deposits above mean spring high-water level. During extreme high tides, sea-water penetrates to the high marshes. In the course of subsequent evaporation, salt crystals precipitate in the fine sediment. This facies has been termed 'sterile' because of its very low biomass production (Borrego *et al.*, 1993). It is very similar to the salt pans of Warmer (1971). The sterile marsh facies occurs in both the Guadiana and the Odiel estuaries (Figs 5 & 7) and has strong continental affinities. It is composed of horizontal layers of alternating brown silty sands and black mud. Individual layers are <0.01 m thick and, with the exception of some root remains, are essentially non-bioturbated. Desiccation cracks and rain drop impressions are frequently found on the muddy surface of this facies. In the lower estuarine and marine domains the sediment of the sterile marsh consists of fine or very fine sand with small wave ripples. Hummocks of sand trapped around halophytes are also very frequent (Fig. 10F). The thickness of the sterile marsh deposits varies from 10 to 15 cm.

Protected beach facies

Silty to sandy beaches are found adjacent to the Neogene–Quaternary cliffs along the Piedras and Odiel estuary mouths. The deposits are composed of accretional laminae which gently dip towards the estuary channel.

Table 1. Depositional facies occurring along the estuaries of the Huelva coast. Critical tide levels as in Fig. 2.

Facies	CTLs	Thickness (m)	Processes	Texture	Physical structures	Biological content
Subtidal channel	MLW	1–3	Fluvial streams Tidal currents	Conglomerate, sand and mud	Ripple lamination Flat bed	Scattered shells and lags
Intertidal channel	MHW MLW	<1	Ebb tidal currents	Mud, shell-lags	Parallel lamination	Plant debris Burrows
Lateral fluvial bars	MHW	>1	Fluvial floods	Poor sorted conglomerate and sand	Megaripples Parallel lamination	–
Channel margin	MHW MLW	>1	Tidal currents Fluvial floods	Well sorted mud and/or silt	Flat bed, plane bed	Scattered shells Burrows
Salt marsh	MSHW MHW	0.5–1	Tidal current Flocculation	Mud (sand)	Flat bed, plane bed Ripple lamination	Roots Plant debris
Sterile marsh	EESHW MSHW	<0.5	Evaporation Deflation	Well sorted mud and fine sand	Flat bed, mud cracks	Plant debris Roots
Protected beach	MHW MLW	2–3	Waves and tides	Poor sorted silty sand	Cross-lamination	Algal mats Burrows

Ridge and runnel systems are not found on the steep foreshore and there is also no beach berm. The distribution of the fine fraction is not directly correlated with tide levels, although grain sizes generally increase towards the estuary channel. This phenomenon is directly related to the number of tidal cycles affecting the different beach sections (Borrego & Pendon, 1989b). The sheltered beaches within the marine domain of the estuaries differ from the exposed outer beaches in both sediment texture and internal structure, the deposits of the former being finer and more steeply inclined (Borrego *et al.*, 1993).

Outer facies association

Because the emphasis of this paper is on the inner estuary, only a brief description of the outer facies association is given. The littoral facies (coastal sand spits and related environments) in the vicinity of the estuary mouths are produced by the interaction of ocean waves and littoral currents (longshore drift, residual tidal currents and rip currents). The exposed beaches have a lower gradient and a wider foreshore than their protected counterparts within the marine domains of the estuaries, where migrating ridge and runnel systems can evolve into berms (Dabrio, 1982). The remnants of former tidal deltas in front of abandoned tidal inlets are reworked into recurved spits (Borrego *et al.*, 1993). The sands of the prograding spits interfinger in a complex manner with the salt marshes in their rear, mainly due to episodic storm-induced overwash processes and aeolian action, the latter having formed a slowly accreting incipient dune belt in the backshore environment of the spits (Flor, 1990).

ESTUARINE DYNAMICS

The variations in lithofacies associations in the various estuarine domains outlined above reflect process variability within each estuary. This, in turn, allows a rational assessment to be made of site-specific sediment dynamic processes, including estimates of sediment supply volumes (Fig. 11). The results of such assessments and estimates are presented below for each estuary and estuarine domain.

Odiel estuary

Fluvial domain

Although short-term dynamic processes in this part of the Odiel estuary are dominated by the tides,

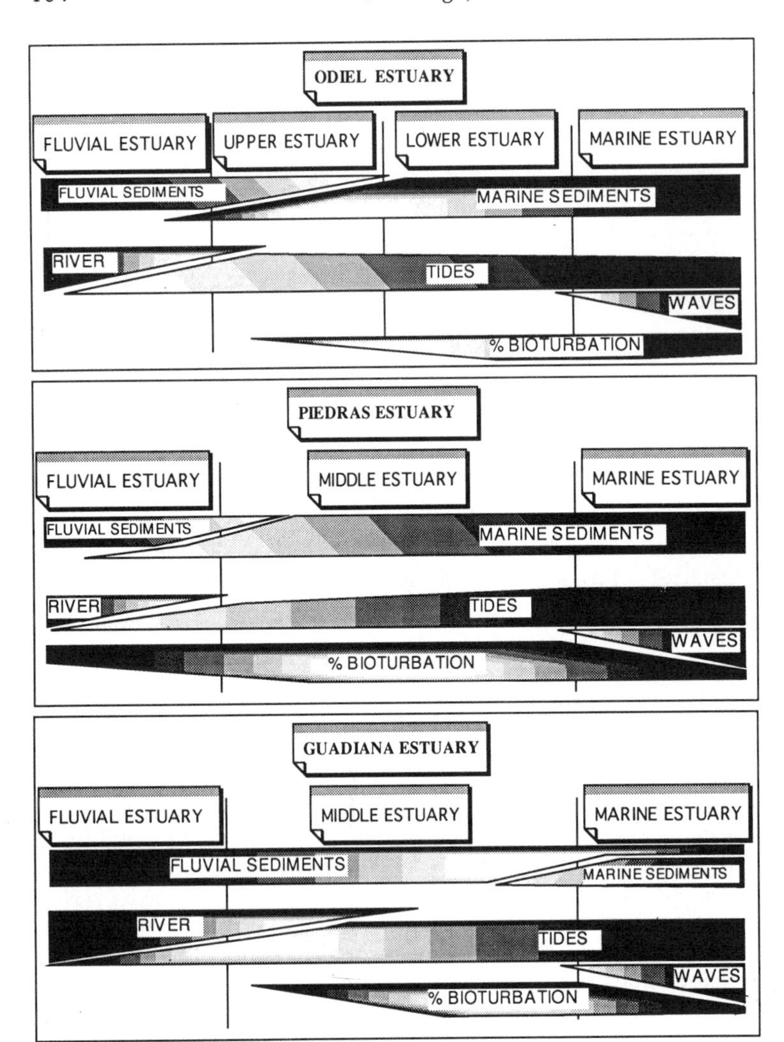

Fig. 11. Comparison of the sediment character as a function of hydrodynamic factors along the main depositional reaches of the three estuaries.

both morphology and deposition are mainly controlled by short-lived, but much more energetic fluvial discharge events. This is reflected in the fluvial character of the sediments. At times of low fluvial discharge the tide is dominated by the ebb cycles which reach current velocities of 0.2–0.5 m s^{-1}. Such velocities are insufficient to significantly modify the larger bedforms developed on the fluvial bars. Tidal deposition is restricted to mud drapes, which are up to 5 cm thick and form during the slack-water period at high tide. However, these have a very low preservation potential, as they are periodically eroded by fluvial flood waters.

Fluvial sediment supply is thus restricted to seasonal floods and/or to extraordinary spring tide events. During such periods there is intense sand transport in form of megaripples on the lateral bars (Borrego, 1992). Under intermediate to high fluvial discharge conditions the sediments supplied by the river consist of yellow silts which cover the lateral bar surfaces and other coarse-grained bedforms with a fine-grained sediment blanket. These silts are normally deposited during the humid winter months from November to March (Borrego, 1992).

The fluvial domain described here is very similar to the fluvial estuary of Fairbridge (1980), its lower part closely corresponding to the fluvial domain as defined by Dalrymple *et al.* (1992).

Upper estuarine domain

This section of the Odiel estuary is formed by a braided channel network which intersects high marsh zones that have a fusiform geometry produced by the narrowness of the old fluvial valley. Some channels have a double function, acting both as tidal channels and fluvial conduits by which excess fresh water is drained during high river stages. Silty to muddy sediments dominate volumetrically, being of both fluvial and tidal origin. Nevertheless, sandy deposits and fluvial gravels also contribute significant volumes to the total sediment in this part of the estuary.

Mean current velocities of the ebb tide, when measured at a height of 50 cm above the channel beds, reach values of 0.3 m s^{-1} during neap tides and 0.65 m s^{-1} during spring tides (Borrego, 1992). Velocities decrease seawards as the channels widen. Sand is transported mainly during the spring-tidal cycles, producing ebb-oriented linguoid megaripple trains. The megaripple crests are reworked under low energy conditions and sediment movement is then dominated by current ripple and lower-plane-bed transport phases. Sandy sediments are subjected to intense tidal reworking, thereby producing very well-sorted, very fine sand that makes up over 40% of the grain size distributions.

As pointed out above, the supply of sand is restricted to fluvial flood events, the interannual fluvial supply being irregular and without any seasonal periodicity. Under low energy conditions the tides deposit black muds with high organic contents in the channels. These are colonized by a dense estuarine fauna which is responsible for intense bioturbation. The alternating layers of sand and mud reflect the energy pulsations which affect this estuarine domain.

The seasonal rhythm of deposition is also reflected in the lower intertidal zone, where fluvial silts alternate with parallel layers of tidally introduced black mud (Fig. 9C). This physical control of the internal sedimentary structures in the lower intertidal zone is replaced by biological activity in the higher intertidal zone, where bioturbation by roots of salt-marsh plants dominate the internal structures of the sediment column.

The sterile marsh forms a transitional zone between the intertidal and the supratidal facies. It also experiences the energetic oscillations that occur in the other parts of this estuarine domain, especially by acting as a flood plain during severe fluvial flood events, in the course of which the brown muddy silt layers with scattered plant remains are deposited. At times of low river stages the sterile marsh is flooded only by extreme spring tides which deposit the black, organic-rich mud layers that are >5 cm thick. These alternating depositional sequences are subsequently disturbed by the effects of prolonged periods of emersion, in the course of which a variety of dessication marks are produced.

This upper estuarine domain corresponds to the central estuary as defined by Fairbridge (1980) or the estuarine zone with fluvial domination as classified by Dalrymple *et al.* (1992).

Lower estuarine domain

This estuarine sector of the Odiel River is composed of various salt marsh bodies separated by distributary tidal channels. The dynamics is controlled by the tide which imposes a typical tidal physiography. Intensely bioturbated silty muds form the volumetrically most abundant sediments. Although the tides are practically identical in duration, current speeds are nevertheless higher during the ebb cycle (0.25 m s^{-1} during neap tides and 0.54 m s^{-1} during spring tides). An important difference manifests itself in the duration of the slack water period, which is longer during the ebb cycles (40 min, speed >0.10 m s^{-1}) than during the flood cycles (25 min) (Borrego, 1992). Fluvial as well as local wave effects are either absent or very scarce. During spring tides the currents promote the deposition of coarse lag deposits (mainly shell fragments), although muddy sediments deposited under low energy conditions are generally more abundant, filling hollows or cavities between the shell fragments. Such conditions are favourable for many benthic organisms which are responsible for intense bioturbation of the fine sediment. Low energy conditions dominate the entire intertidal zone (channel margins and salt marshes), fine sediments being deposited during slack-water periods at high tide.

The sterile marsh zones along this estuarine domain are not affected by fluvial activity. Instead, aeolian processes from the southwest now dominate these unvegetated upper intertidal flats. Sediments are imported from deflation zones located along the foreshore and the coastal dune belt. In this manner aeolian deposits are produced, e.g. sand hummocks and undulating fine sand sheets.

This lower estuarine domain corresponds to the

middle estuary of Fairbridge (1980) or to the central estuarine basin of Dalrymple *et al.* (1992).

Marine domain

The marine domain of the Odiel estuary is separated from the open sea by a number of barrier islands composed by recurved spits. Salt marshes have developed in many places on the estuary side between successive spits. The whole environment, from subtidal channels to intertidal flats, is essentially dominated by the tides, although some wave-induced bedforms occur on exposed parts of the intertidal flats. The sediments are exclusively composed of marine sands and sandy silts (Borrego, 1992). Both ebb and flood currents reach velocities of 1.25 m s^{-1} along the outer sectors of the tidal channels, resulting in intense sediment transport. Sediments reaching the open coast are deposited to form ebb-tidal deltas (Borrego, 1992).

Longitudinal sand bars less than 10 m long are formed landwards by the flood tidal currents. These are attached to the channel margins and are reworked by waves to form the foreshore of the protected estuarine beaches. The energy of the tidal currents decreases towards the inner parts of the tidal channels, thus producing fine sediment traps. The growth and stabilization of the barrier islands produce wave-protected intertidal zones in which sandy muds are deposited.

The Odiel marine estuary is similar to the marine/lower estuary of Fairbridge (1980) or to the outer estuary of Dalrymple *et al.* (1992).

Piedras estuary

Fluvial domain

The fluvial domain of the Piedras estuary comprises the lower fluvial valley of the Piedras River up to the actual tide-influenced part. At present there is little freshwater dilution in the estuary because a dam is being constructed upstream. As a result, the entire estuary is currently tide domainated. Most of the sediment input, however, is still of fluvial origin because of a seasonal supply by minor tributaries below the construction site. By contrast, the tidally controlled geomorphic units are clearly dominated by salt marshes, thus reflecting a pronounced marine influence. Sediment transport in the tidal channels is dominated by spring-tidal ebb currents and/or seasonal river floods.

The sediments in the intertidal zone are of mixed origin. Along the channel margins and in the salt marshes, the tide-reworked fluvial sands are associated with fine, organic-rich sediments of tidal origin. Fluvial bars do not occur because the fluvial sediments are supplied in the course of unchannelized, high-energy fluvial discharge events which form small fan-deltas of medium to fine sands that are rapidly reworked by tidal action.

As in the case of the Odiel River, the sediment supply to the Piedras estuary is climatically controlled, both drainage basins receiving strongly seasonal rainfall. The fluvial domain of the Piedras estuary can be compared to the inner estuary of Dalrymple *et al.* (1992).

Central estuarine domain

The central estuarine domain of the Piedras estuary corresponds to the larger part of the Piedras River valley flooded in the wake of the Holocene transgression (Fig. 5). It is dominated by tidal processes which constantly rework large volumes of river-derived sediments, although tidal sediments are more abundant. Salt marshes are well developed. The central domain is dominated by low-energy conditions, as documented by the absence of larger bedforms. The tides are symmetrical, reaching very similar current velocities during both the flood and the ebb phase of the tidal cycle (generally <0.65 m s^{-1}).

During spring tides, sand is transported seawards, as revealed by the orientation of current ripples. Both muddy and sandy sediments are well mixed because the volume of sand supply is low and energy levels are equally low, the Piedras estuary being hypo-synchronic as pointed out earlier.

The central estuarine domain can be correlated with the central estuary basin of Dalrymple *et al.* (1992).

Marine domain

The marine domain of the Piedras estuary is characterized by the development of an elongate spit which occasionally closes the estuary mouth, thereby causing severely restricted hydrographic conditions in the inner estuary at times. The geomorphology of the mouth region reflects wave domination, tidal activity being limited to the inner side of the spit and the formation of ebb-tidal deltas in front of the spit, suggestive of an ebb-dominated system. The flood tide carries sandy sediments of marine origin into the subtidal parts of the channel, most of which are returned to construct the ebb-tidal deltas. These, in turn, are reworked by waves to form intertidal bars along the recurved barrier spit (Borrego *et al.*, 1993).

Some of the sands reach the shallow parts of the subtidal channel, where they are reworked by secondary wave action to be deposited in the intertidal zone of the protected estuarine beaches in form of parallel layers dipping towards the channel. Smaller volumes of sand are transported upstream by the flood currents during spring tides, producing thin, alternating layers of sand and mud.

Along the inner side of the spit, successive supratidal beach ridges, developed on the recurved parts of the prograded spit, alternate with intertidal swales to form a serrated estuarine shoreline. The beach ridges result from aeolian reworking of the recurved spit crests, whereas the swales are the depressions between successive crests. Being protected from wave action, the swales are areas of net tidal deposition (Dabrio & Polo, 1987; Dabrio & Zazo, 1987). These sheltered tidal deposits are mainly sandy, as the material is derived from the adjacent beach ridges by ebb-tide erosion. The swales are preferred habitats of marine phanerogams which can settle there on account of the low energy conditions. These plants are the source of organic-rich layers occasionally observed within the sandy successions (Fig. 8F).

Secondary wave action produces parallel intertidal sand bars along the adjacent channel margin. These migrate onshore to be attached to the west-facing sides of the recurved beach ridges to form small sand spits. Eastward growth of the spits results in further energy reduction in the intertidal depressions on their leeward sides. As a result of this, finer sediments can be deposited to produce an organic-rich muddy sand environment. Marine organisms, mainly *A. ecaudata* and *C. edule*, subsequently modify these deposits by further increasing sediment aggradation. This process continues until the depressions are infilled to the extent of enabling continental halophytes to develop a high salt marsh. These high salt-marsh areas, in turn, are covered by aeolian sands supplied by deflation from the adjacent beach ridges. In this manner, incipient dunes can complete the vertical aggradation process (Borrego *et al.*, unpublished).

This marine domain of the Piedras estuary is equivalent to the outer estuary of Dalrymple *et al.* (1992).

Guadiana estuary

Upper estuarine domain

Although tidal action is still present in this domain of the Guadiana estuary, fluvial sediments and bedforms as well as freshwater vegetation along the high intertidal fringes are characteristic features. The channel bed consists of massive gravel deposits of fluvial origin, the finer fractions of which can be reworked by the tides to form lag deposits further downstream. Lateral fluvial bars occur in the subtidal parts in this section, documenting a fluvial dominance. The only evidence for tidal action is the absence of medium-grained and finer sands, as these size fractions are easily removed by the tidal currents. Furthermore, the occurrence of mud layers on the lateral bars, interlayered with the gravels in the form of mud drapes, is another hint at tidal activity. These mud drapes are formed during the slack water periods at high tide, subsequent ebb currents being too weak to rework the muds at these topographic elevations.

Central estuarine domain

In contrast to the upper estuarine domain, the central domain of the Guadiana River is clearly controlled by the tides. This is evident from the good sorting of the sandy sediments in the deeper parts of the main channel. The main factor controlling the instantaneous position of the freshwater/saltwater interface is the nature of tidal wave propagation, although river discharge exercises a seasonal control over the tidal wave and sand supply.

Alternating deposits of very well-sorted fine sands and organic-rich muds are found along the shallow parts of the channel. Migrating ripples are developed in the fine sand at times of larger fluvial sediment supply. The organic-rich muds result from flocculation processes occurring in this section of the estuary during the months of low river discharge, when the freshwater interface is located further upstream. Intense wave action, generated by northerly winds, periodically acts in the shallows of the main channel.

The central estuarine domain of the Guadiana River is a tide dominated environment corresponding to the middle estuary of Fairbridge (1980) and partly also to the central estuarine basin of Dalrymple *et al.* (1992), the latter being somewhat more energetic (Morales, 1993).

Marine domain

The marine domain of the Guadiana estuary is the most extensive section in this case. It is characterized by a large variety of bedforms and sedimentary facies. As a result, sedimentary processes and products show large vertical, lateral and longitudinal variability. Sediment supply is twofold, as reflected by the mixing of sands of both fluvial and marine origin. The ebb tides transport sand into the estuary mouth, where they are dispersed by wave action to construct sand spits and barrier islands (Borrego *et al.*, 1992). Excess sand is returned to the system by the flood current, to be deposited in the backbarrier lagoons and tidal flats adjacent to the estuary. Across these lagoons an energy gradient is observed, which is reflected by a landward decrease in grain size.

The continuous process of lagoonal infilling has created a laterally prograding depositional system, in which new barrier islands and lagoons are formed to repeat the same cycle over and over again. The system is thus controlled by tidal activity in the backbarrier region and by waves along the open coast. The final result is a wave-dominated deltaic environment, consisting of a large spit that obstructs the estuary mouth and barrier islands that are generated on the swash platforms of the asymmetrical ebb-tidal delta in front of the main inlet (Fig. 7).

This marine domain of the Guadiana estuary thus corresponds to the marine or lower estuary of Fairbridge (1980), or to the outer estuary as defined by Dalrymple *et al.* (1992).

SUMMARY AND CONCLUSIONS

The lithofacies associations observed in the mesotidal estuarine systems along the Huelva coast clearly show the spatial distribution and relationship between sedimentary processes and their products in all three estuaries.

In the vicinity of the estuary mouths the sediments are of marine origin, their dynamics and depositional features being dominated by wave processes. This applies particularly to the Odiel and Piedras estuaries. In the case of the Guadiana estuary, on the other hand, strong tidal currents in the main channel are responsible for the seaward elongation of the spit.

The sediments of the central estuarine domain of the Piedras River, as well as the upper and lower estuarine domains of the Odiel River, are tidally controlled. By contrast, the central estuarine domain of the Guadiana River is dominated by fluvial sediments. Tidal energy is dominant in the central estuarine domains of all three estuaries, as reflected by the intense reworking of fluvially supplied sediments in the upper estuarine domains of the Odiel and the Guadiana Rivers.

The sediments in the fluvial domains of the estuaries are for the most part of mixed origin. In the case of the Odiel and Guadiana Rivers, the development of lateral bars signifies a local domination of fluvial process.

The most dynamic of the three systems is the Guadiana estuary, the others being in a more advanced stage of development, as reflected in the more extensive salt marshes and denser tidal channel networks. The incision of the Guadiana estuarine channel into bedrock promotes sediment bypassing and inhibits deposition in the estuary proper. Instead, deposition is delayed until the marine domain is reached, thereby promoting the development of a wave-dominated deltaic environment.

ACKNOWLEDGEMENTS

The project was financially supported by the Spanish Government, C.I.C.YT. Project PS89-0113, and the Junta de Andalucía, P.A.I. Group 4070. The help of the Puerto Autónomo de Huelva, who furnished hydrodynamic data from the Huelva littoral zone, is gratefully acknowledged. The kind attention given by Mrs. A. M. Mojarro is especially appreciated. Richard A. Davis, Jr (University of South Florida) kindly reviewed an early version of the manuscript. We thank him and two anonymous referees, as well as B. Flemming (Wilhelmshaven), for valuable suggestions.

REFERENCES

Allen, G.P. (1971) Relationships between grain size parameter distribution and current patterns in the Gironde estuary (France). *J. sediment. Petrol.* **41**, 74–88.

Allen, G.P., Bouchet, J.M., Carbonnel, P. *et al.* (1973) *Environments and Sedimentary Processes of the North Aquitaine Coast.* Guidebook, Inst. Géol. Bassin d'Aquitaine, 183 pp.

Ashley, G.M. & Zeff, M.L. (1988) Tidal channel classification for a low mesotidal salt marsh. *Mar. Geol.* **82**, 17–32.

Borrego, J. (1992) *Sedimentología del estuario del Odiel (Huelva-S.O. España).* Ph.D. thesis, Univ. Sevilla.

Borrego, J. & Pendon, J.G. (1988) Algunos ejemplos de influencia de los procesos antrópicos en el medio sedimentario: la ría de Huelva. *HENARES Rev. Geol.* **2**, 299–305.

Borrego, J. & Pendon, J.G. (1989a) Influencia de la actividad humana sobre la evolución sedimentaria de un sector de la ría de Huelva (España). *Geolis* **3**, 125–131.

Borrego, J. & Pendon, J.G. (1989b) Caracterización del ciclo mareal en la desembocadura del rio Piedras (Huelva). *XII Congr. Esp. Sedim., Bilbao, Comunicac.* **1**, 97–100.

Borrego, J., Morales, J.A. & Pendon, J.G. (1992) Elementos morfodinámicos responsables de la evolución reciente del estuario bajo del río Guadiana (Huelva). *Geogaceta* **11**, 86–89.

Borrego, J., Morales, J.A. & Pendon, J.G. (1993) Holocene filling of an estuarine lagoon along the mesotidal coast of Huelva: the Piedras River mouth, southwestern Spain. *J. Coast. Res.* **9**, 242–254.

Clifton, H.E. (1982) Estuarine deposits. In: *Sandstone Depositional Environments* (Eds Scholle, P.A. & Spearing, D.). *A.A.P.G. Mem.* **31**, 179–189.

Clifton, H.E. (1983) Discrimination between subtidal and intertidal facies in Pleistocene deposits, Willapa Bay, Washington. *J. sediment. Petrol.* **53**, 353–369.

Courau, P. (1983) Traces métalliques: principes des méthodes, réduction des contaminations. In: *Manuel des Analyses Chimiques en Milieu Marin.* (Ed. A. Aminot) CNPEO, 395 pp.

Dabrio, C.J. (1982) Sedimentary structures generated on the foreshore by migrating ridge and runnel systems on microtidal and mesotidal coasts of S. Spain. *Sediment. Geol.* **32**, 141–151.

Dabrio, C.J. & Polo, M.D. (1987) Holocene sea-level changes, coastal dynamics and human impacts in southern Iberian Peninsula. In: *Late Quaternary Sea-level Changes in Spain* (Ed. Zazo, C.), pp. 227–247. Museo Nal. de Ciencias Naturales, CSIC.

Dabrio, C.J. & Zazo, C. (1987) Riesgos geológicos en zonas litorales. *I.T.G.E., Serie geológica ambiental.*, 227–250.

Dalrymple, R.W., Knight, R.J., Zaitlin, B.A. & Middleton, G.V. (1990) Dynamics and facies model for a macrotidal sand bar complex, Cobequid Bay–Salmon River estuary (Bay of Fundy). *Sedimentology* **37**, 577–612.

Dalrymple, R.W., Zaitlin, B.A. & Boyd, R. (1992) Estuarine facies models: conceptual basis and stratigraphic implications. *J. sediment. Petrol.* **62**, 1030–1055.

Davis, R.A. Jr. & Hayes, M.O. (1984) What is a wave-dominated coast? *Mar. Geol.* **60**, 313–329.

Doty, M.S. (1946) Critical tide factors that are correlated with the vertical distribution of marine algae and other organisms along the Pacific Coast. *Ecology* **27**, 315–328.

Fairbridge, P.V. (1980) The estuary: its definition and geodynamic cycle. In: *Chemistry and Biochemistry of Estuaries* (Eds Olausson, E. & Cato, I.), pp. 1–36. Wiley, New York.

Flor, G. (1990) Tipología de dunas eólicas. Procesos de erosión-sedimentación costera y evolución litoral de la provincia de Huelva (Golfo de Cádiz occidental, Sur de España). *Estud. Geol.* **46**, 99–109.

Frey, R.W. & Howard, J.D. (1986) Mesotidal estuarine sequences: a perspective from the Georgia Bight. *J. sediment. Petrol.* **56**, 911–924.

Greer, S.A. (1975) Sand body geometry and sedimentary facies at the estuary-marine transition zone, Ossabaw Sound, Georgia: a stratigraphic model. *Senckenbergiana marit.* **7**, 105–135.

Hayes, M.O. (1979) Barrier island morphology as a function of tidal and wave regime. In: *Barrier Islands* (Ed. Leatherman, S.P.), pp. 1–27. Academic Press, New York.

Howard, J.D. & Frey, R.W. (1985) Physical and biogenic aspects of backbarrier sedimentary sequences, Georgia coast, U.S.A. *Mar. Geol.* **63**, 77–127.

Jouanneau, J.M. & LaTouche, C. (1981) The Gironde Estuary. In: *Contributions to Sedimentology 10* (Eds Füchtbauer, H. & Lisitzyn, A.P.), E. Schweizerbart'sche Verlagsbuchhandlung (Nägele v. Ober Miller), Stuttgart, 105 pp.

Knebel, H.J., Fletcher, C.H. & Kraft, J.C. (1988) Late Wisconsinan-Holocene paleogeography of Delaware Bay: a large coastal plain estuary. *Mar. Geol.* **83**, 115–133.

Le Floch, A. (1961) *Propagation de la marée dans l'estuaire de la Seine et en Seine-maritime.* PhD thesis, Univ. Paris.

McLaren, P. (1981) An interpretation of trends in grain size measures. *J. sediment. Petrol.* **51**, 611–624.

Mopu (1987) *Proyecto de trasvase de arenas a la Playa de Castilla.* Servicio de Costas de Huelva, 16 pp.

Morales, J.A. (1993) *Sedimentología del estuario del Guadiana. S.O. España-Portugal.* PhD thesis, Univ. Huelva.

Nichols, M.M., Johnson, G.H. & Peebles, P.C. (1991) Modern sediments and facies model for a microtidal coastal plain estuary: the James Estuary, Virginia. *J. sediment. Petrol.* **61**, 883–899.

Oertel, G.F. (1973) Examination of textures and structures of mud layered sediments at the entrance of Georgia tidal inlets. *J. sediment. Petrol.* **43**, 33–41.

Oertel, G.F. (1975) Ebb tidal inlets of Georgia Estuaries. *Estuarine Res.* **2**, 97–114.

Ojeda, J. (1988) *Aplicaciones de la teledetección espacial a la dinámica litoral (Huelva). Geomorfología y Ordenación del Territorio.* PhD thesis, Univ. Sevilla, 411 pp. (unpubl.)

Rodriguez, J. (1987a) Recent geomorphologic evolution in the Ayamonte-Mazagón sector of the South Atlantic coast (Huelva, Spain). *Trab. Neóg.-Cuatern.* **10**, 259–264.

Rodriguez, J. (1987b) Modelo de evolución geomorfológica de la flecha litoral de Punta-Umbría, Huelva, España. *Cuatern. y Geomorfol.* **1**, 247–256.

Sha, L.P. (1990) Sedimentological studies of the ebb-tidal deltas along the West Frisian Islands, The Netherlands. *Geol. Ultraiect., Univ. Utrecht* **64**, 1–159.

Simmons, H.B. (1955) Some effects of upland discharge on estuarine hydraulics. *Proc. ASCE* **81**, 1–20.

Vanney, M. & Menanteau, L. (1979) Types de relief littoraux et dunaires en Basse Andalousie (de Huelva à l'embouchure du Guadalquivir). *Mélang. Casa Velazquez* **15**, 5–52.

Warmer, J.E. (1971) Paleoecological aspects of a modern coastal lagoon. *Univ. California Publs Geol. Sci.* **87**, 1–131.

Woodroffe, C.D., Chappell, J.M.A., Thom, B.G. & Wallensky, E. (1989) Depositional model of a macrotidal estuary and flood plain, South Alligator River, Northern Australia. *Sedimentology* **36**, 737–756.

Zazo, C. (1980) *El Cuaternario marino-continental y el límite Plio-Pleistoceno en el Litoral de Cádiz.* PhD thesis Univ. Complutense.

Zeff, M.L. (1988) Sedimentation in a salt marsh-tidal channel system, southern New Jersey. *Mar. Geol.* **22**, 33–48.

Spec. Publs int. Ass. Sediment. (1995) **24**, 171–181

The tidal character of fluvial sediments of the modern Mahakam River delta, Kalimantan, Indonesia

R.A. GASTALDO*, G.P. ALLEN† *and* A.-Y HUC‡

**Department of Geology, Auburn University, AL 36849-5305, USA;*
†TOTAL, Centre Scientifique et Technique, Domaine de Beauplan, Route de Versailles, 78470 Saint-Rémy-les-Chevreuse, France; and
‡Institut Français du Pétrole, 1 & 4, av. de Bois-Préau, 92506 Rueil-Malmaison, France

ABSTRACT

The Mahakam River delta, Kalimantan, Indonesia, is a low wave-energy, mixed tide- and fluvially-controlled delta complex, situated at the eastern edge of the island of Borneo. The medium- to fine-grained terrestrial sediment originates from within a 75 000 km^2 drainage area. It is transported through the equatorial basin and debouches into the Makassar Strait. The Mahakam has two active distributary systems, directed north-east and south-east respectively, with an intervening interdistributary area consisting of a series of tidal channels and former fluvial distributary channels which today are no longer connected to the fluvial regime. A non-random sampling strategy was employed during a vibracoring programme conducted in 1988. The vibracores were collected along two transects: (i) cores from the first transect represent depositional environments within the tide-affected fluvial distributaries; and (ii) cores along the second transect were sampled from sites within the tidal interdistributary area. Sample sites of the distributary channel transect included lateral channel bars, distributary-mouth bars, and delta-front settings.

All the sediments recovered from subaqueous sample sites show varying degrees of tidal influence. Mud drapes and couplets of medium–very fine sands and mud are the most commonly encountered sedimentary successions in the active fluvial distributaries, being also characteristic of all tide-dominated distributary channels. Sand/mud ratios are variable, ranging from 90:10 to 30:70. The thickness of clay drapes is also variable. Sedimentary structures include wavy and lenticular bedding composed of asymmetrical ripples and trough cross-stratification. Ripples may be multi-directional within any one core, being inclined upstream, downstream or horizontally disposed. Sand and mud are mixed by bioturbation in the lower delta plain and delta front. Primary sedimentary structures in conjunction with the degree of bioturbation and the presence of phytoclast drapes appear to be useful criteria for the identification of ancient tide-influenced deltaic distributary channels.

INTRODUCTION

The sedimentological features of deltaic regimes have attracted much attention during the past three decades (e.g. Niger delta: Doust & Omatsola, 1990; Colorado River delta: Kames, 1970). It is a well-established fact that coastal delta morphologies are dependent upon the interplay between fluvial and marine processes. However, most data concerning the features of deltas are derived from studies detailing the end-member systems of a tripartite classification: fluvially-dominated, tide-dominated, or wave-dominated deltas (see e.g. Elliott, 1978a; Galloway & Hobday, 1983). Fewer studies have focussed on coastal regimes where this interplay is dominated by a combination of fluvial and marine processes (e.g. Allen *et al.*, 1979).

It is generally agreed that fluvially-dominated distributary channels are characterized by unidirectional flow with periodic stage fluctuations resulting from annual climatic oscillations. Bedload transport is inhibited only during episodic and/or anomalous events such as when waves are associated with strong onshore winds or river discharge is exceptionally low or high. As a result of this, sediments transported in suspension can be deposited within the distributary channels (Wright & Coleman, 1973). It is also believed that when marine influences dominate over fluvial processes, the sediments most affected are those debouched at the river mouth. In areas of moderate to high tidal range, where tidal processes are more effective, sediment transport may be significantly influenced in the lower parts of river courses (Elliott, 1978b). Tidal currents entering the distributaries during spring tides inundate the interdistributary areas. The tidal waters are temporarily stored at high-water slacks, to be released at the onset of the following ebb stage. This situation results in the sedimentation of suspended sediments, thus reflecting the tidal influence.

Vibracores collected within fluvially-dominated distributary channels of the Mahakam River delta, eastern Kalimantan, Indonesia, reveal that the overprinting of tidal features within this low wave-energy, mixed tide- and fluvially-controlled delta complex is not restricted to the mouth of distributaries. The purpose of this contribution is to describe in detail the tidal features and sedimentological variability within these distributary channels.

GEOGRAPHICAL SETTING OF THE MAHAKAM RIVER DELTA

The Mahakam River delta is located at the eastern edge of the island of Borneo (Fig. 1). Headwaters originate in the central highlands of Kalimantan and debouch into the Makassar Strait at the edge of the Kutei Basin (between 0°21′ and 1°10′ South Latitude and 117°40′ East Longitude). Deltaic sedimentation began in the Middle Miocene (LaLouel, 1979) and since then, several major deltaic complexes have accumulated. Each delta complex is separated by marine transgressions (Magnier *et al.*, 1975). The eastward prograding sedimentary wedge is 6000–8000 m thick. The Pliocene–Quaternary history of the delta has recently been reconstructed in the interdisciplinary MISEDOR project (Pelet, 1987).

The modern Mahakam River drainage system encompasses a 75 000 km^2 area. Sediments transported through the equatorial basin form a low wave-energy (mean wave heights are <0.6 m), mixed tide- (semi-diurnal tides with a mean range of 1.2 m and a maximum amplitude of 3 m) and fluvially-controlled delta (Combaz & De Matharel, 1978; Allen *et al.*, 1979). According to the shoreline classifications of Hayes (1979) and Nummedal & Fischer (1978), eastern Kalimantan would be considered a mixed-energy, low mesotidal coastline. The present delta comprises a thin sedimentary sequence, about 50–60 m thick, that overlaps older Pleistocene palaeodeltas (Roux, 1977). At present it is about 50 km ‘long’, as measured from the delta

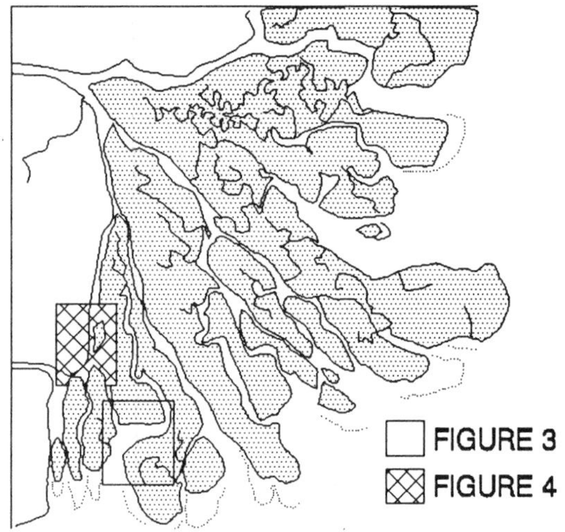

Fig. 1. Generalized locality map delineating the position of the Mahakam River delta on the eastern side of Borneo. Enlargement of the delta showing major channels and areas (Figs 3 & 4).

front to the initial bifurcation of the river at Sanga-Sanga, extending laterally along the coast for nearly 100 km. It is composed of approximately 2000 km^2 of wetlands in the subaerial delta plain and 1800 km^2 of delta front and prodelta sediments.

The Mahakam has two active fluvial distributary systems directed north-east and south-east respectively. An intervening interdistribuary area (Allen *et al.*, 1979) consists of a series of tidal channels that are at present not connected to the fluvial sections of the delta. Channel depths average 7–10 m, maximum depths attaining 15–18 m. The estimated discharge is 1000–3000 m^3 s^{-1} (Allen *et al.*, 1979). The waters carry a high suspended load of silt and clay, whereas medium-fine sands are transported in bedload. The estimated sediment load is 8 $*$ 10^6 m^3 yr^{-1} (Allen *et al.*, 1979). Tidal channels are similar in depth to river channels, with the base of tidal channels often in contact with underlying delta front sediments (personal observation, 1988). Surface tidal currents can exceed 1 m s^{-1} at channel mouths (Allen *et al.*, 1979). Medium to fine sands and silt are localized in the distributary channels and delta front (Allen *et al.*, 1979; Gayet & Legigan, 1987). The remainder of the delta is characterized by mud and clay which are distributed throughout various depositional environments (Allen *et al.*, 1977; Gastaldo & Huc, 1992).

The delta front, an intertidal to subtidal platform 8–10 km in width, fringes the delta plain (Combaz & De Matharel, 1978). Localized sand bars interrupt a monotonous mud sequence in which a marine fauna is often preserved. Laterally extensive deposits of peat accumulate in form of beach ridges at the delta front-delta plain boundary (Allen *et al.*, 1977; Gastaldo *et al.*, 1993). These beach ridges can be up to 2.5 m thick and are composed of river transported phytoclasts. They occur as far as 3 km inland and cover a total surface area of approximately 50 km^2.

Prodelta sediments accumulate on the outer limit of the delta front, where water depth increases to 35 m within 1 km of the delta-front margin (Kartaadiputra *et al.*, 1975). The prodelta is characterized by massive homogenous muds incorporating beds of carbonaceous clay and silt. Decayed phytoclasts (*sensu* Gastaldo, 1994) are common, palynomorphs are rare (Combaz, 1964; Bellet, 1987) and phytoplankton is absent (Combaz & De Matharel, 1978).

STUDY METHODS

Shallow subsurface sampling was conducted by vibracoring within the main organic-rich modern depositional environments (Gastaldo & Huc, 1992). The vibracore sampling pattern was non-random, being arranged in two transects. Cores from the first transect represent depositional environments within the mixed fluvial-tidal distributaries (Fig. 2). The second transect was designed to sample sites from within the tide-dominated inter-distributary setting.

Aluminium irrigation pipes, 6 m in length and 7.5 cm in diameter, were imported from Singapore. Pipe length was the limiting factor for the depth from which subsurface samples could be recovered. Depending on the depositional environment, core lengths varied from <3 m to >5 m. Three specially constructed 12-m tubes were used in fluvial channels where water depths exceeded 8 m (up to 7 m of core recovered). The cores were split longitudinally, one-half of the core being used for sedimentological description, photography and the fabrication of epoxy-resin peels (stored in the core warehouse, TOTAL Indonésie, Balikpapan, Kalimantan). The other half of the core was used for pH, E_H and temperature measurements (see Gastaldo & Huc, 1992) and the recovery of subsamples for phytological and geochemical analyses (see Gastaldo *et al.*, 1993; Huc & Gastaldo, unpublished). Selected phytoclasts were C^{14}-dated.

RESULTS

Channel morphology

The fluvial distributaries form a branching network of channels which radiate from the initial fluvial bifurcation at Sanga-Sanga (Fig. 2). The distributaries exhibit low sinuosity (Fig. 3), consisting of straight segments with channel bifurcations spaced 7–10 km apart. Bifurcations usually occur at sharp bends in the channels which result in deep scouring of the channel thalweg. Although the distributary channels do not exhibit any meandering, the thalweg within the channel exhibits a well-developed meandering pattern, resulting in an intrachannel network of side bars (Fig. 4). Channel cross-sections exhibit asymmetries similar to those observed in meander bends (Fig. 5).

The distributaries are incised into the flat inter-

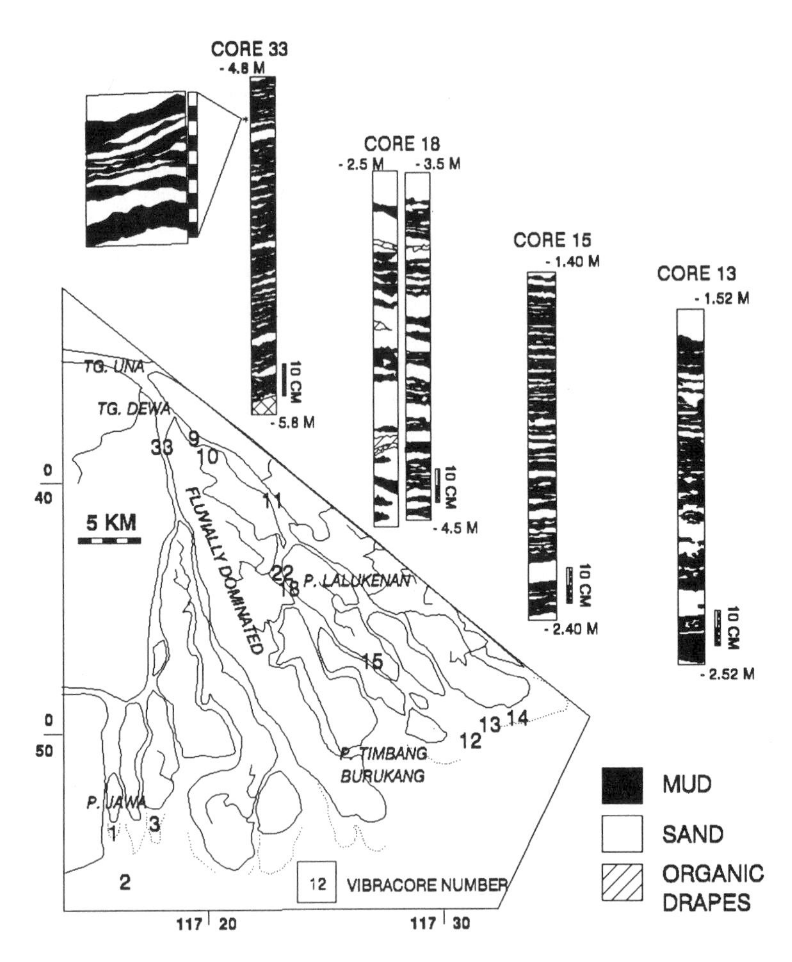

Fig. 2. Fluvially dominated transect in which vibracore sites are indicated by core numbers. Selected vibracore sections illustrate the prevalence of sand–mud couplet facies throughout the delta, including lateral channel bars more than 50 km inland. Facies variability is shown between the delta front (core 13, Fig. 7) and the lateral channel bar south of Tandjung Dewa (core 33; Fig. 8). Composition of organic drapes can be found in Gastaldo & Huc (1992).

tidal wetlands of the delta plain which are covered by a mixed hardwood and tropical palm forest in the interior subzone (Allen *et al.*, 1977) and a monoculture of the mangrove palm *Nypa fructicans* in the subtidal zone. Mangroves colonize newly formed tidal flats along channel margins and emergent channel bars in the subtidal zone, reflecting the upper limits of saline water influence. Channels are 0.5–1.5 km wide and do not exhibit any marked flaring at their mouths, in this respect contrasting to the more trumpet-shaped tidal channels. The positions of the distributaries have been stable during the past 50 years, there being very little or no lateral migration. Distributary thalweg depths vary from 6–>15 m (Fig. 5). The greatest depths occur at channel bifurcations (Fig. 4). In the straight channel reaches, the depth of the thalweg is relatively constant at 6–10 m.

Distributary channels are floored with sand brought in by the Mahakam River. Sand accumulates in form of elongate lateral bars along the channel banks (Figs 4 & 5). The lengths of these lateral bars vary between 2 and 3 km and the resulting sand deposits form elongate pods of similar length. Thicknesses may reach 7–8 m (Fig. 4; Allen *et al.*, 1977).

In the distributary mouths, tidal flow is more pronounced relative to river discharge and mid-channel bars separate ebb- and flood-dominant channels. In these zones the channel thalweg is non-erosive, the sediments being generally more muddy and the sands finer-grained.

A number of bore holes have been sunk within the distributary channels several tens of km landwards from the river mouth. These have shown that the sediments within the distributaries consist of 5–7 m thick erosive-based deposits of medium sand (Allen *et al.*, 1977, 1979). The sand bodies corres-

Fig. 3. Oblique aerial photograph, directed northwards, of the lower delta plain (Fig. 1). Distributary channels are of low sinuosity and consist of straight segments with channel bifurcations occurring about every 7–10 km. Tidal channels exhibit high sinuosity and can be seen to cross swamps. Mangrove swamps can be seen to fringe the monoculture *Nypa* swamps, while juvenile mangroves can be seen to have pioneered tidal flat deposits (lower left).

pond to the side bars mentioned above. As in the case of fluvial point bars, they form deposits roughly equivalent to the thalweg depth, being overlain by 'mud-flat' silty clay and organic-rich clay with abundant plant detritus. The base of the channels commonly incise underlying delta-front sands and muds, the distributary channel sands overlying partly eroded delta-front mouth-bar deposits.

Sediment facies

Six sediment facies can be recognized in the vibracores recovered from the fluvially-dominated channel transect (Gastaldo & Huc, 1992), two of these being particularly distinctive and serving to characterize all channel bars. These are the *sand–mud couplet facies* and the *bioturbated sand–mud couplet facies*. As the names imply, both are composed of alternating layers of sand and mud that appear to have a couplet structure (Fig. 2). Sand grain size is variable throughout the transect. Medium sand dominates in the upper reaches of the distributary channels (vibracores 9, 33), whereas fine–very fine sands are characteristic of the more distal parts of the river (vibracores 15, 18). The sand/mud ratio is variable, ranging from 90: 10 to 30: 70, again dependent on the location of the vibracores (see below). Sand colour varies from yellowish-tan to olive green-grey or dark grey-tan. Primary sedimentary structures include trough cross-bedding, small-scale ripples, and lenticular and wavy bedding. Ripples may be asymmetrical or symmetrical, isolated or in sets (up to 7 cm thick, but averaging at 2 cm) and variable in their inclination (Fig. 6). Thus, ripples are inclined upstream, downstream or are horizontally disposed. Their orientation may be unidirectional, bidirectional or multi-directional within any single core.

Muds up to several centimetres thick and/or bedded phytoclasts up to 2 cm thick overlie the rippled sections, being generally horizontally disposed. Coloration of the mud fraction varies from grey-tan to grey-brown. No primary sedimentary structures are visible. Planar bedding can be seen in the phytoclast fraction, being accentuated by the disposition of plant litter. Plant detritus is composed of entire leaves or leaf fragments, woody and resistant fibrous clasts, as well as dammar (Gastaldo & Huc, 1992). In some instances, phytoclasts may occur scattered within mud drapes. In addition, a variety of insect remains have been recovered from all allochthonous assemblages and bivalve fragments or disarticulated shells occur locally (see below).

The principal difference between the two sediment facies is the considerable bioturbation of the one, ranging from nearly complete homogenization to the presence of discrete, isolated burrows. Homogenization is characterized by the incorporation of sandy sediment into the mud drapes (Fig. 7). Isolated burrows may be vertical, horizontal, or U-shaped. Burrow diameter ranges from <0.4 to 2 cm, and may be lined with faecal pellets. Most burrows are sand-filled and may be cemented by early diagenetic calcite (as verified by XRD).

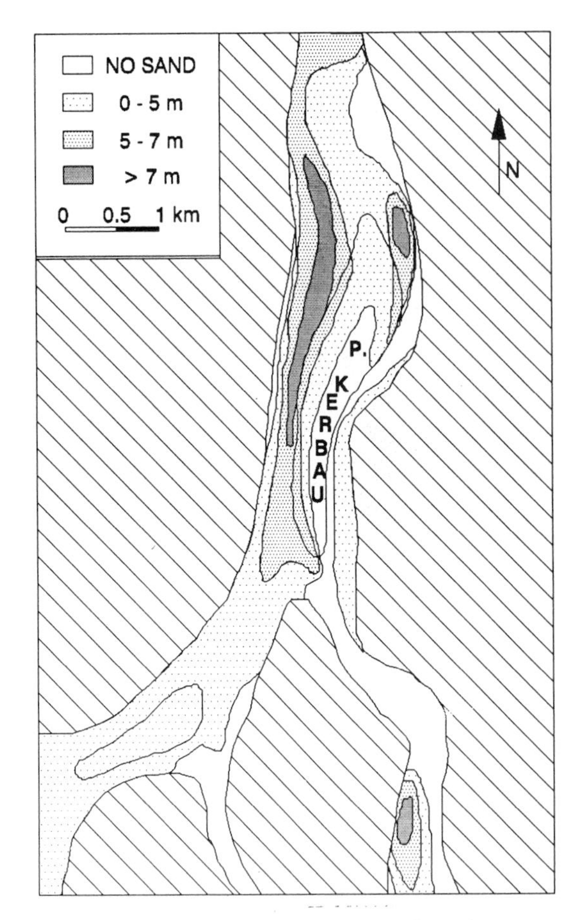

Fig. 4. Isopach of lateral channel bar configuration at P. Kerbau (Fig. 1). Lateral channel bars form a low sinuosity sand ribbon that consists of elliptical pods that result from the coalescence of individual lateral bars.

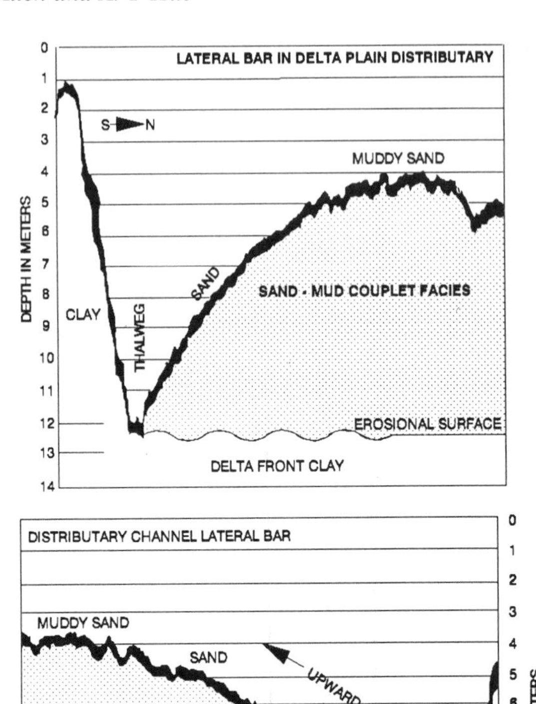

Fig. 5. Cross-sectional morphology of two distributary channels in the delta plain as reconstructed from echo soundings, surface sediment samples and vibracores.

In contrast to the distinctive sand–mud couplet facies and the bioturbated sand–mud couplet facies, the four other facies (i.e. the *bioclastic sand facies*, the *massively-bedded sand facies*, the *grey-black mud facies* and the *organic-rich mud facies*) identified in the vibracores form subordinate constituents, although one geographically localized facies delineates a delta-front setting (see below). The bioclastic sand facies is distinguished by the presence and incorporation of macrofaunal shell fragments (imbricated shell hash in asymmetrical ripples) in a medium-dark grey fine sand. The massively-bedded sand facies is oversaturated and not very common in the core records. Ripples may be present at the base of this facies, but otherwise it is devoid of any primary sedimentary structures. The only visible primary structures in the sulphurous, grey-black mud facies are defined by horizontally bedded phytoclasts. The same is true for the brown-tan to medium-dark grey organic-rich mud facies, but here phytoclasts do not undergo sulphur reduction (see Gastaldo & Huc, 1992, for details of these facies).

Transect variability

All lateral channel bars investigated are composed of the same sand–mud couplet facies. The only visual difference within the delta plain is an increase in the overall sand/mud ratio towards the delta front. Couplets in the upper delta plain (cores

Fig. 6. Core 20: one of 10 vibracores recovered from a lateral channel bar near P. Lalukenan (Fig. 2). Note the interbedding of sand and mud, with sand disposed in isolated ripples, inclined strata and trough cross-bedded structures. Inclination of beds may be variously oriented. Core barrel is cut into 1 m sections. Top of core (sediment–water interface) is in the upper left; bottom of core is in the lower right. DNAG scale is in cm.

Fig. 7. Core 13: one of seven recovered from the delta front (Figs 2 & 9). Bioturbation from sand into mud is evident in the middle setion of the core. Entire and fragmentary invertebrate bioclasts are interspersed within the facies of the lowermost delta plain and delta front. DNAG scale is in cm.

9, 33) reflect minimum sand and maximum mud components, with sand/mud ratios lying between 15:85 and 25:75 (Figs 2 & 8). Sediments in vibracore 11, taken approximately 25 km downstream of the bifurcation at Tg. Una, reflect a marked increase in the amount of sand, the sand/mud ratio approaching 50:50. Ratios of the sand component in the lateral channel bar deposited adjacent to P. Lalukenan, approximately 38 km from Tg. Una, vary considerably depending on the positions of the vibracores within the bar. Ten vibracores (cores 18–23, 29–32) were extracted in close proximity to each other from the bar top (Fig. 6). Sand/mud ratios within these cores ranged from 80:20 (cores 18, 21) to 40:60 (core 20), with higher sand/mud ratios near the bar top. The higher proportion of sand in this bar is not due to an increased number of couplets, but rather to the deposition of cross-bedded and massive (up to 25 cm thick) sand (Fig. 2). Couplets observed in vibracore 15, over 50 km from Tg. Una are composed mainly of alternating silt and mud layers. Where sand–mud couplets exist (> −4.0 m depth), the sand/mud

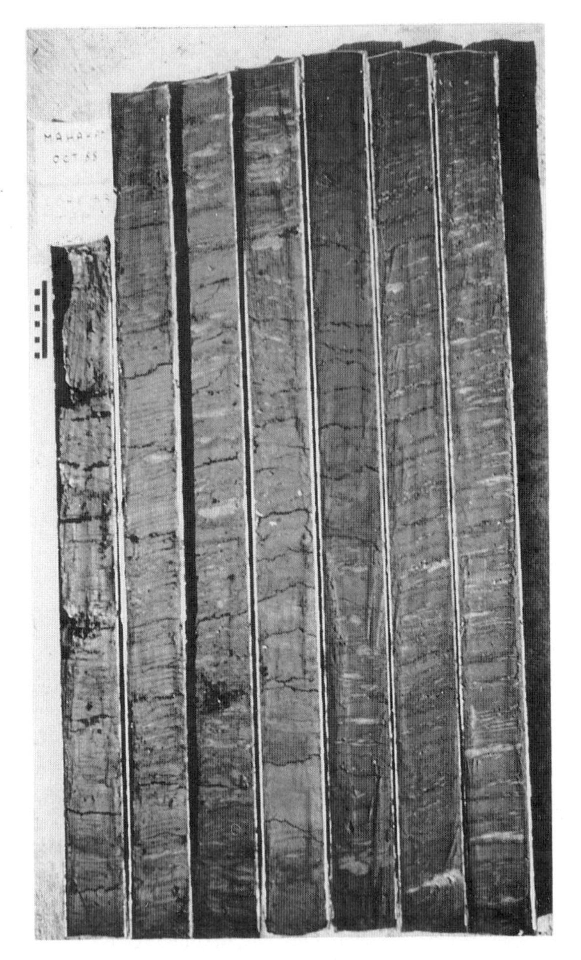

Fi.g 8. Core 33: recovered from the lateral channel bar immediately south of Tandjung Dewa (Fig. 2). Note the presence of sand and mud couplets in which thinner and thicker packages of sand–mud can be seen to alternate throughout the length of the core. Sand is concentrated in laminar, ripple and flaser-bedded structures primarily oriented downcurrent. DNAG scale is in cm.

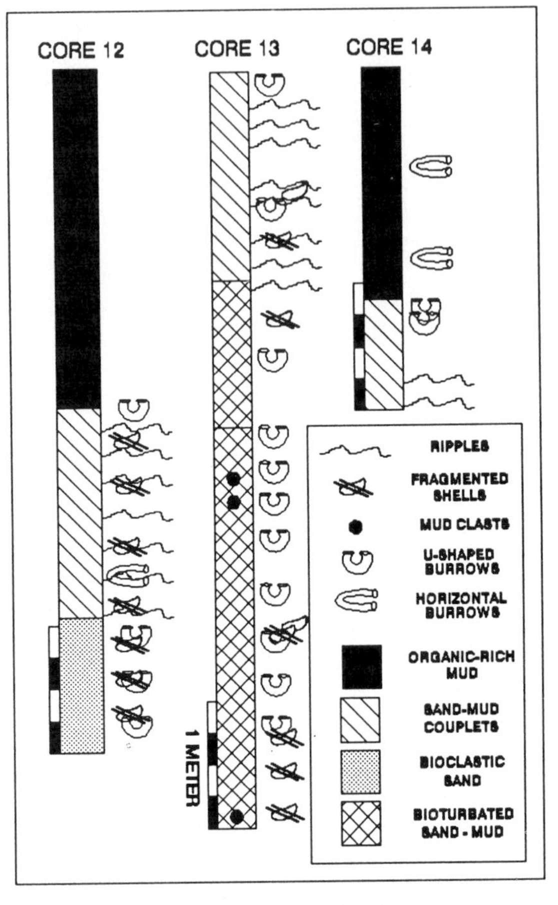

Fig. 9. Schematic representation of variability in the delta front environment between three cores recovered in close proximity to each other (Fig. 2). Organic-rich mud is characteristic of a tidal flat environment, whereas bioclastic sands characterize distributary mouth bars. For details of sediment facies see Gastaldo & Huc (1992).

ratio is 90:10. The upsection change to silt–mud couplets is due to the channel margin location of the core. Bioturbation first appears in these lower delta plain sites.

Facies diversity increase in the lower delta-plain channels, where there is more interaction between marine and fluvial processes. The delta front (cores 1–6, 12–14) is the depositional environment exhibiting the greatest diversity of sediment facies (Gastaldo & Huc, 1992; Fig. 9). The sand–mud couplet facies is still distinctive, but may be severely bioturbated. Macrofaunal shell fragments are often present and semi-lithified mud clasts can be found. The sand–mud couplet facies often overlies a bioclastic sand that is representative of distributary mouth bar deposits. It may be overlain by organic-rich muds signalling a tidal flat environment.

DISCUSSION

There is general agreement that couplets of alternating sand and mud result from the regular succession of currents and slackwater characteristic of tidal regimes (Reineck, 1972; Terwindt & Breussers,

1972; Visser, 1980; Boersma & Terwindt, 1981). The alternation between deposition from bedload and suspension is most effective in meso- to macrotidal environments. It is controlled by the relative volumes of sand and mud, current velocity, as well as the duration of the bedload and suspension load depositional phases (Elliott, 1978a; Klein, 1985). When the amount of sand is small relative to that of the mud (Reineck & Wunderlich, 1968), a large variety of wavy, lenticular and flaser bedding occurs. Their presence are but some of the criteria used to distinguish sedimentation originating from tidal processes (Klein, 1985). Such processes have been ascribed to the lower parts of distributary pathways (Elliott, 1978b).

The sedimentology of the Mahakam distributaries is controlled by the diurnal tidal current reversal and the lunar neap-spring cyclicity in amplitude. Field measurements in the southernmost distributary (Allen *et al.*, 1979) indicate that the duration of slackwater varies with tidal amplitude. During neap tide, for example, the period of zero bottom current can last for more than 1 h. Tidal current reversal occurs throughout the entire delta plain. During periods of heavy rainfall, however, the flood currents are strongly reduced at neap tide in the proximal reaches of the distributaries.

The hydrology of the distributaries is also affected by the existence of a pronounced salt wedge which can extend landwards up to the proximal delta plain (approximately 20–30 km from the distributary mouth bars). Salinities recorded in these zones range from 0‰ at the surface to 20–30‰ near the bottom (Allen *et al.*, 1979, p. 68). The resulting density stratification creates a stratified density circulation regime comprising landward moving bottom water that converges with seaward moving riverine water landwards of the salt intrusion. This creates a density current node that traps suspended sediment within a turbidity maximum in the zone of salt intrusion. The turbidity maximum migrates with the tides, affecting practically the entire delta plain. Suspended mud concentrations within the turbidity maximum can exceed 1000 mg l^{-1} (Allen *et al.*, 1979), such that a mud lamina can be deposited during a current slack lasting several hours.

The combination of tidal current reversal with high suspended sediment concentrations results in the accumulation of clay laminae during tidal slack. The potential for slackwater mud sedimentation is greatly enhanced during neap tides when the tidal currents are too weak to resuspend the deposited mud, particularly during waning tidal amplitudes. The presence of strong bottom currents would probably impede the preservation of slackwater mud laminations during spring tides. Therefore, as in most tide-dominated estuarine or distributary channel deposits, the sand–mud cycles that represent semi-diurnal slacks would be preserved mainly during neap tides.

The predominance of sand–mud couplets in all vibracores recovered from fluvial distributaries, even at sites over 50 km inlands from distributary mouth bars, indicates that tidal influences play a significant role in the development of within-channel deposits in such mesotidal settings. The distribution and quantity of sand that accumulates within any particular channel bar appear to be related to location within both the delta and the bar. For example, the sedimentological features observed in the ten vibracores (cores 18–23, 29–32; Fig. 6) extracted in close proximity to each other are highly variable. Sediments accumulate in bars along the margins of straight channels. One might expect that under such circumstances single depositional events (e.g. neap tide-generated mud laminae) should be traceable throughout the bar. However, it was not possible to find any correlation between particular beds, couplets or stratigraphical intervals, not even between two cores taken within 5 m of each other. Furthermore, these cores displayed a higher percentage of thick, clean sand. This feature may be explained by, amongst other things, the bar being located immediately downstream of a bifurcation and/or bar development occurring in a wide and straight part of the distributary (e.g. at the upstream end of a relatively linear channel).

Alteration of the sedimentological fabric within the bars occurs where conditions favour infaunal colonization. Bioturbation is restricted to the lower part of the delta, particularly within the delta front. This is probably related to the salinity gradient along the channel bottom and salinity tolerances of the burrowing organisms. Bottom salinities seawards of zones located about 10 km from the channel mouths are always greater than 10–20‰, whereas salinities measured landwards of a line about 30 km from the mouth are very low or zero (Allen *et al.*, 1979). The degree to which the sediments are homogenized may also be a function of increased substrate availability. The absence of bioturbation in the upper reaches of fluvial channels is primarily due to chemical barriers to infaunal colonization.

Because of the limitation imposed by the restricted number of vibracores, it is not possible to fully explain sediment variability and distribution within the bars, nor to accurately reconstruct bar development. Couplet generation is probably sporadic, related to neap–spring or equinoxial cyclicity rather than to daily fluctuations in tidal regime. The bundles of couplets observed within the vibracores do not exhibit the same rhythmic features as those described from daily deposition in other estuarine settings (Tessier, 1992; Tessier *et al.*, 1992, in press). In addition, a C^{14}-date of 765 ± 200 yr BP was recorded in an organic bed at -2.0 m depth (core 32—Krueger sample GX-14772 described as woody; Gastaldo & Huc, 1992). Leaves recovered from this organic drape were mostly intact. There was no indication that they had undergone mechanical fragmentation, a feature commonly associated with re-entrainment (see Gastaldo *et al.*, 1993; Gastaldo, 1994). However, since the litter bed was analysed in a bulk sample, it is possible that the C^{14}-date was biased by the presence of woody detritus in the litter. Gastaldo & Huc (1992) note that C^{14}-dated wood recovered near the top of this bar was at least 5200 years older than the organic drape that was dated at 765 ± 200 yr BP. The wood recovered near the top of the bar thus represents recycled organic material of high mechanical resistance. With this in mind, the C^{14}-date of the organic drape may not be an accurate reflection of the actual age of the horizon in the bar and may hence provide a misleading estimate of sedimentation rate. Even so, if these couplets had resulted from daily tidal cyclicity, we would have expected to find much younger leaf litter at this depth. In spite of the fact that the C^{14}-date may be biased towards an older age, the presence of entire leaves as organic drapes supports the contention that couplets are deposited intermittently and that deposition must be related to as yet unknown changes in flow conditions.

CONCLUSIONS

One of the characteristics of tidal deposits is their cyclic nature and, in particular, the rhythmic alternations of sand and mud deposited in response to current velocity cycles that occur with different periodicities. Lateral channel bars up to 15 m thick form a low sinuosity sand ribbon that consists of coalesced elliptical pods representing individual lateral bars. The sedimentary structures within these bars would generally be ascribed to tidal processes but, contrary to common opinion, these characteristic ‘tidal features’ are not restricted to the lower parts of deltaic distributary channels. These features may be the primary, and even unique, sedimentary structures in distributary channel deposits where tidal ranges are of at least mesotidal amplitude. Such sedimentary structures (e.g. alternating sand/mud layering, trough cross-stratification, asymmetrical ripples, and wavy, lenticular and flaser bedding) can be found in thick accumulations within lateral channel bar deposits some tens of kilometres inland. In the Mahakam, these features are found just below the first bifurcation of the river at Tg. Una. In the upper delta plain, these deposits are unaffected by bioturbation. Bioturbation associated with lateral channel bars is restricted to the lower reaches of distributary channels and distributary-mouth bars. The absence of bioturbation within wavy-, lenticular- and/or flaser-bedded sediments, along with the presence of thick organic drapes composed of phytoclasts within a thick (10 m or more) ‘tidal’ accumulation may represent useful criteria in ascribing deposits in the rock record to tidally-influenced fluvial distributary channels rather than to tidal flat environments.

Distributary mouth bars (Mayor Jawa, Tandjung Bukan and Mayor Bujit) are not as thick as lateral channel bars. They exhibit facies similar to lateral channel bars, except that there is more bioturbation and that fragmented macro-invertebrate shells occur together with phytoclasts. Bivalves, gastropods and echinoids are of allochthonous origin, transported into these shallow-water areas by waves and tides. The delta front is the most complex setting of fluvial sediment deposition. Rapid facies changes occur in response to fluctuations in the balance between fluvial discharge and marine processes, resulting in an interplay of facies that may characterize either fluvial, transitional or fully marine depositional settings.

ACKNOWLEDGEMENTS

We would like to thank Mr Choppin de Janvry and Mr Yves Grosjean, TOTAL INDONESIE, for their support and assistance with logistics during our field work in 1988. Dr William A. DiMichele is thanked for his tireless efforts as chief field-assistant. Dr Bruce Purser, University of Paris-Sud, Orsay, and Dr Bernard Durand are thanked for their support and assistance to the senior author during a 1988–89 sabattical in Paris. Support for this project was provided by TOTAL INDONESIE. In addition, it was supported in part by a grant from the National Science Foundation (EAR 8803609) to RAG. The Petroleum Research Fund, as administered by the American Chemical Society, is also acknowledged for partial support of this project (ACS PRF 20829-AC8).

REFERENCES

ALLEN, G.P., LAURIER, D. & THOUVENIN, J. (1977) Sediment distribution patterns in the modern Mahakam delta. *Indon. Petrol. Ass., Proc. 5th Ann. Conv. (Jakarta, 1976)*, 159–178.

ALLEN, G.P., LAURIER, D. & THOUVENIN, J.P. (1979) Etude sédimentologique du delta de la Mahakam. TOTAL, *Compagnies Françaises des Pétroles, Paris, Notes Mém.* **15**, 1–156.

BELLET, J. (1987) Palynofaciès et analyses élémentaires de la matière organique. In: *Le Sondage Misedor: Géochimie Organique des Sédiments Plio-Quarternaires du Delta de la Mahakam (Indonésie)* (Ed. Pelet, R.), pp. 183–196. Editions Technip, Paris.

BOERSMA, J.R. & TERWINDT, J.H.J. (1981) Neap–spring tide sequences of intertidal shoal deposits in a mesotidal estuary. *Sedimentology* **28**, 151–170.

COMBAZ, A. (1964) Les palynofaciès. *Rev. Micropaléont.* **7**, 205–218.

COMBAZ, A. & DE MATHAREL, M. (1978) Organic sedimentation and genesis of petroleum in Mahakam Delta, Borneo. *Am. Ass. Petrol. Geol. Bull.* **62**, 1684–1695.

DOUST, H. & OMATSOLA, E. (1990) Niger Delta. *AAPG Mem.* **48**, 201–238.

ELLIOTT, T. (1978a) Deltas. In: *Sedimentary Environments and Facies* (Ed. Reading, H.G.), pp. 97–142. Blackwell, Oxford.

ELLIOTT, T. (1978b) Clastic shorelines. In: *Sedimentary Environments and Facies* (Ed. Reading H.G.), pp. 143–177. Blackwell, Oxford.

GALLOWAY, W.E. & HOBDAY, D.K. (1983) *Terrigenous Clastic Depositional Systems: Applications to Petroleum, Coal, and Uranium Exploration.* Springer Verlag, New York, 423 pp.

GASTALDO, R.A. (1994) The genesis and sedimentation of phytoclasts with examples from coastal environments. In: *Sedimentation of Organic Particles* (Ed. Traverse, A.). Cambridge University Press, Cambridge, pp. 103–127.

GASTALDO, R.A. & HUC, A.Y. (1992) Sediment facies, depositional environments, and distribution of phytoclasts in the Recent Mahakam River Delta, Kalimantan, Indonesia. *Palaios* **7**, 574–590.

GASTALDO, R.A., ALLEN, G.P. & HUC, A.Y. (1993) Detrital peat formation in the tropical Mahakam River delta, Kalimantan, eastern Borneo: formation, plant composition, and geochemistry. *Geol. Soc. Am. Spec. Publs.* **286**, 107–118.

GAYET, J. & LEGIGAN, PH. (1987) Etude sédimentologique du sondage MISEDOR (delta de la Mahakam, Kalimantan, Indonésie). In: *Le Sondage Misedor: Géochimie Organique des Sédiments Plio-Quaternaires du Delta de la Mahakam (Indonésie)* (Ed. Pelet, R.) pp. 23–72. Editions Technip, Paris.

HAYES, M.O. (1979) Barrier island morphology as a function of tidal and wave regime. In: *Barrier Islands from the Gulf of St. Lawrence to the Gulf of Mexico* (Ed. Leatherman, S.), pp 1–27. Academic Press, New York.

KAMES, W.H. (1970) Facies and development of the Colorado River delta in Texas. In: *Deltaic Sedimentation, Ancient and Modern* (Eds. Morgan, J.P. and Shaver, R.H.). *Soc. Econ. Paleontologists and Mineralogists, Spec. Publ.* **15**, 78–106.

KARTAADIPUTRA, L., MAGNIER, P., & OKI, T. (1975) The Mahakam Delta, Kalimantan, Indonésie. *Proc. 9th World Petrol. Congr. (Totajv)* **2**, 239–250.

KLEIN, G. DEVRIES (1977) *Clastic Tidal Facies.* Continuing Education Publ. Co., Champaign, IL, 149 pp.

KLEIN, G. DEVRIES (1985) Intertidal flats and intertidal sand bodies. In: *Coastal Sedimentary Environments* (Ed. Davis, R.A. Jr), pp. 185–224. Springer-Verlag, New York.

LALOUEL, P. (1979) Log interpretation in deltaic sequences. *Indon. Petrol. Ass., Proc. 8th Ann. Conv.* **1**, 247–290.

MAGNIER, P., OKI, T. & KARTAADIPUTRA, L. (1975) The Mahakam Delta, Kalimantan, Indonésie. *Proc. 9th World Petrol. Congr. (Tokyo)* **2**, 239–250.

NUMMEDAL, D. & FISCHER, I. (1978) Process-response models for depositional shorelines: the German and Georgia bights. *ASCE, Proc. 16th Coastal Eng. Conf.*, 1215–1231.

PELET, R. (Ed.) (1987) *Le Sondage Misedor: Géochimie Organique des Sédiments Plio-Quaternaires du Delta de la Mahakam (Indonésie).* Editions Technip, Paris, 383 pp.

REINECK, H.-E. (1972) Tidal flats. In: *Recognition of Ancient Sedimentary Environments* (Eds Rigby, J.K. & Hamblin, W.K.). S.E.P.M. *Spec. Publs* **16**, 146–159.

REINECK, H.-E. & WUNDERLICH, F. (1968) Classification and origin of flaser and lenticular bedding. *Sedimentology* **11**, 99–104.

ROUX, G. (1977) The seismic exploration of the Mahakam Delta — or — 'Nine years of shooting in rivers, swamps and very shallow offshore'. *Indon. Petrol. Ass., Proc. 6th Ann. Conv.* **2**, 109–142.

TERWINDT, J.H.J. & BREUSSERS, H.N.C. (1972) Experiments on the origin of flaser, lenticular, and sand–clay alternating bedding. *Sedimentology* **19**, 85–98.

TESSIER, B. (1992) Upper intertidal cyclic deposits in the Bay of Mont-Saint-Michel (Normandie, NW France). In: *Tidal Clastics '92, Abstr. Vol.* (Ed. Flemming, B.W.). *Cour. Forsch.-Inst. Senckenberg* **151**, 82–84.

TESSIER, B., ARCHER, A.W. & FELDMAN, H.R. (1992) Comparison of Carboniferous tidal rhythmites (Eastern and Western Interior Basins, USA) with modern analogues (The Bay of Mont-Saint-Michel, NW France). In: *Tidal Clastics '92, Abst. Vol.* (Ed. Flemming, B.W.). *Cour. Forsch.-Inst. Senckenberg* **151**, 84–85.

VISSER, M.J. (1980) Neap-spring cycles reflected in Holocence subtidal large-scale bedform deposits: A preliminary note. *Geology* **8**, 543–546.

WRIGHT, L.D. & COLEMAN, J.M. (1973) Variation in morphology of major river deltas as functions of ocean waves and river discharge regimes. *Am. Ass. Petrol. Geol. Bull.* **47**, 370–398.

Spec. Publs int. Ass. Sediment. (1995) **24**, 183–191

Tidal lamination and facies development in the macrotidal flats of Namyang Bay, west coast of Korea

Y.A. PARK*, J.T. WELLS†, B.W. KIM* *and* C.R. ALEXANDER‡

**Department of Oceanography, Seoul National University, Seoul, 151–742 Korea;*
†Institute of Marine Science, University of North Carolina, 28557, USA; and
‡Skidaway Institute of Oceanography, Savannah, GA 31416, USA

ABSTRACT

Most of the west coast of Korea is fronted by broad intertidal sand- and mud-flats formed in a macrotidal environment that is seasonally subjected to monsoonal winds and intense winter storm surges. Field studies, conducted in about 60 km^2 of Namyang Bay, Korea, were undertaken for the dual purpose of describing tidal flat sedimentary facies (surficial and vertical) and tidal lamination structures (core slabs). Surficial sediments coarsen both seawards (from 7.3 phi to 4.5 phi) and with depth, resulting in a decrease in fine silt and clay-sized material. Furthermore, the data highlight the environmental significance of primary fine lamination in this high-tide-range intertidal depositional environment. Three types of primary fine lamination can be recognized on the basis of grain-size and lamina thickness. Variability of lamina thickness appears to be related to the spring–neap tidal cyclicity and high–low turbidity variations.

INTRODUCTION

Tidal flat sedimentation is either associated with major rivers and their downdrift estuarine and deltaic deposits or occurs restricted to coastal embayments and the lee of barrier islands. The broad and extensive system of tidal flats along the west coast of Korea in the south-eastern Yellow Sea is developed on an open, exposed coast and thus grades laterally into a shelf depositional system. The tidal flats of Korea are of general scientific interest because: (i) there is an apparent lack of a major sediment source; (ii) the tidal flats are developed in a macrotidal environment (tidal range 4–9 m) that is periodically subjected to strong winter storm surges due to monsoonal winds; and (iii) the tidal flats serve as large (summer) reservoirs for fine-grained sediments (mainly silts) that can undergo exchange with the adjacent shelf at frequencies ranging from tidal to at least annual cycles.

This paper describes and discusses (i) the spatial distribution of textural facies on the surface and down-core within the tidal flats; (ii) the flow characteristics and suspended sediment variability in the major tidal channels; and (iii) the lamination structures that are unique to this tidal flat system.

GEOGRAPHICAL SETTING

In general, the west coast of the Korean peninsula is fronted by extensive mud- and sand-flats (Wells & Huh, 1979; Park, 1987) and by a number of large subtidal sand bodies (Off, 1963; Klein *et al.*, 1982). It was during the Holocene transgression that many embayments were filled with mud and that wide tidal flats developed from clastic sediments derived, at least in part, from local sources. Much of the locally derived sediment is dispersed by intense monsoonal surges (cold air outbreaks) which move down the axis of the Yellow Sea, beginning in mid–late October and ending in April (Murakami, 1979; Huh, 1982). Throughout this period of winter monsoons, winds blow from the north for a duration of 1–2 days every 5–7 days at speeds that can exceed 40 knots. These surges are particularly effective at destratifying the water column, resuspending and mixing sediments, and driving large water masses to the south in a seasonal flow, referred to as the South Korean Coastal Current (Uda, 1966; Lee, 1968; Wells & Huh, 1984; Yoo, 1986; Wells, 1988). On an annual time-scale, sediment supply to the coast by rivers is low during winter, when north winds generate large waves in the Yellow Sea, and at a peak in summer, when winds from the south produce little wave activity (Fig. 1). It is reasonable to assume, then, that coastal muds are eroded in winter and carried south by the South Korean Coastal Current into or through the Korea Strait. During high river discharge in summer, deposition may occur over previously eroded surfaces, thus replenishing the coastal belt of intertidal sediments.

As a generalization, intertidal sediments coarsen seawards (Lee *et al.*, 1985; Park, 1987). Dominant tidal flat organisms are crabs, polychaetes and suspension-feeding molluscs, including oysters (Park, 1987). Korean tidal flats have been described as 'gelatinous' or 'oozing' mud-flats at the one extreme (Bartz, 1972; Wells, 1983) and at the other,

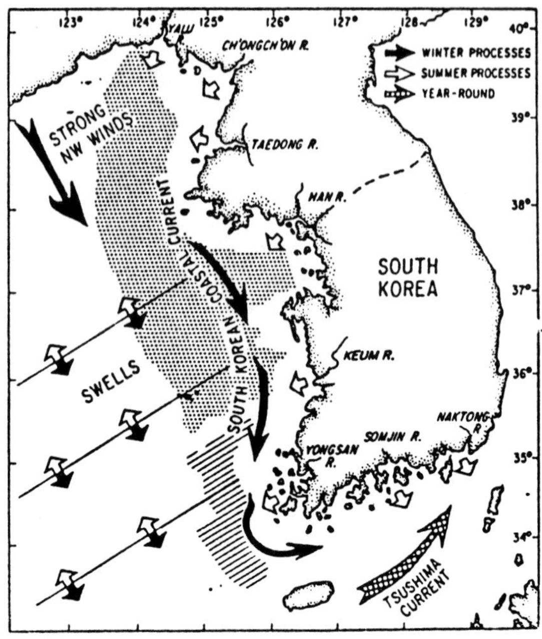

Fig. 1. Locality map with summary of physical processes that control sediment dispersal and accumulation along the west coast of the Korean peninsula.

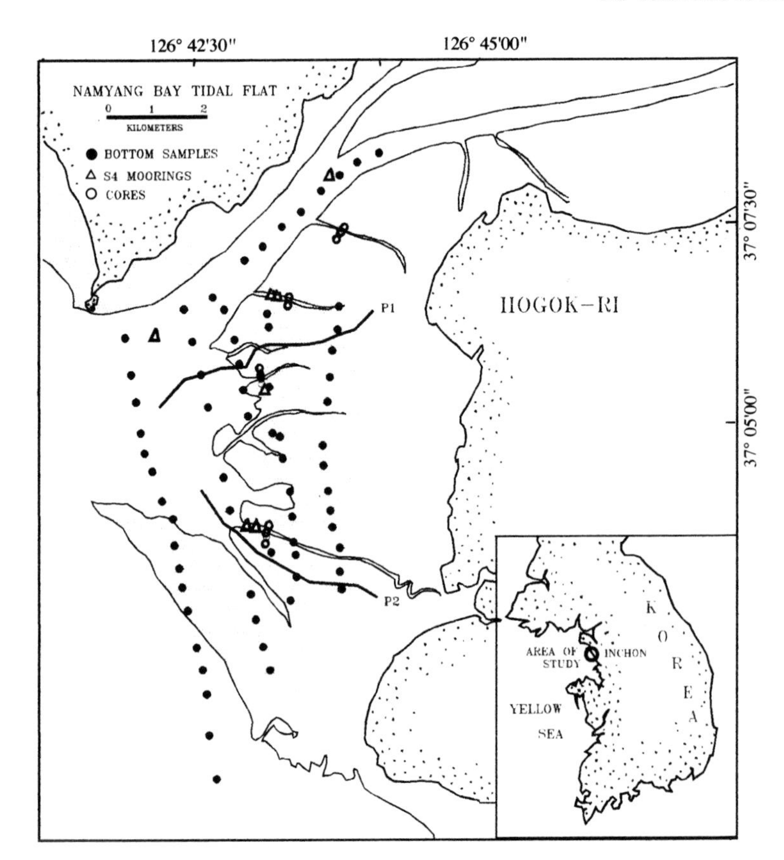

Fig. 2. Index map of the Namyang Bay tidal flats, west coast of Korea, showing transect lines, sample and core positions, anchor stations, and the 4-day moorings.

as sedimentary environments that are sufficiently firm to support ox carts and tractors (Frey *et al.*, 1989). A similar range in sediment consistency was observed even within the areally restricted study area of Namyang Bay near Inchon. In addition to grain-size effects, the variability in short-term accumulation rate undoubtedly affects the state of sediment consolidation. Accumulation rates (100 yr time-scale) are reported to range from 0.15 to >2.0 cm yr^{-1} on the Namyang Bay tidal flats and 0.2–0.6 mm yr^{-1} for the few other intertidal areas along the Korean coast from which cores have been obtained (Alexander *et al.*, 1990).

The field site for our study, the tidal flat region of Namyang Bay, is located in the southern part of Kyeonggi Bay, approximately 30 km south of Inchon (Fig. 2, inset). The predicted spring and neap tidal ranges average 7.7 m and 4.9 m respectively. The highest astronomical tide exceeds 9 m. Maximum intertidal exposures, measured normal to the shoreline, are on the order of 5 km. However, recent land-reclamation projects have cut off many small embayments and the area of intertidal exposure has decreased considerably over the last decade. Ten large drainage channels and dozens of small tributary channels incise the tidal flat, giving it a lobate appearance (Fig. 2). The larger intertidal channels are 2–5 km long and appear as 'permanent' features on most nautical charts; however, the channels lose their definition once they reach the subtidal zone. Although there is no measurable freshwater discharge to the system, substantial currents and highly variable sediment concentrations occur within the tidal channels.

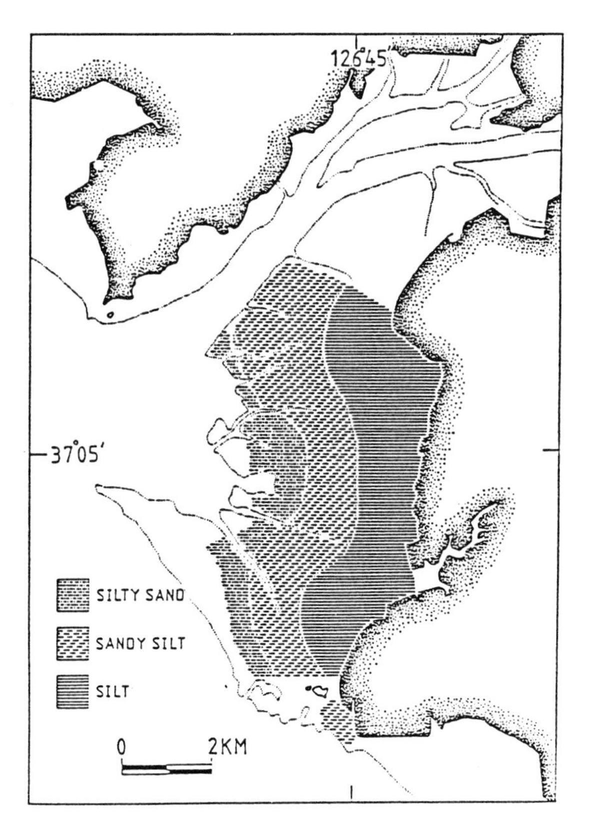

Fig. 3. Distribution of sedimentary facies zones in the Namyang Bay tidal flats. Sediment nomenclature is according to Folk (1954).

STUDY METHODS

Surficial sediments were sampled by means of an Ekman-type grab sampler, covering all but the inner third of the tidal flat area (Fig. 2). In each case about 200 g of sediment were removed from the centre of the scoop. In addition, 15 short cores (50–60 cm) were collected in aluminium conduit tubing along lines across four tidal channels (Fig. 2). Cores were split lengthwise, subsampled, photographed and logged.

Current speed and direction, depth, beam transmission and water samples were obtained from 57 profiles and 54 mid-depth deployments in major tidal channels and over the tidal flat surface (Fig. 2). Deployments were made from a fishing vessel that could rest directly on the tidal flat surface at low tide. The duration of measurements at each station (6–12 h) was unfortunately limited by military restrictions (sun-up to sun-down working hours) and

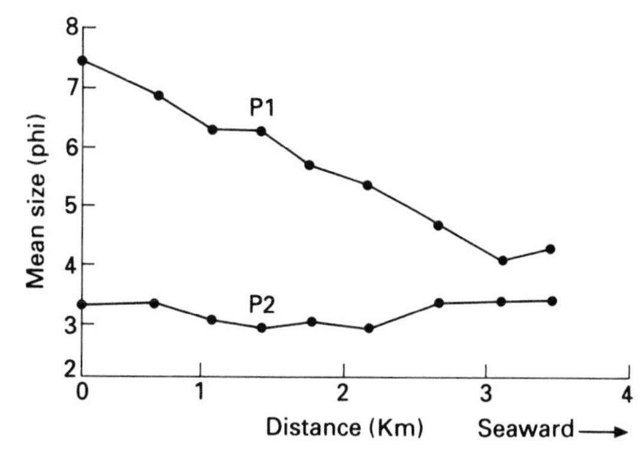

Fig. 4. Cross-flat trends in mean grain size and sediment sorting. Sediments generally get coarser seawards. Note the differing trends along profile lines P1 and P2.

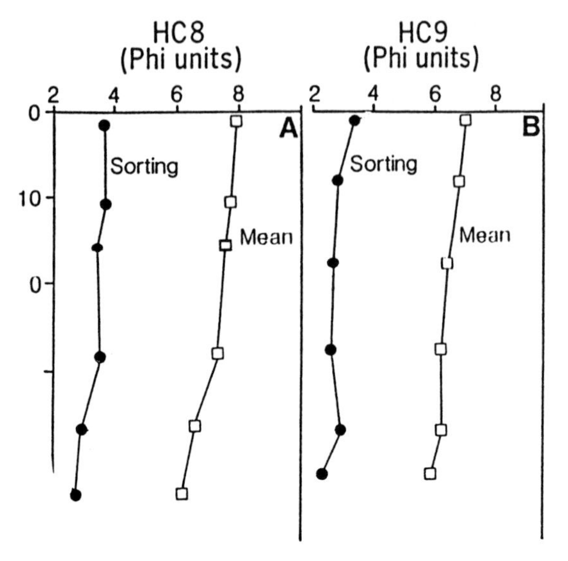

Fig. 6. Down-core trends in mean grain size and sorting. Cores HC8 (A) and HC9 (B) illustrate the typical, down-core sedimentary facies succession. Note the coarsening in grain size and better sorting of sediment with depth.

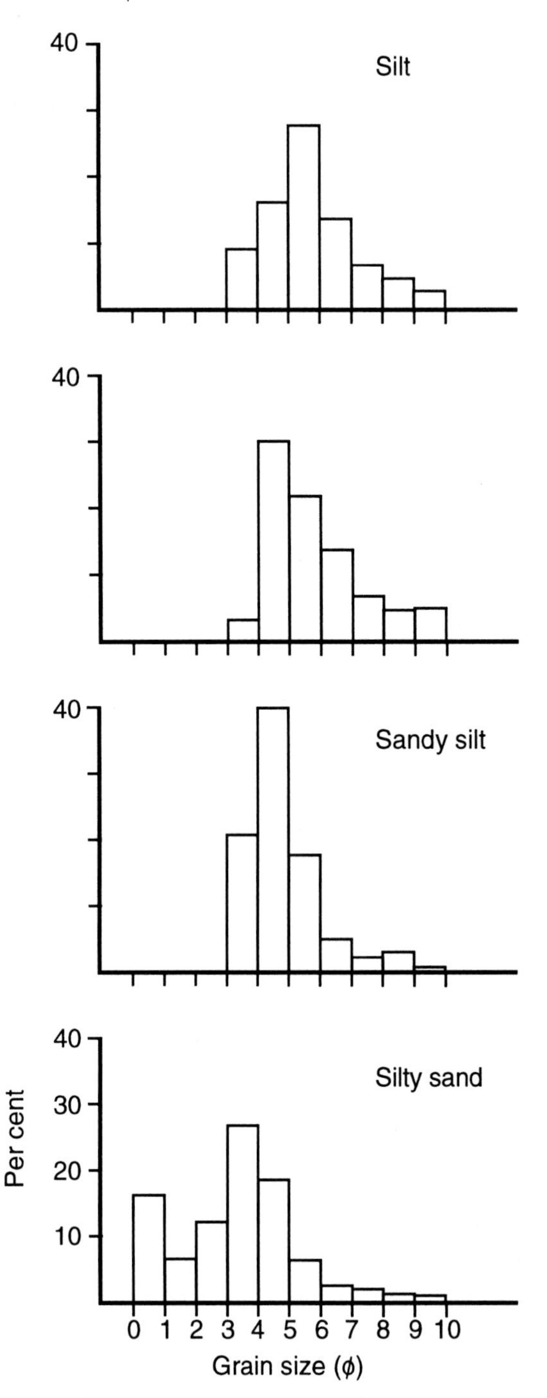

Fig. 5. Grain-size histograms from the Namyang Bay tidal flats which illustrate grain-size distribution trends within typical samples from the high flats (top) to the low flats (bottom).

by the enormous tidal range, which required ending work while water depths were still sufficiently large to regain the shore by boat. Measurements of current speed, current direction and beam transmission were also obtained from a 4-day mooring that was deployed in the centre of a large channel along the north-west margin of the tidal flat (Fig. 2). The mooring consisted of an Endeco 174 current meter and a Sea-Tech transmissiometer with a 5-cm path length that were combined in a single housing and attached 1 m above the bed to a steel pipe. Data were acquired continuously but stored internally as 2-min averages on a magnetic cassette tape recorder.

Water samples taken at each station were filtered in the field on pre-weighed 0.45 μm Millipore membrane filters using a pressure filtration system. Sediment concentrations were then determined as dry mass per unit volume. Beam transmission was calibrated to suspended sediment concentration using these values ($r^2 = 0.66$ over the range 20–225 mg l^{-1}). Small variations in absolute sediment concentration, which could not be resolved with the transmissiometer, were of less interest than the overall magnitude and timing of major turbidity events in the water column (e.g. interfacial waves; see Adams *et al.*, 1990). Bottom samples and core

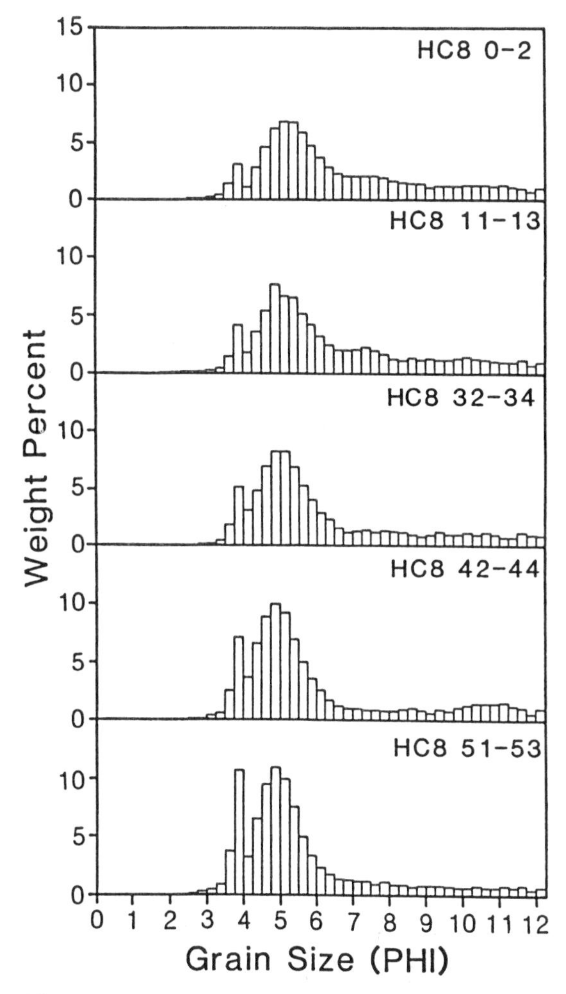

Fig. 7. Down-core progression of selected grain-size frequency distributions from core HC8. The coarsening with depth clearly results from a decrease in silt- and clay-sized material and an increase in very fine sand and coarse silt.

subsamples (at 10-cm intervals) were separated into sand and mud fractions in the laboratory using a 0.063-mm sieve. Samples with greater than about 25% sand (total of 80) were sieved at 1/4-phi intervals following standard techniques (e.g. Folk, 1954).

The resin-impregnated core slabs of representative short cores (50–60 cm) from the tidal flats were cut vertically to the lamination and polished to observe grain size and sedimentary structures under the polarizing microscope. Grain-size frequency (based on 200 counts) and lamina thickness were

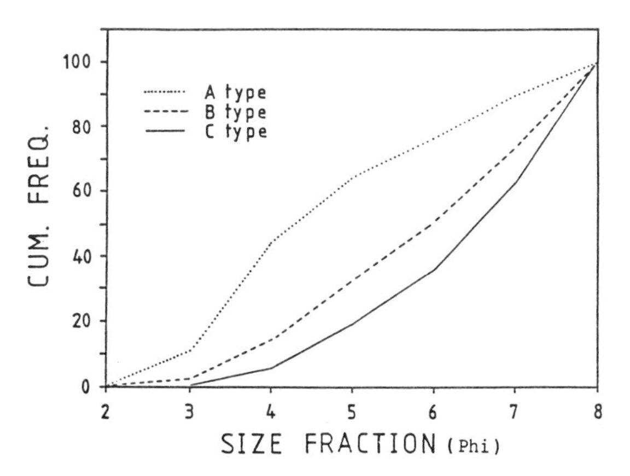

Fig. 8. Three types of tidal lamination in the Namyang Bay tidal flat. Type-A laminae show the coarsest grain-size texture.

established under the microscope. In particular, the thickness of lamina layering was also measured on X-ray radiographic prints (enlarged print).

RESULTS

Sediment properties

Surficial sediment distribution

Surficial sediment samples from the tidal flats were analysed to delineate patterns in surficial sedimentary facies or types based on textural parameters following the method and nomenclature of Folk (1954). Three shore-parallel sedimentary facies were identified by this procedure: (i) a *silt facies* occupying the zone nearest to the mean high water line; (ii) a *sandy silt facies* on the central flats; and (iii) a relatively narrow and restricted *silty sand facies* along the outer edge of the intertidal flat (Fig. 3).

As shown in Fig. 3, the zonation patterns of the three sedimentary facies are oriented parallel to each other along the shore (from north to south). As a consequence, the sediments on the intertidal flats coarsen in a seaward direction, mean grain size increasing from about 7.3 phi near the shoreline to about 4.5 phi on the outer tidal flats (Fig. 4). Typical grain-size frequency distributions of the individual facies are illustrated in Fig. 5.

The seaward coarsening towards the lower tidal flats is attributed to a general increase in wave and

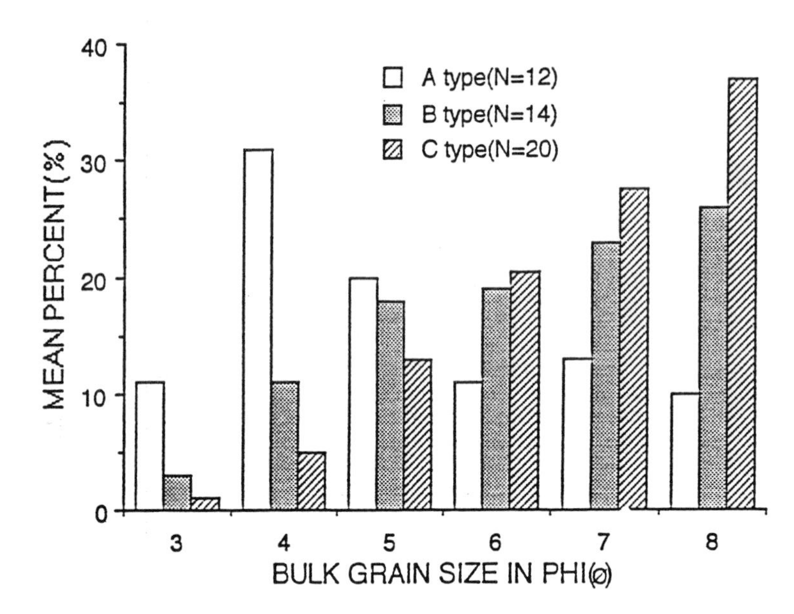

Fig. 9. Grain-size histograms from three different lamina types observed in the sediments of the Namyang Bay tidal flats.

tidal current energy in a seaward direction, causing progressive winnowing of finer-grained sediments. The winnowed material is carried both landwards and seawards, some of it depositing subsequently in subtidal areas adjacent to the intertidal flats. Reineck & Singh (1973) reported a similar trend from the tidal flats of the southern North Sea.

Vertical sediment distribution

The textural trends within individual cores were determined by analysing subsamples from the surface (0–2 cm), from the middle (14–16 cm) and from the lower (27–30 cm) sections of each core. In general, mean grain size coarsens down-core, while the sorting coefficient decreases (Fig. 6). Accordingly, vertical changes from top to bottom in core sediment texture resemble lateral changes from the nearshore across the tidal flat. The down-core coarsening suggests that these tidal flats are prograding seawards, the finer-grained nearshore sediment overriding coarser-grained mid- and outer-flat sediments in the process (Fig. 7).

Sedimentary structures

X-radiographs of representative core slabs from the intertidal flats exhibit mottled sedimentary structures (bioturbated) with varying degrees of preservation of primary physical stratification (fine lamination). The proportion of preserved lamination seems to increase from the middle tidal flats to the high tidal flats, whereas it decreases towards the lower tidal flats.

The positions of 11 representative cores are shown in Fig. 2. The upper 5–7 cm of these cores are mostly characterized by interlayered bedding comprising three types of lamination, here labelled as type-A, type-B and type-C. The lamina types were classified according to their sand content and thickness.

A-type laminae are the coarsest-grained, containing 30% sand and having a median grain size of 4–5 phi. By contrast, type-B laminae contain 10–20% sand and have a median grain size of 5–6 phi, whereas type-C laminae contain <10% sand and have a median grain size of 6–7 phi (Figs 8, 9).

The thicknesses of the three lamina types range from 0.23 to 2.7 mm (Fig. 10A, B). In fact, lamina thickness and sediment texture (sand content and median grain size) appear to be closely correlated, thicker laminae being coarser-grained. This is illustrated by photomicrographs (Fig. 11A, B) which show that the thicker laminae contain coarser grains (very fine sand and silt), whereas thinner laminae are finer-grained (mostly silt and clay).

A conceptual diagram showing the relationship between texture (grain size) and thickness of lamination in the intertidal flat deposits of Namyang Bay is presented in Fig. 12.

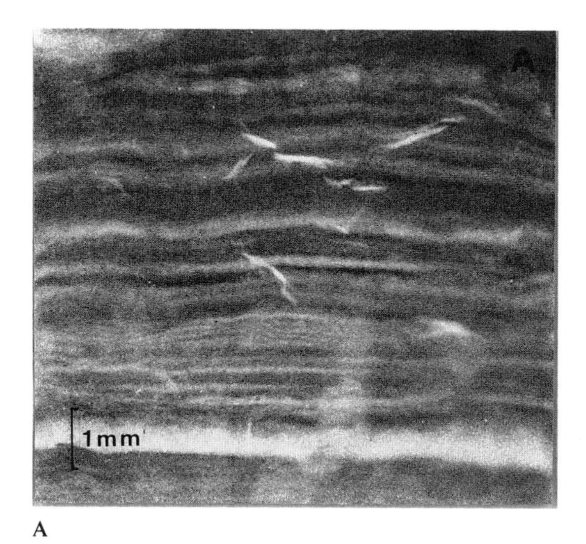

A

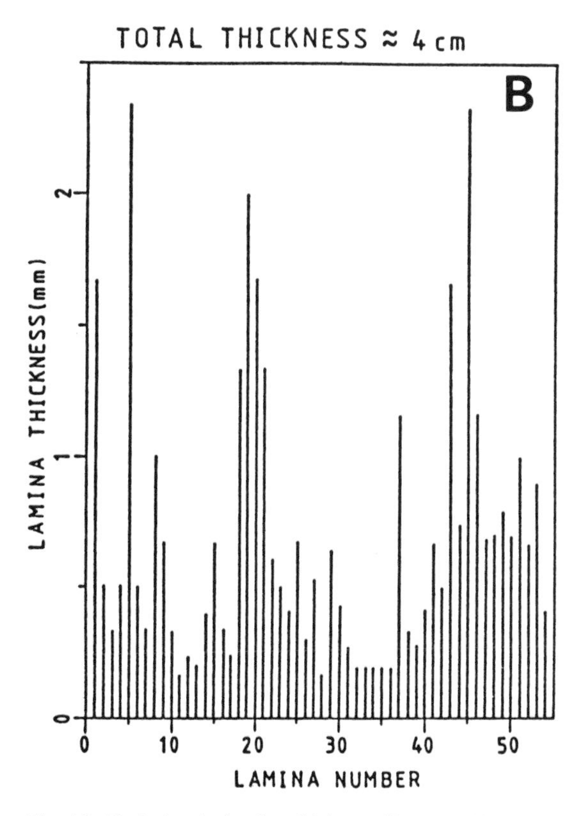

Fig. 10. Variation in lamina thickness illustrated by a representative core slab. (A) X-radiograph of the core slab. (B) Measured lamina thickness versus measured lamina frequency.

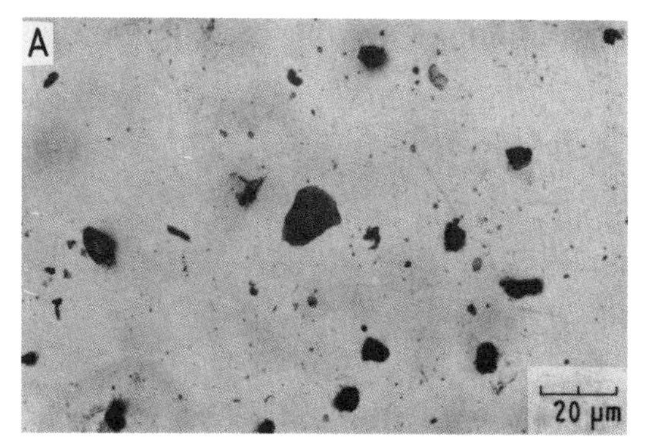

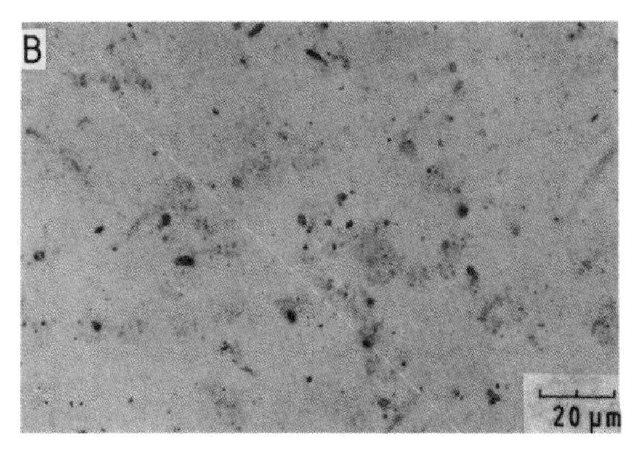

Fig. 11. Photomicrographs showing (A) silt grains from a type-A lamina, and (B) clay-rich texture from a type-C lamina.

Modes of tidal lamination

The alternation between ebb and flood tidal currents, wave action and slack water are considered to be responsible for the origin and mode of tidal lamination and bedding (Reineck, 1960). On the other hand, Klein (1977) has presented a modified model for the formation of tidal bedding (mainly intertidal) based on Reineck & Wunderlich (1968).

The regular formation of tidal bedding under fairweather (spring and neap tide) conditions seems to be well recognized and understood in general. In fact, the study area along the northern margin of Namyang Bay provides convincing evidence for the effects of individual (summer) storms on the tidal flat surface. Following upon a specific storm (Fig. 13), the authors noted that the upper 5–10 cm of soft, oxidized fluid-like mud had been resuspended. It is thus reasonable to associate the low transparency values (high suspended sediment concentrations, Fig. 13) with the effect of resuspension

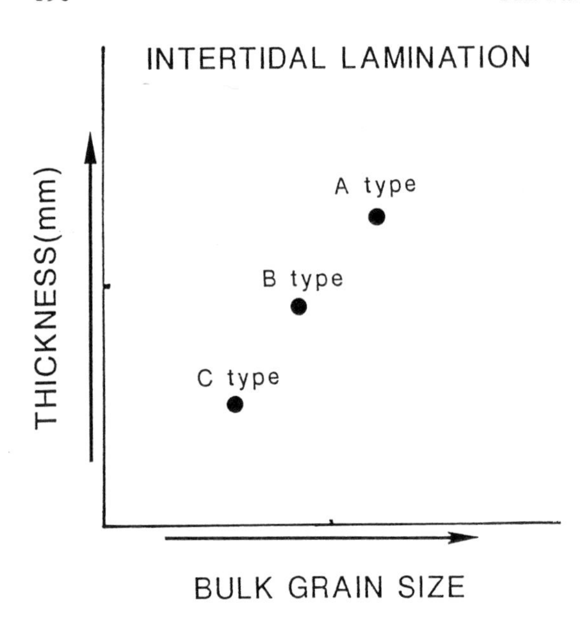

Fig. 12. Conceptual diagram showing the three different lamina types as defined by grain size and thickness.

by high current velocities measured during and immediately after the storm.

One of the unresolved problems in tidal clastic research is the question of how tidal bedding or lamination, developed over full spring–neap cycles (regular daily processes), can be distinguished from storm-generated structures (episodic processes). In the present study, type-A laminae (the thickest and coarsest of lamina types) can be interpreted as the result of enhanced deposition from punctual storm events or similar high energy conditions. On the other hand, it might be speculated that the general high energy (spring tidal current and storm-wave induced current, etc.) would also produce a greater thickness and a coarser texture in the tidal bedding on the tidal flats.

CONCLUSIONS

The unique sedimentary structures, i.e. tidal bedding (lamination), characterizing the intertidal de-

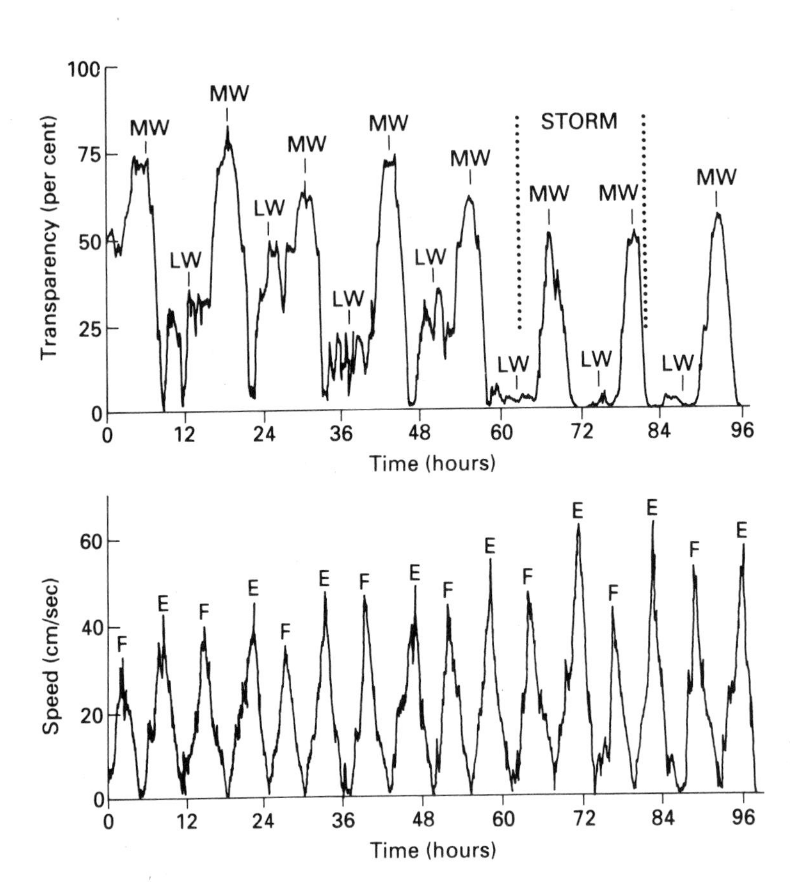

Fig. 13. Transparency (beam transmission) and current speed taken at a 4-day mooring station in the Namyang Bay tidal flat region. Storm passage beginning at hour 62 coincided with an increased attenuation of the beam transmission signal. Note the dramatic change during low-water periods.

posits of the macrotidal Namyang Bay along the west coast of Korea show three types of lamination, characterized by different sediment textures and laminae thicknesses, that are related to fairweather tidal cyclicity and/or storm events.

However, the mechanism for a possible role of punctual high-energy events in promoting thicker and coarser bedding or lamination warrants closer examination. Surficial sediments of the Namyang Bay tidal flat coarsen seawards, i.e. towards the lower tidal flats (from 7.3 phi to 4.5 phi). Furthermore, the vertical trends of sediment textures in cores suggest that the tidal flat system of Namyang Bay has not accreted in place, but has prograded in a seaward direction.

ACKNOWLEDGEMENTS

We are grateful to the hard-working and ocean-going graduate students of the Department of Oceanography, Seoul National University, Seoul, for their diligent assistance in the field work. The research was supported through Y.A. Park and J.T. Wells by the Korea Science and Engineering Foundation (KOSEF) and the National Science Foundation (NSF) of the USA, respectively. We also thank the anonymous reviewers who provided critical and helpful comments to our manuscript.

REFERENCES

ADAMS, C.E., WELLS, J.T. & PARK, Y.A. (1990) Internal hydraulics of a sediment-stratified channel flow. *Mar. Geol.* **95**, 131–145.

ALEXANDER, C.R., NITTROUER, C.A., DEMASTER, D.J., PARK, Y.A. & PARK, S.C. (1990) Macrotidal mudflats of west Korea: a model for interpretation of intertidal deposits. *J. sediment. Petrol.* **61**, 805–824.

BARTZ, P.M. (1972) *South Korea.* Clarendon, Oxford, 203 pp.

FOLK, R.L. (1954) The distinction between grain size and mineral composition in sedimentary rock nomenclature. *J. sediment. Petrol.* **62**, 334–359.

FREY, R.W., HOWARD, J.D., HAN, S.J. & PARK, B.K. (1989) Sediments and sedimentary sequences on a modern macrotidal flat, Inchon, Korea. *J. Sediment. Petrol.* **59**, 28–44.

HUH, O.K. (1982) Satellite observations and the annual cycle of surface circulation in the Yellow Sea, East China Sea and Korea Strait. *La Mer* **20**, 210–222.

KLEIN, G.D. (1977) *Clastic Tidal Facies.* Continuing Edu. Pub. Co., Illinois, U.S.A.

KLEIN, G. DEVRIES, PARK, Y.A., CHANG, J.H. & KIM, C.S. (1982) Sedimentology of a subtidal, tide-dominated sand body in the Yellow Sea, northwest Korea. *Mar. Geol.* **50**, 221–244.

LEE, S.W. (1968) Monthly variation of mean sea level along the coasts of Korea. *J. Oceano. Korea*, **2**, 24–33.

LEE, C.B., PARK, Y.A. & KOH, C.H. (1985) Sedimentology and geochemical properties of intertidal surface sediments of the Banweol area in the southern part of Kyeonggi Bay, Korea. *J. Oceanol. Korea* **20**, 20–29.

MURAKAMI, T. (1979) Winter monsoonal surges of East and Southeast Asia. *J. Meteorol. Soc. Japan* **57**, 134–159.

OFF, T. (1963) Rhythmic linear sand bodies caused by tidal currents. *Am. Ass. Petrol. Geol. Bull.* **47**, 324–341.

PARK, Y.A. (1987) Coastal sedimentation. In: *Geology of Korea* (Ed. Lee, D.S.), pp. 389–405. Geol. Soc. Korea, Kyohak-Sa Publ., Seoul.

REINECK, H.E. (1960) Ueber die Entstehung von Linsen und Flaser-schichten. *Abh. deutsch. Akad. Wiss.* Berlin, H1, 369–374.

REINECK, H.-E. & SINGH, I.B. (1973) *Depositional Sedimentary Environments.* Springer-Verlag, New York, 439 pp.

REINECK, H.-E. & WUNDERLICH, F. (1968) Zeitmessungen und Gezeitenschichten. *Natur u. Museum* **97**, 193–197.

UDA, M. (1966) Yellow Sea. In *Encyclopedia of Oceanography* (Ed. Fairbridge, R.W.) Van Nostrand and Reinhold, New York, 1021 pp.

WELLS, J.T. (1983) Dynamics of coastal fluid muds in low-, moderate-, and high-tide-range environments. *Canadian J. Fisheries and Aquatic Sciences*, **40**, 130–142.

WELLS, J.T. (1988) Distribution of suspended sediment in the Korea Strait and southeastern Yellow Sea: onset of winter monsoon. *Mar. Geol.* **83**, 273–284.

WELLS, J.T. & HUH, O.K. (1979) Tidal flat muds in the Republic of Korea: Chinhae to Inchon. *Sci. Bull. Off. Nav. Res., Tokyo* **4**, 21–30.

WELLS, J.T. & HUH, O.K. (1984) Fall-season patterns of turbidity and sediment transport in the Korea Strait and southeastern Yellow Sea. In: *Ocean Hydrodynamics of the Japan and East China Sea* (Ed. Ichiye, T.), pp. 387–397. Elsevier, Amsterdam.

YOO, H.R. (1986) Remotely sensed water turbidity pattern on the Korean side of the Yellow Sea. *Ocean Res.* **8**, 49–55.

Spec. Publs int. Ass. Sediment. (1995) **24**, 193–211

Patterns of sedimentation in the macrotidal Fly River delta, Papua New Guinea

E.K. BAKER*, P.T. HARRIS*‡, J.B. KEENE* *and* S.A. SHORT†

**Ocean Sciences Institute, University of Sydney, Sydney NSW 2006, Australia; and*
†Australian Nuclear Science and Technology Organisation, Lucas Heights Research Laboratories, Menai NSW 2234, Australia

ABSTRACT

Cores and grab samples provide preliminary information on the character of sedimentary deposits in the macrotidal Fly River delta of southern Papua New Guinea. ^{210}Pb activities in the cores and surface sediments show a wide variability, suggesting differences in the initial specific activity (dpm g^{-1}) of the sediments being deposited. As a consequence, estimates of sediment accumulation rate using the ^{210}Pb technique are not always possible. The available data suggest sediment accumulation rates in one part of the study area of about 8–10 cm yr^{-1}

The sediments consist of laminated, very well sorted, heavy mineral rich sands and poorly sorted silts with generally less than 30% clay. Biological activity is restricted by the extreme current speeds (up to 2.4 m s^{-1}), high suspended sediment load (near bed turbidity levels reach 40 g l^{-1}) and low salinities (<5‰). The distributary channel deposits are sheltered by islands from the large swell waves that propagate landwards from the Coral Sea. This and the paucity of organisms contribute to the preservation of ubiquitous sand/mud laminations. The laminae do not exhibit direct evidence of tidal cyclicity (tidal bundles). Deposition of mud–sand couplets is interpreted as occurring during neap tides, possibly coinciding with times of relatively low surface swell wave energy. This is based on an average couplet thickness of 4.2 mm and an accumulation rate of around 8–10 cm yr^{-1} derived from ^{210}Pb analysis.

INTRODUCTION

Studies of sedimentation in deltas have in the past focused mainly on 'river dominated' and 'wave dominated' types (as classified in the tripartite schemes of e.g. Miall, 1984 and Wright, 1985). Previous studies of 'tidally dominated' deltas include that of the Ganges/Brahmaputra in India (Coleman, 1969; Barua, 1990), the Orinoco in Venezuela (van Andel, 1967; Eisma *et al.*, 1978), the Mahakam in Indonesia (Allen *et al.*, 1979) and the Amazon in Brazil (e.g. Kuehl *et al.*, 1986; Nittrouer & DeMaster, 1986). These depositional environments are characterized by fine-grained (<0.1 mm), rapidly accumulating sediments which prograde offshore over the adjacent continental shelf.

‡*Correspondence address*: Australian Geological Survey Organisation, Antarctic CRC, University of Tasmania, GPO Box 252C, Hobart, Tasmania 7001, Australia.

The Fly delta has been considered to idealize a tidally dominated delta in terms of general geomorphology (Fisher *et al.*, 1969; Wright, 1977, 1985). The funnel shape of the delta does not appear to be controlled by any underlying bedrock morphology (Harris *et al.*, 1992). High sedimentation rates measured on the delta front indicate a net offshore progradation rate of about 6 m yr^{-1} (Harris *et al.*, 1992).

A feature common to ancient and modern tidally influenced deposits is the cyclicity in the spacing and thickness of sand–mud couplets, related to spring–neap variations in tidal current speeds (e.g. Visser, 1980; Dalrymple & Makino, 1989; Dalrymple & Zaitlin, 1989; Kvale *et al.*, 1989; Williams, 1989). It is difficult at present to compare the Fly delta with other locations, as little is known about the nature of the deposits. The question we ask is

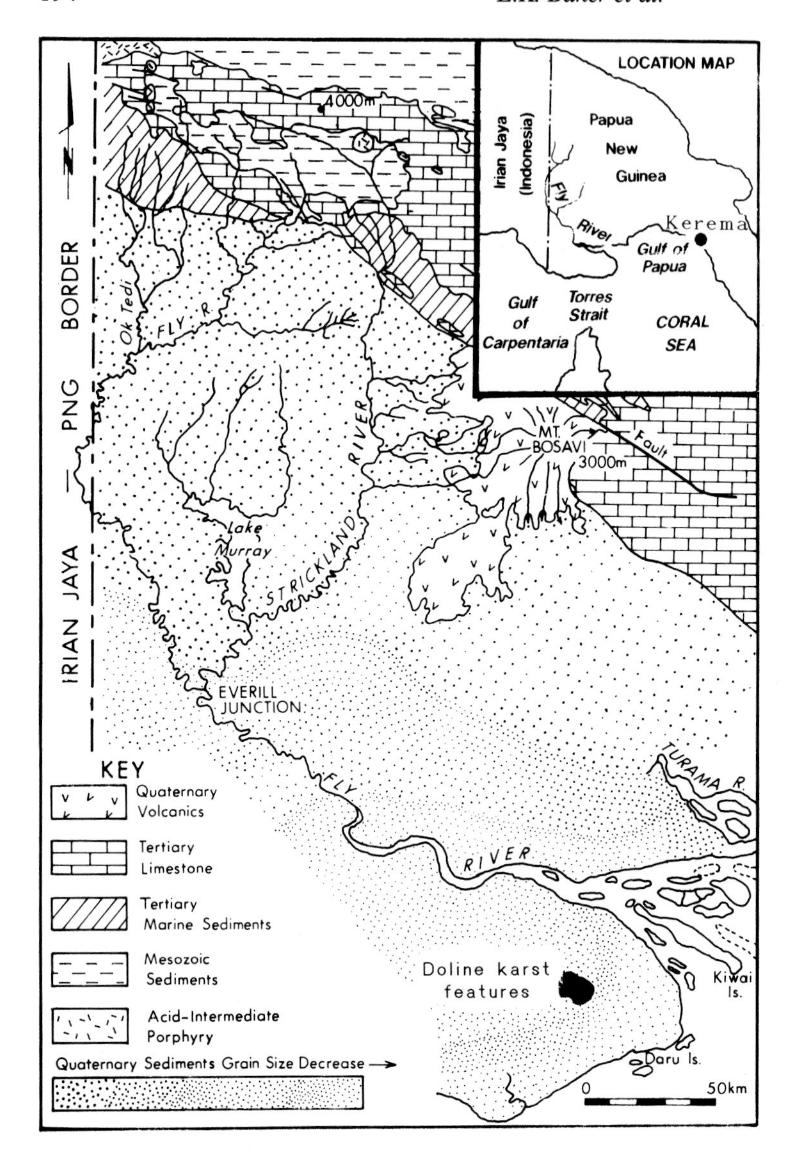

Fig. 1. Location map of the Fly delta in south-western Papua New Guinea.

'What characteristics do Fly delta deposits exhibit as an indicator of tidal influence in their deposition?'.

This paper presents the results of an exploratory expedition to the southern Fly delta, carried out in September 1991. Sediment cores and other field data are assessed to consider the above question in the context of a preliminary study.

REGIONAL SETTING

The Fly River

Papua New Guinea is the northern extension of the Australian continental plate (Jaques & Robinson, 1977). The Fly region is part of a foreland basin which formed following the continent–island arc collision that occurred during the middle Oligocene (Pigram *et al.*, 1989; see Fig. 1).

The Fly River originates in the highlands of Papua New Guinea at an elevation of approximately 4000 m. It flows 1000 km to the Gulf of Papua, the last 800 km of which lie in a flat alluvial plain. The catchment is relatively small (76 000 km^2), considering that the river has an estimated mean annual discharge of 7000 m^3 s^{-1} and a total sediment load of 125 million t yr^{-1} (Ok Tedi Mining Ltd, 1988). In terms of water discharge, it ranks among the 10 largest tropical rivers in the world (Robertson *et al.*, 1990). The combination of steep slopes and high rainfall make landslides are a common occurrence and have been reported to supply many hundreds of millions of tonnes of material into the river (Eagle & Higgins, 1990). The climate is continually wet in the mountains, where the annual precipitation is 10 m, becoming monsoonal towards the coast, with a distinct dry season and an annual rainfall of 2–3 m (Eagle & Higgens, 1990).

Mining and deforestation in the upper part of the catchment have dramatically altered the sediment load of the River. Prior to the 1984 commencement of operations at the Ok Tedi copper mine, the sediment load of the river was 85 million t yr^{-1} (Ok Tedi Mining Ltd, 1988). Of this, 80% was supplied by the Strickland River which joins the Fly at Everill Junction. By 1992 the sediment load is estimated to have increased by almost 50%, with nearly half of the sediment now being contributed by the western, i.e. the Fly River part of the catchment (Eagle & Higgins, 1990).

The Upper Fly is essentially a boulder and gravel bedded, single channel river that has cut deep gorges into the karst topography (Pickup, 1984). A dramatic change in slope occurs at Kiunga, which is only 20 m above sea-level but 800 km from the river mouth. From here the middle Fly becomes a sandy meandering river, with numerous ox-bow lakes and scroll bar complexes (Blake & Ollier, 1969, 1971). The floodplain, which is only 4 km wide at Kiunga, widens to about 14 km at Everill Junction. The lower Fly, downstream of Everill Junction, is tidally influenced. As most of the coarse sediment is deposited in the meandering channels of the floodplain, predominantly fine-grained sand and mud reach the delta (Higgens *et al.*, 1987).

Tidal and wave regime of the delta

The Fly River possesses a highly dynamic funnel shaped delta system which extends for approximately 30 km offshore. The maximum spring tidal range at the mouth is about 4 m, increasing to around 5 m at the apex of the funnel. The tides are semi-diurnal with a large diurnal inequality (Fig. 2). Spring tidal ranges are affected by the moon's declination and exhibit variation over a 28-day period, producing one set of larger spring ranges and another set of smaller ranges (Fig. 2). Neap tides exhibit strongly diurnal characteristics with intervals of around 10 h during which a change in sea-level of less than 0.5 m occurs. As a result there are long periods of slack water during neaps.

Water within the deltaic distributary channels at times exhibits estuarine circulation patterns and is salinity stratified. Near-bed saline water flows landwards with a tidally averaged velocity of up to 20 cm s^{-1} (Wolanski *et al.*, 1992). Strong wind shear homogenizes the water. In general, fresh water tends to exit from the southern distributary channel (Wolanski *et al.*, 1992) whilst saline water intrusion is greatest in the northernmost channel.

Surface wave data from the region are scarce. Wave rider buoy observations from the location nearest to the Fly are from Kerema, located on the coast about 220 km east of the delta (Fig. 1). These data indicate that, on an annual basis, wave height exceeds 0.25 m for 90% of the time, 0.7 m for 50% of the time, 1.5 m for 10% of the time and 2 m for 1% of the time (Thom & Wright, 1983). The significant wave height at Kerema during the south-east trade wind season is 1.3 m, whereas during the south-west monsoon it is only 0.3 m (Thom & Wright, 1983).

Delta sediments

The four main distributary channels of the delta are shallow (typically <8 m deep) and are separated by rapidly migrating (up to 30 m yr^{-1}) linear sand–mud islands (Spencer, 1978). Deltaic sediments are mostly composed of very well sorted fine-grained sand and mud. It is estimated that 90% of the sediment entering the delta distributary channels is carried as suspension load and only 10% is transported as bedload (Ok Tedi Mining Ltd, 1988). Wolanski & Eagle (1991) and Wolanski *et al.* (1992) have carried out studies on the spatial and temporal variability of suspended sediment concentration (SSC). Within the distributary channels, SSCs typically measures 4–6 g l^{-1} and may reach peak concentrations of 40 g l^{-1}, forming a near-bed 'fluid mud' layer. SSCs are found to be highest during the accelerating phase of the flood tidal current over a spring tidal cycle. Lower SSCs occur during the weaker ebb tidal flows and show a variation due to the diurnal inequality. During neap tides the SSC is much reduced as sediment settles to the bed (settling velocities of around 0.1 to 0.3 cm s^{-1} were measured for Fly delta silts). Sediment dewatering and

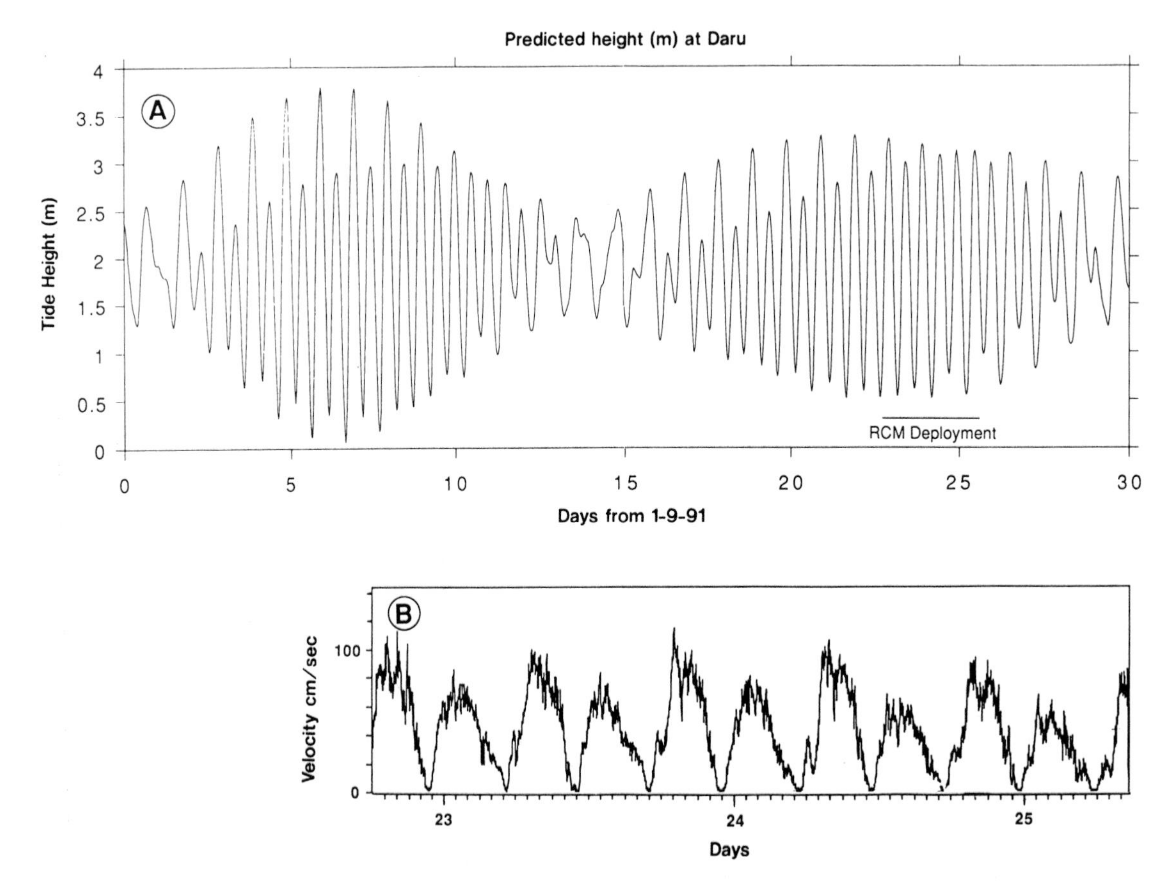

Fig. 2. (A) Time series plot of predicted tidal height at Daru for September 1991 (provided by the Royal Australian Navy Hydrographic Service). (B) Time series plot of tidal current velocity measured at station 85 (see Fig. 3B for location).

compaction are restricted by turbulent spring tidal currents in which the escape of water is inhibited by the strong fluid shear. Dewatering does occur on neap tides, however, and after only 2 h of compaction the critical erosion velocity may be tripled (Wolanski *et al.*, 1992).

Within the distributary channels, coarse sandy sediments are winnowed to form mouth bars and elongate shoals; the latter are quickly colonized by mangroves forming elongate islands. Most of these deltaic islands are covered by a dense growth of mangrove forest and are subjected to periodic inundation during spring tides. In the area where salinity (at high tide) is >10‰, the species *Rhizophora apiculata, Bruguiera parviflora* and *B. gymnorhiza* predominate, whereas the mangrove palm *Nypa fruticans* dominates at locations with salinities <10‰ (Robertson *et al.*, 1990). Meiofaunal and macrofaunal densities are low on the island mud-flats, although variability is exhibited in the level of bioturbation observed in box cores (Alongi, 1991).

Previous investigators of intertidal mud-flat sediments obtained 10–15 box cores at four sites (Alongi, 1991; Alongi *et al.*, 1992; Fig. 3A). Sediment analyses and X-radiographs of these box cores indicate the occurrence of three facies. The first facies (site AIB) comprises laminated sands and muds containing 1.4% $CaCO_3$. Water salinity was 2.5‰. Facies 2 was identified at one site located at the northern end of Ura Island (site URA, Fig. 3A) where salinity at the time of collection was 15‰; it is characterized by numerous biogenic structures. Facies 3 was identified at two sites located at the western end of Umuda Island (site UMA, Fig. 3A) where salinity was 17‰, and at the eastern end of

Sobuwabuda Island (site SOB, Fig. 3A) where salinity was 14.8‰. This facies is characterized by sandy compacted sediments containing wood fragments together with other organic matter and exhibiting both ripple laminated and bioturbated units. For example, the upper 3 cm of sediment in one core comprised a laminated sequence overlying an erosional unconformity below which sediments were extensively bioturbated (Alongi, 1991, fig. 4).

Alongi *et al.* (1992) described cores from three subtidal environments (two cores at each of the seven stations). Cores from the mid-channel environment comprised compacted, fine black sand nearly devoid of benthic infauna (stations, D1, D2 and D10). The black colour is related to a high Fe content of around 4–7% dry weight. Other cores obtained from the lee side of islands comprised laminated sands and muds (stations D4, D5 and D7). Examination of the X-radiograph from one 24 cm long core (station D5) indicated that the mean thickness of sand laminae was 7.2 mm, whilst mud laminae had a mean thickness of 14.2 mm (mean couplet thickness is 21.4 mm); the laminae did not exhibit any discernible rhythmic spacing. A third subtidal facies found at the entrance to a channel (station D9) comprised bioturbated and laminated silt and clay.

METHODS

During a cruise in September 1991 on the RV *Sunbird*, 28 stations were occupied in the southernmost distributary channel of the Fly delta (Fig. 3B; Harris *et al.*, 1992). Surficial grab samples (Shipek) were obtained at each station and piston cores were collected at four of the stations. Short 10-cm long 'mini-cores' were obtained from eight of the undisturbed grab samples. A 1.5 m swell generated by strong south-easterly Trade Winds made the collection of samples extremely difficult.

Sedimentary structures were described from the cut cores and X-radiograph positives. Grain size was determined using a 2600 Series Malvern Laser Particle Size Analyser. Water content was calculated from the weight loss of wet sediment dried at 80°C for 24 h and bulk grain density was measured using a standard pycnometer. Sedimentation rates were calculated by ^{210}Pb isotopic analyses carried out at the Lucas Heights Research Laboratories (for methods see Harris *et al.*, 1993). Surface and near bed water samples were obtained at each station using Niskin bottles; suspended sediment concentration was determined by filtration through preweighed 0.45 μm filter papers.

The percentages of sand and mud in sections of the piston cores were estimated using a Tracor Northern (TN-8500) image analysis system (IAS). The X-radiograph positives were transferred to the IAS via a Sony colour video camera to produce grey-scale maps of the sand and mud densities. The total surface area of the sand (high density areas) inside a frame was estimated. To calibrate the technique, grain sizes of approximately 100 samples of the high density material were determined with a laser particle size analyser. In addition, the IAS was used to measure the thickness (in μm) of the sand and mud layers. Thicknesses were determined also by visual estimation.

Bathymetric data were collected from echosounding profiles produced by a Raytheon precision depth recorder. Current measurements were obtained at 1-min intervals over a 62-h period at station 85 (Fig. 3B) using an Aanderaa self-recording current meter moored approximately 100 cm above the channel bed. The locations of the stations were determined by Global Positioning System, with an estimated accuracy of ± 15 m.

RESULTS

Hydrography

Echosounding profiles reveal a shallow, flat-bottomed channel with a maximum water depth of 7 m at high tide (Fig. 3B). Water depths (Table 1) are corrected to Lowest Astronomical Tide (LAT) using predicted tidal heights derived for Daru. These were provided by the Royal Australian Navy Hydrographic Service. Errors in depth correction due to tidal phase shift between Daru and the study area are expected to be less than 0.5 m.

Current measurements showed that the flood and ebb tides are of comparable duration in this part of the delta (Harris *et al.*, 1992). The flood tide is stronger, however, reaching a maximum current speed of 1.2 m s^{-1} compared to 0.8 m s^{-1} achieved by the ebb tide. Suspended sediment concentration (SSC) varied over the tidal cycle with a maximum concentration of approximately 5 g l^{-1} in the near-bottom water observed during the accelerating stage of the flood tide. The SSC also showed marked vertical stratification with a maximum reading of

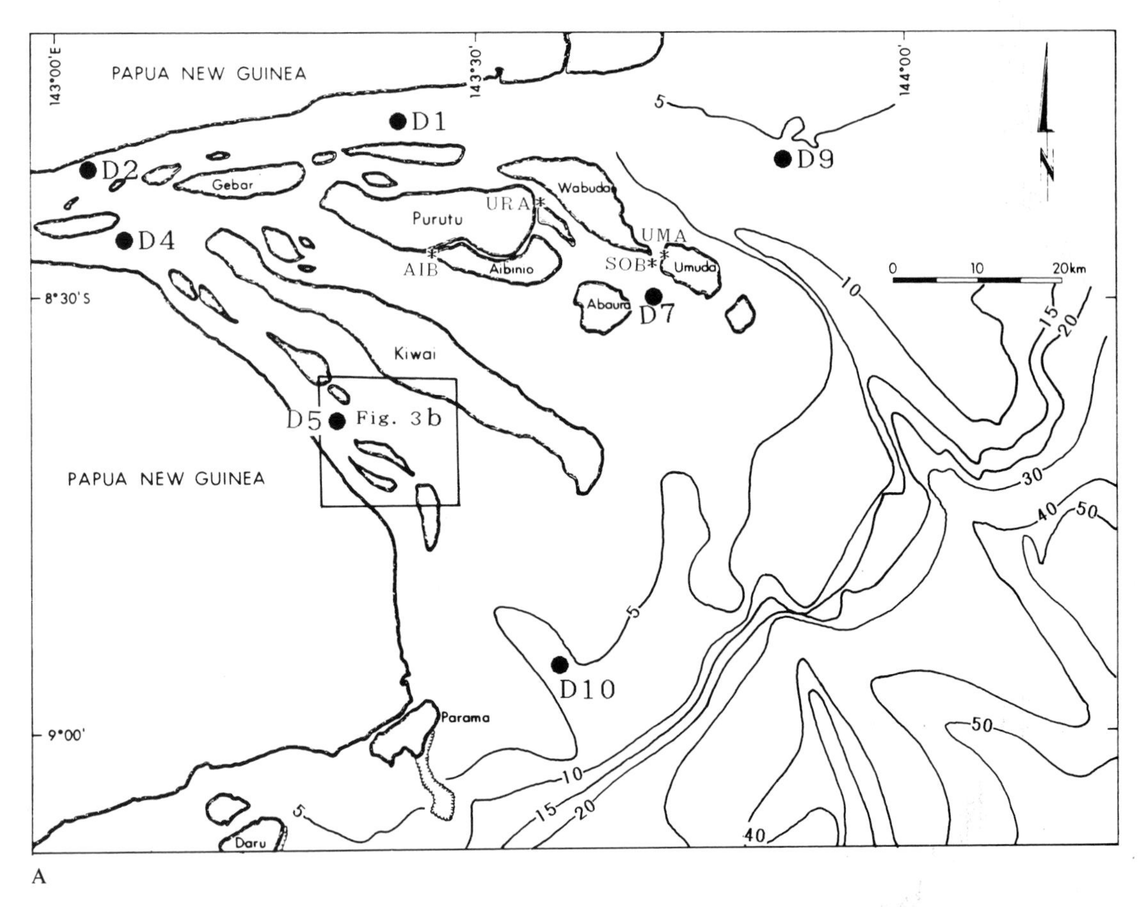

Fig. 3. (A) Map of the Fly delta showing location of study area in the southern channel. Marked sample sites are those of Alongi (1991) and Alongi *et al.* (1992). (B) Southern distributary channel in the Fly delta, showing locations of surficial grab samples (Nos 84–104 and 108–109) and core samples (Nos 105–107 and 111). Transect A–A′ across the channel (1.2 km) illustrates the positions of the sample sites and the relative water depth.

$1.3\,g\,l^{-1}$ and a mean of $0.46\,g\,l^{-1}$ in the surface water. For bottom water, the mean SSC value was $1.3\,g\,l^{-1}$.

Sediments

Surficial sediments obtained at the 28 stations (Fig. 3B) show considerable variation in percentages of sand, silt and clay (Table 1). Sand content varies from 0 to 78% with a mean of 29.4%, silt varies from 17 to 77% with a mean of 53.2%, and clay varies from 0 to 33% with a mean of 17.4%. Bulk samples have a mean water content of 41.1% dry weight ($n = 8$). Spatially, there is a weak inverse correlation between sand content and water depth, such that sand content increases with decreasing water depth, indicating that sand is concentrated along the channel margins and the intertidal zone.

At station 109 (Fig. 3B), collected near the centre of the channel, a grab sample recovered sediment composed of muddy sand with numerous ‘mud balls’, water-saturated wood fragments and occasional gastropod shells (Fig. 4). Mud balls are defined here as rip-up clasts that have been rounded through transport; they are elongate to elliptical in shape and range in size from 1 to 10 cm.

An intertidal flat, approximately 500 m in width, borders the southern side of Meamibu Island (Fig. 5A). It is drained by numerous small (0.3–0.5 m wide and 0.2 m deep) tidal channels that cut down into the wavy and parallel laminated sand and mud (Fig. 5B). There is some bioturbation, mostly

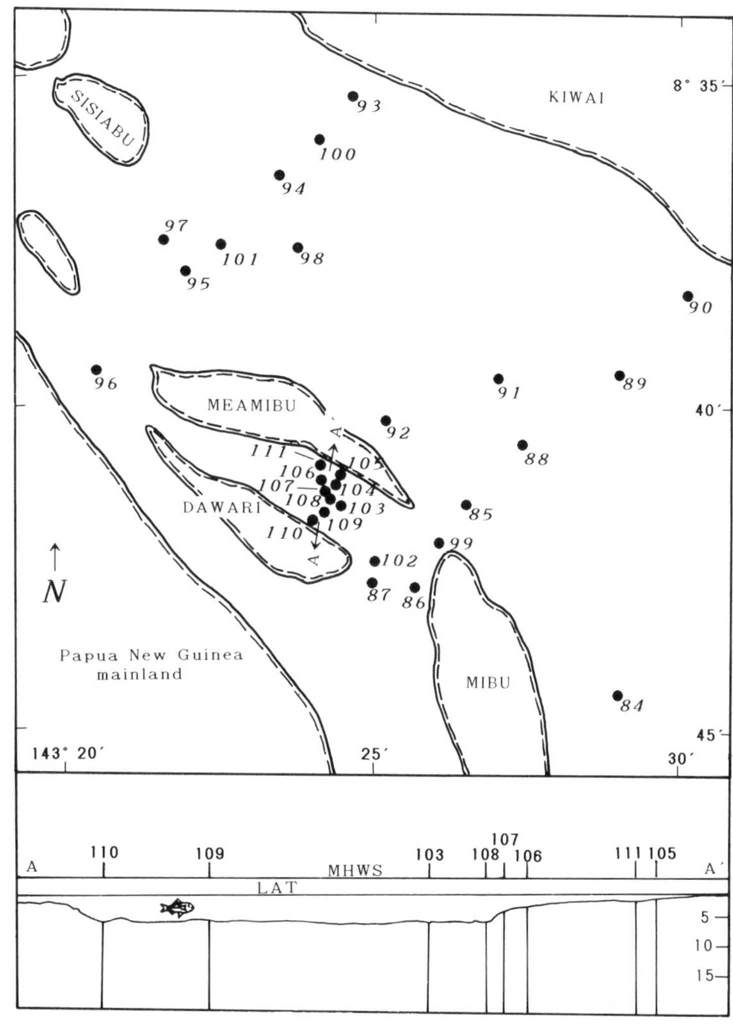

Fig. 3. *(Continued.)* B

due to the burrowing of small crabs, but in general biological activity is low. A 0.5–1.0 m high cliff separates the mud flat from the dense mangrove and *Nypa* palm forest. Large blocks of sediment, many rounded into balls, form an apron at the base of the cliff, indicating recent undercutting and collapse (Fig 5C). On the opposite side of the channel there is no intertidal flat development and the steep foreshore appears to be erosional (possibly due to the southward migration of the main channel).

Four cores were collected between the small islands of Meamibu and Dawari in a shallow water area extending from the intertidal mud flats into deeper, mid-channel waters (Fig. 3B). Subtidal deposits are predominantly composed of alternating layers of sand (0.5–22 mm thick) and mud (0.5–27 mm thick) with some layers exhibiting slump features. Sand content generally increases with depth through the cores, although there are some muddy intervals. An underlying sandy lag deposit was encountered in two cores (Nos 106 and 107).

The upper parts of the cores show evidence of soft sediment deformation in the form of dispersed laminae, microfaulting and asymmetrically folded layers (Fig. 6). Water content measured in the surficial samples was high (35–55%, see above) but generally decreased with depth in the core (e.g. for core 111, correlation between depth and water

Table 1. Station numbers, latitude, longitude, corrected water depth, surficial sediment water content, sand–silt–clay mean grain size and description of surficial (0–10 cm) sediments at Fly delta stations

Station number	Lat.	Long.	Depth (m)	H_2O (%)	Sand (%)	Silt (%)	Clay (%)	Mean size (μm)	Comments
84	8°44.62	143°29.25	2.2		58	36	6	83.8	Massive bedded muddy sand
85	8°41.54	143°26.67	3.2		67	27	6	98.7	Massive bedded muddy sand
86	8°42.03	143°25.77	4.2	41	4	77	19	14.0	Mud beds with sand stringers
87	8°42.51	143°25.00	2.5		42	56	2	46.9	Laminated sand and mud
88	8°40.67	143°27.59	5.1	38	3	77	20	14.3	Mud with sand stringers
89	8°39.52	143°28.95	5.1	41	0	72	28	6.9	Laminated mud and sand
90	8°38.33	143°30.23	2.2		68	27	5	78.2	Massive bedded sand with mud balls
91	8°39.48	143°27.04	5.3		0	73	27	7.9	Massive bedded 'jelly mud'
92	8°40.35	143°25.17	+0.5		75	24	1	95.4	Massive bedded sand
93	8°35.35	143°24.52	5.5		75	21	4	99.5	Massive sand with mud balls and wood
94	8°36.71	143°23.56	4.2	54	2	73	25	8.8	Mud with sand stringers
95	8°38.12	143°21.90	2.6		79	17	4	106.9	Massive bedded silty sand
96	8°39.52	143°20.69	4.0		26	56	18	13.9	Laminated mud and sand
97	8°37.78	143°21.72	3.2		1	68	31	5.9	Mud beds with sand stringers
98	8°37.52	143°24.09	4.4		0	71	29	6.9	Mud beds with sand stringers
99	8°42.10	143°26.05	2.4		66	28	6	111.0	Massive bedded silty sand
100	8°35.93	143°24.07	6.1		0	67	33	6.0	Massive bedded mud
101	8°37.39	143°22.58	1.0		0	72	28	7.6	Mud beds with sand stringers
102	8°42.48	143°25.06	0.2		78	18	4	106.2	Massive bedded sand
103	8°41.51	143°24.46	4.2		12	63	25	8.0	Laminated sand and mud
104	8°41.22	143°23.99	2.7		46	42	12	38.7	Massive bedded sandy mud
105	8°41.00	143°24.29	2.8	35	56	18	26	5.3	Laminated mud and sand
106	8°41.10	143°24.18	2.5	35	38	49	13	13.6	Disturbed at top
107	8°41.26	143°24.26	2.7	32	12	61	27	8.1	Laminated mud and sand
108	8°41.34	143°24.30	2.9		0	72	28	7.6	Massive bedded mud
109	8°41.61	143°24.20	6.2		13	72	15	39.9	Organic matter, mud balls
110	8°41.72	143°24.03	6.2		1	76	23	10.2	Mud balls
111	8°40.91	143°24.15	1.9	53	0	77	23	9.6	Laminated mud and sand

content is $r^2 = 0.467$, $n = 12$). The sands are very well sorted with mean grain sizes of around 87.2 μm (Fig. 7). The mud layers are composed of poorly sorted silts with generally less than 30% clay and an average grain size of 14.5 μm (Fig. 8).

Four main styles of laminated sediment are apparent, including planar and wavy laminations, lenticular bedding and ripple cross laminations. The planar and wavy laminated sections are characterized by thin sand layers that range in thickness from only a few grains to 2–3 mm (Fig. 9). Generally, when the sand layers are more than 1 mm thick, they are loaded onto the underlying mud layers. The load structures can be up to three times thicker than the sands. There is a sharp contact at the base of most of the sand layers, but in many cases the tops are gradational into the overlying mud. The muddy layers are much thicker than the sandy layers and may be up to 27 mm thick. Faint grading can be distinguished within some of the thicker layers, others appearing to be made up of compressed mud clasts or occasional mud clasts in a muddy matrix.

Lenticular bedding is common especially where the sand beds are thicker than 5 mm (Fig. 10). The sand lenses are generally well defined with sharp tops and bases. The lenticular sands are interspersed with ripple or cross-laminated sands. The latter are up to 20 mm thick, bidirectional and often display sharp tops and bases (Fig. 10). Climbing ripples can be observed in some of the thicker sand layers.

The sandy deposits encountered at the bases of cores 106 and 107 contain numerous laminated

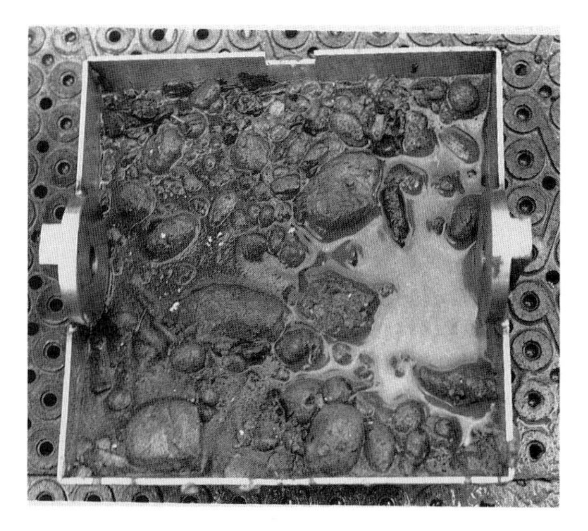

Fig. 4. Photograph showing mud balls obtained in a grab sample (station 109; bucket is 30 cm across).

sand and mud rip-up clasts (Fig. 11). In places the clasts are diffuse and more difficult to distinguish, but in some parts the originally laminated nature of the ripped-up sediment is exhibited and appears little disturbed.

Sediment accumulation rates

Measurements of excess ^{210}Pb were obtained for all four cores (Nos 105, 106, 107 and 111) and for the upper 10 cm of 'mini-cores' (Table 2; Fig. 3B). The five points obtained for core 105 closely followed a standard logarithmic decay curve ($r^2 = 0.837$) giving an accumulation rate of 9.9 cm yr^{-1} (Fig. 12). Excess ^{210}Pb from core 106 ($n = 5$; Fig. 13) gave a sediment accumulation rate of 2.6 cm yr^{-1} with a good correlation ($r^2 = 0.891$). In contrast with these two cores, five measurements from core 107 (Fig. 14) gave a relatively high sediment accumulation rate of 34.4 cm yr^{-1} with a correlation coefficient r^2 of only 0.244 (i.e. no significant correlation exists). Similar variability in ^{210}Pb activity with depth was also observed in core 111 (Fig. 15).

The results from core 111 indicate the presence of at least two sediment facies with different ^{210}Pb signatures. The mean excess ^{210}Pb value, averaged over the nine uppermost points (top 75 cm of core 111), is 0.818 ± 0.115 dpm g^{-1}. This is statistically different to the mean value derived from the lowermost nine points (below 75 cm depth) which is 0.475 ± 0.100 dpm g^{-1}. The two facies are separated by a change in lithology at 75 cm (Fig. 15). The upper part of the core is composed of mixed (slumped?) sediments which overlie laminated sediments. Based on all 18 points, an accumulation rate of 8.0 cm yr^{-1} is estimated ($r^2 = 0.538$) although the variability in excess ^{210}Pb exhibited in the core may preclude the calculation of a reliable accumulation rate.

Excess ^{210}Pb was measured in four 'mini-cores' (Table 2) in order to evaluate its spatial variability within the surficial sediments. The results suggest that mean ^{210}Pb in surface sediments ranges from around 0.53 to 0.94 dpm g^{-1}, with an overall mean value of 0.68 dpm g^{-1}. This range of ^{210}Pb values is found throughout the top 2 m of the cores (Figs 12–15), except for the the lower activity observed at depth in core 106 (Fig. 13).

Lamination thickness and periodicity

Image analysis and visual measurement of individual sand–mud layer thicknesses and couplet thicknesses (e.g. Fig. 16) revealed no apparent rhythmic pattern in any of the four piston cores. Plots of mud lamina and couplet thicknesses versus lamina number were analysed using spectral analysis techniques (cf. Williams, 1989) which indicated a random distribution in all four cores. Analysis of smoothed data sets, produced by a running average method, gave the same result as the 'raw' data.

The statistical test of de Boer *et al.* (1989) was applied in order to check the possible occurrence of a thick–thin lamina sequence caused by diurnal inequality (Fig. 2). The largest number of sequential thick–thin laminae found within any core section was six; such a small number indicates that no statistically significant correlation exists for the cores studied.

DISCUSSION

Variability in excess ^{210}Pb

The extreme variability in excess ^{210}Pb observed in surficial sediments (Table 2) and with depth in the cores (Figs 12–15) suggests differences in the initial specific activity (dpm g^{-1}) of the sediments being deposited (e.g. Kuehl *et al.*, 1986). If ^{210}Pb is preferentially absorbed on to, e.g. clay particles, then differences in sediment texture may give rise to spatial variability. However, comparison of surficial

A

B

Fig. 5. (A) Photograph showing intertidal mud-flat (footprints in the foreground provide a scale). (B) Close-up view showing truncation of horizontally bedded sediments outcropping on the side of a tidal creek (jar is 6 cm high). (C) Photograph showing retreating cliff face with mud balls developed at the base of the slope (jar is 6 cm high).

sediment grain size and excess ^{210}Pb showed no significant correlation between these variables in the present study (Tables 1 and 2).

Other factors contributing to variability in initial activity include (i) the mixing of ^{210}Pb-rich ocean water with ^{210}Pb-depleted river water; and (ii) ^{210}Pb scavenging by particle reactive species such as organic carbon and Fe/Mn oxides (Kuehl *et al.*, 1986). Analyses of excess ^{210}Pb from delta front cores (Harris *et al.*, 1993) show that the highest values are found in the more seaward parts of the delta. The range in ^{210}Pb activities for delta front surficial sediments is 0.3–2.2 dpm g^{-1}, whereas the range in distributary channel surficial sediments is 0.5–0.9 dpm g^{-1}. This may suggest that the high levels of ^{210}Pb observed at offshore sites is due to the presence of sea water. However, the actual cause of the observed variability in initial ^{210}Pb activities is to date unknown.

Tidal signature in Fly delta sediments

The results of our preliminary investigation have shown that rhythmically spaced, sand–mud couplets are not observed in the subtidal deposits of Fly delta distributary channels. This may be explained by, amongst other things: (i) sediment accumulation rate being too low at the study site; (ii) current speeds and sediment types not being conducive to the preservation of tidal rhythmites; and/or (iii) the

Fig. 5. *(Continued.)* C

erosion of sediments by waves and/or currents (Dalrymple & Makino, 1989; Nio & Yang, 1991).

Studies in other modern depositional environments have shown that deposits displaying tidal cyclicity occur on the lee slopes of migrating bedforms (tidal bundles) and on accreting banks associated with migrating channels (rhythmites; Nio & Yang, 1991). In the area studied, no large-scale bedforms were observed in echosounding profiles. Cores were obtained from what is interpreted to be an accreting bank on one side of a migrating channel. The best estimate of sedimentation rate in the cores is 8–10 cm yr^{-1}. Such an accumulation rate appears to be too low for the deposition of a continuous sequence of couplets with a measurable thickness during each tide or even within a neap–spring period. The laminae thicknesses observed in cores from the present study averaged 2.8 mm for mud and 1.4 mm for sand, without any significant variation between the four cores. Given that two couplets are deposited during each tidal cycle (i.e. four per day in this semi-diurnal subtidal setting),

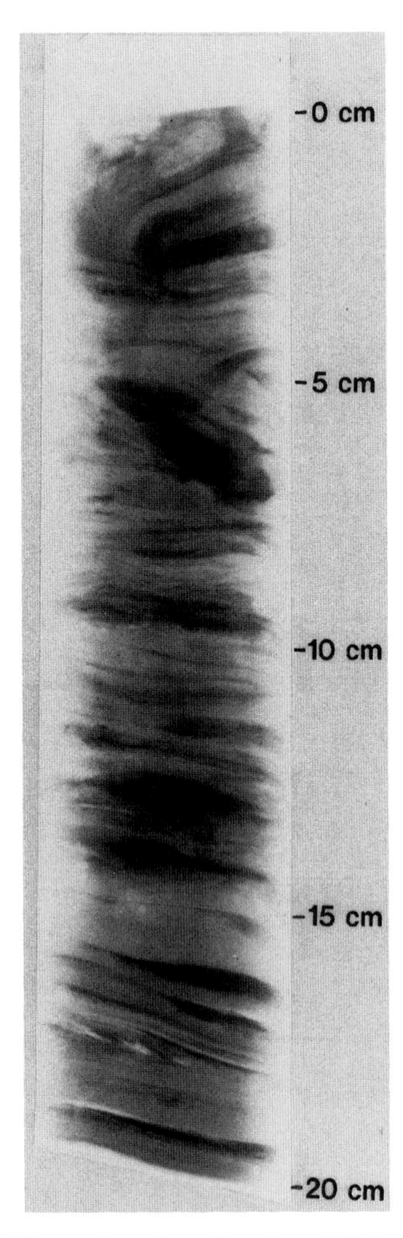

Fig. 6. X-ray radiograph positive of core 105 showing soft sediment deformation in the form of asymmetrical folding and dispersed laminae (mud is the light colour, sand is the dark colour).

an accumulation rate of 8.4 mm d^{-1} (~3 m yr^{-1}) is required to accommodate the observed laminae thicknesses (if a couplet is deposited and preserved during every tidal cycle).

The suspended sediment concentration, grain

Table 2. Excess ^{210}Pb in cores and surface samples obtained in the Fly delta

Station number	Depth (cm below surface)	Excess ^{210}Pb	Standard deviation	Mean excess ^{210}Pb (surface)
86	1–2	0.631	0.053	0.662
	3–5	0.692	0.062	
88	1–2	0.699	0.054	0.594±0.237
	3–5	0.760	0.054	
	6–8	0.322	0.032	
89	1–2	1.028	0.076	0.943±0.075
	3–5	0.887	0.072	
	6–8	0.915	0.067	
94	1–2	0.866	0.062	0.794±0.221
	3–5	0.969	0.064	
	6–8	0.754	0.071	
105	1–2	0.578	0.044	0.578
	37–39	0.470	0.050	
	72–74	0.486	0.050	
	110–112	0.445	0.048	
	145–147	0.336	0.039	
106	3–4	0.593	0.063	0.593
	34–36	0.501	0.046	
	82–84	0.537	0.046	
	133–135	0.131	0.031	
	110–112	0.058	0.024	
107	3–5	0.527	0.042	0.527
	44–46	0.794	0.058	
	127–129	0.335	0.041	
	170–172	0.508	0.055	
	230–232	0.454	0.041	
111	0–3	0.619	0.066	0.715±0.089
	5–7	0.795	0.052	
	10–12	0.730	0.059	
	15–17	0.866	0.076	
	20–22	0.603	0.056	
	25–27	0.893	0.080	
	30–32	0.841	0.045	
	45–47	0.844	0.072	
	55–57	0.948	0.066	
	70–72	0.406	0.053	
	80–82	0.630	0.051	
	90–92	0.643	0.049	
	100–102	0.443	0.048	
	120–122	0.433	0.034	
	140–142	0.492	0.048	
	145–147	0.338	0.037	
	170–172	0.439	0.052	
	190–192	0.450	0.047	

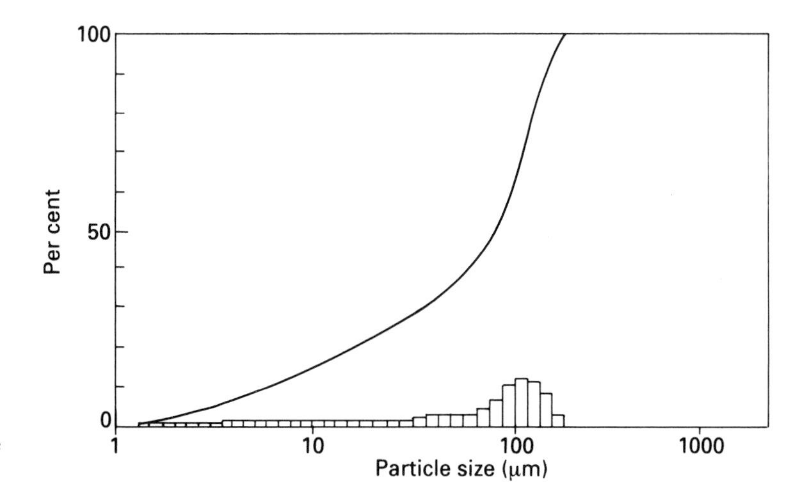

Fig. 7. Laser particle sizer plot showing a typical cumulative curve and histogram for a 'sand' layer.

size and maximum tidal current speeds observed in the Fly delta are not dissimilar to those reported for the Bay of Fundy by Dalrymple & Makino (1989). In the Bay of Fundy, average thicknesses of sand and mud laminae are 1–38 mm and 1–18 mm respectively, which corresponds to the range of laminae thicknesses observed in the present study. The main difference between the Fly and Bay of Fundy is that vertical accretion rates, which coincide with the occurrence of rhythmites in the latter area, are of the order of several metres per year (i.e. much greater than those observed in the present study).

The current speeds and sediment types occurring in the study area may not be conducive to the deposition of tidal rhythmites. Based on the current meter data obtained at station 85 during neap–spring tides, tidal currents in the general area of interest reach a peak speed of about 1.2 m s^{-1} at 1 m above the bed and are flood asymmetric (symmetry index = 0.7). Such speeds, when combined with measured grain sizes and water depths, correspond to upper-plane-bed conditions in bedform stability diagrams (e.g. Leeder, 1982, p. 85). Spring tidal currents probably erode the sand/mud laminae deposited during the previous tidal cycle (which may not have dewatered significantly; see above). Couplets may be deposited and preserved only during neap tides, when tidal currents are weakest and dewatering of the sediment is facilitated by the long periods of slack water (Fig. 2). Mean thicknesses of mud and sand laminae for the present

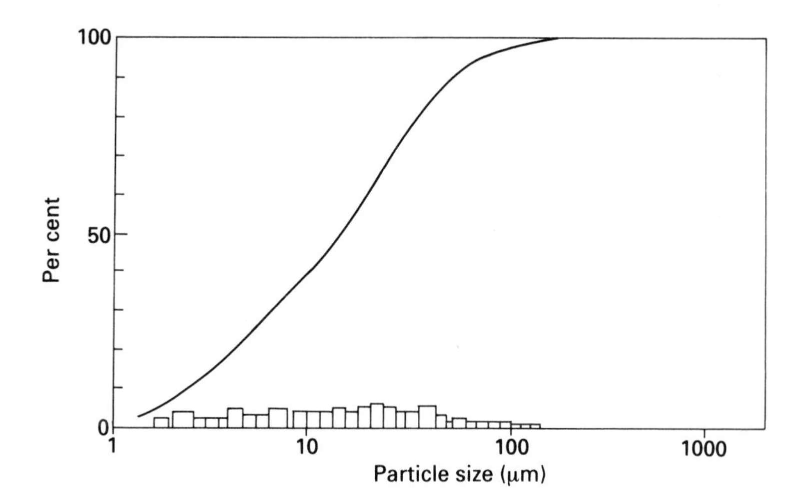

Fig. 8. Laser particle sizer plot showing a typical cumulative curve and histogram for a 'mud' layer.

study (2.8 mm for mud and 1.4 mm for sand), taken in combination with the maximum estimated sedimentation rate (9.9 cm yr^{-1}), suggest the deposition of around 24 sand–mud couplets per year or two per month (or one couplet per neap cycle).

Alternatively, such a pattern might be explained by the seasonal variation in surface wave energy, related to the large waves produced during the south-easterly Trade Winds and small waves produced during the north-west monsoon. Although tidal currents are strong in the distributary channels of the Fly, surface waves propagating landwards from the Coral Sea impinge on the coast, producing strong near-bed oscillatory currents in the channels. These oscillatory currents may preclude dewatering and/or the deposition of a mud drape during tidal slack water periods (particularly during the short slack water periods accompaning spring tides). Many tidal cycles may thus pass before conditions are conducive to mud drape deposition and preservation. In this case, the observed mud/sand layers may relate to episodes of lower wave energy, rather than to any specific tidal cyclicity. This would suggest that both tidal and wave conditions must be favourable if mud couplets are to be

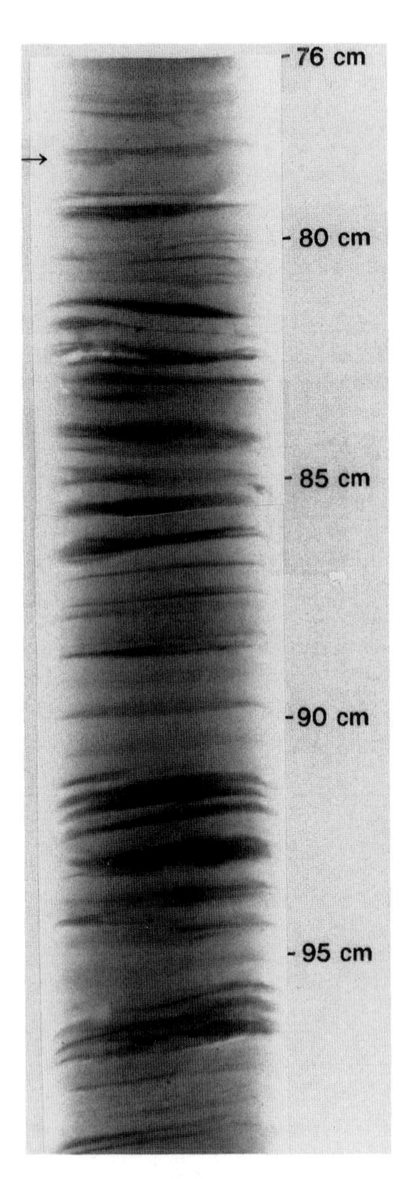

Fig. 9. X-ray radiograph positive of core 111 showing wavy and planar laminated sands and muds. The arrow points to load structure formed on the undersurface of a silty sand layer overlying a mud layer.

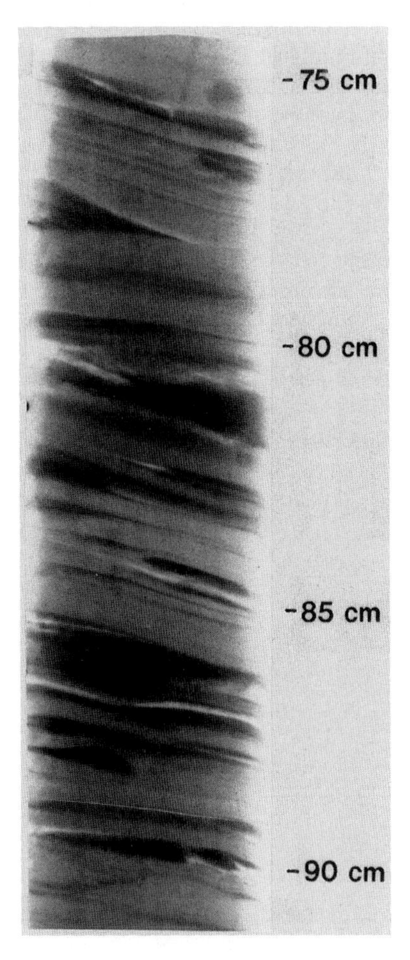

Fig. 10. X-ray radiograph positive of core 111 showing lenticular bedding and cross-laminated sands.

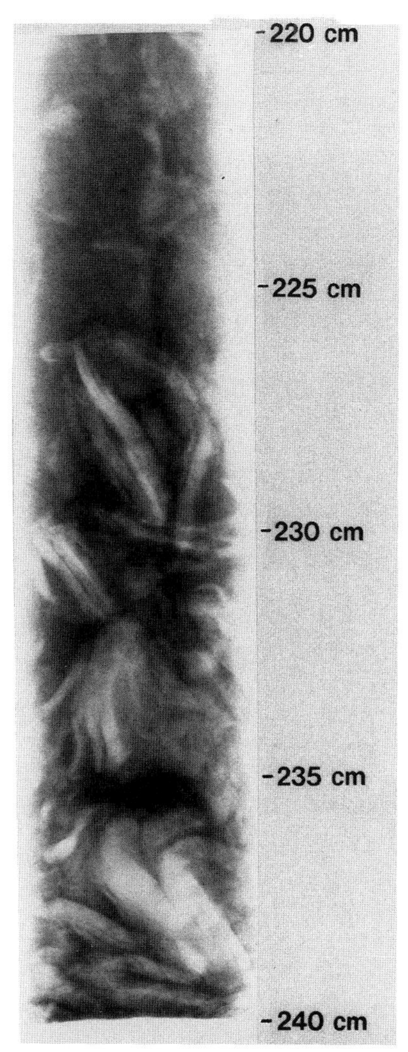

Fig. 11. X-ray radiograph positive of core 106 showing laminated rip-up clasts in a sandy silt matrix.

preserved. In the present case, couplets are only preserved during the low wave energy monsoon season, which lasts about 3 months during any year, suggesting that about eight couplets are deposited and preserved (during neap tides?) for each of these months.

Harris *et al.* (1993) concluded that laminae observed in cores obtained from the offshore prograding edge of the Fly delta (Figs 3A & B) contained seasonal varves related to the variations in surface wave activity. The main differences between delta front and distributary channel deposits are that: (i) the distributary channel deposits have a higher sand content (40–60% versus 10% on the delta front); (ii) ripple cross-bedding occurs in cores from the distributary channel deposits but not in those from the delta front; and (iii) rip-up clasts occur in cores from the distributary channel deposits but not in those from the delta front (slump features are observed in cores from both environments).

The limited core data available from the present study plus those of Alongi (1991) and Alongi *et al.* (1992) indicate that tidal cyclicities are absent from intertidal and subtidal deposits located in the seaward parts of the delta. Thus, tidal processes that have produced the funnel-shaped geomorphology of the Fly delta are not represented as rhythmic tidal bundles in subtidal channel deposits or on the intertidal flats. Alongi *et al.*(1992) noted that laminated sediment deposits were found in more sheltered areas behind islands (the cores from the present study were also from this protected environment). It may be that in the upper sections of the delta, where lateral channel migration rates may also be higher, tidal rhythmites are deposited in sheltered locations. However, further investigations are required to determine whether or not this is the case.

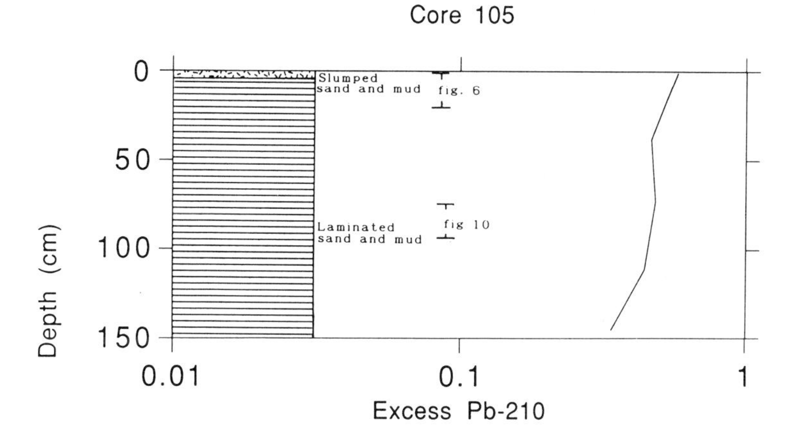

Fig. 12. Plot of excess ^{210}Pb versus depth in core 105 showing the decay curve which gives a sediment accumulation rate of 9.9 cm yr^{-1}.

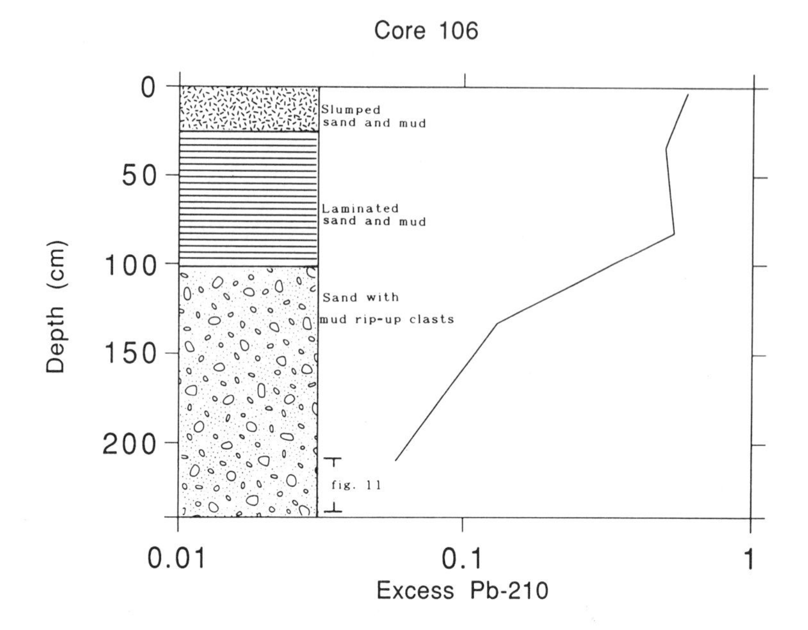

Fig. 13. Plot of excess ^{210}Pb versus depth in core 106 showing the decay curve which gives a sediment accumulation rate of 2.59 cm yr^{-1}.

CONCLUSIONS

Profiles of ^{210}Pb activities in cores together with surface measurements show a wide variability in initial activities, suggesting differences in the initial specific activity (dpm g^{-1}) of the sediments being deposited. This is similar to the results achieved by Kuehl *et al.* (1986) for the Amazon delta. The establishment of a geochronology for some sequences is therefore not always possible by the

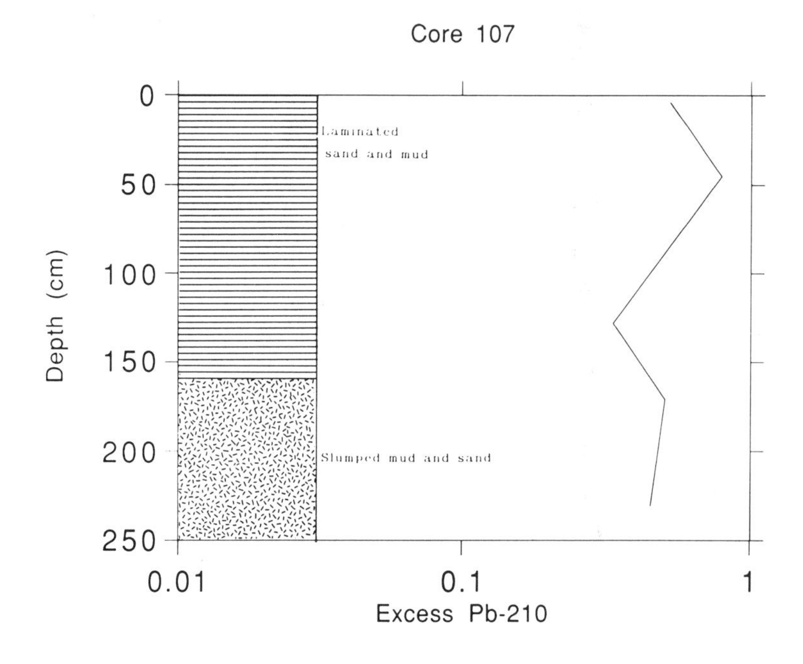

Fig. 14. Plot of excess ^{210}Pb versus depth in core 107 showing the variation in excess ^{210}Pb down the core. The apparent accumulation rate is 34.4 cm yr^{-1}; however, this core appears to be unsuitable for ^{210}Pb geochronology.

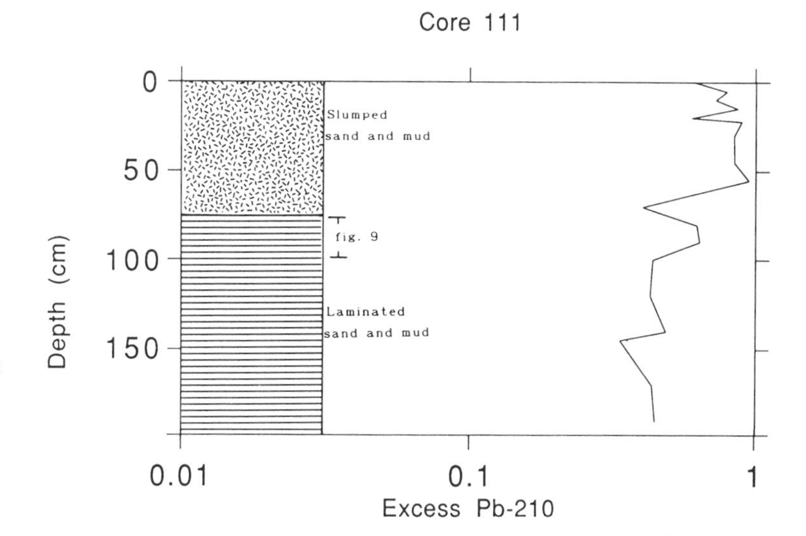

Fig. 15. Plot of excess ^{210}Pb versus depth in core 111 showing the change in lithology at 75 cm and the highly variable excess ^{210}Pb. The decay curve in the lower laminated section gives a sediment accumulation rate of 18.5 cm yr^{-1}.

^{210}Pb technique. The available data suggest sediment accumulation rates of about 8–10 cm yr^{-1} in one part of the study area.

Sand and mud laminae deposited in subtidal distributary channel deposits of the Fly delta do not exhibit a strong regular cyclicity in either laminae spacing or thickness (i.e. tidal bundles are absent). Rather, the sedimentation rate and laminae thicknesses may be explained by a seasonal variation in surface wave reworking in combination with deposition during slack water at neap tide. The sand content increases along a transect extending from the delta front to the distributary channel deposits. Distributary channel deposits are distinguished by the occurrence of rip-up mud clasts and ripple cross-bedding in sand layers which are absent from delta front deposits.

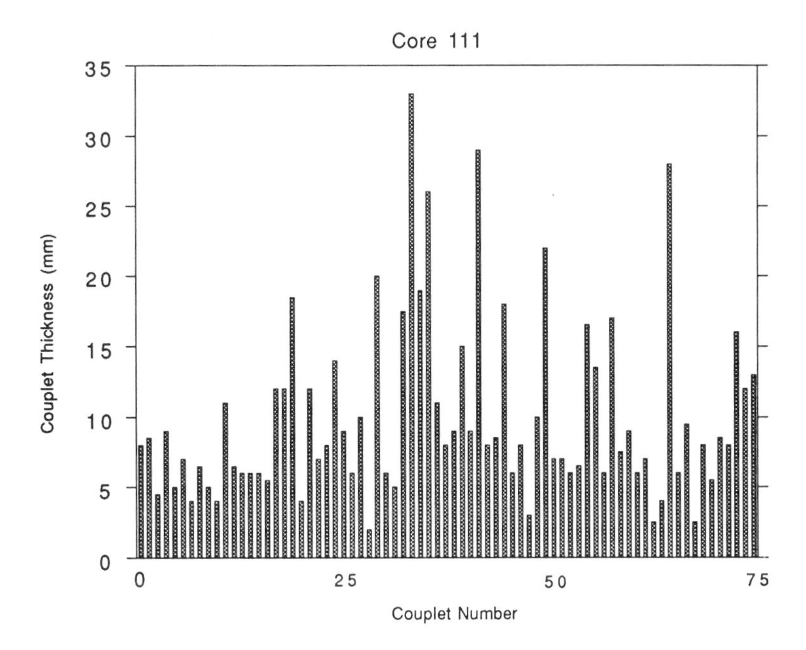

Fig. 16. Couplet thickness versus couplet number for core 111 (from 75 to 180 cm depth).

ACKNOWLEDGEMENTS

Dr Bob Dalrymple (Queens University, Ontario) provided valuable discussion and a critical review of an earlier version of this paper for which we are grateful. Thanks are due to the captain and crew of RV *Sunbird*, T. Ford and L. Wilson, and to Dr P. Gibbs, D. Mitchell and A. Cole for their assistance with the field work. We would also like to thank F. Chee for assistance with the image analyser and A. Cole for help with drafting. Data collection and analyses were supported by a grant from the Australian Research Council (Ref. A38830391).

REFERENCES

ALLEN, G., LAURIER, D. & THOUVENIN, J. (1979) Etude sédimentologique du delta de la Mahakam. *Comp. Franç. Pétroles Notes Mém.* **15**, 115 pp.

ALONGI, D.M. (1991) The role of intertidal mudbanks in the diagenesis and export of dissolved and particulate materials from the Fly Delta, Papua New Guinea. *J. exp. mar. Biol. Ecol.* **149**, 81–107.

ALONGI, D.M., CHRISTOFFERSEN, P., TIRENDI, F. & ROBERTSON, A.I. (1992) The influence of freshwater and material export on sedimentary facies and benthic processes within the Fly Delta and adjacent Gulf of Papua (Papua New Guinea). *Continent. Shelf Res.* **12**, 287–326.

ANDEL, T.H. VAN (1967) The Orinoco Delta. *J. sediment. Petrol.* **37**, 297–310.

BARUA, D.K. (1990) Suspended sediment movement in the estuary of the Ganges-Brahmaputra-Meghna river system. *Mar. Geol.* **91**, 243–253.

BLAKE, D.H. & OLLIER, C.D. (1969) Geomorphological evidence of Quaternary tectonics in southwestern Papua. *Rev. geomorphol. Dyn.* **19**, 28–32.

BLAKE, D.H. & OLLIER, C.D. (1971) Alluvial plains of the Fly River delta. *Z. Geomorph. N.F., Suppl.* **12**, 1–17.

BOER, P.L. DE, OOST, A.P. & VISSER, M.J. (1989) The diurnal inequality of the tides as a parameter for recognizing tidal influences. *J. sediment. Petrol.* **59**, 912–921.

COLEMAN, J.M. (1969) Brahmaputra River: Channel processes and sedimentation. *Sediment. Geol.* **3**, 129–239.

DALRYMPLE, R.W. & MAKINO, Y. (1989) Description and genesis of tidal bedding in Cobequid Bay-Salmon River estuary, Bay of Fundy, Canada. In: *Sedimentary Facies in the Active Plate Margin* (Eds Taira, A. & Masuda, F.), pp. 151–177. Terra Sci. Publ. Comp, Tokyo.

DALRYMPLE, R.W. & ZAITLIN, B.A. (1989) Tidal deposits in the macrotidal, Cobequid Bay–Salmon River estuary, Bay of Fundy. In: *Abst. 2nd Intern. Res. Symp. Clastic Tidal Deposits (Calgary)*, 84 pp.

EAGLE, A.M. & HIGGINS, R.J. (1990) Environmental investigations on the effects of the Ok Tedi copper mine in the Fly River system. In: *Torres Strait Baseline Study* (Eds Lawrance, D. & Cransfield-Smith, T.). Works. Ser. **16**, 97–118.

EISMA, D., GAAST S.J. VAN DER, MARTIN, J.M.A. & THOMAS, A.J. (1978) Suspended matter and bottom deposits of the Orinoco Delta: turbidity, mineralogy and elementary composition. *Neth. J. Sea Res.* **12**, 224–251.

FISHER, W.L., BROWN, L.F., SCOTT, A.J. & MCGOWAN, J.H. (1969) Delta systems in the exploration for oil and gas. *Texas Bur. Econ. Geol. (Austin)*, 78 pp.

HARRIS, P.T., BAKER, E.K., COLE, A.R. & KEENE, J.B. (1992) Sandwave movement, current and sedimentation in Torres Strait. *Ocean Sci. Inst. Rept* **47**.

HARRIS, P.T., BAKER, E.K., COLE, A.R. & SHORT, S.A. (1993) Preliminary study of sedimentation in the tidally dominated Fly River Delta, Gulf of Papua. *Continent. Shelf Res.* **13**(4), 441–472.

HIGGINS, R.J., PICKUP G. & CLOKE, P.S. (1987) Estimating the transport and deposition of mining waste at Ok Tedi. In: *Sediment Transport in Gravel Bed Rivers* (Eds Thorne, C.R., Bathurst, J.C. & Hey, R.D.), pp. 949–976. John Wiley, London.

JAQUES, A.L. & ROBINSON, G.P. (1977) The Continent/island arc collision in northern Papua New Guinea. *BMR J. Austr. Geol. Geoph.* **2**, 2289–2303.

KUEHL, S.A., DEMASTER, D.J. & NITTROUER, C.A. (1986) Nature of sediment accumulation on the Amazon continental shelf. *Continent. Shelf Res.* **6**, 209–225.

KVALE, E.P., ARCHER, A.W. & JOHNSON, H.R. (1989) Daily, monthly, and yearly tidal cycles within laminated siltstones of the Mansfield Formation (Pennsylvania) of Indiana. *Geology* **17**, 365–368.

LEEDER, M.R. (1982) *Sedimentology: Process and Product*, George Allen & Unwin, London, 334 pp.

MIALL, A.D. (1984) Deltas. In: *Facies Models* (Ed. Walker, R.G.), pp. 106–118. Geol. Ass. Can.

NIO, S. & YANG, C. (1991) Diagnostic attributes of clastic tidal deposits: a review. In: *Clastic Tidal Sediments* (Eds Smith, D.G., Reinson, G.E., Zaitlin, B.A. & Rahmani, R.A.). *Can. Soc. Petrol. Geol. Mem.* **16**, 3–28.

NITTROUER, C.A. & DEMASTER, D.J. (1986) Sedimentary processes on the Amazon continental shelf: past, present and future research. *Continent Shelf Res.* **6**, 5–30.

OK TEDI MINING LIMITED (1988) Sixth Supplement Agreement Environmental Study 1986–1988. *Final Draft Report OTML*, PNG.

PICKUP, G. (1984) Landforms, hydrology and sedimentation in the Fly and lower Purari, Papua New Guinea. In: *Channel Processes — Water, Sediment Catchment Controls* (Ed. Schick, A.P.). Catena Suppl., pp. 1–17 Braunschweig.

PIGRAM, C.J., DAVIES, P.J., FEARY, D.A. & SYMONDS, P.A. (1989) Tectonic controls on carbonate platform evolution in southern Papua New Guinea: passive margin to foreland basin. *Geology* **17**, 199–202.

ROBERTSON, A.I., ALONGI, D.M., CHRISTOFFERSEN, P., DANIEL, P., DIXON, P., & TIRENDI, F. (1990) The influence of freshwater and detrital export from the Fly River system on adjacent pelagic and benthic systems, *Aust. Insit. mar. Sci. Rept.* **4**, 199 pp.

SPENCER, L.K. (1978) *The Fly estuari ne delta, Gulf of Papua, Papua New Guinea.* MSc thesis, Univ. Sydney, 278 pp.

THOM, B.G. & WRIGHT, L.D. (1983) Geomorphology of the Purari Delta. In: *The Purari—Tropical Environment of a*

High Rainfall River Basin (Ed. Petr, T.), pp. 47–65. Junk Publishers, The Hague.

VISSER, M.J. (1980) Neap–spring cycles reflected in Holocene subtidal large scale bedform deposits: a preliminary note. *Geology* **8**, 543–546.

WILLIAMS, G.E. (1989) Late Precambrian tidal rhythmites in South Australia and the history of the Earth's rotation. *J. Geol. Soc. London* **146**, 97–111.

WOLANSKI, E. & EAGLE, M. (1991) Oceanography and sediment transport, Fly River Estuary and Gulf of Papua. *Abstr. 10th Australasian Conf. Coastal Ocean Eng. (Auckland, pp. 453–457).*

WOLANSKI, E., RIDD, P., KING, B. & TRENORDEN, M. (1992) Fine sediment transport, Fly River estuary, Papua New Guinea. Rept Ok Tedi Mining Limited, Austr. Inst. Mar. Sci., 31 pp.

WRIGHT, L.D. (1977) Sediment transport and deposition at river mouths: a synthesis. *Geol. Soc. Am. Bull.* **88**, 857–868.

WRIGHT, L.D. (1985) River Deltas. In: *Coastal Sedimentary Environments* (Ed. Davis, R.A., Jr), pp. 1–76. Springer, New York.

Spec. Publs int. Ass. Sediment. (1995) **24**, 213–223

Foraminifers as facies indicators in a tropical, macrotidal environment: Torres Strait–Fly River delta, Papua New Guinea

A.R. COLE, P.T. HARRIS* *and* J.B. KEENE

Ocean Sciences Institute, University of Sydney, Sydney NSW 2006, Australia

ABSTRACT

Foraminifer assemblages are assessed as potential recorders of transportation and redistribution by tidal processes in tropical, tidally dominated carbonate shelf and fluvio-deltaic environments. Four facies are identified on the basis of grain-size distribution, carbonate content and energy: (i) a high energy carbonate sand facies; (ii) a low energy muddy carbonate sand facies; (iii) a transitional mixed terrigenous and carbonate sediment facies; and (iv) a terrigenous delta front facies. Up to 1300 tests were picked from nine representative samples and analysed using a computerized image analysis system. This method allowed the determination of maximum and minimum test dimensions, as well as two-dimensional shape and area, to produce a grain-size histogram of foraminifer tests for each site.

The results show that each of the four facies can be identified from the combination of foraminiferal taphonomic and grain-size properties. Furthermore, such properties identified the mode of transport (i.e. suspension versus bedload) and transport pathways at the sample locations. A small amount of carbonate material is transported landwards in suspension and deposited among the mostly (>95%) terrigenous sediments comprising the prograding Fly delta.

INTRODUCTION

Several studies have been carried out which describe the transportation of foraminifer tests by tidal currents. Most of these have been concerned with tests transported into estuaries (Brasier, 1981; Thomas & Schafer, 1982; Wang, 1983; Wang & Murray, 1983; Wang *et al.*, 1985; Michie, 1987). Other studies have identified test size, sorting, diversity and occurrence of exotic species as factors useful in determining whether transportation of tests over significant distances has occurred (Loose, 1970; Murray *et al.*, 1982; Wang, 1983). Wang (1983) concludes that allochthonous foraminiferal assemblages can be used to recognize tidally influenced estuarine deposits.

There is also an increasing degree of interest focused on the taphonomy of biogenic components of sediments (Brett & Baird, 1986; Fürsich & Flessa, 1987; Meldahl, 1987; Brandt, 1989; Meldahl & Flessa, 1990; Powell, 1992). A study of the taphonomy of skeletal fragments ties in with facies and environmental energy distributions and can provide additional information about the movement and burial history of the sediment. Foraminifers are ideally suited to this kind of study as they contribute significantly to sediments in a wide range of marine environments (Thomas & Schafer, 1982; Martin & Liddell, 1989; Kotler *et al.*, 1992).

For example, Kotler *et al.* (1992) have developed a taphofacies model for Discovery Bay, Jamaica, based on abrasion, dissolution and transportation of tests. The facies are characterized by particular energy regimes that correspond to the degree of abrasion (and hence the amount of destruction) and transportation. In their laboratory studies, Kotler *et al.* (1992) showed that combined dissolution and

**Correspondence address*: Australian Geological Survey Organisation, Antarctic CRC, University of Tasmania, GPO Box 252C, Hobart, Tasmania 7001, Australia.

abrasion is the most effective taphonomic process. Physical destruction of tests is greatly accelerated after dissolution has weakened the test.

The aims of the present study are: (i) to assess the potential of foraminifer assemblages to record transportation and redistribution in tropical, tidally dominated shelf and deltaic environments; and (ii) to determine whether taphonomy can identify the mode of transport (i.e. suspension versus bedload). The question is: 'Can foraminifers be used to distinguish transport pathways and facies by combining taphonomy with grain-size information?'

GEOGRAPHICAL SETTING

Torres Strait is a shallow seaway located between Australia and Papua New Guinea (Fig. 1). It is the site of a major carbonate/terrigenous clastic transition in which the Fly River delta is prograding southwards over the reefal carbonate sediments of the northern part of the Great Barrier Reef (Harris *et al.*, 1993). The slowly accumulating sediments of the shelf are being buried by the rapidly prograding Fly River delta (6 m yr^{-1} laterally), which receives 125 million t yr^{-1} of fine ($D < 100$ μm) fluvially derived sediment (Harris *et al.*, 1993).

Harris *et al.* (1991) have described the distribution of the carbonate and the gravel, sand and mud contents of the surficial sediments based on more than 550 grab samples. In addition, they have presented current-meter data from 12 sites in the area. From the available information, four facies may be identified, based on a combination of tidal current energy, grain size and carbonate content: (i) a high energy carbonate sand facies; (ii) a low energy muddy carbonate sand facies; (iii) a transitional facies; and (iv) a delta front facies (see Fig. 1).

The high energy carbonate sand facies corresponds to an area of Torres Strait which experiences strong tidal currents. Tides in Torres Strait are of the semi-diurnal mixed type (Wolanski *et al.*, 1988) and the tidal range reaches a maximum of 3.8 m (MHWS–MLWS) along the southern coastline of Papua New Guinea. Strong currents are found where the water is accelerated as it flows through constricted, interreef channels. Speeds of over 1.3 m s^{-1} have been measured at 100 cm above the bed in association with coarse-grained, carbonate dunes at 20 m water depth (Harris, 1989, 1991). Coarse-grained bioclastic carbonates are distributed as a patchy veneer of mobile dunes forming a tidally scoured and reworked sand facies which rims the central Torres Strait basin (Fig. 1; Harris & Baker, 1991).

The low energy muddy carbonate sand facies (Fig. 1) is associated with weaker tidal currents. Thus, the coarse-grained mobile dune sediments give way to a low energy facies rich in carbonate mud, located in the back reef 'lagoon' area. Unconsolidated surface sediments of the Strait consist of 40–100% biogenic carbonate debris, derived from skeletal remains of foraminifers, molluscs, bryozoans, corals and algae (Harris, 1988a). Carbonate sediment is accumulating at a rate of about 0.02 cm yr^{-1} (Harris *et al.*, 1993) and dead tests outnumber living ones by a ratio of at least 100 : 1. Foraminifers typically comprise 5–25% of the sediment.

As a result of the large terrigenous sediment input by the Fly River, a strong gradient in the carbonate content of the surficial sediments is exhibited with increasing proximity to the Fly delta (Harris, 1988a). The sediments of the delta consist of laminated terrigenous muds and fine sands, with carbonate contents generally <5% (Harris *et al.*, 1993). On the distal edge of the delta, however, carbonate contents exceed 20%, rising southwards to 60% over a distance of 20 km. This carbonate gradient defines the transitional facies (Fig. 1).

Within the funnel-shaped Fly delta, the tidal wave is amplified and a maximum spring tidal range of about 5 m occurs at the head, near the western end of Kiwai Island (Fig. 1; Wolanski & Eagle, 1991). Strong rectilinear tidal currents, with speeds exceeding 1 m s^{-1}, occur in the deltaic distributary channels. Wolanski & Eagle (1991) have reported extremely high suspended sediment concentrations (SSC), which fluctuate in sympathy with the neap–spring variation in tidal current speeds. Peak SSC values reach 40 g l^{-1} near the bed. Tidal current speeds are much reduced seawards of the distributary channel mouths and reworking is mainly by surface gravity waves on the delta front. The delta front facies (Fig. 1) thus occupies an intermediate tidal energy/high wave energy zone, where deposits are characterized by laminated sands and muds containing less than about 5% $CaCO_3$ (Harris *et al.*, 1993).

METHODS

Samples were selected from over 450 Shipek grab samples obtained by the Ocean Sciences Institute, to represent a range of sediment/energy facies (Fig. 1). The sediments were wet sieved at 63 μm

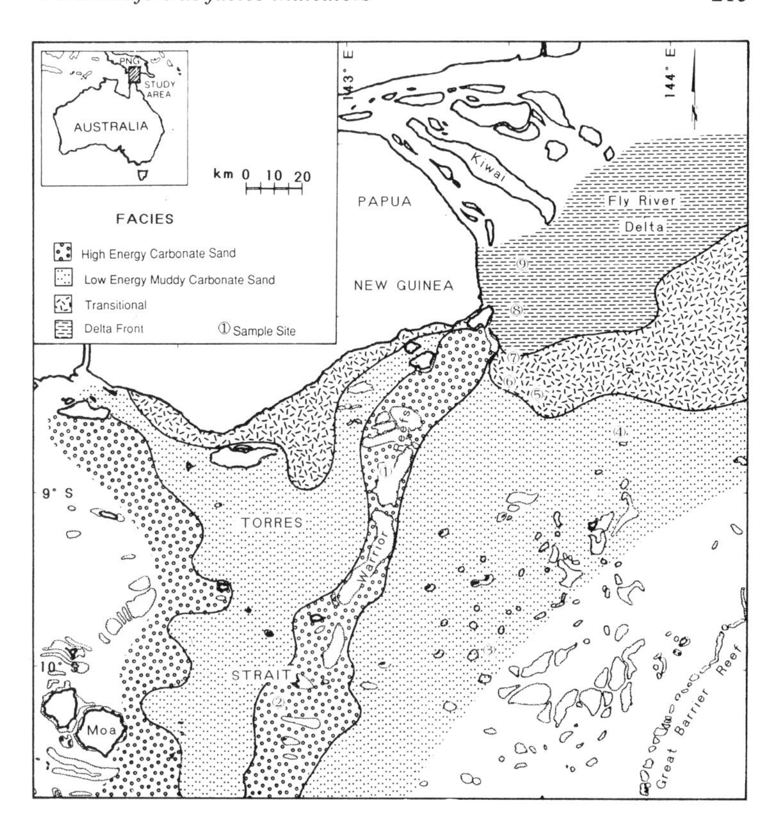

Fig. 1. Location of study area, distribution of facies and location of sample sites.

and foraminifers were picked from 1 to 5 g oven-dried subsamples of the fraction >63 μm to give a total assemblage. Up to 1300 tests were picked from nine representative samples and suborder abundances were determined as defined by Murray (1991, p. 321). Four broad, informal groupings were used, based on wall structure and/or lifestyle (i.e. planktonic forms were counted as a separate group to reflect their different mode of life). The four groups are: (i) miliolids (forms with a porcelainous test); (ii) rotaliids plus buliminids and lagenids (forms with a hyaline test); (iii) agglutinating forms; and (iv) planktonic forms. Surface textures of the foraminifer tests were examined by light microscopy and JEOL 35C scanning electron microscopy.

The size distributions of the four assemblages from each sample site were determined by image analysis, using a Tracor North TN 8502 Image Analysis System. A digitized image was acquired via video from material presented to a light microscope (up to 1200 individual tests were used). The pixel size of the digitized image was 8 μm, approximately 1/10 that of the average smallest test dimension. Thus the maximum and minimum dimensions, as well as two-dimensional shape and area, were determined. The individual results for the maximum projection were then binned into phi categories for each site to give a grain-size distribution.

RESULTS

Samples obtained from nine sites were analysed (see Fig. 1 and Table 1). Based on the four general facies identified above (Fig. 1), a descriptive summary of foraminiferal attributes is presented below.

High energy carbonate sand facies

Sites 1 and 2 are representative of this facies (see Table 1 for a summary of the environmental characteristics of each of the sites). Site 1, located on the Warrior Reefs (Fig. 1), exhibited a low diversity assemblage dominated by rotaliids (Fig. 2), particu-

Table 1. Site details for the nine sites. For sites 7 to 9, where mud content is >80%, the mud is 95% terrigenous (see Fig. 5)

Site	Lat.	Long.	Depth (m)	Mud (%)
1	9 26.6′	143 06.5′	2	1
2	10 06.8′	142 47.0′	19	0
3	9 53.0′	143 32.0′	30	34
4	9 34.9′	143 25.0′	52	24
5	9 12.0′	143 35.1′	31	19
6	9 08.5′	143 29.6′	15	62
7	9 04.9′	143 33.2′	13	80
8	8 59.6′	143 33.6′	7	97
9	8 48.0′	143 31.0′	5	92

larly the genus *Calcarina*. This assemblage is similar to Hallock's (1984) assemblage A, characteristic of exposed reef-flats. The tests from this site are typically dull, bleached white in appearance, a low degree of abrasion having resulted in some of the *Calcarina* spines being broken.

Site 2 is from a dune located in an interreef channel of the Warrior Reefs (Fig. 1). The dune was 4.5 m in height and composed of well-sorted biogenic coarse sand and gravel. The low diversity foraminifer asssemblage here is also dominated by rotaliids (but to a lesser degree than at site 1), with miliolids comprising about 30% of the number counted (Fig. 2). Under the microscope, the rotaliid genus *Amphistegina* is seen to be most abundant (visual estimation), with miliolid genera *Alveolinella* and *Quinqueloculina* making significant contributions. Photomicrographs of tests from this site show iron staining (90% of tests affected) and polished surfaces (Fig. 3). Maximum iron staining is evident in the miliolid species *A. quoyi* and *Q. philippinensis*, miliolids being in general more heavily stained than rotaliids or agglutinating forms. About 30% of the tests exhibit breakage and about 2% have a fresh unstained and unpolished appearance (Fig. 3).

The assemblages at both sites 1 and 2 are dominated by robust forms, with a distinct lack of delicate and small species. Juveniles are present in small numbers only at site 1.

The size distributions (Fig. 4A & B) show different patterns for the two foraminifer populations. Both distributions are moderately well sorted, with mean diameters in the coarse sand range. There are abundant larger foraminifers present, ranging in size up to 10 mm at site 2 (*A. quoyi*) and 3 mm at site 1 (*Marginopora vertebralis*). The distribution curve for site 1 has a positive (fine) skewness and is mesokurtic; the distribution curve for site 2 is also slightly positively skewed, but is at the same time extremely leptokurtic, indicating that the central region of the distribution is extremely well sorted compared to the tails.

Low energy muddy carbonate sand facies

Sites 3 and 4 are from the low energy muddy carbonate sand facies (see Table 1). For both samples the moderate to high diversity assemblages are dominated by rotaliids (*Amphistegina* is the most abundant genus) followed by miliolids as the next most significant suborder (particularly the genus *Quinqueloculina*; see Fig. 2). The assemblages at both sites are dominated by robust forms but there is a significant component (approximately 10%) of smaller delicate buliminids, lagenids, juvenile planktonics and rotaliids.

The tests are white and translucent, with some dark grey staining (5% of tests affected). Abrasion is most evident in miliolid tests in the medium size range (1–2 mm), where solution activity appears to have enhanced the damage. About 25% of these intermediate sized tests have been affected. There is only minimal damage to the smaller forms (<100 μm in size).

The size distribution curves for the foraminifers (Fig. 4C & D) show a shift towards finer grain sizes compared to the high energy facies, with means and medians in the medium (site 4) and fine sand (site 3) ranges; both populations are moderately sorted. The assemblage at site 4 (Fig. 4D) is distinctly bimodal, indicating the existence of two size populations; a similar size range is present at site 3 (Fig. 4C) but in a unimodal population.

Transitional facies

This facies is represented by sites 5, 6 and 7 (Table 1) which are aligned in a transect extending southwards from the delta front. The 10-km transect is characterized by changes in mud content, carbonate content, turbidity and salinity (Fig. 5).

The assemblages along the transect reflect the environmental changes. The high diversity assemblage at site 5 is similar to those at sites 3 and 4 (in the low energy muddy carbonate sand facies) in that rotaliids are the most abundant suborder (Fig. 2), with the genera *Operculina* and *Amphistegina* being

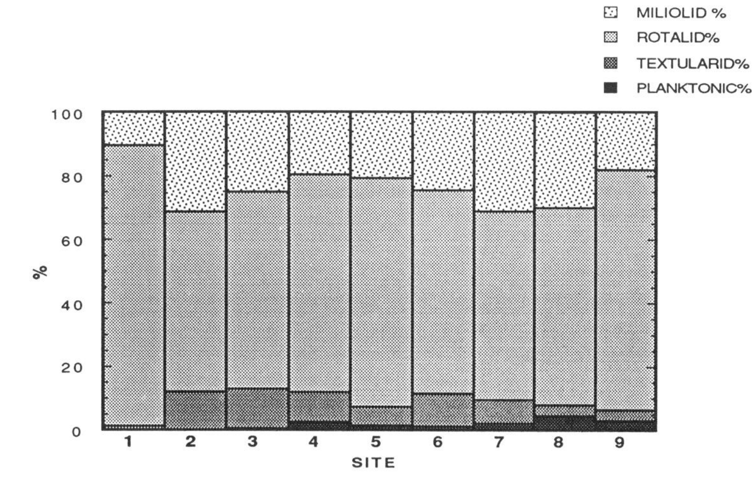

Fig. 2. Suborder composition of the foraminiferal assemblage for each site (see text for explanation of groups).

dominant, and the genus *Quinqueloculina* being the most common of the miliolids. About 10% of the assemblage is comprised of small delicate forms (buliminids and lagenids) and juveniles. There is some minor abrasion damage evident across the entire assemblage, with slightly damaged large robust forms (e.g. *Operculina*) co-existing with the small fragile forms of buliminids, rotaliids and lagenids.

The assemblages at sites 6 and 7 are impoverished (low diversity), reflecting the change in environment along the transect. While the assemblages are still dominated by rotaliids, the tolerant genus *Ammonia* is more important and there is a greater proportion of juvenile forms of all species. Small fragile lagenids and buliminids are also proportionally more important (around 10% of the total assemblage). Total abundances are about 1/5 to 1/10 those of the open marine sites and diversity is consequently lower. Since a brackish water environment is marginal for miliolids and buliminids (their growth is inhibited at salinities lower than 32‰; Murray, 1991, pp. 323–326) these families could be considered exotic forms at sites 6 and 7.

Fig. 3. Photomicrograph of tests from site 2. The mid-grey colour in the black and white photograph is due to the iron staining of the tests. Photograph width is 8 mm.

Mean grain sizes for sites 5, 6 and 7 (Fig. 4E–G) show a shift towards finer grain sizes in comparison to sites 3 and 4. Furthermore, site 5 (mesokurtic) has a moderately sorted, medium sand-sized foraminifer population, sites 6 and 7 (leptokurtic) have well-sorted fine sand populations. The foraminifer populations from this transitional facies appear to correspond to the finer of the two populations from site 4.

Delta front facies

The delta front is represented by sites 8 and 9 (Fig. 1; Table 1). With increasing distance northwards along the transect, the carbonate content decreases to approximately 5%. Correspondingly, salinity decreases from open marine to brackish (9‰) conditions and water turbidity rises (Fig. 5). These environmental parameters are reflected in restricted growth (small size and low diversity) of the foraminifer assemblages. Rotaliids (Fig. 2) are dominated by the genus *Ammonia*; most other species are represented by small delicate juvenile

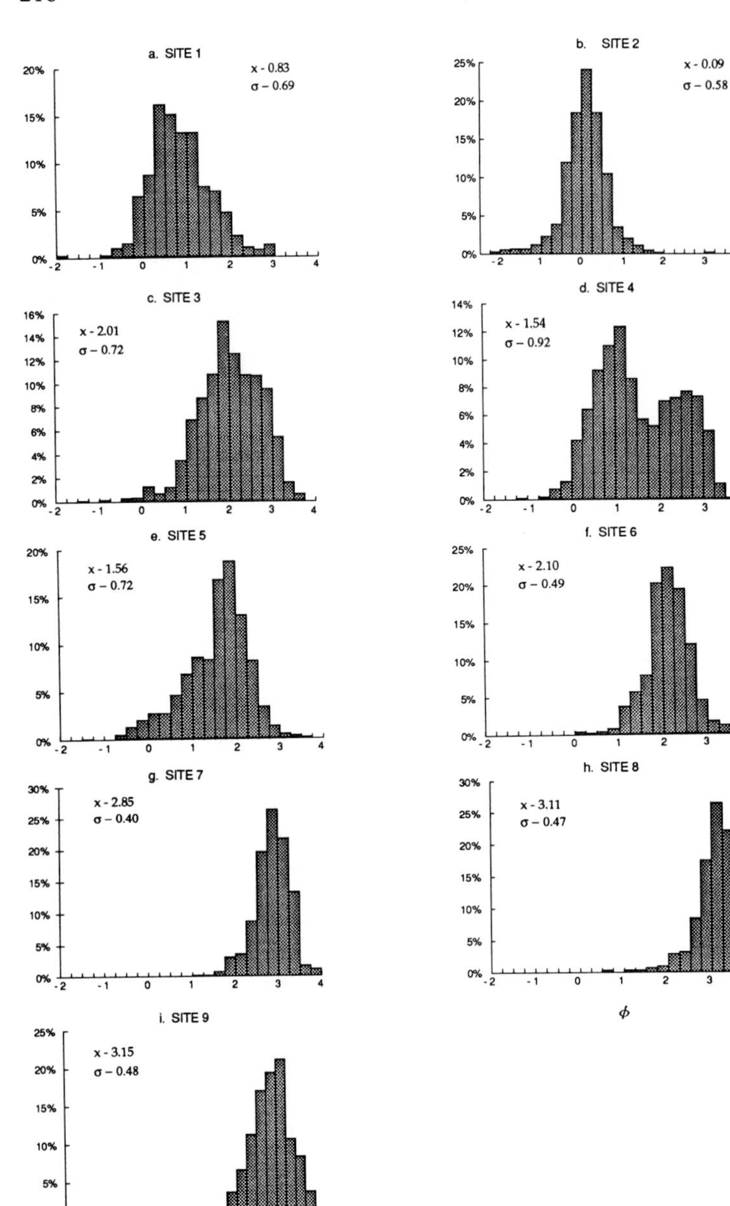

Fig. 4. Size distributions of tests from the nine study sites (mean test sizes (x) and standard deviations (sorting, σ) are shown). A, B: sites 1 and 2 respectively (high energy carbonate sand facies). C, D: sites 3 and 4 respectively (low energy muddy carbonate sand facies). E, F, G: sites 5, 6 and 7 respectively (transitional facies). H, I: sites 8 and 9 respectively (delta front facies).

forms (including miliolids, buliminids, lagenids and planktonics). There is, however, a relative increase in the planktonic component along the transect (Fig. 2).

As salinities are significantly lower than those of the open sea, all miliolid and buliminid species occurring at sites 8 and 9 are considered to be exotic. The surface textures show the combined effects of dissolution and abrasion (the corrasion of Brett & Baird, 1986; also see Fig. 6). Corrosion has affected about 80% of tests at site 9, including the tests of species such as *Ammonia beccarii* which are tolerant of brackish conditions. Corrosion weakens the test which then pits easily (Thomas & Schafer,

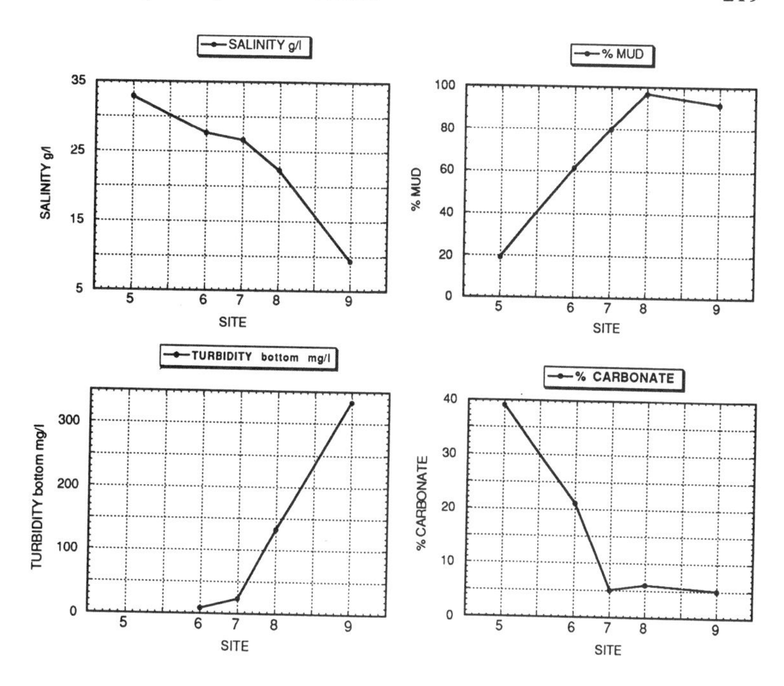

Fig. 5. Salinity (‰), mud (%), turbidity (mg l^{-1}) and carbonate (%) in total sample for sites 5 to 9.

1982; Kotler *et al.*, 1992). The surface textures of exotic miliolid and buliminid tests range from pristine to extensively corroded (Fig. 6). Brackish water tolerant rotaliid and agglutinating species also exhibit a similar range of textures.

DISCUSSION

Examination of the suborder, composition, size distributions and textures of the foraminifer populations suggests that transport in both bedload and suspension occurs. Suitable energy regimes and grain sizes are available in Torres Strait and the Fly delta for both modes of transport. Tidal currents in Torres Strait are strong enough to transport tests of larger foraminifers as bedload, and possibly even by saltation and in suspension. By way of comparison, the threshold velocity measured at 100 cm above the bed (U_{100}) for a 2-mm quartz sphere is 0.77 m s^{-1} (Miller *et al.*, 1977), this speed being regularly exceeded during spring tides at the sample locations in the high energy facies (Harris, 1989, 1991). Given that bedload and suspended load transport occur, it is of interest to consider the effects which such processes might have on the foraminifers.

Effects of bedload transport

A grain size population that has been subjected to bedload transport will become better sorted as a result of finer particles being winnowed away and coarse grains being left behind as a lag deposit. In the case of foraminifer tests, the shape, size and bulk density will determine threshold velocity, mode of transport and transport velocity of the tests (Maiklem, 1968; Kontrovitz *et al.*, 1978). In contrast to, e.g. site 1, the foraminifer assemblage from site 2 exhibits such sorting properties (Fig. 4B) as well as grain surface properties (rounded, polished; Fig. 3), indicating that this material has been subjected to a sorting process resulting from bedload transport. The grain size population for site 2 was determined using both image analysis and settling column, to be sure that hydraulic behaviour is predictable. The mean grain size from the settling column is 0.13 phi smaller than that resulting from image analysis (0.22 compared to 0.09 phi). The size frequency distribution for the settling column data is very well sorted and leptokurtic.

The environment at site 2, comprising mobile dunes, is hostile to all but a few specialized species of foraminifer. It is poor in nutrients, coarse-grained and constantly shifting as it is reworked by

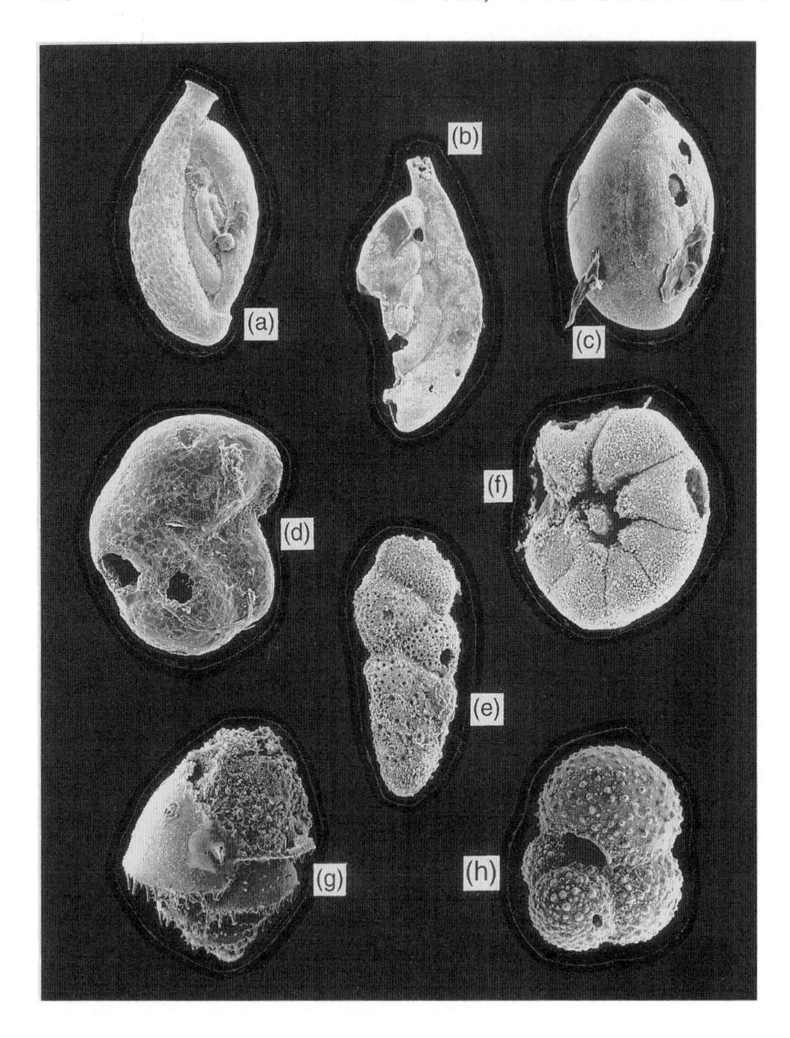

Fig. 6. Scanning electron microscope photographs of tests from site 9 on the delta front. (a) *Spiroloculina cf. rugosa* Cushman & Todd: a medium sand size (0.4 mm) miliolid displaying minimal corrosion (× 195). (b) *Spiroloculina cf. rugosa* Cushman & Todd: extensively corroded and abraded specimen, showing the weakening of the test by corrosion and pitting (× 225). (c) *Quinqueloculina laevigata* d'Orbigny: corrosion and pitting are visible; some organic material is sticking to the test (× 330). (d) *Arenoparella mexicana* (Kornfeld) showing various stages of pitting (note final chamber; × 405). (e) *Loxostomina* sp. corrosion, pitting and enlargement of pores are visible (× 405). (f) *Ammonia* sp. showing corrosion and pitting (final chamber was lost during mounting; × 450). (g) *Sagrina* sp: final chamber is broken off, showing sediment infill (× 495). (h) juvenile planktonic form displaying some breakage (aperture) and pitting (× 405).

strong tidal currents (up to 1.2 m s^{-1}). Only robust forms, for example *Amphistigina lessonii* and *Alveolinella quoyi*, are found at site 2 and almost all tests display abrasion features (polish and breakage; Fig. 3).

This material can be compared to that occurring on coral reefs (site 1) and in the low energy backreef/lagoon sites (sites 3 and 4). The foraminifer populations at these sites have a low to high diversity with variable degrees of dominance depending on species-specific specialization to the subenvironment. The high energy reef material has a low diversity population with only 1 or 2 species dominating. The low energy backreef populations have a high diversity with a greater number of species contributing to each population.

The size distributions at sites 1 to 4 provide information about the degree of modification the foraminifer populations have undergone as a result of bedload transport. The distribution curves for the reef and backreef populations (Fig. 4A, C & D) are representative of assemblages that have been unmodified or only slightly modified by bedload transport. The distributions describe complex, often polymodal (Fig. 4D), essentially biologically controlled populations. Possible minor overprinting by physical sorting processes which winnow/destroy some of the more fragile tests may occur. According to Folk (1980), these would be classified as moderately sorted.

Conversely, the distribution curve for the dune site (Fig. 4B) is well sorted and extremely leptokurtic (indicating an extremely well sorted central region), representing the strong overprinting of physical sorting processes on a biological population. The degree of sorting may be actually greater than indicated by the frequency histogram; for example, the long tails of the distribution are caused by measurement of the maximum dimension of the long, thin cigar-shaped *Alveolinella quoyi.*

Bioturbation and bio-erosion are important processes which have undoubtedly contributed to the mixing and destruction of sediments in Torres Strait. Aspects of bio-erosion likely to affect foraminifer tests include digestion by organisms which ingest grains and subject them to chemical corrosion, and microboring by algae. Holothurians, for example, are abundant in the region and ingest vast quantities of sediment (up to 1 kg of sediment a day for some species; Heezen & Hollister, 1971, p. 185). A sedimentation rate of 0.02 cm yr^{-1}, together with the presence of abundant infaunal sediment feeders, ensure that the sediments are well mixed and processed by bio-eroders. A given foraminifer test population would be expected to contain a component of corroded and possibly abraded tests as a result (e.g. site 4). The relative amounts of mud produced by bio-erosion and mechanical breakdown will vary regionally, but it is the signature of a tidally dominated shelf that such muds are winnowed from the high energy sandy facies and concentrated in the depositional muddy sand facies.

Effects of suspended load transport

Turbidity maxima, indicative of suspended load transport, occur in two parts of the study area: (i) in the central region of Torres Strait, resulting from tidal energy dissipation via friction in the high energy facies (Fig. 1; Harris & Baker, 1991); and (ii) in the Fly delta, related to tidal current reworking and trapping of suspended sediments in the distributary channels (Wolanski & Eagle, 1991). The Torres Strait turbidity maximum is related to the suspension of silt-sized grains derived from bio-erosion and mechanical breakdown of bioclastic material (see above).

The Fly delta turbidity maxima are related to a combination of river sediment input and resuspension by tidal currents and wave energy (Wolanski & Eagle, 1991). Sediment grains are generally <100 µm and are thus preferentially transported in suspension (spherical quartz grains <100 µm in diameter go into suspension as soon as the threshold velocity is exceeded; Leeder, 1982, p. 72). Current meter data, obtained at one location in a southern distributary channel of the Fly delta, indicate that the net landward transport of fine sand takes place integrated over several tidal cycles (Harris *et al.*, 1991).

Trends of decreasing mean size and increasing sorting landwards (observed for sites 5 to 9) can be used to indicate sediment transport pathways (e.g. size distributions become finer and better sorted along a transport path; McLaren & Bowles, 1985). In this case the trends can be accounted for by either suspended load transport and/or environmental constraints. Of course, the size distributions of foraminifers shown in Fig. 4 were derived by direct measurement (image analysis) and the hydraulically equivalent sizes have not been ascertained.

Alternatively, the decrease in mean size landwards could be explained by the observed decrease in salinity (Fig. 5), as most forms do not grow optimally in brackish water. Whereas grain-size data are not uniquely supportive of landward transport, it is the suborder composition of the assemblages (Fig. 2) and their textures which are of interest. The planktonic component shows a small relative increase landwards from sites 5 to 9 (Fig. 2). This landwards increase is the reverse of the expected decrease in the planktonic component away from the open ocean (Murray, 1976). Since planktonic foraminifers are exclusively open ocean forms and are only brought nearshore by currents, their occurrence in such an inshore, brackish water location indicates that they must have been carried landwards in suspension. The contribution of the suspended load to the foraminifera assemblage becomes relatively more important in a landward direction.

The importance of the suspended load contribution is also indicated by the persistence of another exotic component, the miliolids. They contribute nearly 20% of the assemblage at the mouth of the southern Fly delta (Fig. 2), and display textures ranging from pristine to heavily corroded. The nearest suitable marine environment is at least 20 km away, indicating transport pathways of similar (or larger) lengths. The fact that the miliolid tests display textures ranging from pristine to highly corroded indicates that fresh material is constantly being introduced and is subsequently undergoing corrosion. It is unlikely that material is being brought downriver from eroding deposits, as preliminary work on estuarine samples indicate extremely low numbers of foraminifers (most are small *Ammonia* sp., cf. Haig & Burgin, 1982).

Other studies of sediment transport pathways in macrotidal systems have also indicated that the adjacent shelf is typically an important sediment source (e.g. Murray, 1987; Harris, 1988b). Most of these studies refer to estuaries (i.e. partially infilled drowned river valleys). The present study suggests that such an offshore sediment source persists even when the fluvial system supplies a large sediment load to a prograding, tidally influenced estuarine–deltaic system. Although the landward transported calcareous material constitutes only a minor fraction of the total sediment deposited in the delta front facies (i.e. $<5\%$), it is nevertheless present and could be of importance in the interpretation of ancient sequences.

CONCLUSIONS

The present study has shown that foraminiferal taphonomic and grain-size properties can together identify transport pathways. Facies based on physical energy levels and sediment movement can also be identified using the grain-size distributions and taphonomic character of populations. The establishment of transport pathways may be useful in identifying the exotic elements of an assemblage, and also the mode of transport. Foraminiferal taphonomic and grain-size properties may be of use in the interpretation of depositional environments in ancient carbonate deposits.

ACKNOWLEDGEMENTS

This contribution was supported by a grant from the Australian Research Council (Ref. A38830391). We wish to thank David Haig for help and advice in identifying foraminifers and for valuable discussion, as well as Francis Chee, Tony Romeo and Dennis Dwarte of the Electron Microscope Unit, University of Sydney, for their invaluable assistance and time.

REFERENCES

Brandt, D.S. (1989) Taphonomic grades as a classification for fossiliferous assemblages and implications for paleoecology. *Palaios* **4**, 303–309.

Brasier, M.D. (1981) Microfossil transport in the tidal Humber Basin. In: *Microfossils from Recent and Fossil Shelf Seas* (Eds Neale, J.W. & Brasier, M.D.), pp. 314–322. Ellis Horwood, Chichester.

Brett, C.E. & Baird, G.C. (1986) Comparative taphonomy: a key to paleoenvironmental interpretation based on fossil preservation. *Palaios* **1**, 207–227.

Folk, R.L. (1980) *Petrology of Sedimentary Rocks.* Hemphills, Austin.

Fürsich, F.T. & Flessa, K.W. (1987) Taphonomy of tidal flat molluscs in the northern Gulf of California: Paleoenvironmental analysis despite the perils of preservation. *Palaios* **2**, 543–559.

Haig, D. & Burgin, S. (1982) Brackish-water foraminiferids from the Purari River Delta, Papua New Guinea. *Rev. Esp. Micropaleont.* **19**, 359–366.

Hallock, P. (1984) Distribution of selected species of living algal symbiont-bearing Foraminifera on two Pacific coral reefs. *J. Foram. Res.* **14**, 250–261.

Harris, P.T. (1988a) Sediments, bedforms and bedload transport pathways on the continental shelf adjacent to Torres Strait, Australia–Papua New Guinea. *Continent. Shelf Res.* **8**, 979–1003.

Harris, P.T. (1988b) Large scale bedforms as indicators of mutually evasive sand transport and the sequential infilling of wide-mouthed estuaries. *Sediment. Geol.* **57**, 273–298.

Harris, P.T. (1989) Sandwave movement under tidal and wind-driven currents in a shallow marine environment: Adolphus Channel, northeastern Australia. *Continent. Shelf Res.* **9**, 981–1002.

Harris, P.T. (1991) Reversal of subtidal dune asymmetries caused by seasonally reversing wind-driven currents in Torres Strait, northeastern Australia. *Continent. Shelf Res.* **11**, 655–662.

Harris, P.T. & Baker, E.K. (1991) The nature of sediments forming the Torres Strait turbidity maximum. *Austr. J. Earth Sci.* **38**, 65–78.

Harris, P.T., Baker, E.K., Cole, A.R. & Keene, J.B. (1991) Sandwave movement, currents and sedimentation in Torres Strait. *Univ. Sydney, Ocean Sci. Inst. Rept* **47**.

Harris, P.T., Baker, E.K., Cole, A.R. & Short, S.A. (1993) Preliminary study of sedimentation in the tidally dominated Fly River Delta, Gulf of Papua. *Continent. Shelf Res.* **13**, 441–472.

Heezen, B.C. & Hollister, C.D. (1971) *The Face of the Deep.* Oxford University Press, New York.

Kontrovitz, M., Snyder, S.W. & Brown, R.J. (1978) A flume study of the movement of Foraminifera tests. *Palaeogeogr. Palaeoclimatol. Palaeoecol.* **23**, 141–150.

Kotler, F., Martin, R.E. & Liddell, W.D. (1992) Experimental analysis of abrasion and dissolution resistance of modern reef-dwelling Foraminifera: implications for the preservation of biogenic carbonate. *Palaios* **7**, 244–276.

Leeder, M.R. (1982) *Sedimentology.* Allen & Unwin, London.

Loose, T.L. (1970) Turbulent transport of benthonic Foraminifera. *Contrib. Cushman Found. Foram. Res.* **21**, 161–166.

McLaren, P. & Bowles, D. (1985) The effects of sediment transport on grain-size distributions. *J. sediment. Petrol.* **55**, 457–470.

MAIKLEM, W.R. (1968) Some hydraulic properties of bioclastic carbonate grains. *Sedimentology* **10**, 101–109.

MARTIN, R.E. & LIDDELL, W.D. (1989) Relation of counting methods to taphonomic gradients and biofacies zonation of foraminiferal sediment assemblages. *Mar. Micropalaeont.* **15**, 67–89.

MELDAHL, K.H. (1987) Sedimentologic and taphonomic implications of biogenic stratification. *Palaios* **2**, 350–358.

MELDAHL, K.H. & FLESSA, K.W. (1990) Taphonomic pathways and comparative biofacies and taphofacies in a recent intertidal/shallow shelf environment. *Lethaia* **23**, 43–57.

MICHIE, M.G. (1987) Distribution of Foraminifera in a macrotidal tropical estuary: Port Darwin, Northern Territory of Australia. *Aust. J. mar. freshw. Res.* **38**, 249–259.

MILLER, M.C., MCCAVE, I.N. & KOMAR, P.D. (1977) Threshold of sediment motion under unidirectional currents. *Sedimentology* **24**, 507–527.

MURRAY, J.W. (1976) A method of determining proximity of marginal seas to an ocean. *Mar. Geol.* **22**, 103–119.

MURRAY, J.W. (1987) Biogenic indicators of suspended sediment transport in marginal marine environments: quantitative examples from SW Britain. *J. Geol. Soc. London* **144**, 127–133.

MURRAY, J.W. (1991) *Ecology and Palaeoecology of Benthic Foraminifera*. Longman Scientific and Technical, London.

MURRAY, J.W., STURROCK, S. & WESTON, J. (1982) Suspended load transport of foraminiferal tests in a tide- and wave-swept sea. *J. Foram. Res.* **12**, 51–65.

POWELL, E.N. (1992) A model for death assemblage formation: can sediment shelliness be explained? *J. mar. Res.* **50**, 229–265.

THOMAS, F.C. & SCHAFER, C.T. (1982) Distribution and transport of some common foraminiferal species in the Minas Basin, eastern Canada. *J. Foram. Res.* **12**, 24–38.

WANG, P. (1983) Transport of foraminiferal tests in estuaries: a comparison between the East China Sea and the North Sea. In: *Proc. Intern. Symp. Sediment. Cont. Shelf. Spec. Ref. East China Sea*, pp. 517–525. Acta Oceanologica Sinica, China Ocean Press.

WANG, P. & MURRAY, J.W. (1983) The use of Foraminifera as indicators of tidal effects in estuarine deposits. *Mar Geol.* **51**, 239–250.

WANG, P., MIN, Q., BIAN, Y. & HUA, D. (1985) Characteristics of foraminiferal and ostracod thanatocoenosis from some Chinese estuaries and their geological significance. In: *Marine Micropaleontology of China* (Ed. Wang, P.), pp. 229–242. China Ocean Press, Beijing.

WOLANSKI, E. & EAGLE, A.M. (1991) Oceanography and sediment transport, Fly River Estuary and Gulf of Papua. In: *10th Australasian Conf. Coastal Ocean Eng. (Auckland)*, pp. 453–457.

WOLANSKI, E., RIDD, P. & INOUE, M. (1988) Currents through Torres Strait. *J. phys. Oceanogr.* **18**, 1535–1545.

Spec. Publs int. Ass. Sediment. (1995) **24**, 225–236

Submarine cementation in tide-generated bioclastic sand dunes: epicontinental seaway, Torres Strait, north-east Australia

J.B. KEENE* *and* P.T. HARRIS†‡

**Department of Geology and Geophysics, University of Sydney, Sydney NSW 2006, Australia; and*
†Ocean Sciences Institute, University of Sydney, Sydney NSW 2006, Australia

ABSTRACT

Torres Strait forms a shallow (<20 m deep) 150 km wide seaway between Cape York, Australia, and Papua New Guinea. This shallow marine environment is part of a larger active foreland basin forming by subsidence of continental crust as the Indo-Australian plate converges with island terranes to the north. Large subtidal bioclastic dunes (sandwaves) form a sand belt across the shallow sill in this setting. Side-scan sonar and seismic reflection profiles show dunes with an average height of 5.4 m and crests oriented north–south in the study area in central Torres Strait. Current meter data confirm that high tidal energy defines the dominant sediment transport control mechanism. During the north-west Monsoon the tidal flow is asymmetrical, the dominant ebb current flowing eastwards (maximum of 1.15 m s^{-1}) and the subordinate flood current westwards, the latter below the transport threshold of the dune sediments (0.59 m s^{-1}). Reversal of dune asymmetries occurs during winter under the influence of the south-east Trade Winds.

Two vibrocores recovered from dune fields penetrated cross-bedded carbonate sand containing weakly cemented layers at depths >1.6 m below the sea-floor. One vibrocore penetrated 2.67 m of dune deposited cross-bedded sets overlying Pleistocene limestone. The lower metre of this bioclastic sediment was found to be variably lithified forming a skeletal grainstone. Aragonite needles form an isopachous cement in this highly permeable gravelly sand. Peloids are present within the cement and appear to have formed by precipitation. A ^{14}C age of 530 ± 180 yr BP from shell at the base of the sediment indicates the transitory nature of unconsolidated sand deposits in this high-energy environment. The proposed mechanism for cementation is tidal pumping of sea water through the sand body which overlies an impermeable limestone. The cementation is contemporary and not formed at the sediment–water interface. Such internal (subsurface) cementation in carbonate dunes and sand bodies may be more common than previously recognized and enhances the preservation potential of the basal section of dunes by providing resistance to reworking. Hardgrounds elsewhere may have formed in this manner rather than by exposure at or near the sea-floor during periods of quiescence.

INTRODUCTION

Bioclastic sand dunes are a significant sedimentary facies both in mass and areal coverage in the Torres Strait continental seaway. They are particularly extensive in those parts of the seaway between Papua New Guinea and Cape York where high-energy tidal currents are restricted by basement highs and reefs (Fig. 1). Carbonate sand and gravels are the dominant sediment types in this region, together with reefs and basement highs forming two 10–50 km wide linear sand belts across the 150-km wide seaway. These shore-normal sand belts extend northwards from the shoreface at Cape York to within a few kilometres of the Papua New Guinea coast, where there is a lateral facies change to terrigenous

‡*Correspondence address*: Australian Geological Survey Organisation, Antarctic CRC, University of Tasmania, GPO Box 252C, Hobart, Tasmania 7001, Australia.

mud (Harris, 1988). The water depth along and between these belts is everywhere <20 m, thus forming the shallowest sill in the strait linking the Coral Sea in the east with the Arafura Sea to the west. These reefs and associated sand bodies are located more than 120 km to the west of the Great Barrier Reef and the edge of the continental shelf (Fig. 1).

The depositional setting for these coarse-grained skeletal sands does not fit comfortably with existing sedimentary facies models (Tucker, 1985 and others). Neither the carbonate ramp model developed for the Trucial Coast of the Arabian Gulf (Wagner & van der Togt, 1973) nor the shelf-margin models based on studies in the Bahamas (Hine *et al.*, 1981a) are applicable to the Torres Strait carbonate sand bodies.

Previous work on subtidal carbonate sand bodies has concentrated on platform carbonates, particularly in the Bahamas where the emphasis has been on their location at the shelf margin and their important role in offbank transport of sand into deeper water (Hine *et al.*, 1981a). The other depositional setting for subtidal carbonate sands is the carbonate ramp model, where tidal deltas and storms control the facies distribution. The best studied example of this model type is situated off the Trucial Coast of the Arabian Gulf. Both of these models differ significantly from the carbonate deposits in Torres Strait, particularly in the basin setting which, in the case of Torres Strait, is formed by a seaway. This also affects the relationship of the carbonate sands to adjacent facies. The shoreline facies in the north are composed of terrigenous muds and in the south of quartzose sands.

Torres Strait forms part of the present-day marine environments accumulating sediment in the actively subsiding foreland basin formed by the northward convergence of the Indo-Australian plate with the New Guinea orogens (Veevers, 1984). The setting corresponds to a peripheral foreland basin as classified by Dickinson (1974). Pigram & Davies (1987) suggest this collision was initiated in the middle to late Oligocene, and since that time a combination of tectonics and eustatic sea-level fluctuations has determined the extent of the east–west seaway on the leading edge of the Australian continental crust as it converged with island terranes to the north. Today Torres Strait contains most of the shallow marine environments associated with this flexural loading.

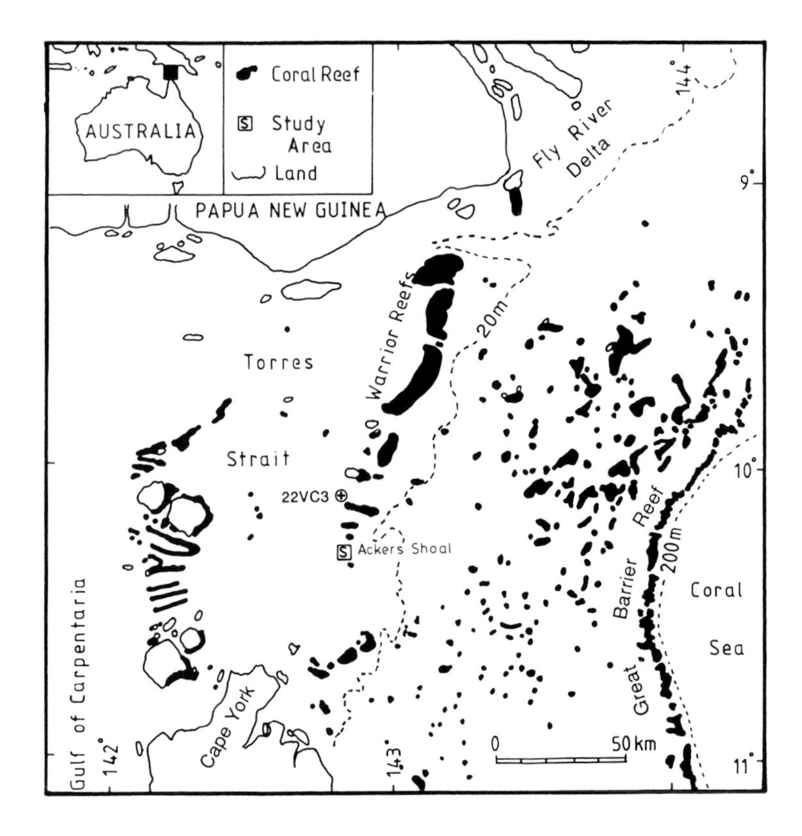

Fig. 1. Location of the study area in Torres Strait. The location of core 22VC3 in another dune field is also shown. Live coral reefs are shown in black.

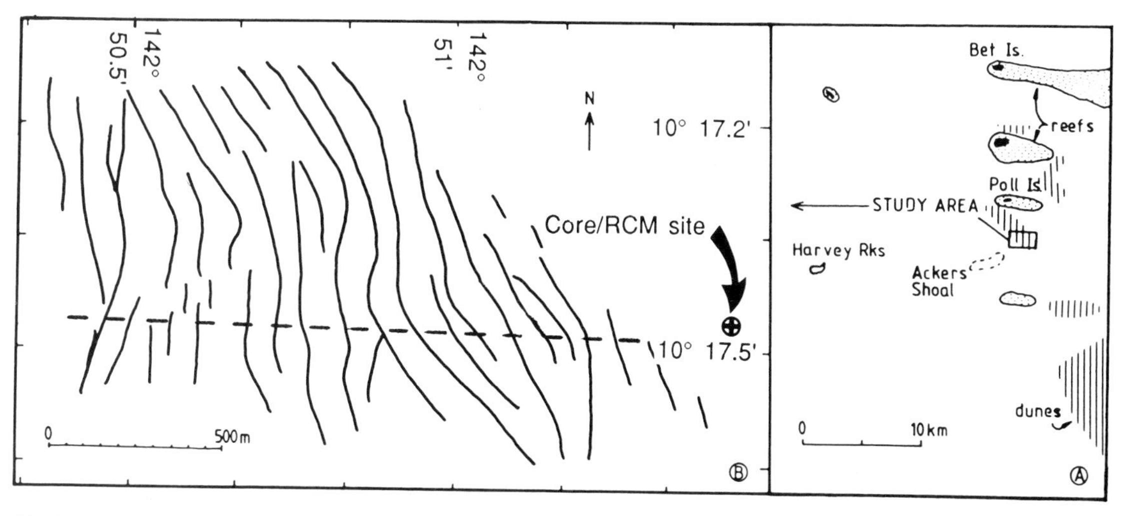

Fig. 2. Location of dunes near Ackers Shoal. The position of core 31VC5 and the current meter deployed in 1990 is indicated with an arrow. The dashed line marks the ship track for the seismic line shown in Fig. 3.

This study aims to link measurements of the physical processes of sedimentation with early diagenesis in carbonate sand dunes found in Torres Strait. It combines measurements of environmental parameters relating to sediment transport (tides, currents) with stratigraphic data from seismic reflection profiles and cores. The dune fields lie between reefs and form part of the eastern sand belt (Fig. 1). A literature survey did not reveal any published data on the internal structure of these, or similar, subaqueous bioclastic sand dunes.

The dune field selected for study is typical of many in the area. It is located between Ackers Shoal, a Pleistocene limestone outcrop on the seafloor, and Poll Island, a small carbonate sand island forming part of a larger reef (Fig. 2). The dunes form an elongated sand body aligned north-west–south-east, covering an area of 5 * 1 km. Within the study area (Fig. 2) the bedforms have an average height of 5.4 m and a wavelength of 41 m with north–south oriented crests (Harris, 1991).

METHODS

Data for this paper were obtained during a research cruise to Torres Strait on board *HMAS Cook* in April 1990. Bathymetric, side-scan sonar and seismic profiling surveys were carried out. An ORE Geopulse Boomer was used together with a Klein model 260 dual-frequency side-scan sonar. Vibrocores were obtained with 6-m long aluminium barrels, 75 mm in diameter. Current speeds were measured using an Aanderaa RCM-4 current meter moored with its rotor located 1 m above the seabed. Current speed and direction were obtained at 10-min intervals. Current velocities were used to estimate bedload transport rates from Bagnold's equation as modified by Hardisty (1983).

Petrographic descriptions of sediments were based on thin-section, scanning electron microscope (SEM), energy dispersive X-ray (EDX) and X-ray diffraction (XRD) analyses. ^{14}C-ages of shell samples were determined by the Macintosh Centre for Quaternary Dating, University of Sydney. The ages were corrected for the oceanic reservoir effect (−450 ± 35 years in eastern Australia; Bowman, 1985) and determined using the calibration curve of Stuiver *et al.* (1986).

RESULTS

Environmental parameters

Tides in the study area are mixed semi-diurnal and have a maximum range of 3.8 m. According to Harris (1991), currents in the region are tidally dominated. Harris also found that dunes in this area reverse their asymmetries between the north-west Monsoon season (December–March) and the south-east Trade Wind season (April–November) due to the slower, but significant wind-driven currents being superimposed on the tidal currents. Side-scan sonographs (Fig. 3) show the dunes to be asymmetrical with north–south oriented crests. The lee side is to the east, and smaller dune bedforms are present on the stoss slopes and at an oblique orientation to the larger bedforms.

The current meter for this study was located on the eastern side of the dune field at a water depth of 21 m (Fig. 2). Useful data were obtained for 43 days. Current speeds and directions were similar to those reported previously by Harris (1991). Sand transport is dominated by one diurnal ebb tide reaching velocities of 1.15 m s^{-1}. However, longer records obtained in 1990 show some other interesting effects, particularly with regard to the neap–spring cycle. The sand comprising the dunes has a mean diameter of 0.61 mm, which has a threshold speed measured at 1 m above the bed (U_{100}) of 0.59 m s^{-1}, based on the curve of Miller *et al.* (1977). A plot of velocity/direction versus time (Fig. 4) shows that threshold velocities for the grains comprising the dunes are exceeded only at spring tides, there being effectively no bedload transport during neap tides. Bedload transport, averaged over 28 days (two neap–spring cycles), is indeed towards the east. During the stronger spring tides, the tidally averaged transport rate increases and all of the transport occurs under the influence of the single strong ebb current. Thus, dune migration occurs over a 6-h period each day during spring tides. The estimated dune celerity (using the method outlined by Harris, 1991) is 0.34 m d^{-1} during springs (0.15 m d^{-1} if neap–spring averaged).

The current speed/direction time series plot also indicates that during neaps (days 17–22 in Fig. 4) the current speed is reduced to less than about 0.15 m s^{-1} and the direction of flow is towards 040°, showing *no reversal* over the semi-diurnal tidal frequency. Whether such a constant north-eastward flowing current is wind-driven or possesses some other components has as yet not been determined. A progressive vector plot of the raw current data indicates a net eastward drift at a rate of about 0.1 m s^{-1} towards 080° (Fig. 5). The data were filtered to remove tidal frequency currents using Munk's 'tide-killer' filter (Thompson, 1983). The filtered data show a periodicity in current speed at a frequency of about 14 days, with residual current velocities ranging between 0.05 and 0.4 m s^{-1}. The direction of the flow varies between 060° and 240° but is mainly oriented towards 140°. The peaks in residual current speed correspond with the spring tidal currents. It can be concluded, therefore, that the residual currents do contain some tidally generated component, perhaps generated by the local bathymetry (such as the presence of a sand bank).

The significance of storm-related processes remains to be investigated in the area. Satellite altimetry data used by McMillan (1982) to estimate global surface wave heights over a year suggest that significant wave heights in Torres Strait rarely, if ever, exceed 3.5 m, and during the north-westerly Monsoon (December–March), they are nearly always less than 1.5 m. Surface wave activity is most constant during the south-east Trade Wind season. It is likely, however, that periodic tropical cyclones during the Monsoon months have a far greater effect on sediment transport.

Pleistocene limestone

Ackers Shoal to the south-west of the study area (Fig. 2) is a sea-floor outcrop of Pleistocene limestone. The limestone forms an unconformity throughout the central region of Torres Strait and can be traced as a strong reflector on seismic profiles (Fig. 3). Core 31VC5 sampled 5 cm of this brown-coloured limestone. It is very well lithified and in thin section consists of sparse bioclasts in a micritic matrix with minor amounts of quartz silt. The bioclasts have recrystallized to spar and X-ray analysis indicates low-Mg calcite as the only phase present. This sparse biomicrite or wackestone was deposited in a lower energy environment when compared with the overlying mud-free sand being deposited today. It was probably lithified during subaerial exposure when sea-level was lower. The brown iron oxide staining and recrystallization of aragonite and high-Mg calcite support the suggestion of a subaerial environment for lithification.

The limestone is encrusted with shallow water organisms (oysters, algae, bryozoans) and is extensively bored. This happened either during the Holocene transgression or in times when the limestone was exposed on the sea-floor. Boring by organisms and mechanical erosion has generated identifiable, but relatively minor, lithoclasts for the overlying sediment. The limestone is impermeable to pore waters in the overlying Holocene sands.

Bioclastic sand dunes

The dunes near Ackers Shoal form a 3–8 m thick

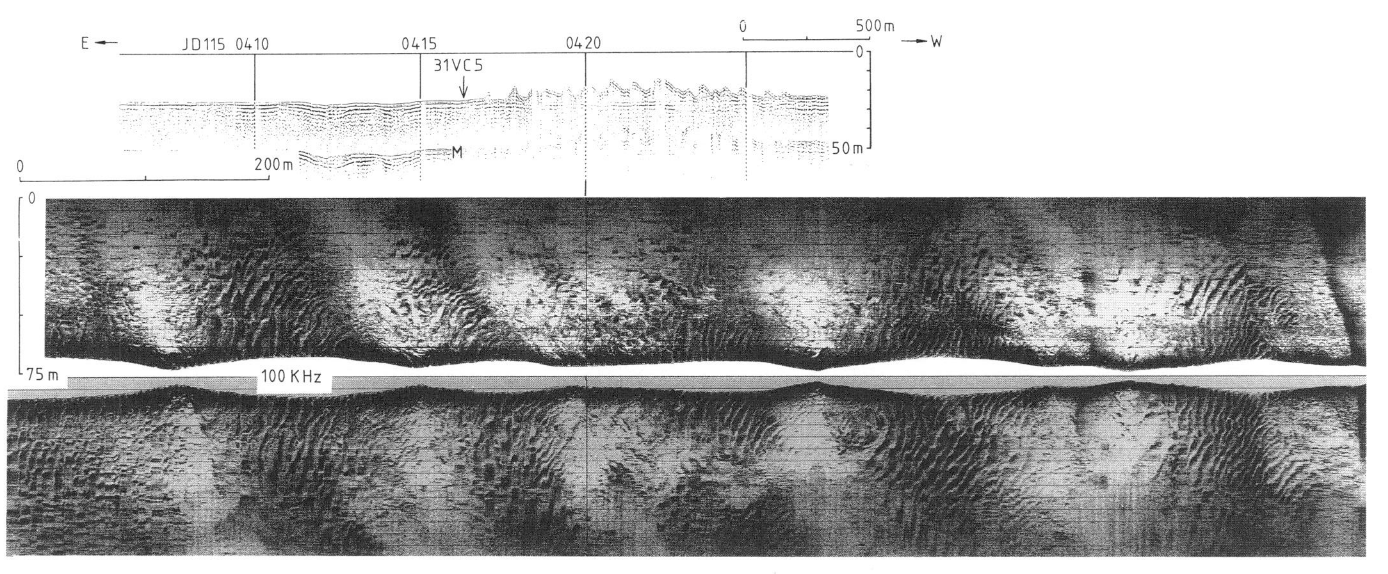

Fig. 3. An east–west seismic profile and side-scan sonograph showing the dunes near Ackers Shoal and the location of core site 31VC5. The location of the ship track is shown in Fig. 2. The seismic profile and sonograph overlap for 500 m centred on 0420 (note that the seismic image and sonograph are not at the same scale). The erosional unconformity between the Holocene sands and the underlying Pleistocene limestone is marked by a strong reflector at about 3 m below the seabed (reflector M is a multiple of the sea-floor). Smaller dunes in the troughs and on the flanks of larger dunes have crests oriented north–south.

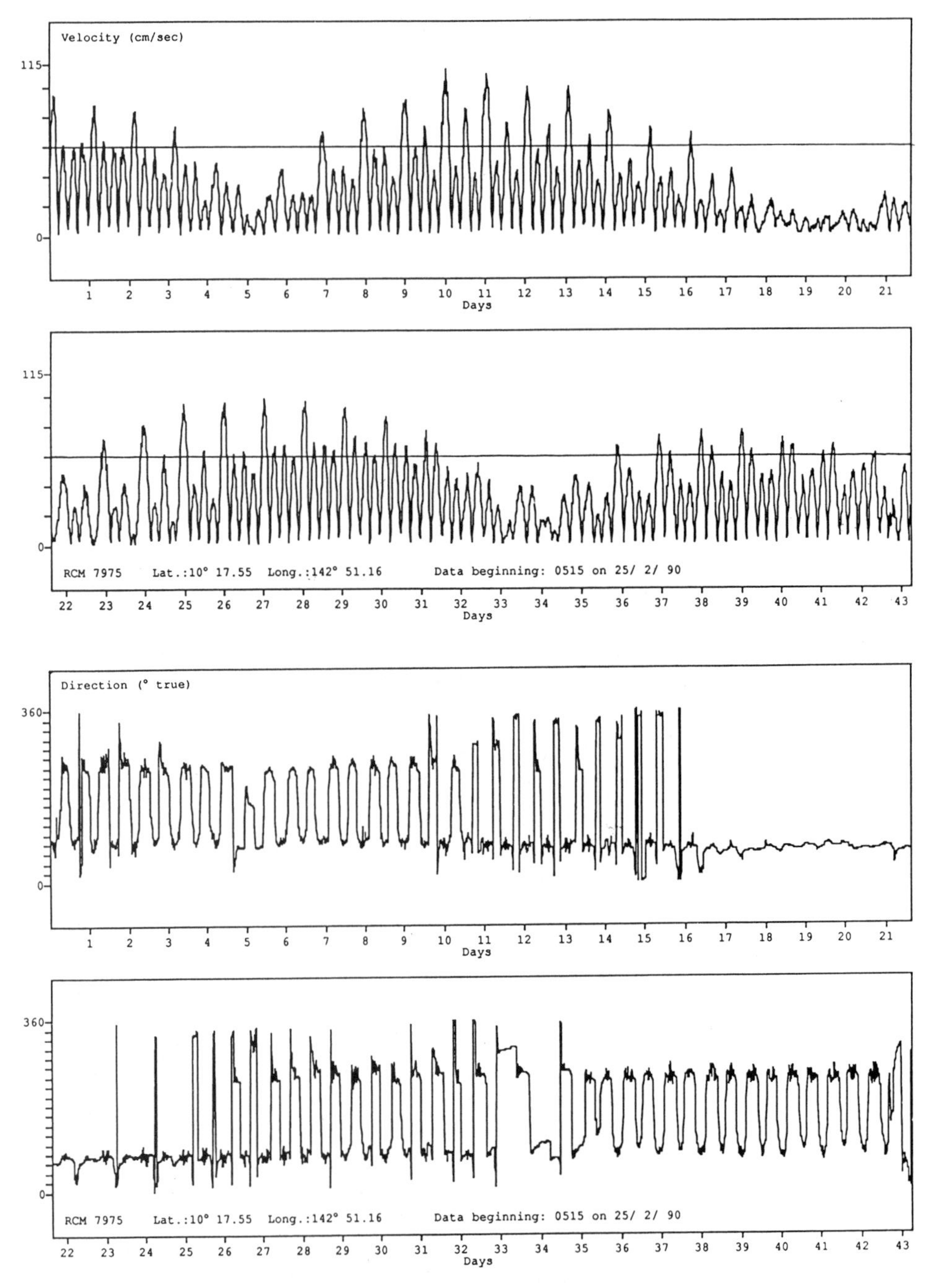

Fig. 4. Time series plots of current speed and direction (raw data) from the Ackers Shoal deployment in February–March 1990. Note the steady current direction for days 17–23. The horizontal line marks the threshold velocity (0.59 m s^{-1}) related to the mean grain size of sand comprising the dunes at Ackers Shoal, which is exceeded only during spring tides (e.g. days 7–16).

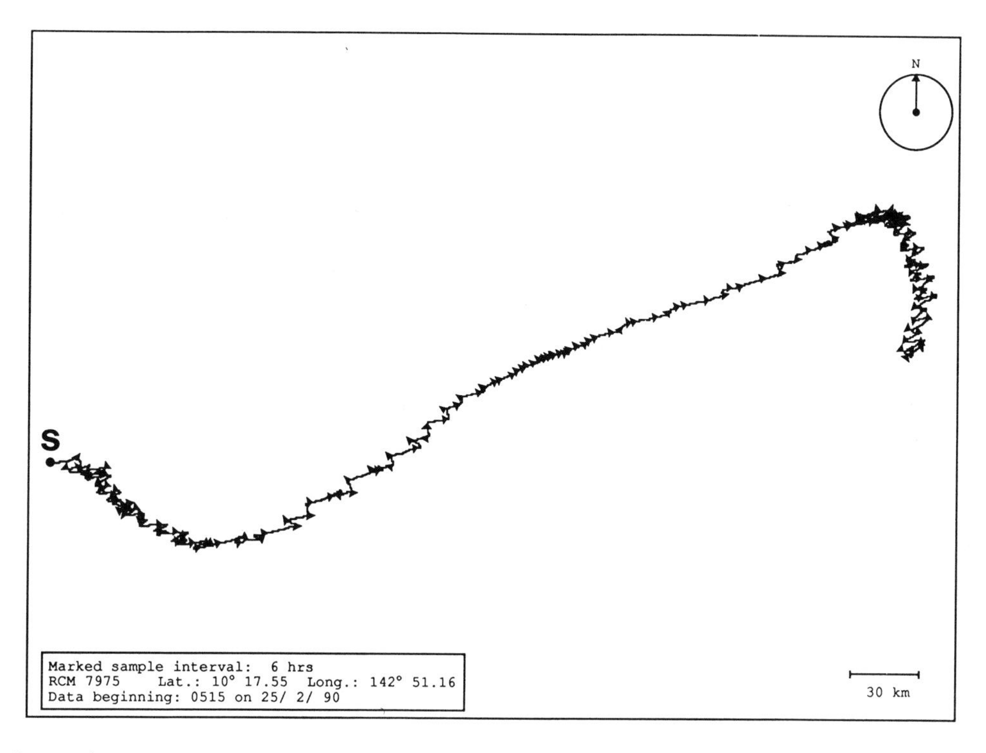

Fig. 5. Progressive vector plot of Ackers Shoal current data.

late Holocene sand deposit overlying Pleistocene limestone. Core 31VC5 in 21 m of water was taken on the eastern margin of the present-day dune field at the same location as that of the current meter (10°17.5′S; 142°51.5′E, Figs 2 & 3). The current meter data and cross-bedding in the core confirm dune migration in this area.

The core sampled 2.67 m of carbonate sand and gravel and also penetrated 5 cm into the underlying limestone. The sediment consists of dark, yellowish-orange skeletal sand containing fragments of coralline algae, molluscs, coral, *Halimeda*, foraminifers, echinoderms and bryozoans with minor limestone lithoclasts and peloids. The carbonate content is greater than 95%, the texture ranging from sand to gravelly sand. There is no mud present. Settling-column analysis revealed the sediment to be very poorly to extremely poorly sorted throughout, most bioclasts being 0.25–2 mm in size. The most noticeable sedimentary structure is a strong fabric to the platey and elongate grains (Fig. 6). This fabric, or grain alignment, represents foreset bedding preserved from migrating small dunes. Horizontal bedding is weakly defined by variation in sorting. The lower metre (1.60–2.67 m) of sand in this core is variably lithified. Porosity is substantial (35–40%) and permeability is very high even in the weakly cemented layers.

A shell sample from near the base at 2.65 m depth gave a ^{14}C-age of 530 ± 180 yr BP, indicating the relatively young age of this deposit. A second core (core 22VC3; 10°06.8′S, 142°47.3′E, Fig. 1), located in 17 m of water approximately 20 km to the north-west of core 31VC5, sampled the flank of a dune and contains poorly sorted gravelly sand down to a depth of 3.5 m. Below this depth there is a facies change to muddy gravelly sand which has not been reworked. This lower facies has a ^{14}C-age of 4970 ± 140 yr BP. The upper facies is distinguished by foreset cross-bedding and has incipient cementation at 1.9 m and 3.0 m. Both cores were taken in active dune fields and contain similar sediment and cementation.

In terms of dune size and wavelength, the sand dunes examined in this survey can be termed full-bedded, implying that the supply of sand in the region is plentiful (Harris, 1991). The dunes rework the surface of an underlying sand body. Sediment reworking by tidal currents is a feature of these sand bodies and the individual clasts are frequently well rounded and polished, although they have a very low sphericity. The relatively young age of the dune sands in core 32VC5 gives an average sedimentation rate of over 3 m per 1000 years, a rate which cannot be sustained for any length of time. This sedimentation rate does, however, indicate the transitory nature of unconsolidated sand deposits in this high-energy environment.

Submarine cementation

Core 31VC5 contains early diagenetic cements from a depth of 1.6 m to the base of the deposit at 2.67 m. Two weakly cemented layers were studied by thin-section, XRD, SEM and EDX analyses. A 4-cm thick layer (2.12–2.16 m, Fig. 6) is the most lithified, with aragonite cement coating most grains. However, the amount of cement is volumetrically very small as can be estimated from thin sections (Figs 7 & 8), the cemented layers being identical in composition to the sediment above and below. Texturally they contain slightly more shell gravel which may have enhanced their initial permeability. The layers are friable and grade both up and down into almost identical unconsolidated sediment.

In thin sections of samples from this core, aragonite forms isopachous fringes on grains (Figs 7 & 8). The radially oriented acicular crystals (30–100 μm in length and 1–2 μm in width) coat most grains and form a mesh cement at point contacts of grains. This is clearly seen in the SEM photographs (Figs 9 & 10). The aragonite precipitates on all types of particles, from the large single crystals of echinoid spines to micritic peloids and lithoclasts. The few grains lacking a coating of cement have no particular features in common. Micritic rims caused by algal coatings are present on some grains. Many of the aragonite-lined cavities are 0.5–1 mm in size, testifying to the high initial porosity in these sediments.

A sample of the weakly lithified layers was disaggregated using an ultrasonic bath and the resulting fine fraction, consisting mostly of cement, was analysed by XRD. Aragonite is the dominant phase present with only a minor amount of high-Mg calcite. Small crystals of high-Mg calcite are shown in Fig. 11.

Unlike cemented reef facies, this sand body lacks fine-grained internal sediment deposits. In thin

Fig. 6. Photograph of core 31VC5 showing grain alignment interpreted as foreset bed deposits of migrating bioclastic dunes. This strong fabric is present throughout the core. The section on the left is the 0.50–0.85 m core depth interval and the section on the right is from the 1.90–2.25 m interval. The upper section (left) is unconsolidated, whereas the lower section (right) is weakly cemented, particularly below 2.10 m. The arrow indicates a cemented layer sampled for SEM analysis.

section and under the SEM, however, peloidal textures do occur in some pore spaces (Figs 7, 8 & 12). The occurrence of these peloids only in the cemented horizons of the core raises the possibility of a diagenetic origin. Direct precipitation of high-Mg calcite has been proposed by, amongst others, MacIntyre (1985) as being one of the origins for peloids. The peloids illustrated in Figs 7, 8 & 12 from the cemented layers in core 31VC5 are composed of a micritic nucleus of carbonate (high-Mg calcite?) coated with radiating aragonite needles. Their uniform size of 25–50 μm and

Fig. 7. Thin section photomicrograph showing acircular crystals of aragonite forming isopachous fringes on bioclastic and micritic grains (Core 31 VC5, depth 2.12 m; crossed nicols; photo width is 0.5 mm).

spherical–oval shape, together with their occurrence in pores where aragonite cementation is pervasive, support their formation via precipitation in this environment.

DISCUSSION

Most highly mobile sand bodies, e.g. those formed by tidal currents, do not show evidence of submarine cementation and thus their preservation in the rock record is dependent upon high rates of sediment supply and/or subsidence or eustatic sea-level changes. Exceptions to this principle are found in bioclastic sediments where modern submarine cements have been recorded in the Arabian Gulf (Shinn, 1969), from the Bahama Platform (Dravis, 1979) and in other subtidal carbonate environments (Marshall, 1986).

Shinn (1969) put forward three essential conditions for submarine cements to form: low rates of sedimentation, sediment stability and high initial permeability of the sediments. Where these conditions are met in tropical waters, cements are precipitated at or near the sediment–water interface leading to the formation of hardgrounds. Hine *et al.*

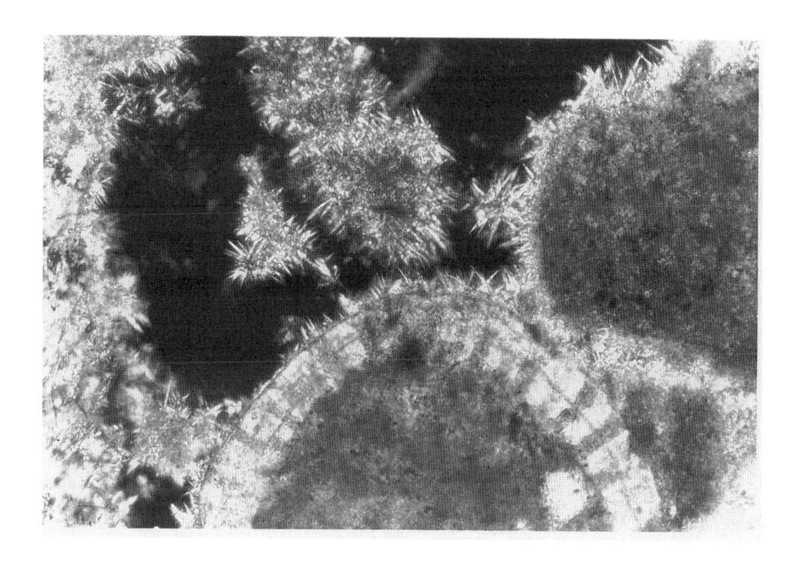

Fig. 8. Thin section photomicrograph showing aragonite crystals forming a weak cement and aggregates in pore spaces. These precipitated aggregates appear to develop into peloids (core 31VC5, depth 2.12 m; crossed nicols; photo width is 0.5 mm).

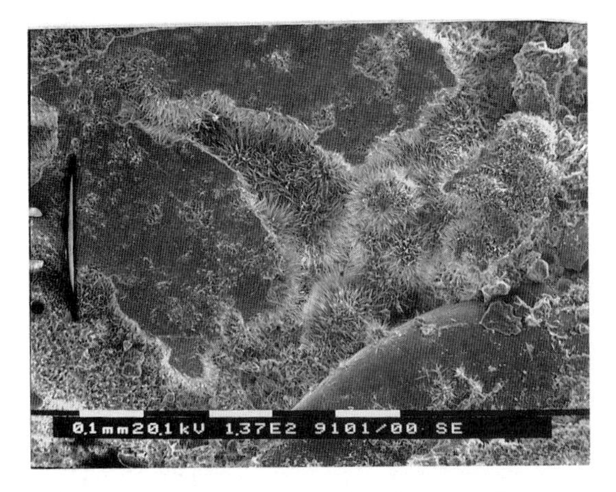

Fig. 9. SEM photo of fracture surface of a weakly cemented layer in core 31VC5 (2.15 m sub-bottom depth). Aragonite cement lines most pore spaces.

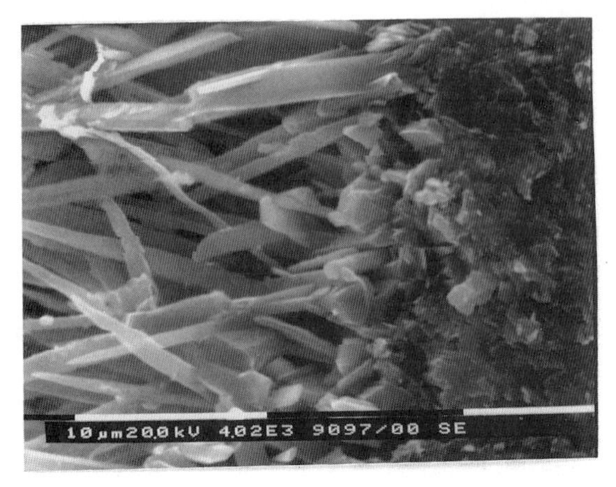

Fig. 11. SEM photo of aragonite needles precipitated onto a bioclast (right). Small subhedral crystals are high-Mg calcite.

(1981b) suggested that these hardgrounds formed carapaces on the surface of dune fields during periods of quiescence and used this to explain the cemented horizons observed within shelf margin sand bodies. In both the Arabian Gulf and Bahamas Platform, hardgrounds are forming under relatively low-energy conditions but in quite different depositional settings, i.e. lower shoreface and shelf margin dunes respectively. Dravis (1979) reported surface cementation forming in a few months or less in oolitic tidal sands on Eleuthera Bank in the Bahamas under high energy conditions. Cementation at this location was probably enhanced by algae binding the sediment. The high sedimentation rate of 72.6 cm per 1000 years, calculated from radiocarbon dating, is probably not sustained uniformly over the entire oolite shoal.

In general the formation of submarine cements requires the flow of substantial volumes of seawater through the sediment. Most occurrences of marine cements in carbonate sands are associated with reefs, shelf margins or shorelines where high-energy waves and currents prevail. This has led to the belief that cementation is restricted to shorelines and shelf margins where there is strong current activity (Tucker & Wright, 1990). High-energy con-

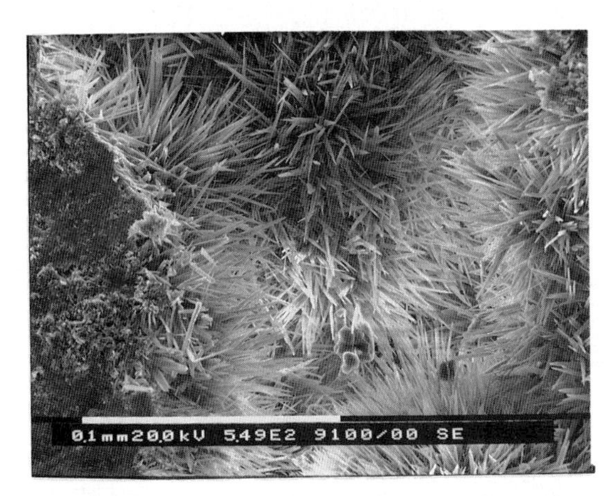

Fig. 10. Enlargement of Fig. 9 showing aragonite mesh cement.

Fig. 12. SEM photo of a radial aggregate of aragonite needles precipitated in a pore space. These precipitated aggregates have the appearance of peloids in thin section.

ditions have been proposed as a prerequisite for moving sea-water through the pore system in reefs, being used to explain why the seaward margins of reefs are better cemented than the inner and mid-shelf reefs (Marshall, 1986). The lack of a penetrating pumping mechanism is thought to explain why hardgrounds are restricted to the upper few centimetres of the surface and are self-limiting in thickness (Shinn, 1969; Dravis, 1979).

In the case of the cemented horizons observed in the Torres Strait sand dunes, no such periods of quiescence are required. There are no hardground indicators such as bored or well cemented surfaces that would indicate whether these layers formed at or near the sediment–water interface. This precludes their interpretation as buried hardgrounds. Cementation is post-burial, occurring once the sediment has been buried beneath a particular thickness of sediment which represents the frequently reworked layer for that environment. In the Torres Strait dune field this critical thickness amounts to about 2 m of sediment. Early diagenesis is taking place below this depth in sediment which was last reworked not less than 500 years ago. It is important to distinguish between the fast short-term sedimentation rate (derived from this age) and the longer-term sedimentation rate which is much slower (less than 0.5 m per 1000 years). It may be that some previously reported hardgrounds are in fact horizons which were initially cemented within the sand body and then exposed by erosion. Early cementation, if extensive, will enhance the preservation potential of such formations.

Contemporary aragonite cementation in the marine phreatic zone is possible because, first, the sediment is stable for long enough below the reworked layer in this high energy environment and, second, high permeability allows sea-water to flow along connecting pathways, possibly bedding planes between sets of crossbeds. The underlying Pleistocene limestone acts as an impermeable barrier and the mechanism which allows sea-water to penetrate into the sand is the tidal flow over the dune field. The asymmetry of the tidal flow enables a net flow through the dune field, albeit on a much longer time frame (i.e. months).

Extensive cementation in this setting will occur provided that tidal action and permeability are maintained. Periodic reworking below that level due to fairweather dune migration processes could occur with storms. Erosion by storms would create lithified clasts and slabs and expose the cemented horizons for boring and encrusting by sessile organisms characteristic of hardground development.

CONCLUSIONS

1 Tide-generated bioclastic dunes contain cemented horizons forming today at depths of 2–3 m below the dune troughs in active, high-energy dune fields. These occurrences differ from hardgrounds previously described which form at or near the surface and are subsequently buried during periods of low energy.

2 The proposed process leading to cementation is the pumping mechanism of the tides which moves sea-water along connected permeability pathways. The pressure differential due to the tides flowing over the dune field leads to a slow but net flow through the permeable sands above the impermeable Pleistocene limestone. The effectiveness of this process determines both the thickness and extent of cementation.

3 The depositional environment in Torres Strait, where tidal sand bodies are found, does not fit into existing carbonate platform models. The dunes are not related to the shelf edge or shoreline but form along the bathymetric sill of an epicontinental seaway. The sedimentary structures reflect currents parallel to the length of the seaway, whereas the overall sand body is aligned shore-normal.

ACKNOWLEDGEMENTS

The authors are grateful to the Australian Defence Science and Technology Organisation and the Australian Research Council (Ref. No. A39131170) for financial support. The Royal Australian Navy provided ship time on *HMAS Cook* in support of this project. Radiocarbon dating was carried out by the Macintosh Centre for Quaternary Dating, University of Sydney, reference numbers SUA 295Q0 and SUA 2951. Vibrocoring and boomer seismic equipment was purchased with a grant awarded by the University of Sydney. Special thanks are due to D. Mitchell for assistance with the field work.

REFERENCES

Bowman, G.M. (1985) Revised radiocarbon oceanic reservoir correction for southern Australia. *Search* **16**, 164–165.

DICKINSON, W.R. (1974) Plate tectonics and sedimentation. In: *Tectonics and Sedimentation* (Ed. Dickinson, W.R.). *Soc. Econ. Paleontol. Miner. Spec. Publ.* **22**, 1–27.

DRAVIS, J. (1979) Rapid and widespread generation of recent oolitic hardgrounds on a high energy Bahamian Platform, Eleuthera Bank, Bahamas. *J. sediment. Petrol.* **49**, 195–208.

HARDISTY, J. (1983) An assessment and calibration of formulations for Bagnold's bedload equation. *J. sediment. Petrol.* **53**, 1007–1010.

HARRIS, P.T. (1988) Sediments, bedforms and bedload transport pathways on the continental shelf adjacent to Torres Strait, Australia–Papua New Guinea. *Continent. Shelf Res.* **8**, 979–1003.

HARRIS, P.T. (1991) Reversal of subtidal dune asymmetries caused by seasonally reversing wind-driven currents in Torres Strait, northeastern Australia. *Continent. Shelf Res.* **11**, 655–662.

HINE, A.C., WILBER, R.J., BANE, J.M. & NEUMANN, A.C. (1981a) Carbonate sand-bodies along contrasting shallow-bank margins facing open seaways, northern Bahamas. *Am. Ass. Petrol. Geol. Bull.* **65**, 261–290.

HINE, A.C., WILBER, R.J., BANE, J.M., NEUMANN, A.C. & LORENSON, K.R. (1981b) Offbank transport of carbonate sands along leeward bank margins, northern Bahamas. *Mar. Geol.* **42**, 327–348.

MACINTYRE, I.G. (1985) Submarine cements—the peloidal question. In: *Carbonate Cements* (Eds Schneidermann, N. & Harris, P.M.). *Soc. Econ. Paleontol. Miner. Spec. Publ.* **36**, 109–116.

MARSHALL, J.F. (1986) Regional distribution of submarine cements within an epicontinental reef system: central Great Barrier Reef, Australia. In: *Reef Diagenesis* (Eds Schroeder, J.H. & Purser, B.H.), pp. 8–26. Springer, Heidelberg.

MCMILLAN, J.D. (1982) A Global Atlas of GEOS-3 significant wave height data and comparison of the data with national buoy data. *NASA Contractor Rept* **156882**. Wallops Flight Center, Virginia, 97 pp.

MILLER, M.C., MCCAVE, I.N. & KOMAR, P.D. (1977) Threshold of sediment motion under unidirectional currents. *Sedimentology* **24**, 507–527.

PIGRAM, C.J. & DAVIES, H.L. (1987) Terranes and the accretion history of the New Guinea orogen. *BMR J. Austr. Geol. Geophys.* **10**, 193–211.

STUIVER, M., PEARSON, G.W. & BRAZIUNAS, T. (1986) Radiocarbon age calibration of marine samples back to 9000 cal yr B.P. *Radiocarbon* **28**, 980–1021.

SHINN, E.A. (1969) Submarine lithification of Holocene carbonate sediments in the Persian Gulf. *Sedimentology* **12**, 109–144.

THOMPSON, R. (1983) Low pass filters to suppress inertial and tidal frequencies. *J. phys. Oceanogr.* **13**, 1077–1083.

TUCKER, M.E. (1985) Shallow-marine carbonate facies and facies models. In: *Sedimentology: Recent Developments and Applied Aspects* (Eds Brenchley, P.J. & Williams, B.P.J.). *Spec. Publs. Geol. Soc. London* **18**, 139–161.

TUCKER, M.E. & WRIGHT, V.P. (1990) *Carbonate Sedimentology*. Blackwell, London, 482 pp.

VEEVERS, J.H. (1984) *Phanerozoic Earth History of Australia*. OUP, Oxford, 418 pp.

WAGNER, C.W. & TOGT, C. VAN DER (1973) Holocene sediment types and their distribution in the Southern Persian Gulf. In: *The Persian Gulf* (Ed. Purser, B.H.), pp. 124–156. Springer, Heidelberg.

Ancient Tidal Processes and Sediment Dynamics

Spec. Publs int. Ass. Sediment. (1995) **24**, 239–258

Reconstruction of tidal inlet and channel dimensions in the Frisian Middelzee, a former tidal basin in the Dutch Wadden Sea

A.J.F. VAN DER SPEK

Geological Survey of The Netherlands, PO Box 157, 2000 AD Haarlem, The Netherlands

ABSTRACT

The Middelzee, a Late Holocene tidal basin in The Netherlands Wadden Sea, was created by marine erosion in the Boorne Valley. It reached its maximum extension around AD 1000. The basin was rapidly filled in and supratidal salt marshes were formed which were subsequently secured with dikes. By AD 1600 the landward part of the Middelzee had been reclaimed. This caused partial infilling of the remaining channels in the Wadden Sea and the tidal inlet.

The cross-sectional surface area and depth of the tidal channels and tidal inlet were calculated from the basin surface area and the tidal discharge, using empirical relationships derived in this paper as well as published in the literature. The calculations show that, given the maximum thickness of inlet channel deposits, the tidal prism was distributed over two inlet channels until AD 1700. This is in agreement with historical nautical charts.

The calculated maximum channel depths for tidal channels in the Middelzee are in agreement with the channel sediment thicknesses found in borings. Locally, these calculated maximum depths were significantly less than the maximum thickness of the channel deposits. This indicates that Middelzee channels have cut down into channel sediments deposited during earlier transgressions in the Boorne Valley. This provides valuable information on the evolution of the tidal basin.

The minimum annual sediment import between AD 1000 and AD 1600, estimated from the decrease in basin surface area and the infilling of the channels, is $1.9 * 10^6$ m^3. This rate falls well within the range of the estimated present-day sediment input into the Wadden Sea tidal basins.

INTRODUCTION

The present-day Netherlands Wadden Sea can be subdivided into several tidal basins, separated by tidal watersheds, each of them flooded and drained through its own inlet. This subdivision was even more explicit in the past, when the orientation of these basins was governed by north–south running Pleistocene valleys. The long-term development of these tidal basins is poorly understood.

The changes in palaeomorphology of a part of the former Wadden Sea in north-west Friesland during the medieval period are reconstructed in this paper. Marine erosion between AD 800 and AD 1000, created a tidal basin reaching 30 km inland. This basin is called the Middelzee (Fig. 1). The expansion and the subsequent silting up and reclamation of this basin have been described by contemporary authors (see Boeles, 1951 and Halbertsma, 1955 for overviews). The stepwise reclamation of the Middelzee over 600 years illustrates the rate of silting up of the embayment. Publications and maps dealing with aspects of archaeology, geology (Cnossen, 1958; Ter Wee, 1976; de Groot *et al.*, 1987), geomorphology and soil composition (Cnossen, 1971; Kuijer, 1974, 1976a,b, 1981) complement the historical descriptions.

Tidal inlets and channels are the most mobile elements in a tidal basin. Determination of their positions and dimensions is essential in the analysis

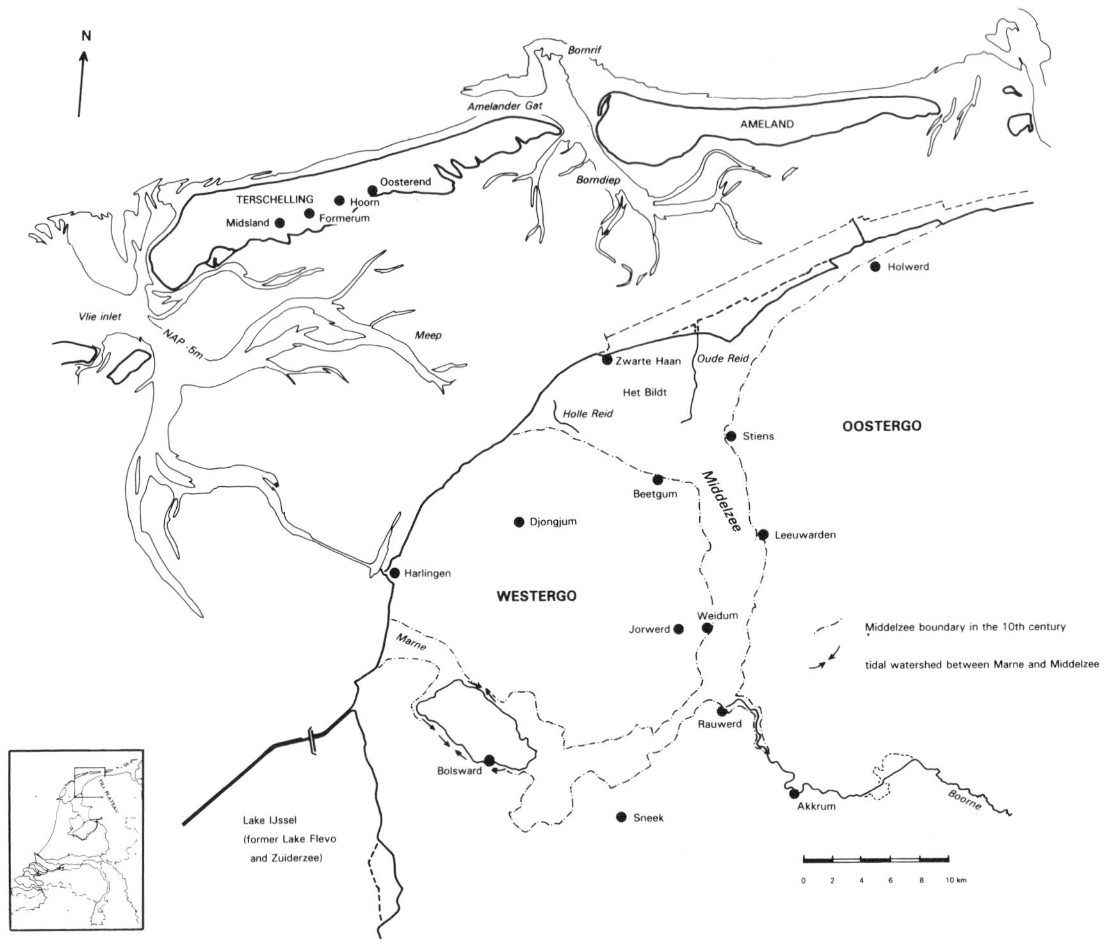

Fig. 1 Location map of NW Friesland and the Middelzee. The 5-m isobath is given relative to NAP, the Dutch Ordnance Level, which is about present-day mean sea-level. See inset for location.

of tidal basin evolution. The positions of the tidal inlet and channels of the Middelzee, however, are poorly known. Shoreface retreat and lateral inlet migration destroyed most of the evidence of former Holocene tidal inlets and barrier islands. Tidal channel and tidal flat deposits in the backbarrier area are bound to be reworked by lateral migration of younger channels. Only the major channels that scoured deep into the Pleistocene subsurface can still be traced. Reduction of the tidal prism caused by silting up of the tidal basin must have resulted in a decrease in width and/or depth of the channels in the basin and in the inlet, since the cross-sectional area and the tidal prism in channels and inlets are directly related.

Channel and inlet reconstructions can be improved using the empirical relationships between the cross-sectional area of a channel and its tidal prism. This relationship was introduced for tidal inlets by Le Conte in 1905 (FitzGerald *et al.*, 1984) and has been quantified by O'Brien (1931, 1969) and many others since. Gerritsen & de Jong (1985) established a relationship for the tidal inlets of The Netherlands Wadden Sea. A similar relationship between the cross-sectional areas of channels in the tidal basin and their tidal prisms has been quantified by Eysink (1979, 1991) and Gerritsen (1990). Using these relationships, the dimensions of tidal inlets or tidal channels can be estimated if the surface area of the tidal basin is known, since the tidal prism of a basin can be approximated by the product of the surface area of the basin times the tidal range. Thus, the decrease of the inlet and channel dimensions of the Middelzee caused by the silting up of the basin with time can be estimated using the reduction in surface area indicated by land reclamation.

FORMATION AND SILTING UP OF THE MIDDLEZEE

The Boorne Valley and the formation of the Middelzee

The Middelzee was the final stage of the Holocene transgression in the Boorne Valley. The Boorne was a small river draining the Saalian glacial till plateau in Friesland, Groningen and Drente (Fig. 1) from the Late Saalian onwards. It originated as a glacial stream that ran to the south-west. Its course became blocked by aeolian sands during the Weichselian. The lower reach of the river shifted gradually to the north-west (Cnossen & Zandstra, 1965). Finally, the Boorne ran from Akkrum to the north-north-west (Fig. 1).

During the Holocene the sea penetrated into the Boorne Valley from the north, changing it into an estuary. The oldest preserved estuarine deposits in the Boorne Valley are found west of Jorwerd (Fig. 1) at 8.0–10.5 m below present mean sea-level (Ter Wee, 1975, 1976). They were formed in the Late Atlantic (6400–5300 BP).

The sea-level rose rapidly in the coastal plain of The Netherlands before 7000 BP, with rates reaching 0.75 m per century. Between 7000 BP and 5500 BP the sea-level rise decelerated. After 5500 BP the average rate of sea-level rise was only 0.15 m per century. This allowed sediment supply in the western part of The Netherlands to fill in the tidal basins which had been created there. Eventually the tidal inlets closed between 5500 BP and 3300 BP (Beets *et al.*, 1992; Van der Spek & Beets, 1992). The situation in the northern part of The Netherlands was different. The tidal basins were not completely filled in with sediment, and consequently the inlets remained open.

The slowly but continuously rising mean sea-level caused the sea to penetrate further south in the Boorne Valley until it reached into the area south of Westergo (Fig. 1) in the Late Subboreal. Initially, this caused expansion of the tidal basin and scouring of the tidal channels (Ter Wee, 1976). However, sediment transport into the basin resulted in infilling of the landward part of the basin with time. The levees along the channels were built up to form high-lying, supratidal areas which became inhabited around 2600 BP (Ter Wee, 1976).

Further away from the channels, clay was deposited predominantly and was compacted with time, creating depressions in the landscape. The Boorne diverted its course to such a depression east of the former estuary (Fig. 2). With the continuing rise in sea-level, the tidal action gradually penetrated landwards during the Post-Roman and Early-Medieval periods (Cnossen, 1958). At the end of the 9th century the basin expanded rapidly and tidal channels scoured deep into the subsurface (Kuijer, 1974, 1976a; Ter Wee, 1976; de Groot *et al.*, 1987). The expansion of the basin was most likely enhanced by human activities such as peat digging. In the 10th century the lower reach of the Boorne had changed into the Middelzee (Ter Wee, 1975, 1976). Seawater penetrated as far inland as Sneek (Fig. 1).

The Marne, a channel of the Vlie tidal inlet, penetrated Westergo from the west (Fig. 1). It joined the Middelzee near Bolsward, where a tidal watershed developed (Kuijer, 1974).

The inlet of the Middelzee was situated between the islands of Terschelling and Ameland (Fig. 1). Radiocarbon-dating of peat layers indicates that the central part of Terschelling must have been present by AD 360–620. In addition, the pollen content of the peat suggests human occupation by then (de Jong, 1984). Information on the age of Ameland is sparser, the oldest peat layers having been dated at AD 670–850 (de Jong, 1984).

Channels in the Middelzee and Het Bildt

The course and dimension of former channels and creeks in the Middelzee south of Weidum (Fig. 1) are known from geological mapping (Ter Wee, 1976; de Groot, *et al.*, 1987). For areas north of Weidum and in the wider part of the Middelzee, later on known as Het Bildt (Fig. 1), as well as for the adjacent Wadden Sea, this information must be inferred from other sources. Written sources, going back to early medieval times, give rough indications only and neither accurate maps nor charts are known for the period of maximum extent of the Middelzee. Isbary (1936) and Rienks & Walther (1954) suggest that the two major marsh creeks which are today recognizable in the landscape, the Holle Rijd and the Oude Rijd (Fig. 1), are the remnants of medieval tidal channels.

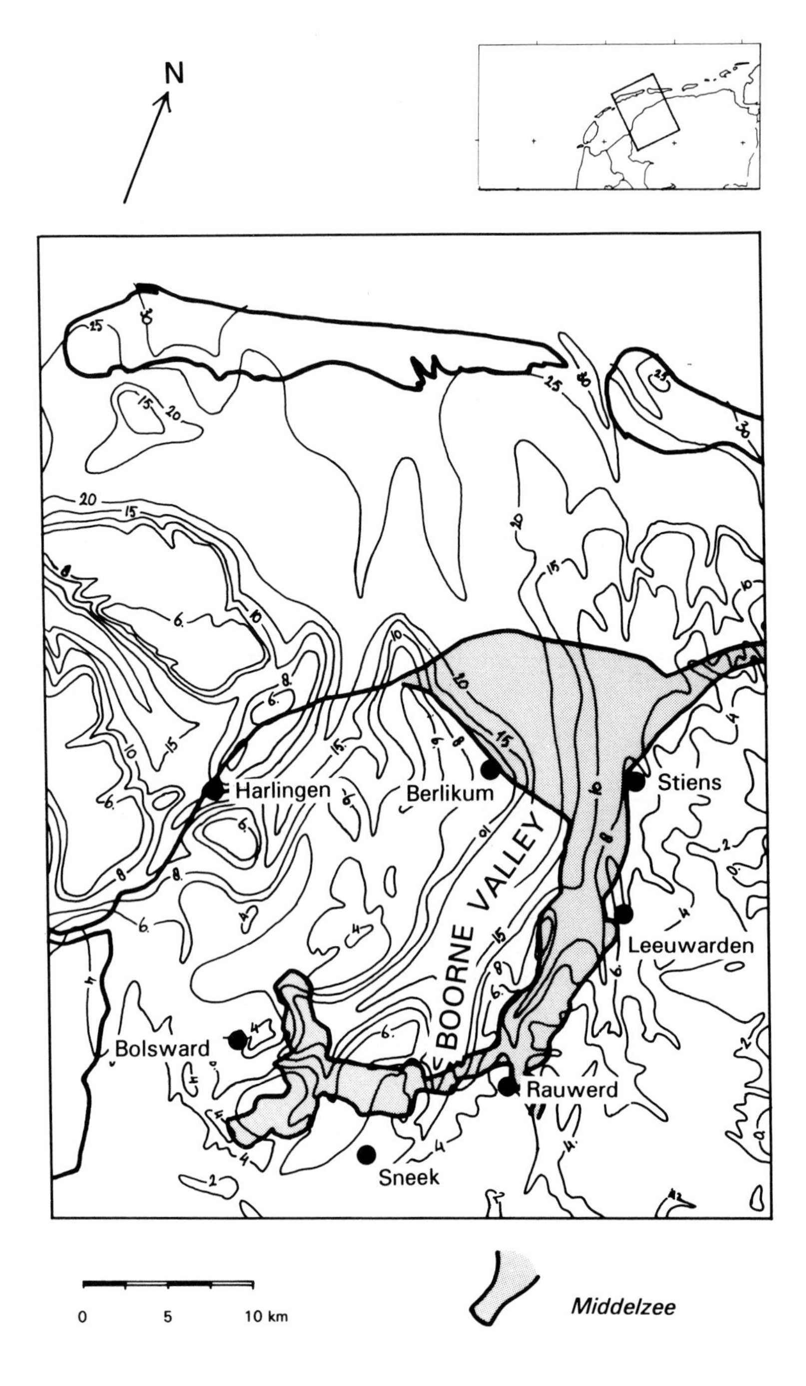

Fig. 2. Base Holocene map of NW Friesland. Depth contours are in metres below NAP. See inset for location.

Siltation and reclamation of the Middelzee

Increased flooding and marine erosion associated with rising sea-level resulted, according to Kuijer (1976a,b), in the damming of channels and the construction of dikes along the high-lying reaches of land. Improved social organisation also contributed appreciably to these activities (Borger, 1985). Reconstitution of events linked to land reclamation and dike construction in the Middelzee is mainly based on the historical study of Rienks & Walther (1954). Figure 3 gives an overview of the dikes along the Middelzee and their dates of construction.

The oldest dikes were constructed at the beginning of the 10th century by raising the levees and high-lying marsh bars of northern Westergo and Oostergo (Fig. 3). At the end of the 10th century, Westergo and the branch of Middelzee between

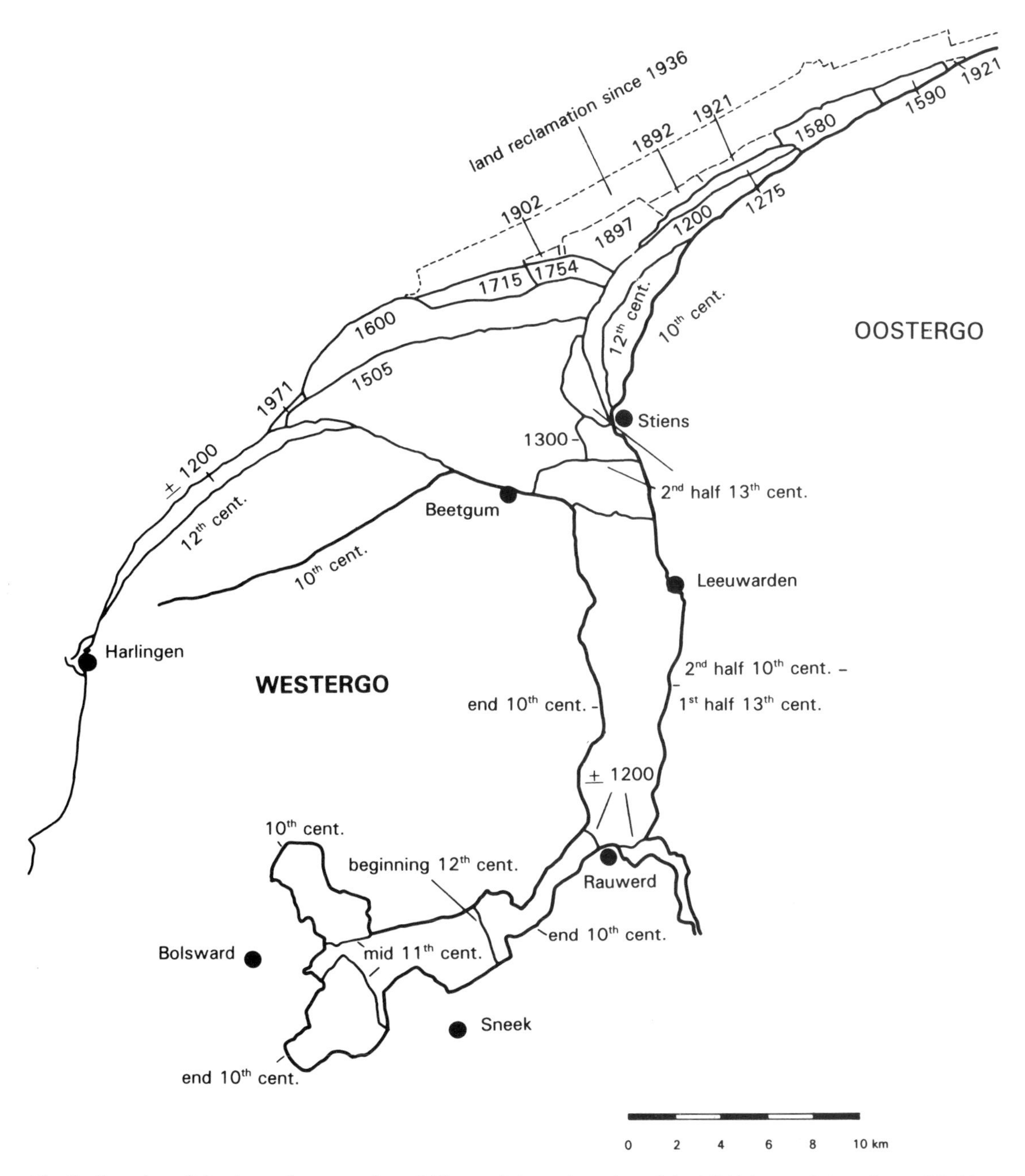

Fig. 3. Overview of the dates of construction of dikes and the reclamation of the Middelzee and the adjacent Wadden Sea (mainly after Rienks & Walther, 1954).

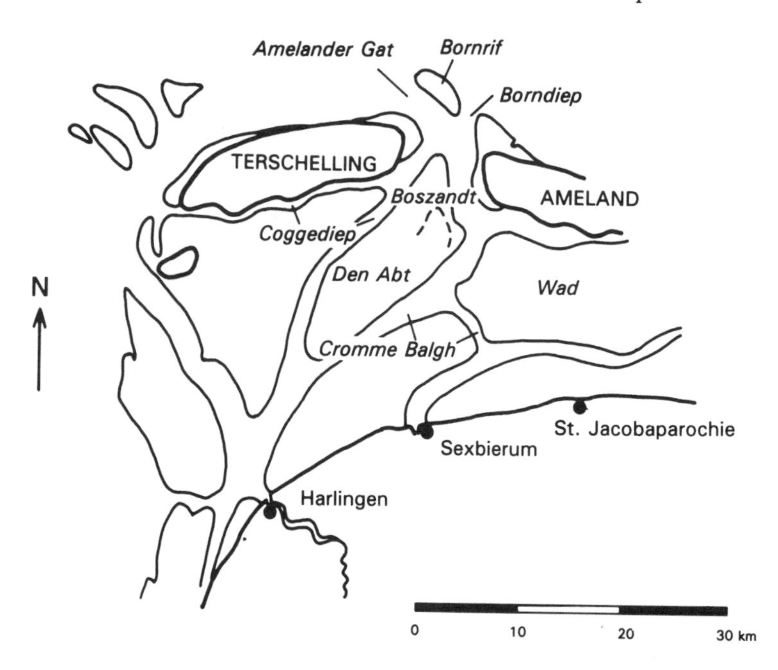

Fig. 4. Simplified nautical chart of the Amelander Gat around 1585 after Waghenaer ('Tvlie ende Tmaersdiep', in Spieghel der Zeevaerdt, part 1, 1584).

Rauwerd and Bolsward were completely encircled by a system of dikes (Fig. 3). The western part of Oostergo was protected by a single, more or less north–south running dike. The silting up of the tidal watershed near Bolsward to supratidal level, followed by the construction of a dike across it, transformed this branch of the Middelzee into a dead end. Calm conditions caused rapid silting up of the intertidal (mud) flats. The latter evolved into salt marshes which were protected with dikes. Thus, the south-western branch of the Middelzee was reclaimed stepwise from west to east

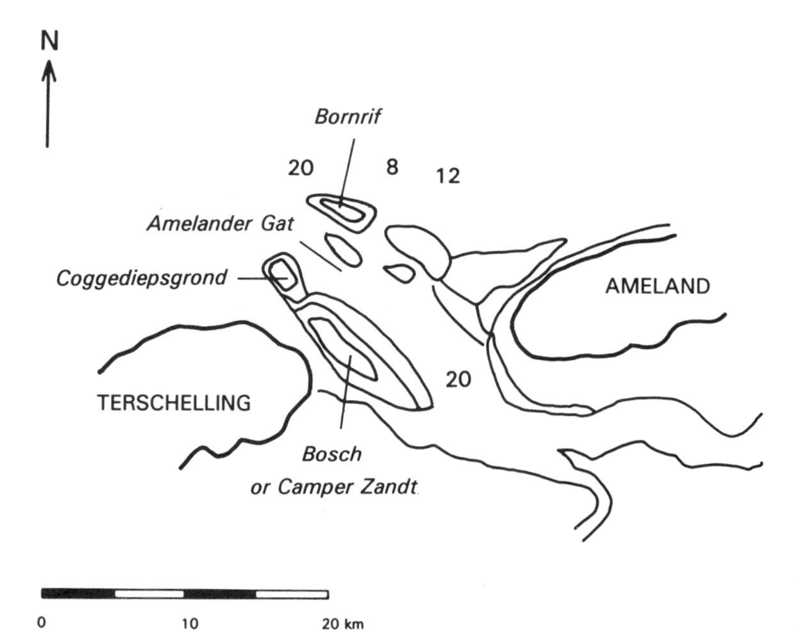

Fig. 5. Simplified nautical chart of the Amelander Gat around 1585 after Haeyen (1585). Depths are given in metres.

(Fig. 3). By AD 1200 it was closed off from the estuary.

The remaining Middelzee silted up rapidly and by AD 1300 it was reclaimed up to Beetgum and Stiens (Fig. 3), leaving a coastal embayment now known as Het Bildt. Het Bildt silted up to supratidal level within a century. In 1505 a dike was raised on the seaward marsh bar (Schotanus & Sterringa, 1664). With the construction of this dike, which connected the dikes of Oostergo and Westergo, an almost linear coastline was formed. Salt marshes that accreted at the toe of these dikes were reclaimed in the following centuries (Fig. 3).

Maps and charts of the Wadden Sea

The first charts of the North Sea coasts and the Wadden Sea were published by Waghenaer (1584; Fig. 4) and Haeyen (1585; Fig. 5) when the Middelzee had already been reclaimed. In the 16th century the Amelander Gat (the inlet between Terschelling and Ameland) had two channels, the Coggediep in the west and the Borndiep in the east. A shoal that was called Bosch or Camper Zandt separated these channels. The Amelander Gat was the main channel in the ebb-tidal delta. It was bordered by shallow grounds to the north called Bornrif.

The channels and shoals in the tidal inlet show a distinct evolutionary pattern. At the end of the 16th century Waghenaer and Haeyen depicted a wide tidal inlet with a relatively small shoal surface (Figs 4 & 5). In 1608 Blaeu's map showed an extended shoal, with three distinct bars at its seaward side (Fig. 6). The chart published by Blaeu in 1623 shows an even larger shoal in the inlet, with the Camper Zandt silted up to supratidal level (Fig. 7). From this map sequence it can be concluded that the inlet width was reduced over this period.

Haeyen (Fig. 5) indicated a depth of '12 vadem', (20.4 m) for the Borndiep. He did not give a depth for the Coggediep, apparently this being not considered important for shipping. Blaeu's chart from 1608 gives identical information on channel depth (Fig. 6). Winsemius (1622) reported that in 1610 the Coggediep had a depth of '9 vadem' (15.3 m).

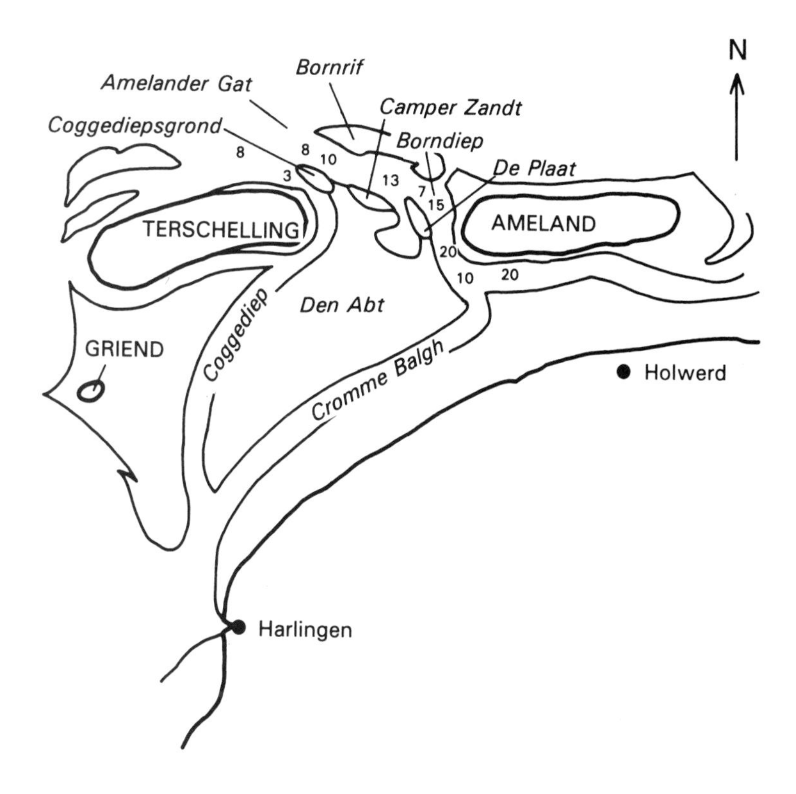

Fig. 6. Simplified nautical chart of the Amelander Gat after Blaeu's 'Licht der Zeevaert' (1608). Depths are given in metres.

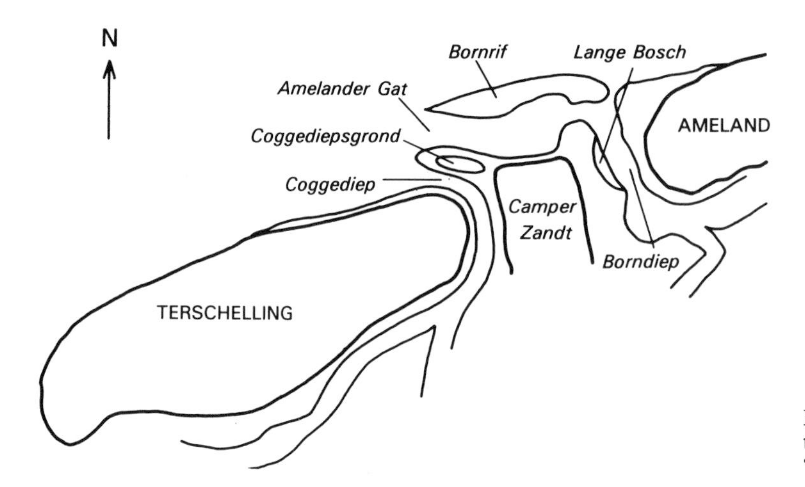

Fig. 7. Simplified nautical chart of the Amelander Gat after Blaeu's 'Zeespieghel' (1623).

THE AMELANDER GAT AFTER THE 17TH CENTURY

At the beginning of the 18th century the Coggediep was still in existence, according to the charts of Guitet and Witsen dating from 1710 and 1712. Guitet indicated a depth of '10 vadem' (16.8 m) in 1710 (Fig. 8). From his map it can be seen that the Coggediep at that time was a wide inlet channel. This suggests that the discharge must have been partitioned between the two channels. Historical reconstructions by Schoorl (personal communication), however, show that Coggediep was very small in 1695. According to the hydrographical chart of the Amelander Gat produced by Buyskes, the Coggediep had disappeared completely in 1798 and the shoals Camper Zandt and Bosch were connected to the island of Terschelling. The Wadden Sea south of the eastern part of Terschelling had accreted to intertidal level without being dissected

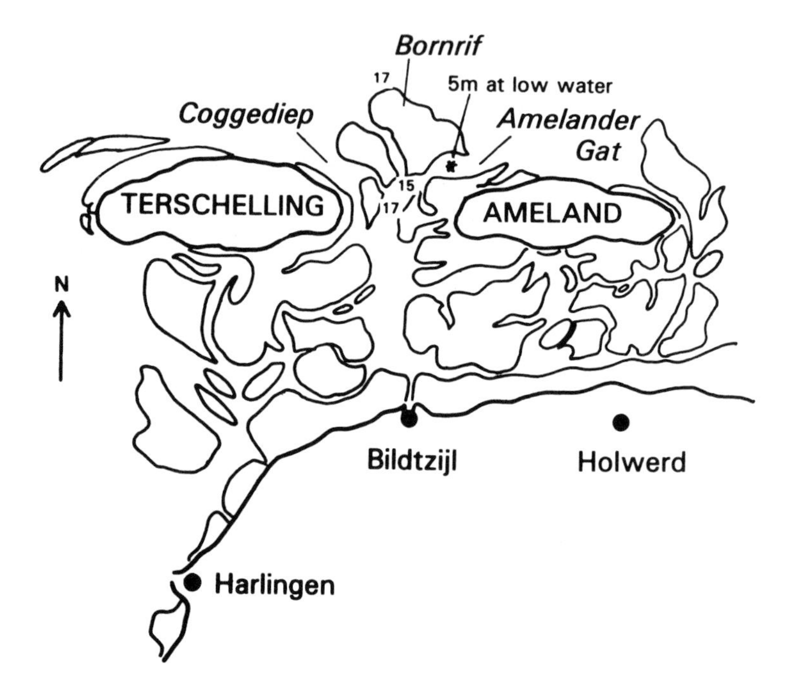

Fig. 8. Detail of Guitets map of the Wadden Sea ('Wad en Buytenkaart', 1710), after the original in the library of Leiden University. Depths are given in metres.

by channels. Buyskes measured a depth of '15 vadem' (25.5 m) in the Borndiep.

The silting up of the Coggediep cannot be a consequence of the reclamation of the Middelzee because, as discussed above, it persisted for two centuries after the reclamation of the Middelzee and Het Bildt. It is not plausible that, all other factors being equal, the silting up of a channel in a tidal inlet lags two centuries behind such a major reduction of its tidal prism. The dimensions of the Coggediep should have rapidly followed this reduction in drainage area, since the wave-driven shoreward sediment flux that provides the sand for inlet sedimentation was not diminished. It is inferred that the Coggediep must have been (partly) connected to another tidal basin to gain sufficient discharge to remain open. Indeed, historical evidence suggests that it was connected to the Zuiderzee (see Fig. 1 for location of the latter) and silted up only after the establishment of a fixed connection between the Vlie tidal inlet and the Zuiderzee.

Reduction in channel depth of the Coggediep delayed the arrival of the flood tide on the Terschelling tidal watershed via the Amelander Gat. The flood tide arrived earlier at the watershed via the westerly tidal inlet since the tidal wave runs from south-west to north-east along The Netherlands coast. Thus a water-level gradient was created on the watershed that resulted in a net eastward water and sediment displacement and, consequently, migration of the tidal watershed. This process would have been amplified by the prevailing westerly winds (FitzGerald & Penland, 1987). The tidal watershed south of Terschelling, although still dissected by channels, started to shift eastwards in the first half of the 18th century (Isbary, 1936). This coincides very well with the final silting up of the Coggediep. Simultaneously, the tidal watershed accreted to intertidal level, consolidating the separation between the Vlie and Borndiep tidal basins.

After the silting up of the Middelzee, tidal currents started to penetrate further eastwards between Ameland and Friesland. Southward running channels like the Cromme Balgh (Fig. 4) were abandoned, while eastward running channels gained more discharge and consequently became deeper. This, as explained above, caused the tidal watershed south of Ameland to move to the east. The tidal watershed south of Ameland migrated 4 km to the east between 1831 and 1950. The Borndiep reacted to this eastward shift by rotating its course from north–south to north-west–south-east and started to erode the island of Ameland. Erosion resulted in the south-western shore of Ameland moving 1.4 km to the east before the process was counteracted by the implementation of protective measures.

Harle inlet, between the islands of Spiekeroog and Wangerooge in the German Wadden Sea, has a similar history of inlet narrowing caused by partial basin reclamation (FitzGerald *et al.*, 1984; Flemming & Davis, 1994). The tidal watershed south of Spiekeroog, the island west of the inlet, moved to the east concomitantly with the shallowing of the western inlet channel.

BASIN RECONSTRUCTION

From historical and geological data it is clear that, since approximately AD 600, the inlet of the successor of the river Boorne must have been situated somewhere between the village of Formerum on the island of Terschelling and the present-day west coast of Ameland (Fig. 1). The maximum depth of Holocene channel erosion here is 25–28 m below MSL (Fig. 9). The maximum depth can be up to 31 m below MSL if reworking of Eemian deposits by Holocene channels is considered.

Map reconstructions of the Middelzee have been made on the basis of centennial maps in order to assess reduction in basin surface area with time. After estimation of the tidal prism from the surface area, the dimensions of the tidal inlet can be derived from the relationships between tidal volume and cross-sectional profile of the inlet. Thus a reconstruction of the evolution of inlet dimensions can be made.

The Middelzee basin

Assessments of the maximum extension of deposits associated with the Middelzee are based on the distribution of Duinkerke-III sediments on geological maps (Ter Wee, 1976; de Groot *et al.*, 1987). The distribution in the area north of Weidum (Fig. 1) was inferred from maps of soil composition. The maximum distribution of these sediments, however, is related to storm floods and will therefore not be incorporated in the calculations which are based on average tidal activity.

The landward boundaries of the Middelzee were formed by dikes from *c.* AD 1000 onward. These boundaries have been determined for each century

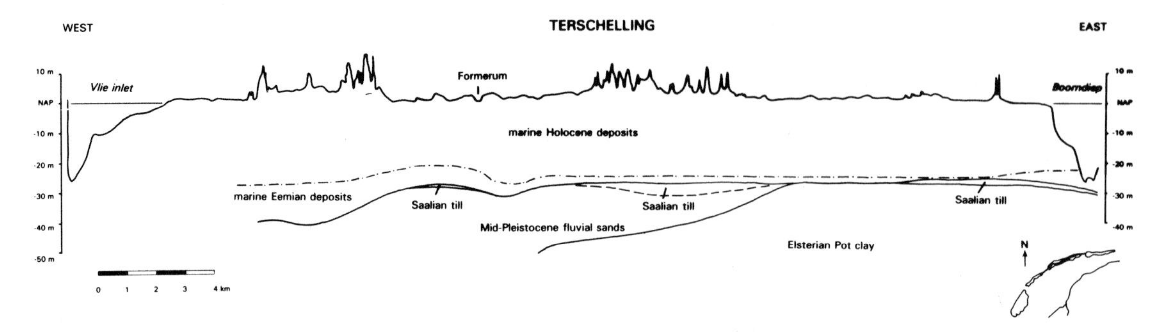

Fig. 9. West–east cross-section through the island of Terschelling (after Van Staalduinen, 1977).

from AD 1000 to AD 1900 (Fig. 10). Dating of dike construction is mainly based on the study of Rienks & Walther (1954).

The tidal watersheds in the Wadden Sea are also considered to be basin boundaries, although the watershed south of Terschelling was initially dissected by channels that ran to the Zuiderzee. The position of this watershed moved towards the east with time. The position before AD 1300 is assumed to have been situated between Midsland and Dongjum (Figs 1 & 10), roughly along the western contour of MSL-20 m shown in Fig. 2. In AD 1300, when the Middelzee was reduced to a funnel-shaped embayment, the divide was situated between Midsland and Minnertsga (Fig. 10) where it remained until AD 1600 (Isbary, 1936). By 1800 it had shifted to a position between Hoorn and Zwarte Haan. In 1900 the divide was at its present position between

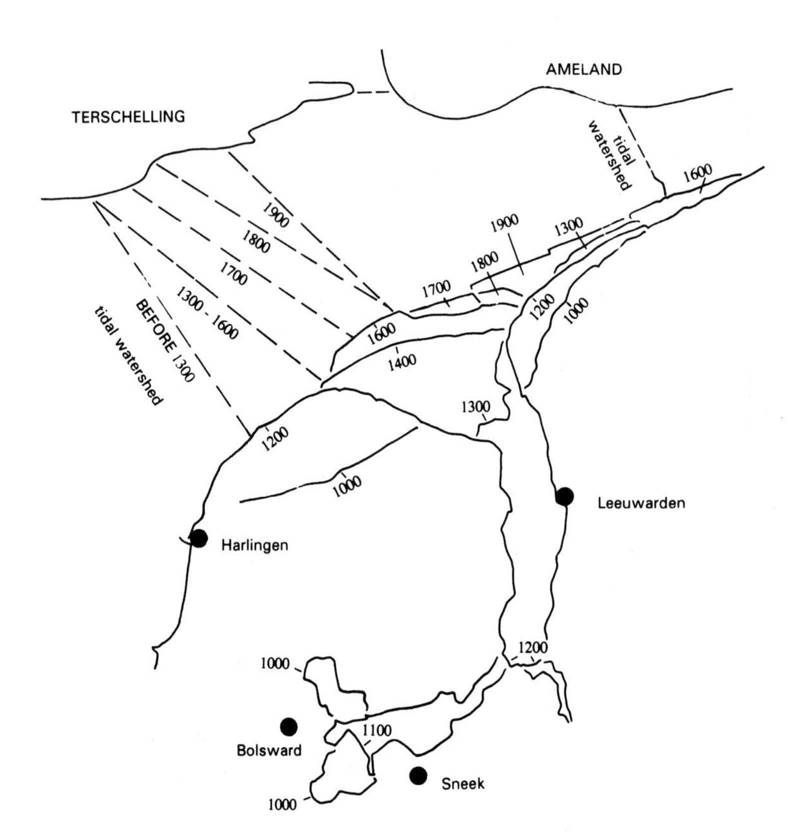

Fig. 10. Centennial reconstruction of the dimensions and positions of the Middelzee and tidal watersheds behind the islands Terschelling and Ameland since AD 1000. Dikes raised within 20 years after reconstruction dates have been considered as being present at these dates, since the salt marsh they were built on must already have accreted to supratidal level by that time.

Oosterend and Zwarte Haan (Fig. 10). The tidal watershed south of Ameland did not shift eastwards until 1831. Its position is therefore assumed to coincide with the position of the dam between Holwerd and Ameland that was built in 1870 (and which was destroyed within a couple of years). The migration of this divide as from 1831 was omitted from the reconstruction for 1900.

The seaward boundary of the basin is assumed to be at the present-day southern shores of Terschelling and Ameland, along the shortest line between the two islands across the tidal inlet.

Surface areas of the reconstructions shown in Fig. 10 are given in Table 1. A graph of the areal reduction of the Middelzee tidal basin with time is shown in Fig. 11. The surface area of the Middelzee decreased almost linearly from 737.5 km^2 in AD 1000 to 275.5 km^2 in AD 1900, a reduction of 63%.

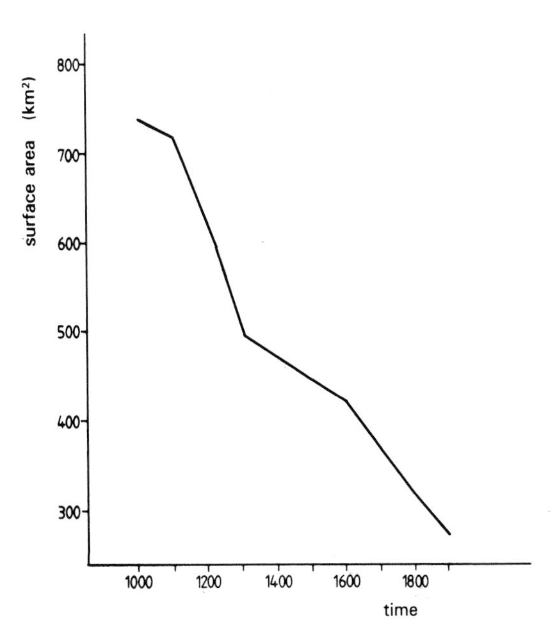

Fig. 11. Plot of the reduction in surface area of the Middelzee with time.

Inlet dimensions

The tidal prism can be estimated on the basis of reconstructed basin surface areas, since tidal prisms are determined by the size and shape of the basin and the tidal range (Van Veen, 1950; O'Brien, 1969). If the tidal prism is known, the cross-sectional area of the tidal inlet can be calculated, using the empirical relationship between these two parameters. The maximum inlet depth can also be derived from the tidal prism.

In Fig. 12 the tidal prisms of recent tidal basins in The Netherlands Wadden Sea are plotted against basin surface areas at mean high-water level (data from de Glopper, 1967; Anonymous, 1974; Endema, 1979). A linear regression function of the form

$$P = 3.66 + (1.34 * SA) \quad (1)$$

(where P is tidal prism in 10^6 m^3, SA is surface area at mean high water in 10^6 m^2) can be calculated.

Table 1. Surface area in km^2 of the Middelzee tidal basin for reconstructed 100-year periods since AD 1000. The tidal prisms, cross-sectional areas and inlet depths have been derived using the relationships given in the text

Reconstructed period AD	Surface area (10^6 m^2)	Tidal prism (10^6 m^3)	Cross-sectional area below MSL (10^3 m^2)	Maximum inlet depth (m)
Maximum	923.4			
1000	737.5	992	64	50
1100	716.3	964	62	49
1200	625.0	841	55	44
1300	495.1	667	44	38
1400	444.9	600	40	35
1600	421.5	569	38	34
1700	367.2	496	34	31
1800	316.7	428	30	29
1900	275.5	373	26	27
Equation		1	3	4

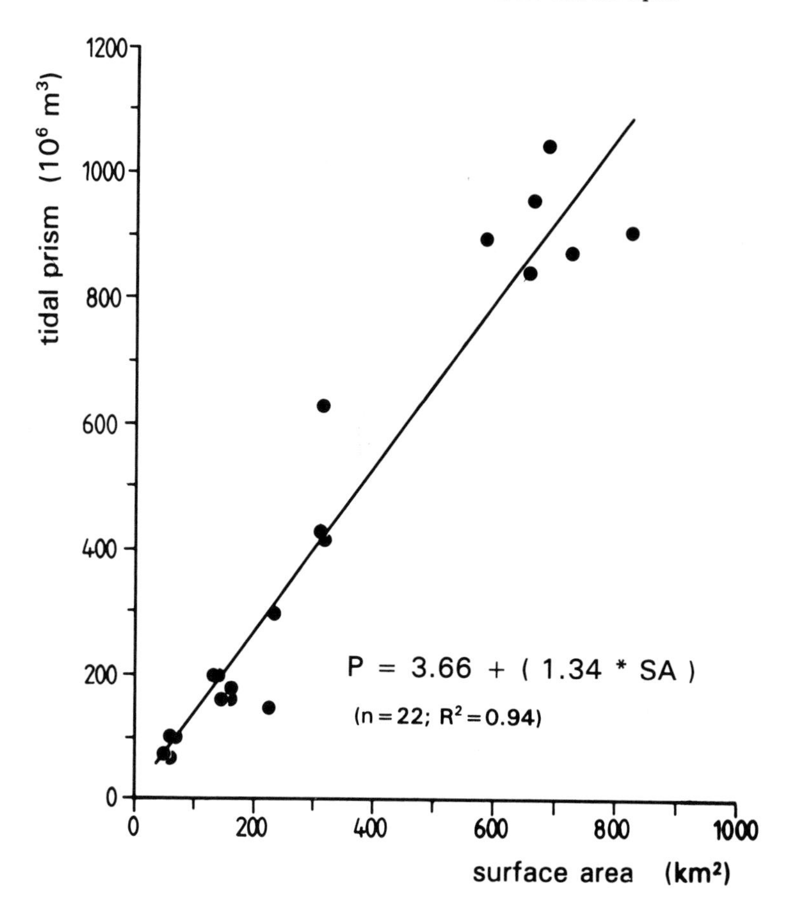

Fig. 12. Plot of tidal prism versus basin surface areas for the tidal basins of the present-day Netherlands Wadden Sea (based on data from de Glopper, 1967; Anonymous, 1974; Endema, 1979).

This relationship is valid for surface areas from $50 * 10^6$ m^2 to $820 * 10^6$ m^2. The tidal prisms that were calculated for the reconstructions of the Middelzee range from $373 * 10^6$ m^3 to $992 * 10^6$ m^3 (Table 1). Now that the tidal prisms are known, the corresponding cross-sectional areas of the tidal inlet can be calculated. Gerritsen & de Jong (1985) gave a statistical relationship between tidal volume and cross-sectional inlet area below MSL for the present tidal inlets in the The Netherlands Wadden Sea:

$$TV = (33198 * A_{c,msl,inlet}) - 127.6 * 10^6 \text{ m}^3 \quad (2)$$

(where TV is tidal volume in 10^6 m^3 and $A_{c,msl,inlet}$ is the inlet cross-sectional area below mean sea-level in m^2). From a hydraulic point of view, the closest correlation is expected to exist between the cross-sectional inlet area and the largest of the ebb or flood volumes, the so-called dominant volume. However, in this case the tidal prism, i.e. the mean value of the ebb and flood volumes, is used since the dominant volume cannot be estimated from the basin surface area. Since the tidal prism is equal to one-half of the tidal volume (Gerritsen, 1990, 1992), equation 2 can be rewritten in the form:

$$A_{c,msl,inlet} = 60.2 * 10^{-6} * P + 3844 \quad (3)$$

The cross-sectional areas calculated using this formula range from $26 * 10^3$ m^2 to $64 * 10^3$ m^2 (Table 1).

However, cross-sectional areas are difficult to assess from geological information. Tidal inlets very often migrate laterally along the coast, leaving a sequence of point-bar deposits. This prevents a reliable determination of the inlet width. Maximum inlet depth, however, can be derived from the thickness of channel-fill sequences. The relationship between tidal prism and maximum inlet depth will not be as strong as that between cross-sectional area and tidal prism since inlet width and shape are not incorporated (de Glopper, 1967). Sha (1990, p. 38, fig. 16) shows a linear relationship between tidal prisms of present-day tidal basins in the Wadden Sea and maximum inlet depths. This relationship was quantified using data from Anonymous (1974) and Endema (1979) (see Fig. 13), resulting in the relationship:

$$h_{NAP,max,inlet} = 37 * 10^{-9} * P + 13 \quad (4)$$

where $h_{NAP,max,inlet}$ is maximum inlet depth in m below NAP (NAP = Dutch Ordnance Level, which is about present-day mean sea-level) and P is the tidal prism in 10^6 m^3, the relationship being valid for prisms from $70 * 10^6$ m^3 to $1050 * 10^6$ m^3. On this basis, the inlet depths calculated for the Middelzee decrease from 50 m in AD 1000 to 27 m in AD 1900 (Table 1).

From borings it is known that Holocene erosion between Oosterend in Terschelling and the present-day west coast of Ameland, the location of the inlet since the formation of the Middelzee, reaches at most 31 m below MSL (Fig. 9). The calculated inlet depths indicate that before AD 1700 the inlet was too large to be accommodated in a single channel, suggesting that the tidal discharge must have been spread over two channels. This is in agreement with historical maps. Around AD 1700 the channel Coggediep silted up and the Borndiep became the only main channel in the Amelander Gat. It can also be concluded that in AD 1000, when the Middelzee reached its largest dimension, the inlet must have consisted of two channels with maximum depths of about 30 m below MSL.

The historical nautical charts of the Amelander Gat do not give sufficient information to verify calculated maximum inlet depths. In most cases depth values were only given for the Borndiep. Besides, calculated maximum depths would hardly correspond to historical chart data since, from a nautical point of view, navigation concerns itself with minimum depths.

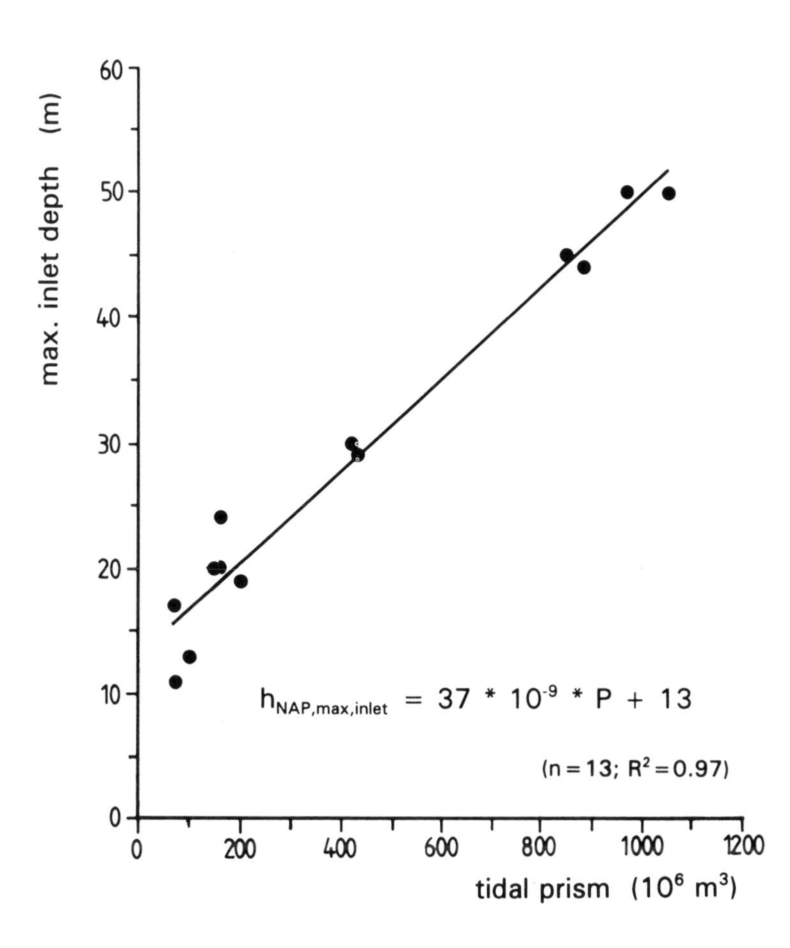

Fig. 13. Plot of maximum inlet depth versus tidal prism for the tidal basins of the present-day Netherlands Wadden Sea (based on data from Anonymous, 1974 and Endema, 1979).

Channel dimensions

Similar relationships between tidal discharge and cross-sectional channel area are derived by Eysink (1979) and Gerritsen & de Jong (1985) for channels in the Wadden Sea tidal basins (see Table 2). However, since channel depth is the only meaningful variable in the study of fossil channel deposits (channel width cannot be established because of lateral channel migration), it is geologically more meaningful to use a relationship between tidal discharge or prism and maximum channel depth. Figure 14 shows 69 combinations of maximum channel depth and observed discharge for the present Wadden Sea tidal basins (data from de Glopper, 1967; Endema, 1979; Gerritsen & de Jong, 1985). The variation in channel depth for a given discharge is large. This is a consequence of variations in channel depths to width ratios due to, amongst other things, sediment composition,

Table 2. Calculated discharge volumes, cross-sectional areas and maximum channel depths for five cross-sections through the Middelzee (see Fig. 16) for the situation AD 1000. These values have been calculated with the formulae given in the text. The cross-sectional areas were calculated with the equation for channels in the Wadden Sea tidal basins: $A_c = 71.6 * 10^{-6} * P + 134.6$ (Gerritsen & de Jong, 1985)

Cross-section	Surface area (km^2)	Discharge volume (10^6 m^3)	Cross-sectional area (10^3 m^2)	Maximum channel depth (m)		
1	37	53	3.9	10	9	9
2	45	64	4.7	11	10	9
3	66	92	6.7	13	12	11
4	93	129	9.4	14	14	13
5	161	220	15.9	18	18	16
Equation		1		5	6	7

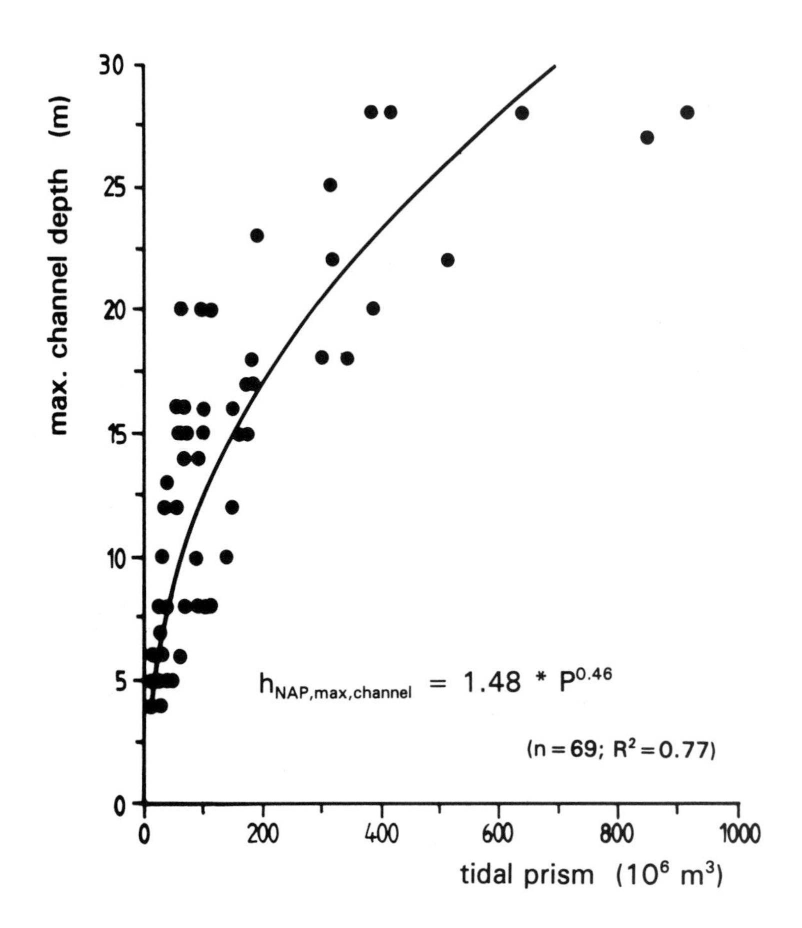

Fig. 14. Plot of maximum channel depth versus tidal prism for the present-day Netherlands Wadden Sea (based on data from de Glopper, 1967; Endema, 1979; Gerritsen & de Jong, 1985).

branching of channels and scouring of channels caused by constructions.

Van Bendegom (1949) described channel width as a function of depth for an idealized Wadden Sea channel with an approximately rectangular profile as $w = 5 * h^2$ (w is channel width in m, h is depth in m) From this it can be estimated that the cross-sectional area A_c of a channel equals $h * w = 5 * h^3$. This means that A_c varies with the cubic power of the channel depth. Since the tidal prism has a linear relationship with the cross-sectional area (see above), the prism will also vary with the cubic power of the depth. When the cube root of the prism (in 10^6 m^3) is related to the maximum channel depth, a linear relationship of the form

$$h_{NAP,max,channel} = 3.49 * P^{1/3} - 3.09 \quad (5)$$

can be calculated (Fig. 15). Direct correlation of both data sets yields:

$$h_{NAP,max,channel} = 1.48 * P^{0.46} \quad (6)$$

(with P in 10^6 m^3; see Fig. 14). Geyl (1976) correlated maximun channel depth and the drained surface area behind the cross-section for the tidal basins of the Wadden Sea. This yields

$$h_{NAP,max} = 1.98 * SA^{0.41} \quad (7)$$

with SA in 10^6 m^2. Approximation of the maximum channel depth can be calculated with these equations if the discharge, or the surface area of a part of a basin, are known.

Maximum channel depths were calculated from surface areas behind a number of cross-sections through the Middelzee (Fig. 16) using equations 1, 5 and 6. Depths were calculated for the reconstruction of the basin in AD 1000, when the basin had its widest extension and thus the deepest channels. Direct calculation of the maximum channel depth

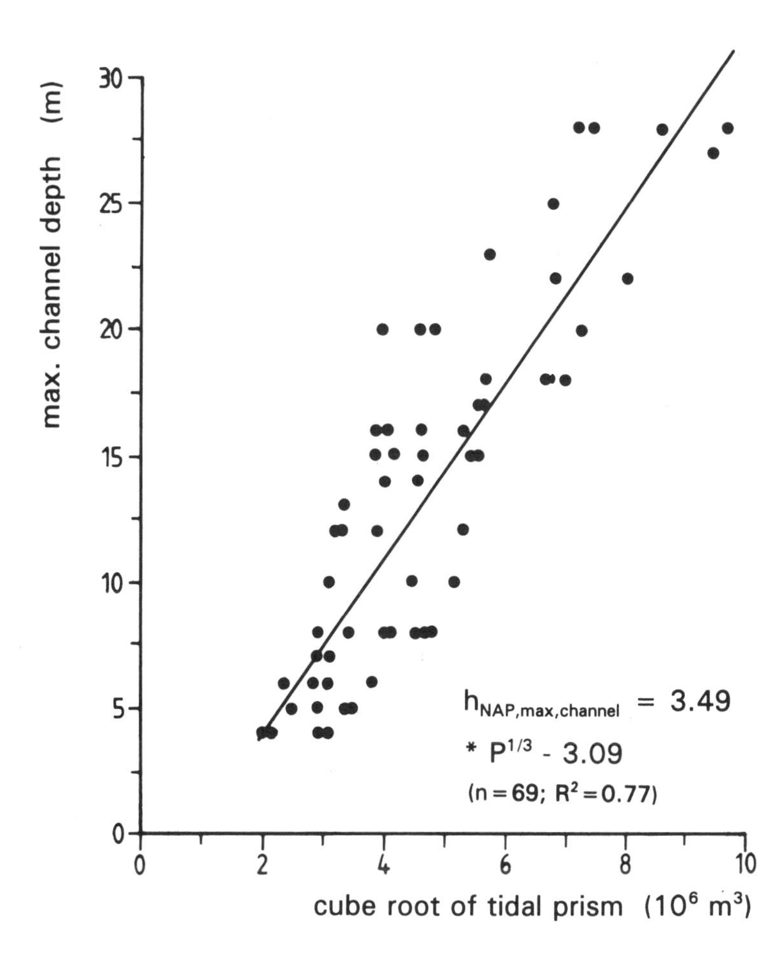

Fig. 15. Plot of maximum channel depth versus the cube root of the tidal prism for the present-day Netherlands Wadden Sea (based on the same data set as in Fig. 14).

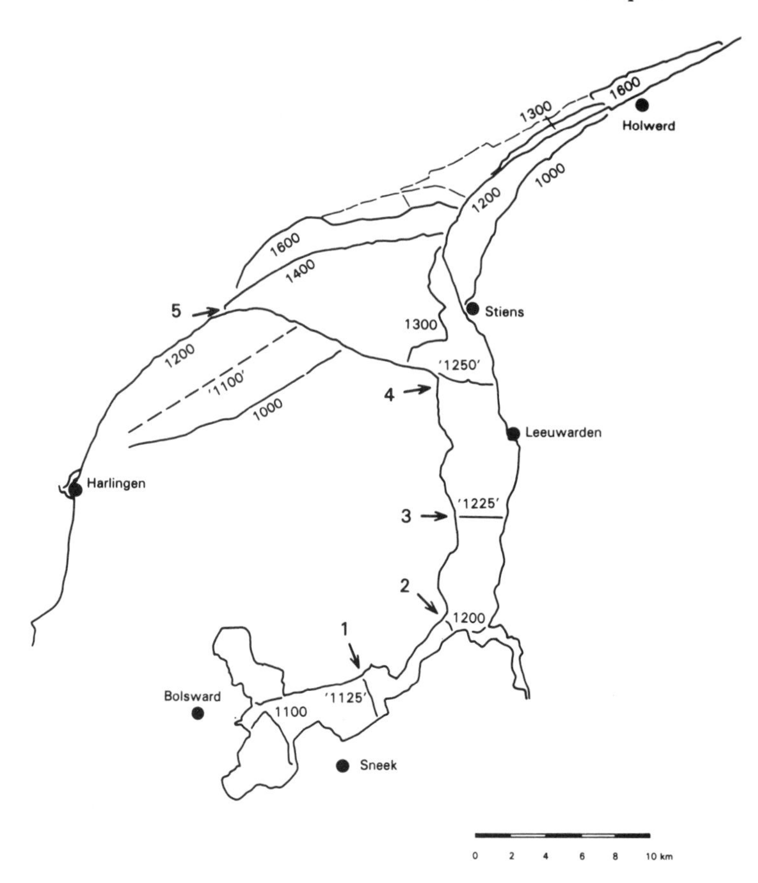

Fig. 16. Positions of several cross-sections through the Middelzee (1 to 5) for which maximum channel depths were calculated. The dates indicate the steps of sediment budget calculation. Dates in quotation marks are hypothetical.

from the surface area with equation 7 gives almost identical results. The maximum channel depth below MSL ranges from 9–10 m north of Sneek to 16–18 m in the northern part of Het Bildt (Fig. 16; Table 2).

The bases of the channel deposits in the cross-sections 2 and 3 were found in boreholes at 12.5 m and 10.7 m below NAP respectively. Channel depths of 9–11 m below NAP (cross-section 2) and 11–13 m below NAP (cross-section 3) were calculated (Table 2). The channel in cross-section 2 is confined to a narrow passage. This might have caused the channel to scour deeper than commonly observed. The calculated depth for cross-section 1 seems to be in agreement with geological information from the southern part of the Middelzee provided by Ter Wee (1976).

According to the Base Holocene map (Fig. 2), a calculated maximum channel depth of 14 m (Table 2) for cross-section 4 can have been reached in the most western part. The calculated channel depth for cross-section 5 is smaller than the greatest erosional depth of the Holocene situated at >20 m below NAP (Fig. 2). However, the maximum erosional depth of channels connected with the Middelzee is about 18 m. This suggests that the deepest erosion was not caused by Middelzee channels but by an earlier marine transgression in the Boorne Valley. Thus, at least two different generations of channel deposits are superimposed here.

BUDGET CALCULATIONS

A sediment budget for the Middelzee was calculated using the decrease in surface area with time, as illustrated by the history of dike construction, and the thickness of the deposited sediments. For the southern part of the Middelzee the distribution and thickness of the deposits are known from the geological map. The deposits were subdivided into channel deposits and mud-flat/marsh deposits,

Table 3. Calculated ranges of sediment volumes deposited until AD 1600 in the Middelzee basin and in the salt marshes along its borders and to the north (see text for further explanations)

Time interval AD	Middelzee (10^6 m^3)	Salt marshes (10^6 m^3)	Total (10^6 m^3)
<1000	44–56	39–64	83–120
1000–1100	26–36	13–15	122–177
1100–1125	25–42	–	147–213
1125–1200	74–76	20–24	241–313
1200–1225	55–69	–	296–382
1225–1250	91–115	–	387–497
1250–1300	31–51	2	421–550
1300–1400	99–158	–	520–708
1400–1600	70–80	3	590–788

Table 4. Time history of the decline of the total channel volume of the Borndiep tidal basin and inferred sedimentation in the channels

Reconstructed period AD	Channel volume (10^6 m^3)	Sedimentation (10^6 m^3)	Total sedimentation (10^6 m^3)
1000	2030	–	–
1100	1944	86	86
1200	1586	358	444
1300	1120	466	910
1400	955	165	1075
1600	881	74	1149
1700	717	164	1313
1800	576	141	1454
1900	468	108	1562

based on sediment thickness and composition. Average thicknesses were estimated for both categories. The deposited sediment volumes were calculated for the intervals shown in Fig. 16, based on the ages of the dikes and a number of inferred intermediate stages.

The southern part of the Middelzee had been completely silted up and reclaimed by *c.* AD 1225 (Fig. 16). A volume of 224–279 * 10^6 m^3 of sediment had been deposited (Table 3).

North of cross-section 3 (Fig. 16), thicknesses of Middelzee deposits formed in the period AD 1225–1600 were calculated on the basis of a few borings. This means that they have only been approximated for this period. An amount of 292–404 * 10^6 m^3 of sediment was deposited in the Middelzee between AD 1225 and AD 1600 (Table 3).

In the period AD 1000–1600 the salt marshes north of Oostergo and Westergo accreted from intertidal to supratidal levels and were secured by dikes (Figs 3 & 16). The stepwise history of this process is illustrated with Fig. 16. Based on Bakker (1954) and Bakker & Wensink (1955), the average thickness of these deposits is estimated to be 0.5–0.6 m. A sediment volume of 77–108 * 10^6 m^3 was deposited in the salt marshes before AD 1600 (Table 3).

The total sediment volume deposited in and along the Middelzee between AD 1000 and AD 1600 was 507–668 * 10^6 m^3. The average annual sedimentation was 0.9–1.1 * 10^6 m^3. This amount is about the same as the present-day annual import of sand into the Borndiep tidal basin amounting to 0.9 * 10^6 m^3 (Stive *et al.*, 1991). The sediment is estimated to comprise 75% fine to very fine sand and 25% clay for the channels, and 50% sand and 50% clay for the other deposits. As the volume of the channel deposits is 267–399 * 10^6 m^3, this means that about 65% of the sediment deposited in the Middelzee was sand. Figure 17 illustrates the total deposition in the period under consideration.

The sediment budget calculated above relates to the Middelzee. The silting up of this part of the basin also caused sedimentation in the adjacent Wadden Sea. The reduction of the tidal flow between the inlet and the Middelzee due to the silting up of the latter will have caused sedimentation in the channels. To complete the sediment budget for the Borndiep tidal basin, these amounts must also be estimated. Eysink (1979, 1991) derived an equation relating the total channel volume below mean sea-level to the tidal prism through the inlet:

$$I_{msl} = 65 * 10^{-6} * P^{3/2} \quad (8)$$

(where I_{msl} is the channel volume below MSL in 10^3 m^3 and P is the tidal prism in m^3). The sedimentation in the channels was calculated with this equation from the reduction of the tidal prism with time (Table 1). Between AD 1000 and AD 1900, 1562 * 10^6 m^3 of sediment were deposited in the channels of the Middelzee tidal basin and the adjacent Wadden Sea (Table 4; Fig. 17). This corresponds to an annual sedimentation of 1.7 * 10^6 m^3. For the period AD 1000–1600, the annual sedimentation in the channels amounts to 1.9 * 10^6 m^3. The sedimentation in the channels in the Middelzee proper until AD 1600 was 267–399 * 10^6 m^3. This is 17–26% of the total sedimentation in the channels in that period.

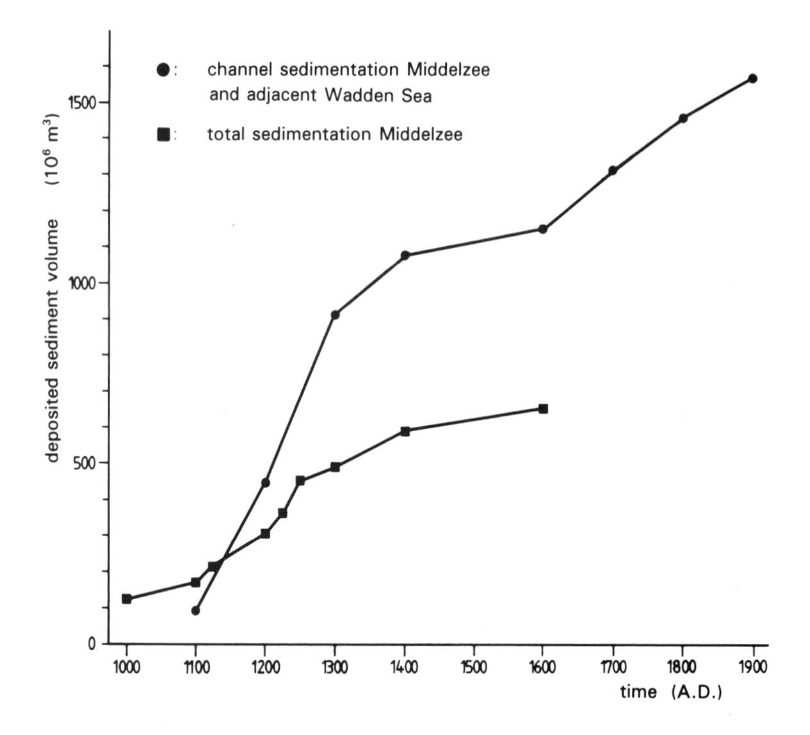

Fig. 17. Plot of sediment volumes deposited in the Middelzee and the adjacent Wadden Sea since AD 1000.

The calculated sedimentation rate of $1.9 * 10^6$ m^3 per year falls well within the range of yearly sand import into the present tidal basins of The Netherlands Wadden Sea. The Vlie tidal basin, with a surface area of $668 * 10^6$ m^2 and a tidal prism of $1078 * 10^6$ m^3 (Kool *et al.*, 1984), which is comparable to the Middelzee at its largest size, has an annual sand import of $2.6 * 10^6$ m^3 (Stive *et al.*, 1991).

CONCLUSIONS

The Middelzee, the latest stage of marine activity in the Pleistocene Boorne Valley, was formed during the medieval period. The subsequent silting up of the basin is illustrated by the age of the reclamations. At least $1562 * 10^6$ m^3 of sediment was deposited in the basin in the period AD 1000–1900. This caused a decrease in the tidal prism from $992 * 10^6$ m^3 to $373 * 10^6$ m^3, a reduction of 62%. The reduction of the tidal prism caused a decrease in the dimensions of the tidal channels in the inlet and the basin, as can be illustrated by a series of old nautical charts. The calculated annual sediment import into the basin falls well within the range of present-day values for The Netherlands Wadden Sea. This means that the rapid silting up of the Middelzee can be explained without the exceptionally high rates of sediment import that were suggested in historical reviews (Halbertsma, 1955; Schoorl, 1980).

The coast of Friesland had a more or less linear shape after the silting up of the Middelzee and its final reclamation in AD 1600. The geometry of the Borndiep tidal basin had changed from elongate to square. The morphodynamics of the basin changed considerably with the new geometry. When the basin was dominantly north–south orientated the main channels were running parallel to this direction, the position of the inlet being relatively stable. After AD 1600 the tidal flow penetrated further to the east, between the island of Ameland and the Frisian coast. This resulted in migration of the tidal watershed from south of Ameland towards the east. The inlet channel rotated to a more NW–SE orientation and eroded the western end of Ameland.

The empirical relationships established for the present-day tidal basins of the Wadden Sea proved to be a valuable tool in the analysis of the evolution of fossil tidal basins. Besides geological information, such as thickness and age of deposits, other channel features can be used to elaborate and refine basin reconstructions. This is an important contribution to the understanding of tidal basin development and the behaviour of coastal systems. This paper gives examples using the surface area of (part of) the basin to estimate channel depths. Thus, two different generations of Holocene channel deposits could be distinguished in the sediment fill of the mouth of the Boorne Valley. Conversely, if the depth of a channel is known from geological data, the dimension of the adjoining basin can be estimated.

The relationship between the basin surface area and the tidal prism varies greatly. Although the tidal prism is approximated by the basin surface area multiplied by the tidal range, the three-dimensional geometry of the basin which determines, for example, the extent of the intertidal area, also plays an important role in controlling the magnitude of the tidal prism. The relationship given in this paper holds for the present-day tidal basins in The Netherlands Wadden Sea, although the mean tidal range increases from 1.4 m in Texel inlet in the west to 2.2 m in the mouth region of the Ems estuary in the east. The intertidal area ranges from 20% of the basin surface area for the large basins to 80% for the small basins (de Glopper, 1967).

So far it is not clear why the tidal prism of $992 * 10^6$ m^3 in the Amelander Gat in AD 1000 was distributed over two channels with a maximum depth of 31 m, while in the present-day situation the tidal prisms in the Texel and Vlie inlets, amounting to $965 * 10^6$ m^3 and $848 * 10^6$ m^3 respectively, are contained in single channels with maximum depths of 50 m and 45 m respectively. For the Texel inlet the extensive protection works on the southern shore inhibit lateral migration of the Marsdiep channel and force it to scour deeply into the subsurface (Sha, 1990). This offers an explanation for the Texel inlet. For the Vlie inlet there is no such explanation. However, it is clear that basin length relative to basin width is small for present-day tidal basins in the Wadden Sea (with exception of the Marsdiep basin) when compared to the Middelzee. The estuaries in the south-western Netherlands, with large basin length to width ratios, all have more than one tidal channel in the inlet. There seems to be a complicated interaction between basin topography, including subsoil resistance, and hydrodynamic forces that accounts for this difference. This interaction needs further study.

ACKNOWLEDGEMENTS

This research was funded by The Netherlands Organisation for Scientific Research (NWO), grant Co2-77.125, and the Ministry of Internal Affairs, file number DUO 789612. I thank Dirk Beets, Jan Rik van den Berg, Poppe de Boer, Doeke Eisma, Kiek Jelgersma, Henk Schoorl and Marcel Stive for the valuable discussions and careful examination of earlier drafts of this paper. Furthermore, I would like to thank the Oosterwolde Department of the Geological Survey for conducting borings in the former Middelzee and James Baker for correcting the English. Finally, thanks to Burg Flemming for his patience.

REFERENCES

ANONYMOUS (1974) *Rapport van de Waddenzeecommissie; advies inzake de principiele mogelijkheden en de voor- en nadelen van inpolderingen in de Waddenzee.* The Hague, 326 pp.

BAKKER, J.P. (1954) Relative sea-level changes in Northwest Friesland (Netherlands) since pre-historic times. *Geol. Mijnb.* **16,** 323–346.

BAKKER, J.P. & WENSINK, J.J. (1955) Overzicht van de Holocene reliefgeneraties en sedimentopvolging in Barradeel. In: *Barradeel, rapport betreffende het onderzoek van het Landskip-genetysk wurkforbân fan de Fryske Akademy*, pp. 16–42. Laverman, Drachten.

BEETS, D.J., VALK, L. VAN DER & STIVE, M.J.F. (1992) Holocene evolution of the coast of Holland. *Mar. Geol.* **103,** 423–443.

BENDEGOM, L., VAN (1949) *Beschouwingen over de grondslagen van kustverdediging.* Rijkswaterstaat, 84 pp.

BLAEU, W.J. (1608) *Licht der Zeevaert.* Amsterdam.

BLAEU, W.J. (1623) *Zeespiegel.* Amsterdam.

BOELES, P.C.J.A. (1951) *Friesland tot de 11e eeuw*, 2nd edn. M. Nijhoff, The Hague, 598 pp.

BORGER, G.J. (1985) De ouderdom van onze dijken. *Hist. Georg. Tijdschr.* **3,** 76–80.

CNOSSEN, J. (1958) Enige opmerkingen omtrent het ontstaan van het Beneden-Boornegebied en de Middelzee in verband met de subatlantische transgressie. *Boor en Spade* **9,** 24–38 (with English summary).

CNOSSEN, J. (1971) De bodem van Friesland; toelichting bij blad 2 van de bodemkaart van Nederland, schaal 1:200,000. Sticht. Bodemkart. Wageningen, 132 pp.

CNOSSEN, J. & ZANDSTRA, J.G. (1965) De oudste Boorneloop in Friesland en veen uit de Paudorftijd nabij Heerenveen. *Boor en Spade* **14,** 62–87 (with English summary).

ENDEMA, D. (1979) *Grootheden/gegevens Waddenzee, Waddeneilanden en Noordzeekust.* Rept **WWKZ-79.H003,** Rijkswaterstaat, 20 pp.

EYSINK, W.D. (1979) *Morfologie van de Waddenzee, gevolgen van zand- en schelpenwinning.* Rept **R 1336**, Delft Hydraulics Lab., 92 pp.

Eysink, W.D. (1991) Morphological response of tidal basins to changes. The Dutch coast: paper no. 8. In: *Proc. 22nd Coast. Eng. Conf., Delft, 1990* (Ed. Edge, B.L.), pp. 1948–1961. ASCE, New York.

FitzGerald, D.M. & Penland, S. (1987) Backbarrier dynamics of the East Friesian Islands. *J. sediment Petrol.* **57**, 746–754.

FitzGerald, D.M., Penland, S. & Nummedal, D. (1984) Changes in tidal inlet geometry due to backbarrier filling: East Friesian Islands, West Germany. *Shore and Beach* **52**, 2–8.

Flemming, B.W. & Davis, R.A. Jr. (1994) Holocene evolution, morphodynamics and sedimentology of the Spiekeroog barrier island system (southern North Sea). In: *Tidal Flats and Barrier Systems of Continental Europe: A Selected Overview* (Eds Flemming, B.W. & Hertweck, G.). *Senckenbergiana marit.* **24**, 117–155.

Gerritsen, F. (1990) *Morphological stability of inlets and channels of the western Wadden Sea.* Rept **GWAO-90.019,** Rijkswaterstaat, 86 pp.

Gerritsen, F. (1992) Morphological stability of inlets and tidal channels of the Western Wadden Sea. In: *Proc. 7th Internat. Wadden Sea Symp., Ameland, 1990* (Eds Dankers, N., Smit, C.J. & Scholl, M.). *Neth. Inst. Sea Res. Publs Ser.* **20**, 151–160.

Gerritsen, F. & Jong, H. De (1985) *Stabiliteit van door- stroomprofielen in het Waddengebied.* Rept **WWKZ-84.V016,** Rijkswaterstaat, 53 pp.

Geyl, W.F. (1967) Tidal neomorphs. *Z. Geomorph. N.F.* **20**, 308–330.

Glopper, R.J. De (1967) Over de bodemgesteldheid van het Waddengebied. *Van Zee tot Land* **43**, W.E.J. Tjeenk-willink Publ., Zwolle, 67 pp. (with English summary).

Groot, T.A.M. de, Adrichem Boogaert, H.A. van, Fischer, M.M. *et al.* (1987) Toelichtingen bij de geologische kaart van Nederland 1 : 50.000, Blad Heerenveen West (11W) en Blad Heerenveen Oost (11O). Neth. Geol. Surv., 251 pp. (with English summary).

Haeyen, A. (1585) *Amstelredamsche Zee-caerten.* Christoffel Plantyn, Leiden.

Halbertsma, H. (1955) Enkele oudheidkundige aantekeningen over het ontstaan en de toeslijking van de Middelzee. *Tijdschr. Kon. Ned. Aardr. Gen.* **72**, 93–105 (with English summary).

Isbary, G. (1936) Das Inselgebiet von Ameland bis Rottumeroog. Morphologische und hydrographische Beiträge zur Entwicklungsgeschichte der friesischen Inseln. *Arch. Deutschen Seewarte* **56 (3)**, Hamburg, 55 pp.

Jong, J. de (1984) Age and vegetation of the coastal dunes in the Frisian Islands, the Netherlands. *Geol. Mijnb.* **63**, 269–275.

Kool, G., Peereboom, P., Lieshout, M.F. & Boer, M. de (1984) *Verloop natte en droge oppervlakten en kombergingen.* Rept **WWKZ-84.H009**, Rijkswaterstaat.

Kuijer, P.C. (1974) Bodemkaart van Nederland schaal 1 : 50.000, Toelichting bij de kaartbladen 10W Sneek en 10O Sneek. Sticht. Bodemkart. Wageningen, 126 pp.

Kuijer, P.C. (1976a) Bodemkaart van Nederland schala 1 : 50.000, Toelichting bij de kaartbladen 5W Harlingen en 5O Harlingen. Sticht. Bodemkart. Wageningen, 95 pp.

Kuijer, P.C. (1976b) Bodemkaart van Nederland schaal 1 : 50.000, Toelichting bij het kaartblad 11W Heerenveen. Sticht. Bodemkart. Wageningen, 140 pp.

Kuijer, P.C. (1981) Bodemkaart van Nederland schaal 1 : 50.000, Toelichting bij de kaartbladen 6W Leeuwarden, 6O Leeuwarden en het vaste land van de kaartbladen 2:W Schiermonnikoog en 2:O Schiermonnikoog. Sticht. Bodemkart. Wageningen, 181 pp.

O'Brien, M.P. (1931) Estuary tidal prisms related to entrance areas. *Civil Eng.,* **1**, 738–739.

O'Brien, M.P. (1969) Equilibrium flow areas of inlets on sandy coasts. ASCE, *J. Waterw. Harbors Div.* **95**, (WW1), 43–51.

Rienks, K.A. & Walther, G.L. (1954) *Binnendiken en slieperdiken yn Fryslan.* A.J. Osinga Publ., Bolsward, 555 pp. (with English summary; charts are published in a separate atlas).

Schoorl, H. (1980) The significance of the Pleistocene landscape of the Texel–Wieringen region for the historical development of the Netherland coast between Alkmaar and East Terschelling. In: *Transgressions and the History of Settlement in the Coastal Areas of Holland and Belgium* (Eds Verhulst, A. & Gottschalk, M.K.E.). Belg. Centr. Land. Geschied. **66**, 115–153.

Schotanus À Sterringa, C. (1664) *Beschrijvinge van de Heerlyckhydt van Frieslandt.* (Facs. Iss. 1978).

Sha, L.P. (1990) Sedimentological studies of the ebb-tidal deltas along the West Frisian Islands, the Netherlands. *Geologica Ultraiectina* **64,** 160 pp.

Spek, A.J.F. van der & Beets, D.J. (1992) Mid-Holocene evolution of a tidal basin in the western Netherlands: a model for future changes in the northern Netherlands under conditions of accelerated sea-level rise? *Sediment. Geol.* **80**, 185–197.

Staalduinen, C.J. van (Ed.) (1977) *Geologisch onderzoek van het Nederlandse Waddengebied.* Neth. Geol. Surv., 77 pp.

Stive, M.J.F., Roelvink, J.A. & Vriend, H.J. de (1991) Large-scale coastal evolution concept. The Dutch coast: paper no. 9. In: *Proc. 22nd Coast. Eng. Conf., Delft, 1990* (Ed. Edge, B.L.), pp. 1962–1974. ASCE, New York.

Veen, J. van (1950) Eb- en vloedschaar systemen in de Nederlandse getijdewateren. *Tijdschr. Kon. Ned. Aardr. Gen.* **67**, 303–325. (with English summary).

Waghenaer, L.J. (1584) *Spieghel der Zeevaerdt.* Leiden.

Wee, M.W. Ter (1975) Enkele momentopnamen uit de geologische geschiedenis van de Boorne. *It Beaken* **37**, 334–340.

Wee, M.W. Ter (1976) Toelichting bij de geologische kaart van Nederland 1 : 50.000, blad Sneek (10W, 10O). Neth. Geol. Surv., 131 pp. (with English summary).

Winsemius, P. (1622) *Chronique ofte historische geschiedenisse van Vriesland etc.* Franeker.

Spec. Publs int. Ass. Sediment. (1995) **24**, 259–271

Comparison of ancient tidal rhythmites (Carboniferous of Kansas and Indiana, USA) with modern analogues (the Bay of Mont-Saint-Michel, France)

B. TESSIER*, A.W. ARCHER†, W.P. LANIER‡ *and* H.R. FELDMAN§

**Université de Lille I, Laboratoire de Sédimentologie et Géodynamique, URA 719 CNRS, 59655 Villeneuve d'Ascq, France;*
†Department of Geology, Kansas State University, Manhattan, KS 66506, USA;
‡Department of Earth Sciences, Emporia State University, Emporia, KS 66801, USA and
§Kansas Geological Survey, University of Kansas, Lawrence, KS 66047, USA

ABSTRACT

In the upper intertidal area of the Mont-Saint-Michel Bay estuary, the sediment consists of silty and sandy heterolithic facies which commonly display well-developed neap–spring cycles. Such facies have been termed 'tidal rhythmites' because they exhibit rhythmicity directly related to tidal periodicities. The vertically accreted tidal bundles are related to flood-dominant deposition. They constitute a sedimentary record of a single semi-diurnal tidal cycle. The neap–spring cycle is reflected by a vertical evolution in thicknesses of successive 'tidal bundles' and types of bedforms. These neap–spring records range from a few cm to one dm in thickness and can be expressed in planar lamination, flaser to wavy bedding, and climbing-ripple bedding.

Tidal rhythmites are common in mid-continental Carboniferous sequences in the US. They are well developed in the interior coal basins. Because such facies lack marine body fossils and contain few trace fossils, they have commonly been interpreted as non-marine. However, analyses of sequential series of laminae thicknesses indicate well-developed periodicities indicative of tidal influences during deposition. Modern analogues for such rhythmites have been described from upper reaches of estuarine systems such as the macrotidal Bay of Mont-Saint-Michel. Comparison of the Carboniferous tidal rhythmites with these modern analogues indicates many similarities, which extend to both physical and biogenic sedimentary structures.

In the modern environments, similar tidal rhythmites are best developed in high intertidal estuarine settings. In the Mont-Saint-Michel Bay, tidal rhythmites are observed within restricted secondary channels of the inner estuarine domain protected from ocean wave reworking and fluvial energies. However, high concentrations of suspended sediment are required to produce these facies. In the Bay of Mont-Saint-Michel, high suspended sediment concentrations are produced by offshore wave dynamics. The physical similarities observed between the supposed Carboniferous and the modern tidal rhythmites suggest that they have been deposited in comparable environments with similar hydrodynamic conditions. This similarity in structures may be used as a tool to reinterpret some of the Carboniferous facies and to specify their palaeo-environmental conditions.

INTRODUCTION

Tidal rhythmites may be defined as tidal bedding (Reineck & Wunderlich, 1968) which displays a vertical record of tidal cyclicities, such as the semi-diurnal, diurnal, and neap–spring cycles. Sedimentary records of tidal cycles are very well known in migrating bedforms based on the work of Visser (1980) and Terwindt (1981). Such records have also begun to be well understood in vertically accreted deposits. This is especially true within Carboniferous strata of the epicontinental basins of the interior of the US, where rhythmites that appear to be of tidal origin are very common and have been described from a variety of settings (Kvale & Archer, 1990, 1991; Brown *et al.*, 1990; Kuecher *et al.*, 1990; Archer, 1991). These ancient rhythmites show many similarities to modern tidal rhythmites described from upper intertidal areas within estuarine environments. Such modern tidal rhythmites are present within the upper estuarine reaches of the Bay of Fundy, Canada (Dalrymple *et al.*, 1991) and are particularly common within the Bay of Mont-Saint-Michel in France (Tessier *et al.*, 1989; Tessier, 1993). The purpose of the present paper is to report the main similarities between the modern rhythmites from the Bay of Mont-Saint-Michel and those present within the Carboniferous basins of the US. The Carboniferous examples are from the Douglas Group of Stephanian age in eastern Kansas, and from the Mansfield Formation of Westphalian age in south-eastern Indiana. The principal intent of this paper is not to present a modern analogue to Carboniferous coal-bearing settings, but to point out the numerous similarities in the physical and biogenic sedimentary structures and to discuss criteria that may be of use in interpreting ancient macrotidal systems.

Historically, a number of Carboniferous stratigraphical sections in the US, now thought to contain tidal rhythmites, have been previously misinterpreted as non-marine facies. For example, a number of sites in the Carboniferous coal basins of the Central US exhibit rhythmite-bearing sections that commonly contain upright lycopods, have common occurrences of invertebrate trackways, and lack marine body fossils. Such data have traditionally led to interpretations invoking non-marine settings such as fluvial, floodplain, overbank, and limnic (lacustrine) deposits. More recently, detailed sedimentological analyses indicate that many Carboniferous mudstones and siltstones are characterized by fine-scale lamination and that successive laminae thicknesses exhibit a systematic variation in thickness, thus suggesting tidal influences during sedimentation. This interpretation has been applied not only to the study area in Kansas (Archer, 1991; Lanier *et al.*, 1993) and Indiana (Kvale *et al.*, 1989; Kvale & Archer, 1990, 1991) but also to sections in Illinois (Kuecher *et al.*, 1990).

MODERN TIDAL RHYTHMITES IN THE BAY OF MONT-SAINT-MICHEL

Overview of the sedimentary environment

The Bay of Mont-Saint-Michel, located in the western part of the English Channel, is formed within a wedge-shaped geometry bounded by the Cotentin and Brittany peninsulas of north-western France (Fig. 1). This wide embayment is protected from strong wave action and the sedimentary dynamics are controlled almost entirely by tidal currents. The wedge-shaped geometry of the bay further amplifies the high tidal ranges present within the English Channel. Because of the southern deflection of the tidal bulge within the channel, which is related to the Coriolis effect, tidal ranges along the French coast are generally significantly higher than those along the English coast to the north. Moreover, the Contentin Peninsula produces a stationary wave in acting like a barrier against the Atlantic tidal surge. The localized dramatic amplification produces a major macrotidal setting in which the tidal range occasionally reaches 15 m during high spring tides. The Bay of Mont-Saint-Michel is thus one of the foremost macrotidal environments in the world (Larsonneur, 1975, 1994). The tidal periodicities within the bay are typical for the Atlantic Ocean and are characterized by a semi-diurnal system with a very small diurnal inequality.

The modern morphology of the day is, in part, a result of the contained sedimentary prism which has been formed during the Holocene transgression since 10 000 yr BP. Regarding the present shape and orientation of the bay and dynamic factors of sedimentation, two main morphosedimentary environments can be distinguished (Fig. 1):

1 The south-western end of the bay, termed the 'bottom bay', is protected from wave and strong tidal currents by the Grouin de Cancale. This feature is composed of a large muddy intertidal flat on which some cheniers are formed which migrate landwards during exceptionally severe storms.

2 The eastern part of the bay, where three small rivers have produced an estuarine system; sedimentation within this part is dominated by alternating tidal currents. The estuarine environment is characterized by wide salt marshes and a very wide mixed intertidal flat with migrating tidal channels. These flats constitute a very active modern depositional unit.

To date, tidal rhythmites have been observed exclusively in the estuarine domain of the bay and, more precisely, occur within the upper intertidal deposits of the inner estuarine system. The mixed flat of the estuarine domain is the depositional site of a very specific type of sediment, locally termed '*tangue*'. This *tangue* is comprised of grey, silty to

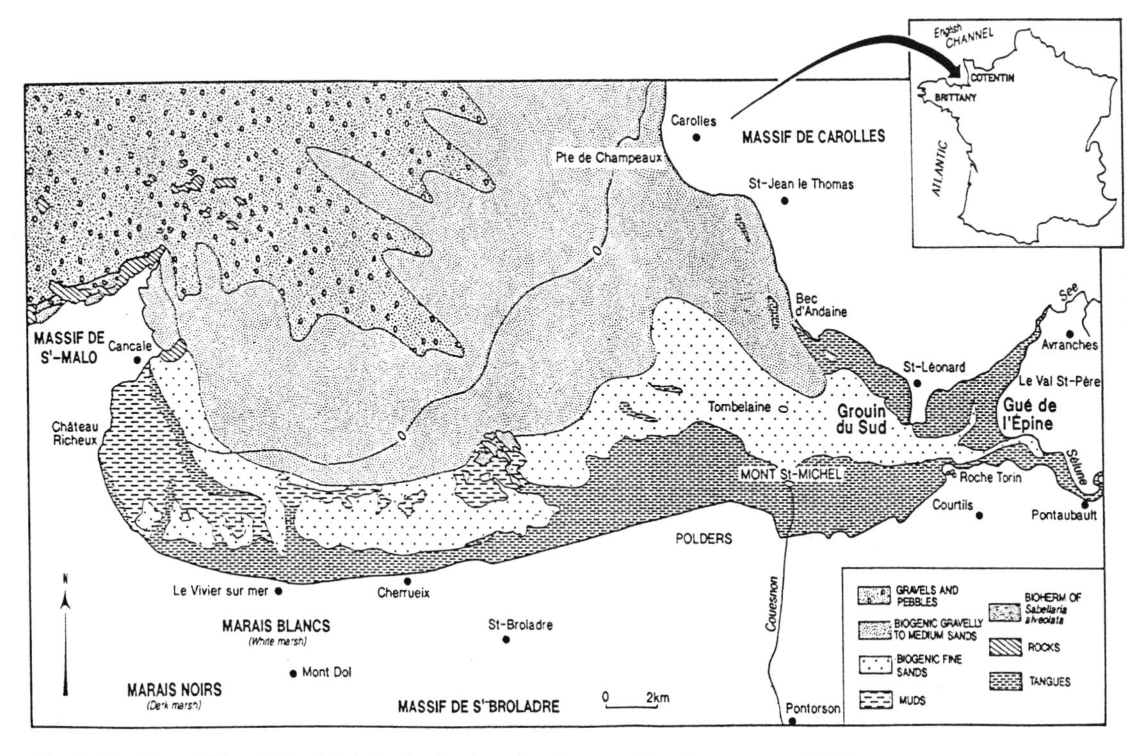

Fig. 1. The Bay of Mont-Saint-Michel: distribution of sediments (after Larsonneur, 1989).

sandy muds that have a mean grain size ranging from 0.03 to 0.09 mm. This sediment is unusual in that 50% of it is composed of carbonate, mostly of biogenic origin. Fragments of molluscs, foraminifers, ostracods, coccoliths, bryozoans, sponge spicules and diatoms constitute the bioclastic fraction. The mineral fraction is represented by quartz, mica and heavy minerals (Larsonneur, 1989, 1994). Because of its physical properties, the *tangue* provides an excellent medium for the formation and preservation of numerous sedimentary structures (Bajard, 1966; Mathieu, 1966).

Occurrence of tidal rhythmites

Tidal rhythmites have been observed primarily in the fine *tangues* of the inner part of the estuarine domain. The muddy sediments in the sheltered south-western part of the bay, where tidal dynamics is controlled by gyratory currents, do not display tidal rhythmites. Similarly, the sand-dominated sediments within the more exposed parts of the bay do not exhibit rhythmites. It is thus apparent that estuarine dynamics play a key role in the occurrence and preservation of these modern rhythmites.

The estuarine morphology controls the tidal dynamics and creates the alternating tidal currents necessary to produce tidal bedding that exhibits well-developed couplets. It also provides an environment protected from open marine processes (e.g. swell) that could mask tidal periodicities. On the other hand, estuaries are favourable to the production of high concentrations of suspended sediment which may be either of marine or fluvial origin. High concentrations are necessary to produce a measurable deposit during each tidal cycle. In the case of the Mont-Saint-Michel Bay, fluvial input of sediment into the system is almost negligible. Most of the sediment is of marine origin and tidal rhythmites are deposited in the inner estuary during or after a period of offshore sediment reworking by waves which produces highly turbid waters that subsequently penetrate into the estuary. This wave-produced turbidity maximum results in the deposition of tidal rhythmites in the upper reaches of the bay.

A final point to emphasize about the estuarine

environments of the bay is the controlling influence of biogenic factors during sedimentation. Because of strong and sudden variations in salinity and temperature linked to the mixed tidal/fluvial dynamics, and because of the unusually high rates of vertical sediment accretion, bioturbation and colonization of the intertidal flats is minimal. In the course of each tidal cycle, the waters within the fluvial channels are fresh during low tide, but achieve near average marine salinities during high tide. Moreover, the higher flats can be subjected to significant freshening from rainwater during low tide and neap tide emergences. This low to extremely low degree of bioturbation results in an excellent preservation of the physical sedimentary structures. Nevertheless, life is present in the upper intertidal, estuarine domain and the surface of the *tangue* can exhibit abundant animal traces (Bajard, 1966). Burrows are produced by the annelid *Nereis diversicolor* and the arthropod *Corpohium volutator*. Some insects and gastropods also produce typical surficial trails and tracks.

It should be noted that the *tangue* cannot be considered as a typical detrital sediment (Bourcart & Charlier, 1959). Its peculiar properties probably control, to a large part, the processes of formation and preservation of the rhythmites.

Although the tidal rhythmites and their biogenic structures are of local significance, they are limited to very specific environments with specific characteristics and can therefore be used for a better understanding of ancient rhythmites and their environmental constraints. Bedforms from supposed Carboniferous tidal rhythmites in Kansas and the associated biogenic structures display a great number of similarities with the modern deposits in the Mont-Saint-Michel Bay. In particular, because factors such as salinity variations and tidal ranges are difficult, if not impossible, to reconstruct in ancient depositional environments, comparisons of ancient sedimentary structures with modern analogues can provide an improved understanding of the processes involved during deposition.

TIDAL RHYTHMITES IN THE CARBONIFEROUS OF NORTH AMERICA

Geographical locations and stratigraphy

The Carboniferous tidal facies discussed below are from the Douglas Group of eastern Kansas and Mansfield Formation of Indiana (Fig. 2). The rocks in Indiana are older and were formed during the Westphalian, whereas the rocks in Kansas were deposited during the Stephanian. None the less, the depositional systems both resulted in similar coal-bearing sequences with virtually identical tidal facies.

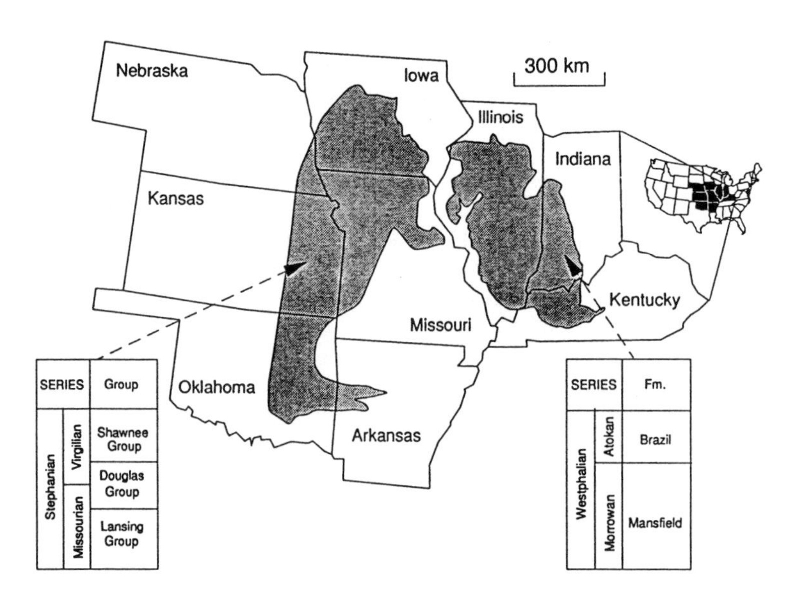

Fig. 2. General outlines of Eastern Interior Coal Basin (Illinois Basin) that includes the study area in Indiana, and of Western Interior Basin that includes the study area in Kansas.

In Indiana, the Mansfield and Brazil Formations consist of thick, cross-bedded sandstones, laminated sandstones and mudstones, and coals and underclays. Within the Mansfield, some of the thicker sandstones occur within palaeovalleys which were incised into lower Carboniferous strata. The valley-fill sands are locally conglomeratic and have been interpreted as representing valley-confined, braided fluvial systems. The upper parts of the Mansfield are more enriched in mudstones, containing thin and relatively discontinuous coal seams. The overlying Brazil Formation contains thicker coal seams; however, these coals too are relatively discontinuous laterally (Kvale & Archer, 1990). Stratigraphical and sedimentological aspects of the Mansfield and Brazil have been described in more detail by Archer & Maples (1984), Kvale *et al.* (1989) and Kvale & Archer (1990, 1991).

The Douglas Group is well exposed in outcrop along a belt that runs from the northern to southern end of the eastern part of the state of Kansas (Fig. 2). Shales probably constitute more than 70% of the Douglas interval, but these have received far less attention than the other lithofacies. As part of a study on upper Palaeozoic shales of Kansas, Cubitt (1979) presented details on, amongst other things, the clay minerals. In general, shales within the Douglas interval are notably siltier and more enriched in quartz than shales in the remainder of the upper Palaeozoic section of Kansas. Soil horizons are common in the shales, but few descriptions have to date been made of these.

The Douglas Group facies can best be understood in the context of the large-scale cycles that divide this group. The Douglas Group contains at least four decametre-scale cycles (Fig. 3). Each cycle starts with a fining-upward succession of fluvial to nearshore facies and is capped with marine facies that include widespread marine limestones. For purposes of discussion these cycles will be referred to as A–D, with 'A' ranging from the base of the Tonganoxie Sandstone up through the Westphalia Limestone, 'B' consisting of the Vinland Shale, Haskell Limestone and Weston Shale, 'C' incorporating everything up to the Amazonia Limestone, including the informally named Ireland Sandstone, and 'D' extending up to and including the Toronto Limestone (the basal member of the Oread Limestone of the Shawnee Group), and the informally named Wathena Shale. There is also a larger scale of cyclicity apparent, in which thick, deeply incised cycles (A and C) alternate with thinner cycles with little or no incisement (B and D).

The basal parts of cycles A and C exhibit large-scale incised palaeovalleys (Fig. 3). The basal part of cycle A consists of the Tonganoxie Sandstone Member, which was initially deposited in palaeovalleys which were incised into the underlying Weston Shale and Lansing Group. Initial valley-fill sequences consist of conglomerates, derived largely from Lansing Group carbonates which were exposed in valley walls and trough-cross-bedded sandstones.

The current understanding of the sequence stratigraphy indicates that cycles A and C represent valleys incised during lowstand and filled during transgression. Cycles B and D represent the highstand and regressive systems.

Occurrence of tidal rhythmites in Carboniferous sections

To date, there has been more published work on the tidal rhythmites of the Mansfield and Brazil Formations in Indiana than on the rhythmites in Kansas. One of the earliest detailed discussions of Carboniferous rhythmites which are thought to be tidal in origin was presented by Kvale *et al.* (1989), who described laminated siltstones of the informally named 'Hindostan Whetstone Member' of the Mansfield Formation. A further description of the tidal facies for this member was presented by Kvale & Archer (1991) and discussion of the various hierarchies of tidal periodicities recorded in such laminated siltstones is provided in Archer *et al.* (1991). Tidal rhythmites occur directly above coal seams in the overlying Brazil Formation (Kvale & Archer, 1990). The coals are of low-sulphur types and, until the recognition of the tidal rhythmites, had traditionally been interpreted as having formed in a setting lacking marine influence.

In the Kansas sections, analyses of lamination patterns within the upper Lawrence Shale indicated the presence of tidal rhythmites (Archer & Feldman, 1990; Archer, 1991). These rhythmites are similar in all aspects to those described by Kvale & Archer (1990) from the Brazil Formation of Indiana. In the Tongonoxie Sandstone of the Douglas Group, a variety of rhythmites has been described by Lanier *et al.* (1993). The Tongonoxie rhythmites include forms identical to those described from the 'Hindostan Whetstone' of Indiana (Kvale *et al.*, 1989; Kvale & Archer, 1991).

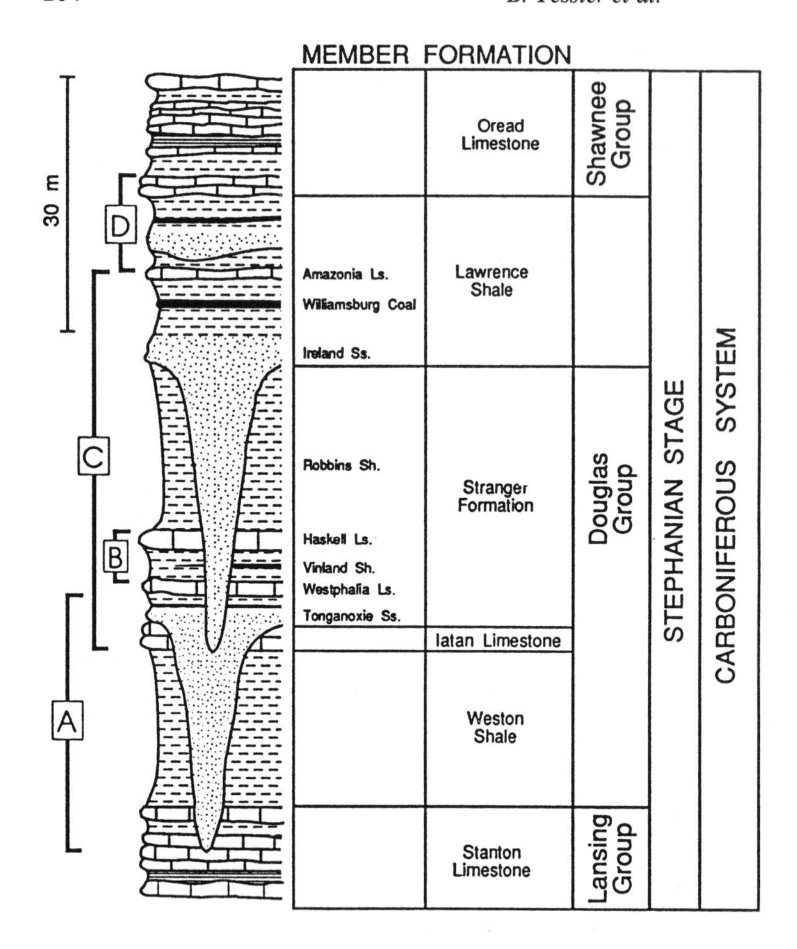

Fig. 3. Schematic stratigraphical section of the Douglas Group based primarily on well-logs and exposures in north-eastern Kansas. The vertical section can be subdivided into four major cycles. Cycles A and C are valley-fill sequences including fluvial, estuarine and marine facies. Tops of cycles A and C are delineated by marine limestones. Cycles B and D are primarily more marine influenced and appear to be progradational deltaic marine sequences.

In general, the Carboniferous tidal facies of both Indiana and Kansas occur within a similar stratigraphical setting. Because of a drier climate and greater availability of geophysical logs related to hydrocarbon exploration, the overall stratigraphical setting of the rhythmite-bearing sequences is better understood in Kansas. However, a general model based on the Kansas section appears to be applicable also to the Indiana sections. In Kansas, tidal rhythmites by and large seem to be restricted to the uppermost parts of the valley-fill sequences of the thicker depositional cycles (cycles A and C, Fig. 3). Although conglomeratic sands occur near the base of the valley-fill sequences, the uppermost parts of the fills become finer grained and exhibit smaller-scale bedforms, including tabular cross-bedding and ripple-scale cross-beds and laminations. As the sequence completely filled and overtopped the palaeovalleys, muddy coal-bearing facies were formed. Within the fine-grained siltstones and mudstones of this interval, systematic variations in laminae thicknesses suggest tidal influences during deposition (tidal rhythmites). Starved ripples and lenticular bedding are common as well. Flat-topped ripples and rain-drop impressions indicate intertidal settings. Palaeosols and coals with well-developed underclays indicate significant periods of exposure (Lins, 1950). Tidal influences are particularly notable above thin coals that occur within the upper phases of the valley-fill sequences (Kvale *et al.*, 1989; Kvale & Archer, 1990; Archer, 1991). The common occurrence of rhythmites directly above coals may in part be related to the generation of accommodation space by peat compaction (Kvale & Archer, 1990). The shales overlying these coals show increasing marine influence up-section and maximum marine conditions resulted in deposition of bioturbated to fossiliferous shales and regionally continuous to discontinuous limestones.

COMPARISON OF MONT-SAINT-MICHEL BAY AND THE CARBONIFEROUS DEPOSITIONAL ENVIRONMENTS

Comparison of tidal rhythmites

Direct comparisons can be made between Douglas Group rhythmites and those that occur within the Mont-Saint-Michel Bay at a variety of scales.

In the Bay of Mont-Saint-Michel, tidal rhythmites have been observed principally in fine-grained *tangues* of the upper intertidal area of the inner estuarine domain. In vertical exposures of excavations, the *tangues* always display alternating bedding composed of sandy beds interlaminated with silty to muddy beds. The sandy units have a planar- or ripple-bedding structure, the latter being gradational to lenticular, wavy, flaser-like or of the climbing-ripple type. In most cases, the silty to muddy beds drape the sandier units, such that the internal structure of the *tangue* consists of a couplet succession comprising a sandy unit overlain by a silty to muddy unit. Such couplets range from 1–2 mm to 1–2 cm in thickness.

Successive couplets are commonly arranged in microsequences, a few cm to about 15 cm in thickness, separated from each other by a dark layer. Each microsequence contains 10–12 couplets and the individual couplet thicknesses thicken and thin progressively from the bottom to the top of the sequence. These microsequences represent a vertical record of neap–spring tidal cycles (Figs 4B, 5A, 6A & 7A).

In a given microsequence, the number of couplets corresponds exactly to the number of tides able to reach the upper intertidal domain which is flooded only during spring tides. Neap tides are characterized by extended periods of subaerial emergence lasting at least several days. A couplet thus represents the depositional record of a single semi-diurnal cycle. Because most of the area is dominated by flood tidal currents, the sand layer is deposited during flood tides. The silty to muddy drape results from settling during the slackwater period of the high tide. In some cases, a second, very thin couplet occurs at the top of a primary couplet, reflecting minor reactivation during the ebb retreat (Tessier, 1993).

Progressive thickening of the couplets is related to the increasing tidal range from neap to spring, progressive thinning representing the decreasing tidal range from spring to neap. The neap emergence is manifested by a dark layer, the colour of which is produced by a diatomaceous film which forms on the *tangue* during the emergence period. At such times no tidally induced sedimentation occurs on the flats and extensive microbial activity develops a thin mucilaginous film.

Measurements of sedimentation rates using vertical stakes have confirmed that the microsequences described above are truly related to deposition

Fig. 4. Comparative examples of planar laminated silt-rich tidal rhythmites from the Tonganoxie Sandstone of Kansas, Douglas Group (A), and the Bay of Mont-Saint-Michel (B). Scale bar on A = 10 cm.

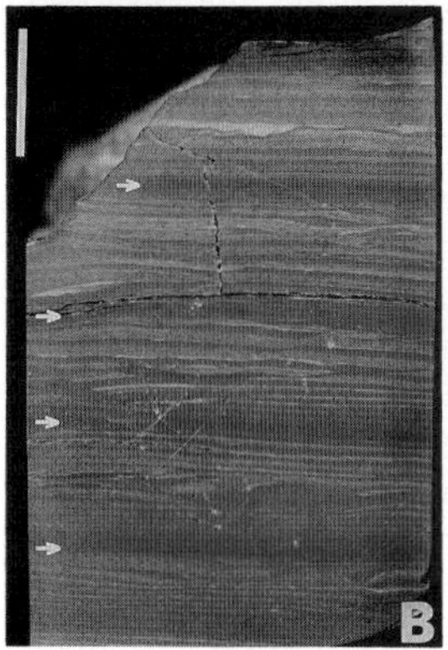

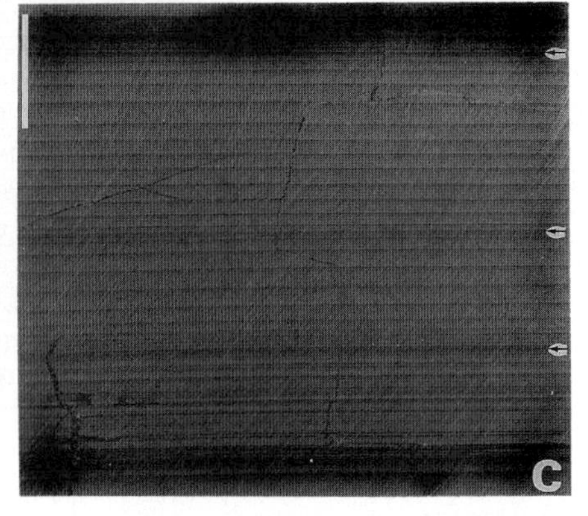

Fig. 5. Comparative examples of planar laminated to lenticular tidal rhythmites from the Bay of Mont-Saint-Michel (A), the Lawrence Shale of Kansas, Douglas Group (B), and the Brazil Formation of Indiana (C). Arrows indicate neap stage laminae. Scale coin on A = 2.8 cm, scale bars on B, C = 1 cm.

during a single neap–spring cycle. These rhythmic tidal microsequences display a number of bedding types that range from planar to flaser and climbing-ripple bedding.

Rhythmites most commonly occur within planar bedding, consisting of sand- or mud-dominated couplets. Couplet thicknesses generally range from a few mm upwards to 1–2 cm (Figs 4B & 5A). Less common are planar rhythmites containing very thin couplets approximately 1–2 mm thick. The planar-bedded rhythmites that occur within the bay are similar in scale and texture to those that occur in the Carboniferous rhythmites (Figs 4A, 5B & C). A lenticular bedded to 'pinstriped' facies, commonly containing well-developed rhythmites, occurs in the Carboniferous Lawrence Shale of the Douglas Group. These rhythmites consist of silt-fine sand layers capped by a clay drape. Similar facies occur within the Mont-Saint-Michel Bay (Tessier *et al.*, 1989). Both of these examples are similar to the 'type 2' tidal rhythmites described by Kvale & Archer (1990) from the Brazil Formation of Indiana.

In flaser to wavy-ripple bedding, modern tidal rhythmites are characterized by the vertical evolution from lenticular and wavy bedding to flaser bedding when the tidal range increases from neap to spring, and from flaser bedding to wavy bedding when it decreases from spring to neap (Fig. 6A). These changes in bedform character reflect an increase in tidal current velocities during spring tides associated with stronger movement of sand-

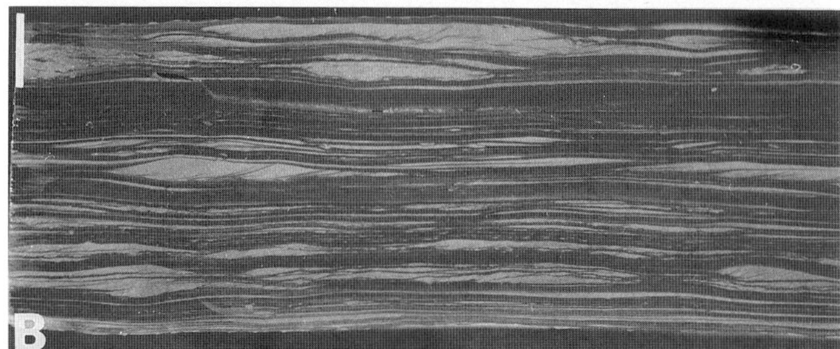

Fig. 6. Comparative examples of tidal rhythmites in ripple bedding from the Bay of Mont-Saint-Michel (A) and the Lawrence Shale of Kansas, Douglas Group (B). Arrows indicate neap stage laminae. Scale bar on A = 5 cm; scale bar on B = 1 cm.

sized particles. A comparison of the modern and ancient lenticular, wavy- and flaser-bedded rhythmites reveals a generally similar scale and texture (Fig. 6B).

In climbing-ripple bedding, two types of record can be distinguished in the Bay of Mont-Saint-Michel (Tessier, 1993). In type I, each sand–mud couplet represents a single ripple train. The successive trains thicken and climb regularly from neap to spring tides. They then become thinner and the angle of climb increases from spring to neap. In type II, one ripple train represents a complete neap-spring cycle. In this type, at the base of the ripple train, the first lamina are moulded over the underlying shape (Fig. 7A). When the tidal range increases, the lamina become progressively sandier and their slope increases to form the lee sides of the migrating ripple. The slope then decreases at the top of the ripple, and the lamina drape the underlying ripple shape as the tidal range decreases to neap-tidal conditions. Similar bedforms occur within the Tongonoxie Sandstone of Kansas and are associated with abundant raindrop imprints, runzel marks, runnel marks, and other types of drain features, all of which indicate development in the intertidal zone. Lanier *et al.* (1993) describe forms similar to the type I described herein as 'type B' and forms similar to the type II described herein as 'type A' (Fig. 7B). Kvale & Archer (1991) have described climbing silt waves, similar to the type II forms described herein, from the 'Hindostan whetstones' of Indiana.

Comparison of biogenic structures

Biogenic structures recorded in the Carboniferous tidal rhythmites are also very similar to those observed in the modern analogues. In general, the sediments of the rhythmites are not bioturbated, but may exhibit abundant surficial biogenic structures which do not disrupt the sedimentary fabric. In rhythmites of the Tongonoxie Sandstone, such surficial tracks and trails include the ichnogenera *Plangtichnus*, *Haplotichnus* and *Treptichnus*. This is an unusual suite of trace fossils which also occurs in the Hindostan Whetstone of the Carboniferous of Indiana (Archer & Maples, 1984; Maples & Archer, 1987). The biogenic structures were originally

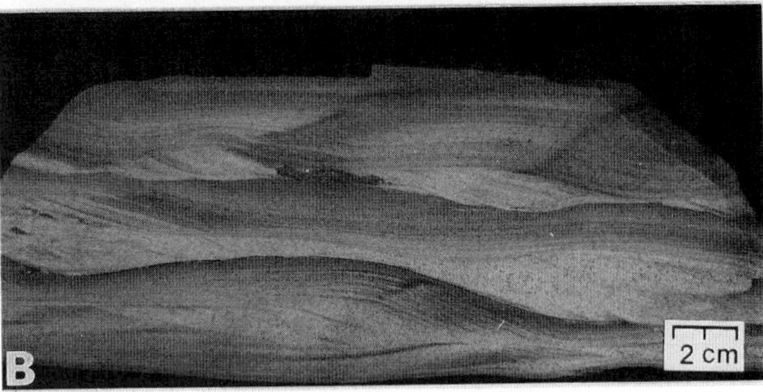

Fig. 7. Comparative examples of tidal rhythmites in climbing ripple lamination of type II from the Bay of Mont-Saint-Michel (A), and the Tonganoxie Sandstone of Kansas, Douglas Group (B). Arrows indicate neap stage laminae. Scale bar on A = 5 cm.

described as having been produced by insects (Miller, 1889). The form referred to as *Plangtichnus* occurs as a convex hyporelief on lower bedding surfaces, or as concave epireliefs on upper bedding surfaces (Figs 8C & D). This trace fossil is similar to modern biogenic structures that are found on the rhythmite-forming intertidal flats in Mont-Saint-Michel Bay (Figs 8A & B), interpreted by Bajard (1966) to have been formed by insect larvae. The insects build superficial burrow galleries in the *tangue* during subaerial exposure of the high flats. Subsequent erosion during the rising tide appears to remove the upper part of the burrow, resulting in the formation of a shallow furrow (concave epirelief) which is then filled with tidally emplaced sediment during flood tides. Both the modern traces of Mont-Saint-Michel Bay and the Carboniferous trace fossils at most disrupt only the upper few mm of the sedimentary laminations on which they occur, the underlying sedimentary structures remaining undisturbed.

INTERPRETATIONS OF THE CARBONIFEROUS RHYTHMITES

At present, the only modern depositional settings that provide useful analogues to the Carboniferous rhythmites, in terms of rapid rates of vertical accretion and characteristic physical and biogenic sedimentary structures, are the upper fluvial to estuarine reaches of macrotidal embayments such as the Mont-Saint-Michel Bay in France and the Bay of Fundy in Canada. Both modern sites have a funnel-shaped geometry that serves to amplify tidal ranges significantly at the fluvio-estuarine transition. They are characterized by high rates of vertical accretion that locally exceeds 1 m yr^{-1} (Tessier, 1990; Dalrymple *et al.*, 1991).

The Douglas Group cycles A and C, for example, are remarkably similar to late Holocene estuarine deposits within incised valleys along the North American and European Atlantic coasts, where studies have revealed similar facies patterns (Rah-

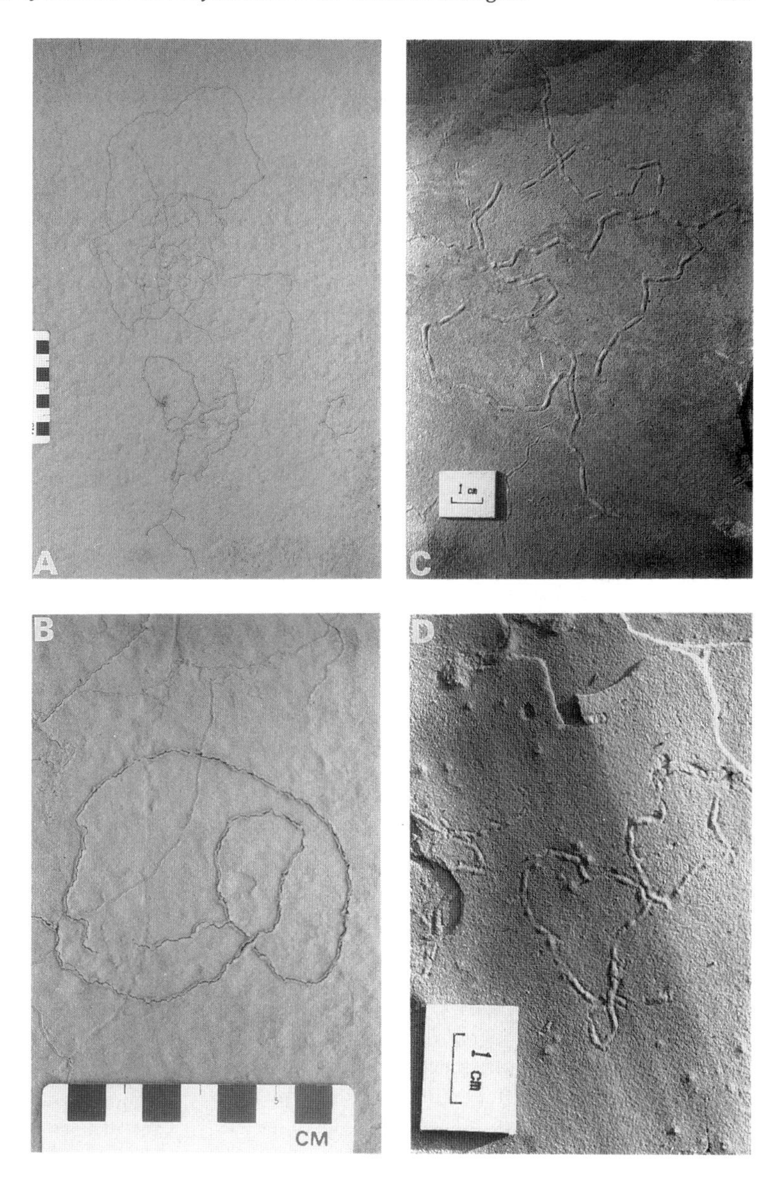

Fig. 8. Comparative examples of biogenic structures. A, B: Surface of the *tangue* in the Bay of Mont-Saint-Michel. These shallow surficial burrows are interpreted by Bajard (1966) as the products of insect larvae of the family Tabanides. C, D: Hyporeliefs of *Plangtichnus* from Hindostan Whetstone, Mansfield Formation, Indiana (Archer & Maples, 1984). These are shallow surficial burrows originally interpreted as having been produced by insect larvae (Miller, 1889).

mani, 1989; Dalrymple *et al.*, 1991; Nichols *et al.*, 1991). The modern estuarine model, based on the James River estuary and the thick Douglas Group cycles, commences with fluvial sand resting unconformably on older rocks. The fluvial sandstone grades upwards into middle estuarine muddy sediment. The Douglas Group cycles are capped by marine shale and limestone, although marine sands also occur within the valley-fill sequence basinwards to the south. Many of the details of the Douglas Group facies model have yet to be worked out, but the similarities with modern estuarine facies are quite impressive.

Our preliminary analyses therefore suggest that many of the so-called fluvial sand deltaic facies, especially within the Carboniferous rocks of Indiana and Kansas, may be related to estuarine depositional processes, although definitive fluvial and probably deltaic successions occur as well. Based upon preliminary surface and subsurface analyses, stratigraphic sections bearing tidal rhythmites appear to be related to incised-valley estuarine successions. These facies are complex and only further work and more access to core and subsurface information can positively identify the facies structure and architecture.

CONCLUSIONS

As demonstrated by this comparison, the modern tidal rhythmites in Mont-Saint-Michel Bay constitute a very good analogue for rhythmites in the Douglas Group of Kansas, the Mansfield and Brazil Formations of Indiana and other rhythmites in the coal-bearing basins of the Carboniferous of the US. They seem to constitute an accurate indicator for estuarine environments.

Despite the strong similarities in primary sedimentary structures, some fundamental differences need to be mentioned. First, the Carboniferous rhythmites are primarily composed of siliciclastics, whereas the sediments that form the rhythmites in Mont-Saint-Michel Bay are predominantly of carbonate origin. Second, the Carboniferous rhythmites are best interpreted as having formed in a semi-diurnal system with a marked diurnal inequality. Today such tidal systems are characteristic of the Pacific Ocean. Conversely, the modern rhythmites in the Bay of Mont-Saint-Michel exhibit periodicities related to a semi-diurnal system that lacks any significant diurnal inequality. The first difference, a mineralogical one, is not particularly important in terms of tidal dynamics. Mont-Saint-Michel Bay is within the temperate realm, whereas the Carboniferous coal-bearing sequences were apparently formed at wet equatorial palaeolatitudes. Despite these strong differences in climate and latitudinal settings, the tidal dynamics, and thus by inference also the tidal ranges, may have been similar during deposition of the modern and ancient analogues. The second difference, which is linked to the tidal system as such, is not significant. Continental geometries were vastly different during the Carboniferous when a single global continent, Pangaea, was being formed. It is therefore likely that many Carboniferous rhythmites will display periodicities more similar to the larger Pacific Ocean than to the relatively smaller Atlantic Ocean. These differences are largely associated with the relative strength of the tropical tidal periodicity and do not significantly affect the development of synodic periodicities, which are the principal driving forces for the development of neap–spring cycles.

REFERENCES

Archer, A.W. (1991) Modelling of tidal rhythmites using modern tidal periodicities and implications for short-term sedimentation rates. In: *Sedimentary Modelling: Computer Simulations and Methods for Improved Parameter Definition* (Eds Franseen, E.K., Watney, W.L., Kendall, C.G. St.C. & Ross, W.). *Kansas Geol. Surv. Bull.* **223**, 185–194.

Archer, A.W. & Feldman, H.R. (1990) Tidal rhythmites within the Lawrence Formation (Pennsylvanian: Virgilian) of Kansas. *Abstr. Kansas Acad. Sci.* **9**, 4.

Archer, A.W. & Maples, C.G. (1984) Trace fossil distribution across a marine to nonmarine gradient in the Pennysylvanian of southwestern Indiana. *J. Paleont.* **58**, 448–466.

Archer, A.W., Kvale, E.P. & Johnson, H.R. (1991) Analysis of modern equatorial tidal periodicities as a test of information encoded in ancient tidal rhythmites. In: *Clastic Tidal Sedimentology* (Eds Smith, D.G., Reinson, G.E., Zaitlin, B.A. & Rahmani, R.A.). *Mem. Can Soc. Petrol. Geol.* **16**, 189–196.

Bajard, J. (1966) Figures et structures sédimentaires dans la zone intertidale de la partie orientale de la baie du Mont-Saint-Michel. *Rev. Géol. dyn. Géogr. phys.* **2(8)**, 39–111.

Bourcart, J. & Charlier, R. (1959) The tangue: a 'non-conforming' sediment. *Geol. Soc. Am. Bull.* **70**, 565–568.

Brown, M.A., Archer, A.W. & Kvale, E.P. (1990) Neap–spring tidal cyclicity in laminated carbonate channel-fill deposits and its implication: Salem Limestone (Mississippian), south-central Indiana, U.S.A. *J. sediment. Petrol.* **60**, 152–159.

Cubitt, J.M. (1979) The geochemistry, mineralogy and petrology of Upper Palaeozoic shales of Kansas. *Kansas Geol. Surv. Bull.* **217,** 117 pp.

Dalrymple, R.W., Makino, Y. & Zaitlin, B.A. (1991) Temporal and spatial patterns of rhythmite deposition on mud flats in the macrotidal Cobequid Bay–Salmon River estuary, Bay of Fundy, Canada. In: *Clastic Tidal Sedimentology* (Eds Smith, D.G., Reinson, G.E., Zaitlin, B.A. & Rahmani, R.A.). *Mem. Can. Soc. Petrol. Geol.* **16**, 137–160.

Kuecher, G.M., Woodland, B.G. & Broadhurst, F.M. (1990) Evidence of deposition from individual tides and of tidal cycles from the Francis Creek Shale (host rock to the Mazon Creek Biota), Westphalian D (Pennsylvanian), northeastern Illinois. *Sediment. Geol.* **68**, 211–221.

Kvale, E.P. & Archer, A.W. (1990) Tidal deposits associated with low-sulfur coals, Brazil Fm. (Lower Pennsylvanian), Indiana. *J. sediment. Petrol.* **60**, 563–574.

Kvale, E.P. & Archer, A.W. (1991) Characteristics of two Pennsylvanian-age semi-diurnal tidal deposits in the Illinois Basin, U.S.A. In: *Clastic Tidal Sedimentology* (Eds Smith, D.G., Reinson, G.E., Zaitlin, B.A. & Rahmani, R.A.). *Mem. Can. Soc. Petrol. Geol.* **16**, 179–188.

Kvale, E.P., Archer, A.W. & Johnson, H.R. (1989) Daily, monthly, and yearly tidal cycles within laminated siltstones of the Mansfield Formation (Pennsylvanian) of Indiana. *Geology* **17**, 365–368.

Lanier, W.P., Feldman, H.R. & Archer, A.W. (1993)

Tidal sedimentation in a fluvial to estuarine transition, Douglas Group, Missourian-Virgilian, Kansas. *J. sediment. Petrol* **63**, 860–873.

LARSONNEUR, C. (1975) Mont-Saint-Michel Bay, France. In: *Tidal Deposits* (Ed. Ginsburg, R.), pp. 21–30. Springer, New York.

LARSONNEUR, C. (1989) La baie du Mont-Saint-Michel. *Bull. Inst. Géol. Bass. d'Aquit.* **46**, 5–74.

LARSONNEUR, C. (1994) The Bay of Mont-Saint-Michel: a sedimentation model in a temperate macrotidal environment. In: *Tidal Flats and Barrier Systems of Continental Europe—A Selected Overview* (Eds Flemming, B.W. & Hertweck, G.). *Senckenbergiana marit.* **24**, 3–63.

LINS, T.W. (1950) Origin and environment of the Tonganoxie Sandstone in northeastern Kansas. *Kansas Geol. Surv. Bull.* **86,** 105–140.

MAPLES, C.G. & ARCHER, A.W. (1987) Redescription of Early Pennsylvanian tracefossil holotypes from the nonmarine Hindostan Whetstone beds of Indiana. *J. Paleont.* **61**, 890–897.

MATHIEU, R. (1966) Structures sédimentaires des dépots de la zone intertidale de la partie occidentale de la baie du Mont-Saint-Michel. *Rev. Géol. dyn. Géogr. phys.* **2(8)**, 113–122.

MILLER, S.A. (1889) *North American Geology and Palaeontology for the Use of Amateurs, Students and Scientists.* Western Methodist Book Concern, Cincinnati, Ohio, 664 pp.

NICHOLS, M.M., JOHNSON, G.H. & PEEBLES, P.C. (1991) Modern sediments and facies model for a microtidal coastal plain estuary, Virginia. *J. sediment. Petrol.* **61**, 883–899.

RAHMANI, R.A. (1989) *Cretaceous Tidal Estuarine and Deltaic Deposits, Drumheller, Alberta.* Guide Book 2nd Internat. Res. Symp. Clastic Tidal Deposits, Calgary, 55 pp.

REINECK, H.E. & WUNDERLICH, F. (1968) Classification and origin of flaser and lenticular bedding. *Sedimentology* **11**, 99–104.

SANDERS, D.T. (1959) Sandstones of the Douglas and Pedee Groups in northeastern Kansas. *Kansas Geol. Survey. Bull.* **134,** 125–159.

TERWINDT, J.H.J. (1981) Origin and sequences of sedimentary structures in inshore mesotidal deposits of the North Sea. *Spec. Publs int. Ass. Sediment.* **5**, 4–26.

TESSIER, B. (1990) *Enregistrement des cycles tidaux en accrétion verticale dans un milieu actuel (la baie du Mont-Saint-Michel), et dans une formation ancienne (la molasse marine miocène du bassin de Digne). Mesure de temps et application à la reconstitution des paléoenvironnements.* PhD thesis, Univ. Caen.

TESSIER, B. (1993) Upper intertidal rhythmites in the Mont-Saint-Michel Bay (NW France); perspectives for paleoreconstruction. *Mar. Geol.* **110**, 355–367.

TESSIER, B., MONTFORT, Y., GIGOT, P. & LARSONNEUR, C. (1989) Enregistrement des cycles tidaux en accrétion verticale; adaptation d'un outil de traitement mathématique: exemples en baie du Mont-Saint-Michel et dans la molasse marine miocène du bassin de Digne. *Bull. Soc. Géol. France* **8(5)**, 1029–1041.

VISSER, M.J. (1980) Neap–spring cycles reflected in Holocene subtidal large-scale bedform deposits. A preliminary note. *Geology* **8**, 543–546.

Ancient Tide-dominated Environments and Facies

Spec. Publs int. Ass. Sediment. (1995) **24**, 275–288

Sequence stratigraphy of the late Pleistocene Palaeo-Tokyo Bay: barrier islands and associated tidal delta and inlet

H. OKAZAKI* *and* F. MASUDA†

**Natural History Museum, Chiba, Chiba 260, Japan; and*
†Institute of Earth and Planetary Science, Osaka University, Toyonaka, Osaka 560, Japan

ABSTRACT

The Kioroshi and Joso Formations, deposited in the late Pleistocene Palaeo-Tokyo Bay, crop out in southern Ibaraki and northern Chiba Prefectures, eastern Kanto, central Japan. They form a depositional sequence which is primarily controlled by glacio-eustatic sea-level changes (*c.* 150–60 kaBP) and characterized by the barrier-island system of Palaeo-Tokyo Bay.

In Palaeo-Tokyo Bay, antecedent valleys were formed by rivers at the low sea-level stage of a glacial period. The subaerial unconformity cut at that time comprises a sequence boundary. During the early transgression associated with the following interglacial period, the drowned valleys were filled with estuarine and fluvial deposits. The transgressive surface relates to an abrupt increase in accommodation, its ravinement surface caps the overlying drowned-valley fill. Subsequently, a transgressive barrier-island system consisting of beach-shoreface, tidal-delta and tidal-inlet systems migrated landwards to form a transgressive systems tract. As the rate of sea-level rise slowed, the retreat of barrier islands ceased and was followed by emergence. During the subsequent sea-level lowering, the coastal plain prograded seawards and formed a highstand systems tract.

The depositional sequence in Palaeo-Tokyo Bay thus documents a transgressive and regressive barrier system. Landward migration of the barrier system in the wake of the rising sea-level resulted in the development of flood-tidal deltas on the lagoonal side of the barrier system and a ravinement surface on the open-sea side. Accordingly, the flood-tidal delta and inlet-fill, which have a high preservation potential, are important as a transgressive system tract of the tidally influenced Palaeo-Tokyo Bay.

INTRODUCTION

The late Pleistocene Kioroshi and Joso Formations form a clearly defined depositional sequence in terms of facies analysis related to sequence stratigraphy (Vail *et al.*, 1977; van Wagoner *et al.*, 1988).

The Kioroshi and Joso Formations were deposited in Palaeo-Tokyo Bay (Narita Research Group, 1962) during the Shimosueyoshi transgression (*c.* 150–120 kaBP) and the subsequent regression (*c.* 120–60 kaBP). Depositional systems that can be recognized include those associated with meandering-river, birdfoot-delta, beach-shoreface, tidal-delta and inlet, and drowned-valley-fill environments (Okazaki & Masuda, 1992). The depositional systems represent a sequence consisting of lowstand, transgressive and highstand systems tracts. We describe the features of each systems tract, based on the recognition of the architecture of individual depositional systems and their bounding surfaces. Correlation between the depositional sequences and glacio-eustatic sea-level changes was made on the basis of interbedded tephras (Machida & Suzuki, 1971; Sugihara *et al.*, 1978) with the help of ESR dating (Nakazato *et al.*, 1993) and identification of the geomorphic surfaces (Sugihara, 1970; Ikeda *et al.*, 1982).

GEOLOGICAL SETTING

The Kioroshi (*c.* 10–30 m thick) and the overlying Joso Formation (*c.* 3–10 m) underlie the Joso Upland (including the Hitachi and Shimosa Uplands), a raised late Quaternary plain north-east of Tokyo (Fig. 1). The upland lies between about 20 m to more than 100 m in altitude and is divided into two geomorphic surfaces, these being the Shimosueyoshi and Musashino Surfaces, in descending order (Sugihara, 1970; Ikeda *et al.*, 1982). The Kioroshi Formation corresponds to, and underlies the upper surface, i.e. the Shimosueyoshi Surface. The Joso Formation corresponds to the lower surface, i.e. the Musashino Surface.

The Kioroshi Formation rests unconformably on the underlying Formation (Shimosa-Daichi Research Group, 1984) and crops out widely over the study area. The Kioroshi Formation can be divided into four facies assemblages, denoted by the letters A, B, C and D, in ascending order. However, the thickness of each facies varies markedly from place to place.

The Joso Formation (Kodama *et al.*, 1981) is divided into three facies assemblages (or facies), namely E, F and G. The Joso Formation mainly occurs in the Hitachi Upland.

The depositional systems and depositional sequence of the Kioroshi and Joso Formations were mapped by facies analysis of more than six hundred outcrops. Continuous tephra layers interbedded in the formations are useful as marker beds (Okazaki *et al.*, 1994).

DEPOSITIONAL SYSTEMS

Facies assemblage A: a drowned-valley-fill system

Description

Facies assemblage A consists of facies A(a) and A(b) (Fig. 2a). Facies A(a) is 2–4 m thick and comprises massive silts to very fine sand, the grain size gradually fining upwards. Bioturbation is intense and *in situ* shells of *Dosinia japonica, Macoma tokyoensis, Lucinoma annulata* and *Crassostrea gigas* are present. Facies A(b) comprises alternations of very fine to fine sand and mud layers, exhibiting lenticular and flaser bedding (Fig. 3a). Plant debris and intense burrowing occur. A channel-shaped erosional surface defines the base of these facies in some places. It contains basal gravel and mud clasts.

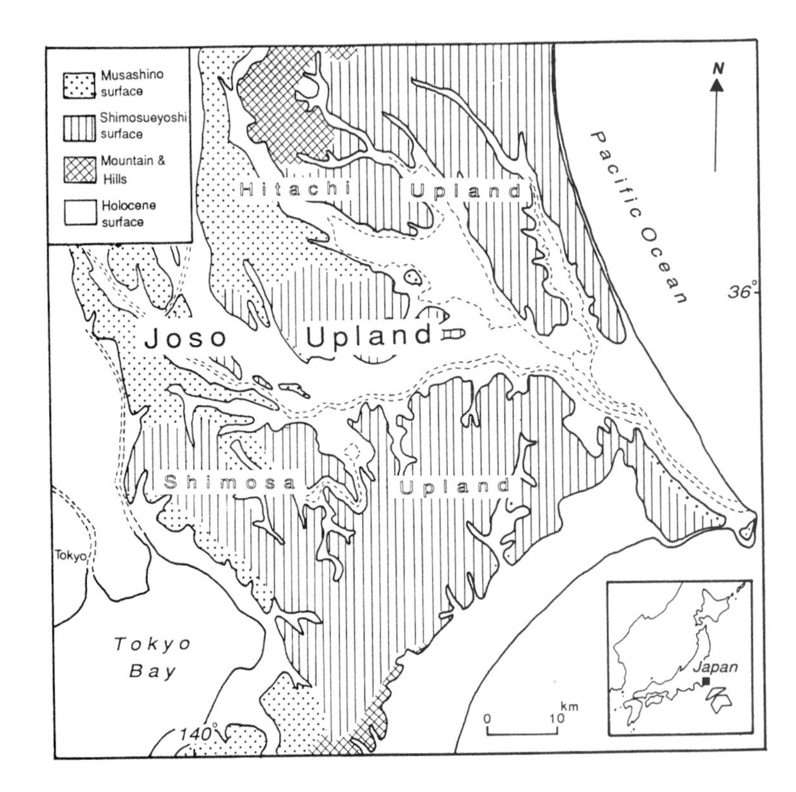

Fig. 1. Index map of the study area, Joso Upland in eastern Kanto, central Japan (division of geomorphic surfaces after Kaizuka & Matsuda, 1982).

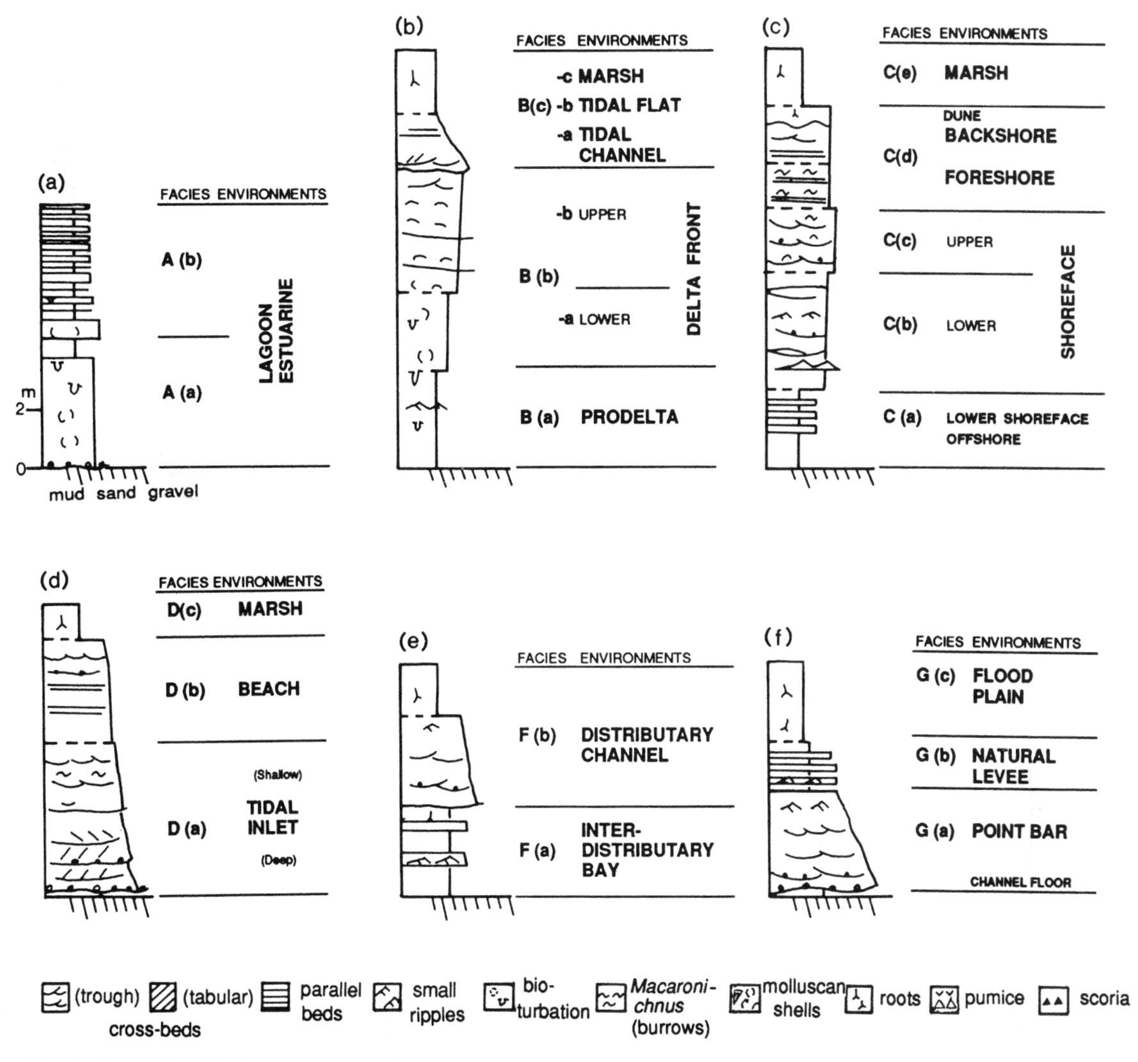

Fig. 2. Generalized facies successions of depositional systems in Joso Upland. (a) A drowned-valley-fill system in the Kioroshi Formation. (b) A tidal-delta system in the Kioroshi Formation. (c) A beach-shoreface system in the Kioroshi Formation. (d) A tidal-inlet-fill system in the Kioroshi Formation. (e) A birdfoot-delta system in the Joso Formation. (f) A meandering-river system in the Joso Formation.

Interpretation

Facies A(a) was probably deposited in a calm bay. Ecological data for molluscs indicate an intertidal to meso-neritic zone. Facies A(b) represents a tidally deposited sand-mud sequence of the type described by Reineck & Wunderlich (1968). These deposits reflect an estuarine or lagoonal environment and, since they fill a valley, are interpreted as drowned-valley-fill deposits.

Facies assemblage B: a tidal-delta system

Description

Facies assemblage B is made up of three facies B(a), B(b) and B(c), in ascending order (Fig. 2b). Facies B(a) consists of mud with thin sheet-sand layers, showing current ripple lamination in lenticular beds and intense bioturbation.

Facies B(b) is divided into two subfacies: B(b)-a

and B(b)-b. The lower subfacies B(b)-a is 3–6 m thick, comprising well-sorted fine to very fine sand with burrows and *in situ* molluscan shells (e.g. *Fluvia mutica, Raeta yokohamensis, Fabulina nitidula, Siliqua pulchella* and *Solen krusensterni*). Parallel lamination and trough cross-stratification are present. The upper subfacies B(b)-b is also 3–6 m thick, but medium grained with very abundant molluscan shells (e.g. *Mactra sulcataria, Tapes variegata, Gomphina neastartoides* and *Glycymeris vestita*). It shows large-scale low-angle, tabular cross-stratification (Fig. 3b). Burrows of *Ophiomorpha* are abundant.

Facies B(c) is divided into three subfacies: B(c)-a, B(c)-b and B(c)-c. Subfacies B(c)-a is 2–3 m thick and consists of fine to medium sand. Trough cross-stratification with angular foresets, mud drapes and herringbone cross-stratification are present (Fig. 3c). Erosive channel-like scours occur in the basal part. Subfacies B(c)-b is 1–2 m thick, comprising fine to medium sand with silt layers. Wave ripples and parallel lamination, convolute bedding and dish structures are present. Burrows and casts of *in situ* shells are common (Fig. 3d). Subfacies B(c)-c is 1–4 m thick and consists of mud with peaty layers and rootlets.

Interpretation

The depositional environment of facies assemblage B has been interpreted using a tidal-delta model (Okazaki & Masuda, 1989a). The mud of facies B(a) indicates a low-energy prodelta environment, where thin sand layers were introduced by storm-generated currents. Facies B(b) corresponds to the delta-front deposits. The large-scale low-angle, tabular cross-stratification indicates the original dip of the deltaic slope. Facies B(c) was formed on the delta plain. The sandy deposits of subfacies B(c)-a and -b were formed by the interaction of waves and tides and represent channel and tidal-flat deposits. Subfacies B(c)-c corresponds to a marsh.

Facies assemblage C: a beach-shoreface system

Description

Facies assemblage C consists of five facies: C(a), C(b), C(c), C(d) and C(e) (Fig. 2c). Facies C(a) is 3–5 m thick and comprises alternations of fine sand and mud. An erosional surface defines the base of the facies. Granules, mud pebbles and molluscan shell fragments occur immediately above the surface. Wave ripples, parallel lamination, and hummocky cross-stratification are present in the sand (Fig. 3e). The trace fossil of *Cylindrichnus* sp. is contained in the mud.

Facies C(b) is 3–5 m thick and comprises very fine to fine sand with minor intercalations of thin silt layers. Low-angle, wedge-shaped cross-stratification and amalgamated hummocky and swaley cross-stratification are the dominant structures (Fig. 3f). Facies C(c) is 1–2 m thick and consists of well-sorted medium to coarse sand. Molluscan remains and mud pebbles are present sporadically. Shallow trough cross-stratification and tabular cross-stratification are the dominant structures. Trace fossils of *Macaronichnus* sp. and *Ophiomorpha* sp. are present.

Facies C(d) is 1–4 m thick and consists of well-sorted fine to medium sand. Shallow dipping parallel lamination is defined by heavy minerals. (Fig. 4a). The trace fossil *Macaronichnus* sp. is prominent. The upper part of the facies contains more heavy minerals (limonite), the sedimentary structures being dominated by small tubular sets and trough cross-laminations. The uppermost part contains plant rootlets and small burrows, being composed of very well sorted fine to medium sand with dish and convolute structures. Facies C(e) is 1–2 m thick and consists of poorly-sorted, massive sandy silt and mud, into which very fine sand layers are intercalated. Plant debris are abundant.

Interpretation

The upward coarsening succession of assemblage C is interpreted as a beach-shoreface system. Swaley and hummocky cross-stratification are formed during storms (Hunter & Clifton, 1982; Nøttvedt & Kreisa, 1987). The interbedded mud and sand in facies C(a) are interpreted to reflect alternating fairweather and storm deposits in the offshore to lower shoreface environments. Facies C(b) corresponds to the lower shoreface under fairweather control. The cross-beds in facies C(c) were formed by migration of asymmetrical ripples and dunes on the upper shoreface (Clifton *et al.*, 1971). The distinct subparallel to planar lamination of facies C(d) is characteristic of beach deposits (Reineck & Singh, 1980). The seaward inclined planar laminations were produced by swash and backwash processes, being roughly aligned parallel to the foreshore surface (Masuda & Okazaki, 1983a;

Fig. 3. (a) Lenticular bedding composed of alternations of fine to very fine sand and mud in the tidal-flat deposits (Loc. 10). The scale is 3 cm long. (b) Large-scale, low-angle tabular cross-stratification (arrowed) of delta-front deposits, dipping gently to the left in the middle part of the cliff (Loc. 8). The cliff is about 17 m high. (c) Trough cross-stratification and mud drapes of tidal-channel deposits. Burrows can be seen in the lower half of the photo. The trowel is 30 cm long. (d) Wave ripples and parallel lamination with bioturbation in tidal-flat deposits. The trowel is 30 cm long. (e) Shoreface deposits. The megaripple is composed of coarse sand and mud. The scale is 10 cm long. (f) Low-angle, wedge-shaped cross-stratification of shoreface deposits. The scale is 15 cm long.

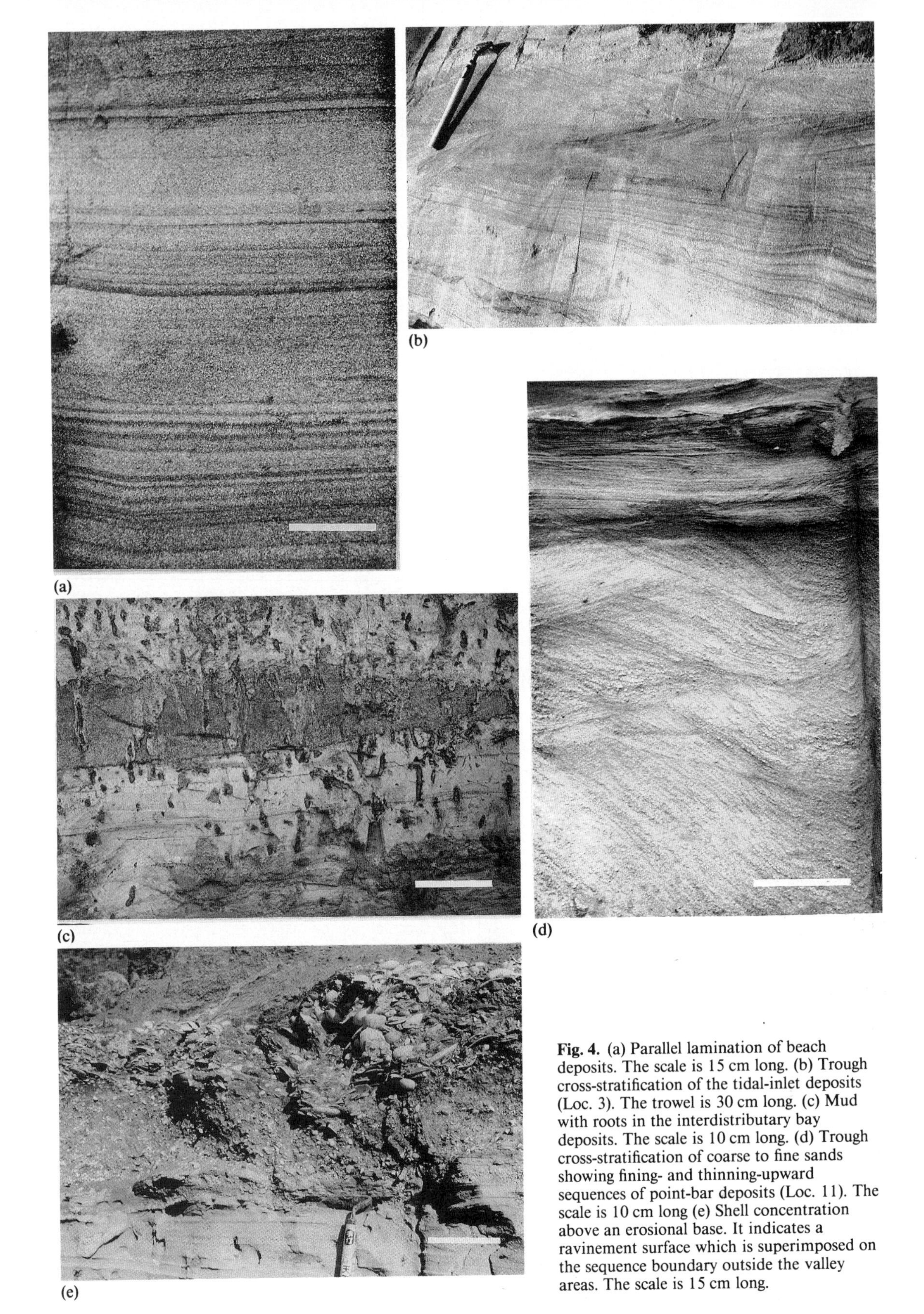

(a) (b) (c) (d) (e)

Fig. 4. (a) Parallel lamination of beach deposits. The scale is 15 cm long. (b) Trough cross-stratification of the tidal-inlet deposits (Loc. 3). The trowel is 30 cm long. (c) Mud with roots in the interdistributary bay deposits. The scale is 10 cm long. (d) Trough cross-stratification of coarse to fine sands showing fining- and thinning-upward sequences of point-bar deposits (Loc. 11). The scale is 10 cm long (e) Shell concentration above an erosional base. It indicates a ravinement surface which is superimposed on the sequence boundary outside the valley areas. The scale is 15 cm long.

Masuda & Yokokawa, 1988). Small- to medium-scale trough cross-stratified units may have been formed in backshore runnels, whereas the well-sorted sands may be of aeolian origin. Facies C(e) probably formed in a backshore salt marsh environment.

Facies assemblage D: a tidal-inlet-fill system

Description

Facies assemblage D is made up of three facies: D(a), D(b) and D(c) (Fig. 2d). Facies D(a) consists of medium to coarse sands which are mostly trough cross-stratified. An erosional base is present. Large-scale trough cross-stratified units, indicating bi-directional currents, are present in the lower part (Fig. 4b). The upper part of the facies is dominated by medium-scale cross-bedding and the trace fossil *Macaronichnus* sp.

Facies D(b) is characterized by low-angle, wedge-shaped cross-stratified medium to fine sands. The upper part of the facies consists of ill-sorted medium to fine sand with small-scale trough cross-stratification. Facies D(c) consists of massive muds containing rootlets and plant debris.

Interpretation

Bidirectional cross-beds, a fining-upward textural trend and an upward thinning of cross-bed set thickness indicate that facies D(a) is a channel-fill sequence resulting from tidal-inlet migration. Gently dipping parallel stratification of facies D(b) reflects beach deposits abruptly capping the channel-fill. Facies D(c) is a marsh deposit.

Facies E: an alluvial fan

Facies E overlies the Kioroshi Formation, being separated from the latter by an erosional surface. The facies comprises medium to very coarse cross-stratified and gravel-bearing tuffaceous sands. The depositional environment of facies E is not clear because of its restricted distribution. It is thought to reflect an alluvial fan environment.

Facies assemblage F: a birdfoot-delta system

Description

Facies assemblage F is made up of two facies: F(a) and F(b) (Fig. 2e). Facies F(a) is 0.5–1.5 m thick and consits of muddy sand with roots and plant debris (Fig. 4c). Thin beds of very fine sand containing sand-streaked laminated mud are common. Facies F(b) comprises 3–5 m thick medium sands dominated by cross-stratification and channel-shaped erosional bases.

Interpretation

The depositional environment of this facies is inferred from both facies and topographic analyses. A birdfoot-delta-plain landform has been identified on the surface of the Tsukuba Upland (Ikeda *et al.*, 1982). The depositional environment of the sandy facies F(b) corresponds to a distributary channel, whereas the muddy facies F(a) was formed in an interdistributary channel on a floodplain and/or in a distributary bay environment (Okazaki & Masuda, 1989a).

Facies assemblage G: a meandering-river system

Description

Facies assemblage G consists of three facies: G(a), G(b) and G(c) (Fig. 2f). Facies G(a) is 3–5 m thick and comprises fine to very coarse sands arranged in a fining-upward sequence. Trough cross-stratification and epsilon cross-stratification are the predominating sedimentary structures. The scale of the cross-stratification decreases upwards (Fig. 4d). Channel-shaped erosional bases are observed, coarse sand and gravel being present above the erosional surfaces. Facies G(b) consists of 1–2 m thick fine sand with thin mud layers. Inverse grading, climbing-ripple lamination, convolute structures and rootlets are locally present. Facies G(c) is present above or lateral to facies G(a) and G(b), whereby the vertical or lateral transition is gradual. The facies is 3 m thick and consists of mud containing abundant plant fragments and roots.

Interpretation

The fining- and thinning-upward cross-beds of facies G(a) are characteristic features of point bars in a meandering river (Allen, 1964). Coarse sediments in the basal part are the result of lag deposits on the channel floor. The alternating sands and muds in facies G(b) are considered to reflect natural levee deposits. Inverse grading is a distinctive indicator of flood deposits (Masuda & Iseya, 1985). The plant-containing muds of facies G(c) are interpreted to represent floodplain deposits.

DEPOSITIONAL ARCHITECTURE OF PALAEO-TOKYO BAY

'Barrier-island system' of Palaeo-Tokyo Bay

The tidal-delta (facies assemblage B), tidal-inlet-fill (facies assemblage D), beach-shoreface (facies assemblage C) and lagoonal deposits (facies assemblage A) recognized in the Kioroshi Formation together define a 'barrier-island system'. The palaeogeography of the barrier-island system has been constructed from the facies distribution and palaeocurrents of Palaeo-Tokyo Bay (Okazaki & Masuda, 1989b; Okazaki & Masuda, 1992) (Fig. 5).

The beach-shoreface depositional systems are distributed in the areas surrounding Palaeo-Tokyo Bay. The position of the open sea, situated towards both the east and the west, is inferred on the basis of the beach deposits in the eastern parts of the Joso Upland. It is thought that emergence was initiated near the centre of these areas, forming geomorphic highs at the time of the Shimosueyoshi period (Masuda & Nakazato, 1988). These geomorphic highs were probably formed by barrier-island chains

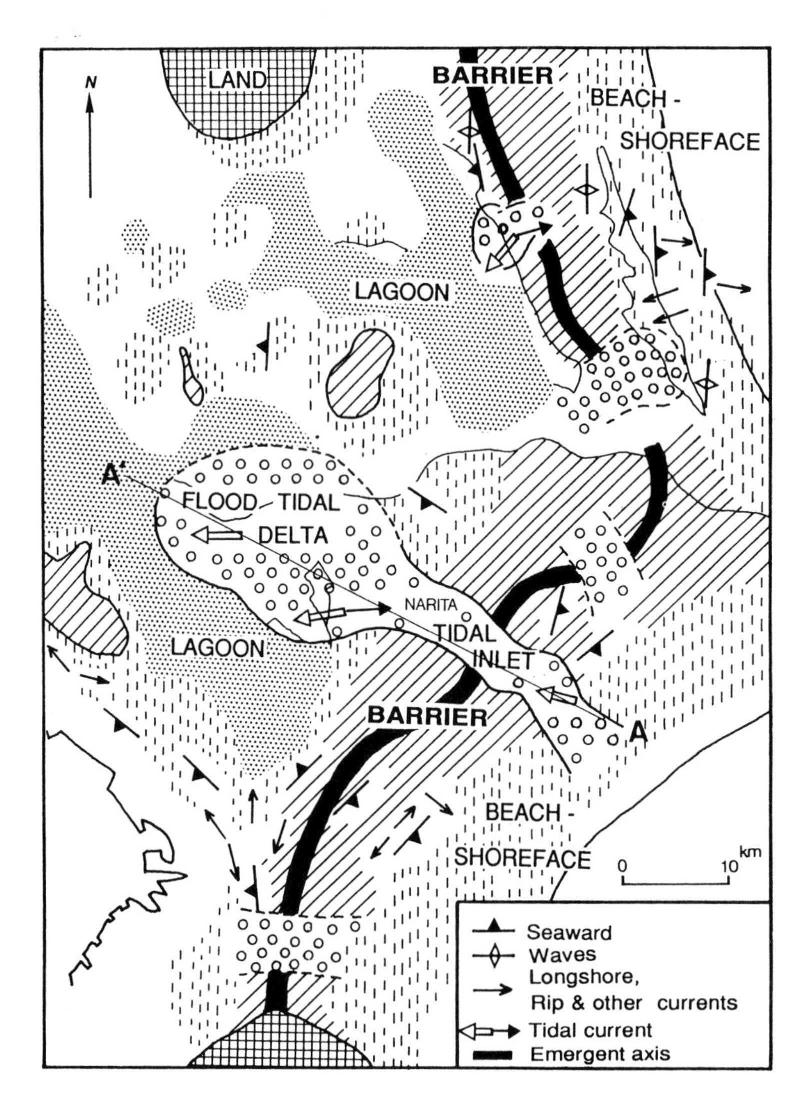

Fig. 5. Reconstruction of the depositional systems and palaeocurrents in the 'barrier-island system' of Palaeo-Tokyo Bay during the Shimosueyoshi sea-level highstand of 130–110 kaBP.

that separated the open sea from an enclosed lagoon. The lagoon was connected to the open sea by tidal inlets. Foreset dips of the cross-bedded sets forming the tidal-inlet fills show bipolar distributions dominated by a westward component. Dip directions are normal to the trend of the barrier islands.

The tidal-delta system developed in the centre of Palaeo-Tokyo Bay. The system contains thick fossil beds of molluscan shells. The geometry is lobate and narrow. The dominant palaeocurrent directions face towards the west in both the delta-front and delta-plain facies. The delta developed at the distal part of a tidal inlet in response to the flood-tidal current that passed through the inlet.

Beneath the flood-tidal delta are large depressions (e.g. in the Shimosa Upland, extending from Yokaichiba to Nagareyama; Fig. 6). These features are interpreted to represent long and narrow river valleys at the base of the Kioroshi Formation. These are filled with drowned-valley-fill deposits that form the delta bottomsets. The drowned valley became shallower due to filling as sea-level rose. Barrier islands or barrier spits developed towards the open-sea margin, while a shallow bay formed on the landward side, the valley mouth acting as a tidal inlet. A flood-tidal delta produced narrow and lobate sand bodies on the lagoonal side of the valley.

Transgressive flood-tidal delta

Figure 7 shows the cross-section indicated in Figs 5 & 6. It corresponds to the valley at the base of the Kioroshi Formation. Flood-tidal delta deposits are developed in this section. Drowned-valley-fill, flood-tidal delta, beach-shore face, birdfoot-delta and meandering-river depositional systems can be observed in ascending order within the cross-section.

The drowned-valley-fill deposits, which are 2–16 m thick, are found in the lowermost part of the area. They are separated from the underlying formation by a distinct erosional surface. The depositional environment of the drowned-valley-fill facies varies from place to place along the valley. The seaward parts to the east consist of sandy fluvial and tidal-channel, muddy flood-plain and tidal-flat deposits (Okazaki, 1991). The landward parts to the west consist of thick muddy flood-plain, tidal-flat, lagoon and open-bay deposits.

The flood-tidal delta deposits are separated from the underlying drowned-valley-fill sequences by a sharp boundary. The delta foresets attain thicknesses of 10–15 m. They are divided into two subfacies, mainly B(b)-a and B(b)-b. The dominant palaeocurrent direction in the delta front is towards

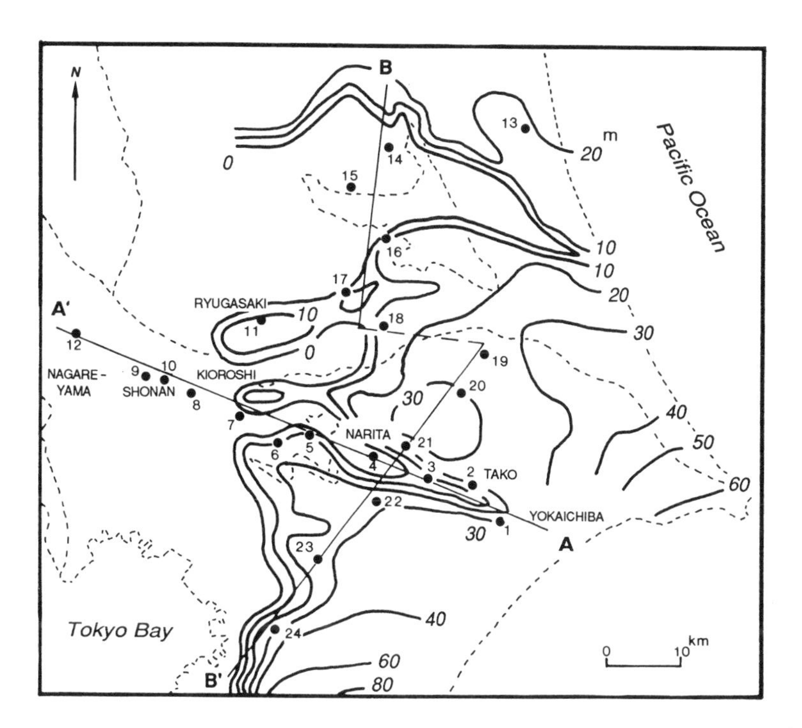

Fig. 6. Contour map of the base of the Kioroshi Formation after Kikuchi (1980). A–A′ and B–B′ show the cross-sections in Figs 7 & 8 (•, location of outcrop of Figs 7 & 8).

the west. Masuda & Okazaki (1983b) reported that the scoria marker (BT) described by Sugihara (1979) occurs in various units. In the eastern seaward part, the tephra are intercalated in the upper delta front of subfacies B(b)-b (Locs 4 and 5 in Fig. 7). In the central area it lies in the upper part of the lower delta front of subfacies B(b)-a (Loc. 6). On the western landward side, it occurs in the lower part of the lower subfacies B(b)-a (Loc. 8) and the prodelta section of facies B(a). This distribution is consistent with a landward migration of the flood-tidal delta. Another tephra marker (TK) was discovered in the eastern Tako area (Loc. 2), but it could not be traced towards the west (Sato, 1993). This suggests that the Tako tidal delta was deposited in a smaller bay at an earlier transgressive stage.

In general, barrier islands and their associated tidal deltas develop during transgressions (Moslow & Heron, 1979; Swift *et al.*, 1991). The transgressive flood-tidal delta of Palaeo-Tokyo Bay was hence most probably formed in the course of barrier-island retreat.

The beach-shoreface system is observed from Narita to Yokaichiba (Locs 1, 2 and 4), where it overlies the flood-tidal delta system. This deposit is 5–10 m thick, becoming thicker eastwards. The base of the deposit rests on a flat or erosional surface, covered by marine shell fragments and coarse-grained sand (Fig. 4e). It is interpreted to represent a ravinement surface. Sediments eroded from the shoreface by storm currents are commonly redeposited further offshore on the ravinement surface (Swift, 1968). The landward shift of the barrier is characterized by shoreface erosion and the deposition of a flood-tidal delta on the lagoon side.

The tidal-inlet fill overlies the delta system at Narita (Loc. 3), where it is preserved at the landward extremes of the barrier-island system.

The opposing dip directions of the beachface deposits between Narita and Yokaichiba suggest that emergence occurred near the centre of both areas. The beach-shoreface deposits in the eastern parts of the area were formed by seaward growth of the barrier island during sea-level fall.

A birdfoot-delta system is present in the Shonan area (Locs 9 and 10). It is separated from the underlying formations by an erosional surface. The 'Kmp' (*c.* 100 kaBP) and 'Pm-1' (*c.* 80 kaBP) tephras are intercalated in the interdistributary-bay mud of the birdfoot delta. The meandering-river system, in

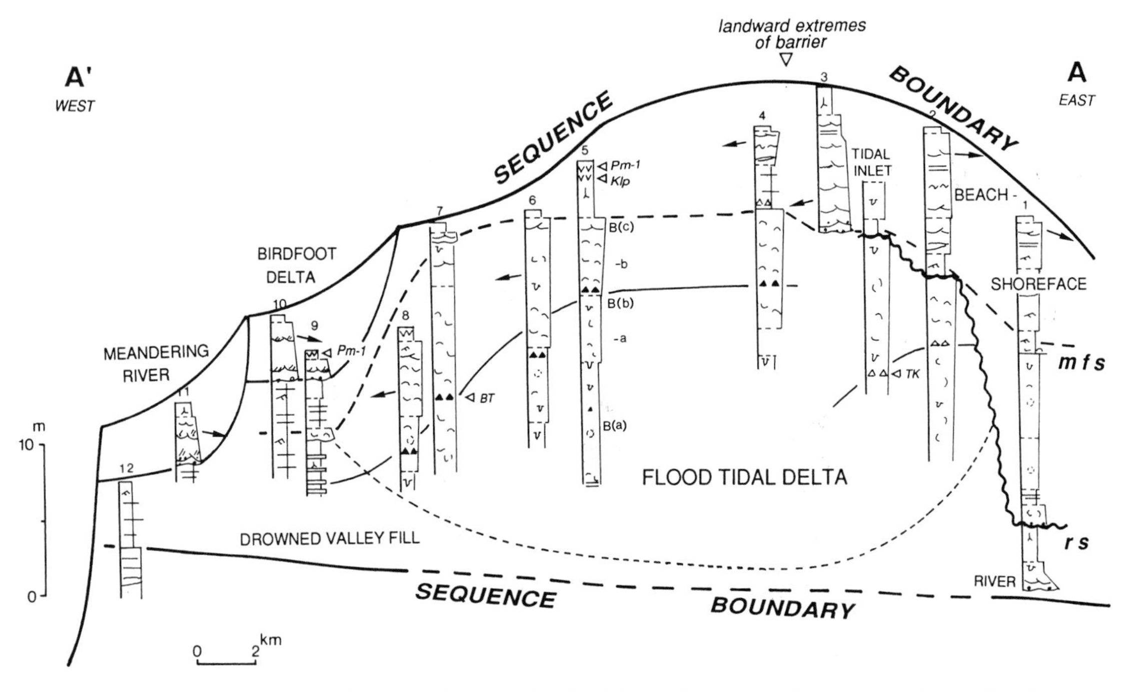

Fig. 7. Valley-parallel east–west section. A–A′ is shown in Fig. 6 (rs, ravinement surface; mfs, maximum flooding surface; other notations: see Fig. 2).

turn, progrades into Ryugasaki (Loc. 11), where it erodes the birdfoot-delta deposits.

DEPOSITIONAL SEQUENCE OF PALAEO-TOKYO BAY

Sequence boundary

Significant erosional relicts are evident at the base of the Kioroshi Formation (Fig. 6). One or two subordinate formations have been removed. This unconformity is interpreted to have been formed by subaerial erosion following a fall in sea-level. This erosional surface is therefore recognized as a sequence boundary (Fig. 8).

Lowstand systems tract

The valley-fill deposits consist of a vertical sequence from fluvial through estuarine-mouth facies, to open-bay muds or oyster reefs. This valley fill was deposited during the late eustatic fall or early eustatic rise. The basal fluvial deposits may correspond to a lowstand systems tract. The lowstand depositional systems are formed and preserved in the deepest entrenched valleys. The top of the facies is bounded by a sharp surface. Above this surface, i.e. at the basal part of the beach-shoreface deposit, there are coarse deposits containing gravels, mud clasts and abundant molluscan shells. This is a first marine flooding surface. The time of the transgression is about 140 kaBP according to ESR dating of the shells on the surface (Nakazato *et al.*, 1993).

Transgressive systems tract

As sea-level rose, the landward shift of the barrier system resulted in the development of a flood-tidal delta on the lagoonal side and a ravinement surface on the seaward side. The transgressive systems tract in Palaeo-Tokyo Bay is hence characterized by a flood-tidal delta and inlet sequence. The ESR age of the tidal delta is 140–120 kaBP (Nakazato *et al.*, 1993).

The ravinement surface can be traced across the margins of the valleys, where it forms a wave-cut terrace in Palaeo-Tokyo Bay (Fig. 9). Accordingly, the surface represents a transgressive event causing an abrupt increase in accommodation following the flooding of the interfluvial areas. Where the ravinement surface truncates previously subaerial interfluves or other coastal headlands, the ravinement surface becomes superimposed on the sequence boundary.

Highstand systems tract

As the rate of sea-level rise slowed, the barrier islands probably stabilized in the area of Narita (Fig. 7), thus constituting an emergent axis. As emergence proceeded, the coastal plain prograded (Murakoshi &

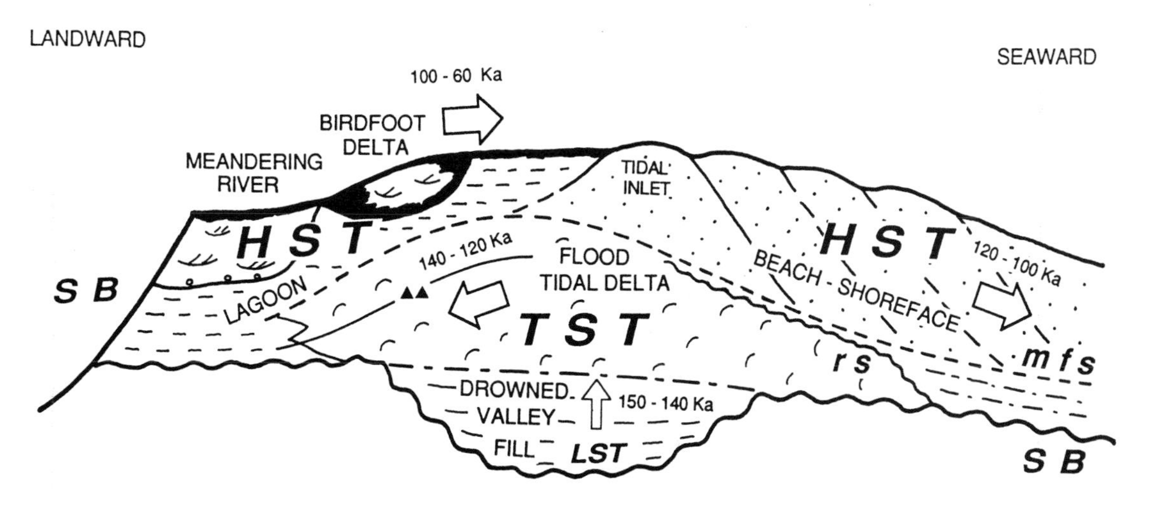

Fig. 8. Schematic depositional cross-section of Palaeo-Tokyo Bay showing the sequence stratigraphical relationships. LST, lowstand system tract; TST, transgressive systems tract; HST, highstand systems tract; SB, sequence boundary.

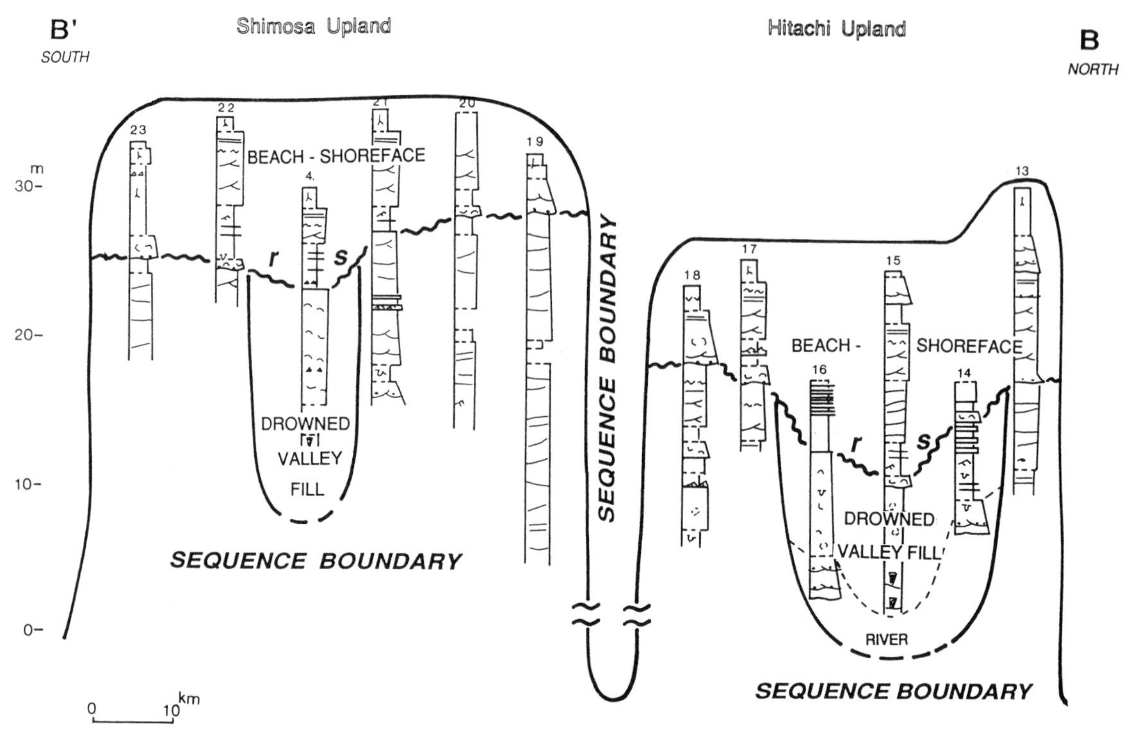

Fig. 9. Valley-normal south–north section. B–B′ is shown in Fig. 6 (notations: see Fig. 2).

Masuda, 1992). As a result, the beach-shoreface slope and tidal-delta plain correspond to the maximum flooding surface just before the emergence. Above the maximum flooding surface, the beach-shoreface deposit prograded towards the open sea, where it formed the highstand systems tract.

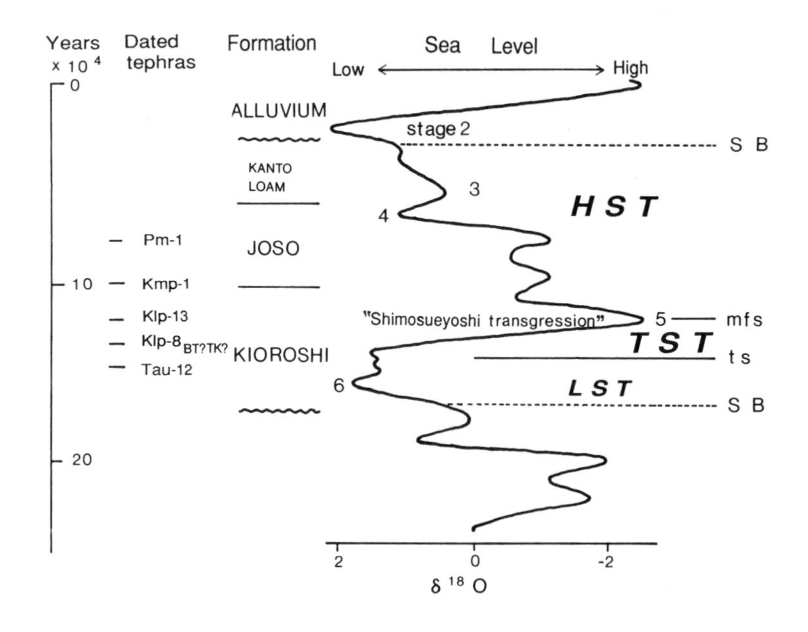

Fig. 10. Sequence stratigraphy of the Kioroshi and Joso Formations. The tephra markers are dated with the fission-track method of Machida & Suzuki (1971) and Sugihara *et al.* (1978). Oxygen isotope curve is after Imbrie *et al.*, (1984) and shows contemporary global sea-level changes (ts, transgressive surface).

Subaerial deposits, comprising the birdfoot-delta and meandering-river deposits, overlie the tidal delta and the beach-shoreface sequences with a distinct channel diastem. These units were deposited at the time of the subsequent fall in sea level, thus forming the late highstand systems tract. Furthermore, the Kioroshi and Joso Formations at the top of the sequence are bounded by the next younger sequence boundary formed later than 60 kaBP.

CONCLUSION

The evolution of the depositional sequence with its associated sea-level history is reconstructed as follows (Fig. 10): A sequence boundary is located at the base of the Kioroshi Formation in valleys that were formed by a river (150 kaBP). This is associated to the period of rapid eustatic fall and sea-level lowstand, corresponding to the glacial period at stage 6 of the oxygen isotope curve of Imbrie *et al.* (1984). During the subsequent transgression (Shimosueyoshi transgression) from stage 6 to stage 5, the valleys were drowned and filled with fluvial-estuarine deposits. The tops of the valleys represent a ravinement surface (140 kaBP) formed in the course of rapid transgression. The barrier-island system, comprising tidal-delta and inlet deposits, retreated to form a transgressive systems tract (140–120 kaBP). The subsequent episode of maximum flooding occurred at stage 5, followed by progradation of the coastal plain towards the open sea in a highstand systems tract. During the next sea-level lowering (100–60 kaBP), a birdfoot-delta and meandering-river systems prograded during the late highstand. At the top of the sequence, the Kioroshi and Joso Formations are bounded by the sequence boundary formed after 60 kaBP. The depositional sequence in Palaeo-Tokyo Bay corresponds to a transgressive and regressive barrier-island system. The flood-tidal delta and inlet-fill, which have a high preservation potential, are of particular importance for the recognition of the transgressive systems tract of the tidally influenced Palaeo-Tokyo Bay.

ACKNOWLEDGEMENTS

We are very grateful to Professor R.M. Carter (James Cook University of Queensland) for his critical review of and instructive comments to an earlier version of this manuscript.

REFERENCES

ALLEN, J.R.L. (1964) Studies in fluviatile sedimentation: six clyclothems from the Lower Old red Sandstone, Anglo-Welsh Basin. *Sedimentology* **3**, 163–198.

CLIFTON, H.E., HUNTER, R.E. & PHILLIPS, R.L. (1971) Depositional structures and processes in the non-barred high-energy nearshore. *J. sediment. Petrol.* **41**, 651–670.

HUNTER, R.E. & CLIFTON, H.E. (1982) Cyclic deposits and hummocky cross stratification of probable storm origin in Upper Cretaceous rocks of Cape Sebastian area, southwest Oregon. *J. sediment. Petrol.* **52,** 127–143.

IKEDA, H., MIZUTANI, K., SONODA, Y. & ISEYA, F. (1982) Geomorphic development of the Tsukuba Upland, Ibaraki Prefecture. *Tsukuba Environ. Stud.* **6**, 150–156.

IMBRIE, J., HAYS, J.D., MARTINSON, D.G. *et al.* (1984) The orbital theory of Pleistocene climate: support from a revised chronology of the marine O^{18} record. In: *Milankovitch and Climate Part 1* (Eds Berger, A.L., Imbrie, J., Hays, J.D., Kukla, G. & Saltsman, B.), pp. 269–305. D. Reidel, New York.

KAIZUKA, S. & MATSUDA, I. (1982) Active tectonics and geomorphic division of the Tokyo metropolitan area; scale 1:200,000. *Naigai Chizu,* 48 pp.

KIKUCHI, T. (1980) Palaeo-Tokyo Bay. *Urban Kubota* **18,** 16–21.

KODAMA, K., HORIGUCHI, M., SUZUKI, Y. & MITSUNASHI, T. (1981) Late Pleistocene crustal movement of the Kanto plain, central Japan. *J. Geol. Soc. Japan* **20**, 113–128.

MACHIDA, H. & SUZUKI, M. (1971) Chronology of the late Quaternary Period as established by Fission Track dating. *Kagaku (Science)* **41**, 263–270.

MASUDA, F. & ISEYA, F. (1985) 'Inverse grading'. A facies structure of flood deposits in meandering rivers. *J. Sediment. Soc. Japan* **22/23**, 108–116.

MASUDA, F. & NAKAZATO, H. (1988) Tectonic movement of Kashima-Boso uplift zone from a sedimentary facies point of view. *Gekkan Chikyu (Earth Mon.)* **112**, 616–623.

MASUDA, F. & OKAZAKI, H. (1983a) Directional structures observed in the Pleistocene strata of Tsukuba Upland. *Environ. Stud. Tsukuba* **7C**, 99–110.

MASUDA, F. & OKAZAKI, H. (1983b) Two types of prograding deltaic sequences developed in the late Pleistocene Paleo-Tokyo Bay. *Ann. Rept Inst. Geosci. Univ. Tsukuba* **9**, 55–60.

MASUDA, F. & YOKOKAWA, M. (1988) Ancient beach deposits. *Gekkan Chikyu (Earth Mon.)* **110**, 523–530.

MOSLOW, T.F. & HERON, S.D. JR. (1979) Quaternary evolution of Core Banks, North Carolina: Cape Lookout to New Drum Inlet. In: *Barrier Islands from the Gulf of St Lawrence to the Gulf of Mexico* (Ed. Leatherman, S.P.), pp. 211–236. Academic Press, New York.

MURAKOSHI, N. & MASUDA, F. (1992) Estuarine, barrier-island to strandplain sequence and related ravinement surface developed during the last interglacial in the Paleo-Tokyo Bay, Japan. In: *Quaternary Coastal Evolution* (Ed. Donoghue, J.F.). *Sediment Geol.* **80,** 167–184.

Nakazato, H., Shimokawa, K. & Iwai, N. (1993) ESR dating for the Pleistocene shell and value of annual dose. *Appl. Radiat. Isot.* **44**, 167–173.

Narita Research Group (1962) The Shimosueyoshi transgression and the Paleo-Tokyo Bay. *Chikyu-Kagaku (Earth Science)* **60/61,** 8–15.

Nøttvedt, A. & Kreisa, R.D. (1987) Model for the combined-flow origin of hummocky cross-stratification. *Geology* **15,** 357–361.

Okazaki, H. (1991) Tidal-influenced deposits in Kashima Upland. *J. Nat. Hist. Mus. Chiba* **2**, 15–23.

Okazaki, H. & Masuda, F. (1989a) Arcuate and birdfoot deltas in the late Pleistocene Paleo-Tokyo Bay. In: *Deltas: Sites and Traps for Fossil Fuels* (Eds Whateley, M.K.G. & Pickering, K.T.). *Geol. Soc. London Spec. Publs* **41,** 129–138.

Okazaki, H. & Masuda, F. (1989b) Paleocurrents in the late Pleistocene Paleo-Tokyo Bay. *J. Sediment. Soc. Japan* **31,** 25–32.

Okazaki, K. & Masuda, F. (1992) Depositional systems of the Late Pleistocene sediments in Paleo-Tokyo Bay area. *J. Geol. Soc. Japan* **98**, 235–258.

Okazaki, H., Sato, K. & Nakazato, K. (1994) Stratigraphy of the Kioroshi and Joso Formations. *J. Nat. Hist. Mus. Chiba* **3**, 16–69.

Reineck, H.-E. & Singh, I.B. (1980) *Depositional Sedimentary Environments.* Springer-Verlag, New York, 549 pp.

Reineck, H.-E. & Wunderlich, F. (1968) Classification and origin of flaser and lenticular bedding. *Sedimentology* **11**, 99–104.

Sato, H. (1993) Stratigraphy of the middle-upper Pleistocene Shimosa Group distributed in the area from Naruto town, Sanbu-gun to Yaka-ichiba city, Chiba Prefecture, Central Japan. *J. Nat. Hist. Mus. Chiba* **2**, 99–113.

Shimosa-Daichi Research Group (1984) The relation between the shape of the base of the Kroroshi Formation and its facies, around Tega-numa, Chiba Prefecture, Central Japan. *Chikyu-Kagaku (Earth Science)* **38,** 226–234.

Sugihara, S. (1970) Geomorphological development of the western Shimosa Upland in Chiba Prefecture, Japan. *Geogr. Rev. Japan* **43,** 703–718.

Sugihara, S. (1979) Stratigraphy and bedrock topography on Narita Formation at Shimosa Upland, Kanto Plain. *Inst. Cult. Sci. Meiji Univ. Mem.* **18**, 1–41.

Sugihara, S., Arai, F. & Machida, H. (1978) Tephrochronology of the Middle to Late Pleistocene sediments in the northern part of the Boso Peninsula, Central Japan. *J. Geol. Soc. Japan* **84,** 583–600.

Swift, D.J.P. (1968) Coastal erosion and transgressive stratigraphy. *J. Geol.* **76**, 444–456.

Swift, D.J.P., Phillips, S. & Thorne, J.A. (1991) Sedimentation on continental margins, IV: lithofacies and depositional systems. In: *Shelf Sand and Sandstone Bodies* (Eds Swift, D.J.P., Oertel, G.F., Tillman, R.W. & Thorne, J.A.). *Spec. Publs int. Ass. Sediment.* **14,** 89–152.

Vail, P.R., Mitchum, R.M.Jr., Todd, R.G. *et al.* (1977) Seismic stratigraphy and global changes in sea level. In: *Seismic Stratigraphy—Applications to Hydrocarbon Exploration* (Ed. Payteon, C.E.). *A.A.P.G. Mem.* **26,** 49–205.

Wagoner, J.C. van, Posamentier, H.W., Mitchum, R.M. *et al.* (1988) An overview of the fundamentals of sequence stratigraphy and key definitions. *S.E.P.M. Spec. Publs* **42**, 39–46.

Spec. Publs int. Ass. Sediment. (1995) **24**, 289–300

Diurnal inequality pattern of the tide in the upper Pleistocene Palaeo-Tokyo Bay: reconstruction from tidal deposits and growth-lines of fossil bivalves

N. MURAKOSHI*§, N. NAKAYAMA†¶ *and* F. MASUDA‡**

**Doctoral Program in Geoscience, University of Tsukuba, Ibaraki 305, Japan;*
†Geothermal Department, Bishimetal Exploration Co., Tokyo 101, Japan; and
‡Institute of Earth and Planetary Science, College of General Education, Osaka University, Toyonaka, Osaka 560, Japan

ABSTRACT

Tide-influenced estuarine sediments have been recognized within the upper Pleistocene Kioroshi Formation of central Japan. The tidal regime and diurnal inequality pattern in an embayment of Palaeo-Tokyo Bay were reconstructed from intertidal and subtidal sandwave deposits and also from the growth-line patterns of fossil bivalve shells which lived in the estuary.

Mud couplets, regular thick–thin alternations and sinusoidal thickness variations with a cyclicity of about 27 bundles were recognized in the tidal bundle sequence. From the analyses of these tide-induced sedimentary structures, semi-diurnal or mixed tidal patterns were reconstructed. Similar analyses were made on bivalve shells. The internal growth lines and growth increments of the shells of *Potamocorbula amurensis* Schrenck show single- and double-spaced growth increments, thickness alternations of growth increments, and thickness inversion of growth lines with a period of about 28 increments. From these analyses, semi-diurnal or mixed palaeotidal patterns with diurnal inequality in the ranges of both low tides and high tides were inferred.

INTRODUCTION

Tidal deposits have recently been identified in the upper Pleistocene Kioroshi Formation of Japan (Masuda & Nakayama, 1988; Masuda *et al.*, 1988; Makino *et al.*, 1989), based on sedimentary structures such as herringbone cross-bedding, reactivation surfaces and mud-draped foresets (see Nio & Yang, 1989). Allen & Homewood (1984), Yang & Nio (1985) and Masuda *et al.* (1988), amongst others, have used such structures to reconstruct the palaeotidal regime of ancient tide-influenced sedimentary deposits.

By the same token, studies on the formative mechanisms of growth lines in shells (e.g. House & Farrow, 1968; Evans, 1972; Richardson *et al.*, 1981) have shown that tidal regimes can also be estimated from the growth patterns observed in the shells of intertidal bivalves. Besides reconstructing palaeotidal regimes, growth lines in recent and fossil shells have been used to estimate environmental conditions such as water temperature and seasonality (e.g. Pannella, 1976; Ohno, 1989). We have applied this palaeontological tool to the study of fossil shells from Palaeo-Tokyo Bay in order to obtain a detailed picture of the palaeotidal pattern and to compare the results with similar estimations derived from the analysis of sedimentary structures of tidal deposits in the same strata.

Correspondence addresses: §Department of Environment Science, Faculty of Science, Shinsyu University, Matsumoto 390, Japan.
¶Geothermal Department, Bishimetal Exploration Co., Tokyo 108, Japan.
**Department of Earth and Space Science, Faculty of Science, Osaka Unversity, Toyonaka, Osaka 560, Japan.

GEOLOGICAL SETTING

Coastal and shallow-marine deposits of the last interglacial period are widely distributed over the

Kanto Plain of central Japan. During interglacial sea-level highstands, which reached a maximum at 125 kaBP, the modern Kanto Plain was occupied by a shallow estuarine embayment, today known as the Palaeo-Tokyo Bay (Narita Research Group, 1962). The deposits in this former tidal bay are included in the Kioroshi Formation, representing the upper part of the Pleistocene Shimosa Group. These deposits and those of the underlying group represent the infill of a forearc basin which existed in this region in Plio-Pleistocene times.

The depositional sequence of the Kioroshi Formation consists of coastal and shallow-marine sediments that reflect the transition from estuarine, through barrier-island to strandplain depositional environments (Murakoshi & Masuda, 1992). The estuarine to barrier island deposits were laid down during the transgressive interval between the glacial at 150 kaBP (oxygen isotope stage 6) and the interglacial at 125 kaBP (isotope stage 5), whereas the strandplain system developed subsequently during the regressive interval between the last interglacial and the last glacial around 80 kaBP (isotope stage 4).

The estuarine valley was incised by fluvial processes during the sea-level lowstand of stage 6. It was subsequently infilled by tidal dune sands, lagoonal and marsh muds, as well as fluvial sediments at the base. The valley is inferred to have extended NW–SE in this area (Fig. 1), the bay being situated in the north-west and the Pacific Ocean in the south-east. The flood current is hence assumed to have flowed north-westward and the ebb current south-eastward.

At Iigurashinden in Youkaichiba, Chiba Prefecture, the deposits of a transgressive estuarine valley-fill are exposed (Fig. 2). The lowermost part of the deposit comprises fine-grained fluvial channel sands. The central part includes massive silts and clays of brackish estuarine origin, fine- to medium-grained tidal flat sands and intertidal to subtidal channel fills. The uppermost part, in turn, is made up of drowned valley deposits comprising sands and silts. In the middle of the estuarine deposits, fine to medium sands displaying herringbone cross bedding are exposed (Fig. 3). From a mud layer and from the cross-bedded section of this deposit some fossil shells of intertidal bivalves (oysters, scallops, cockles, etc.) were collected.

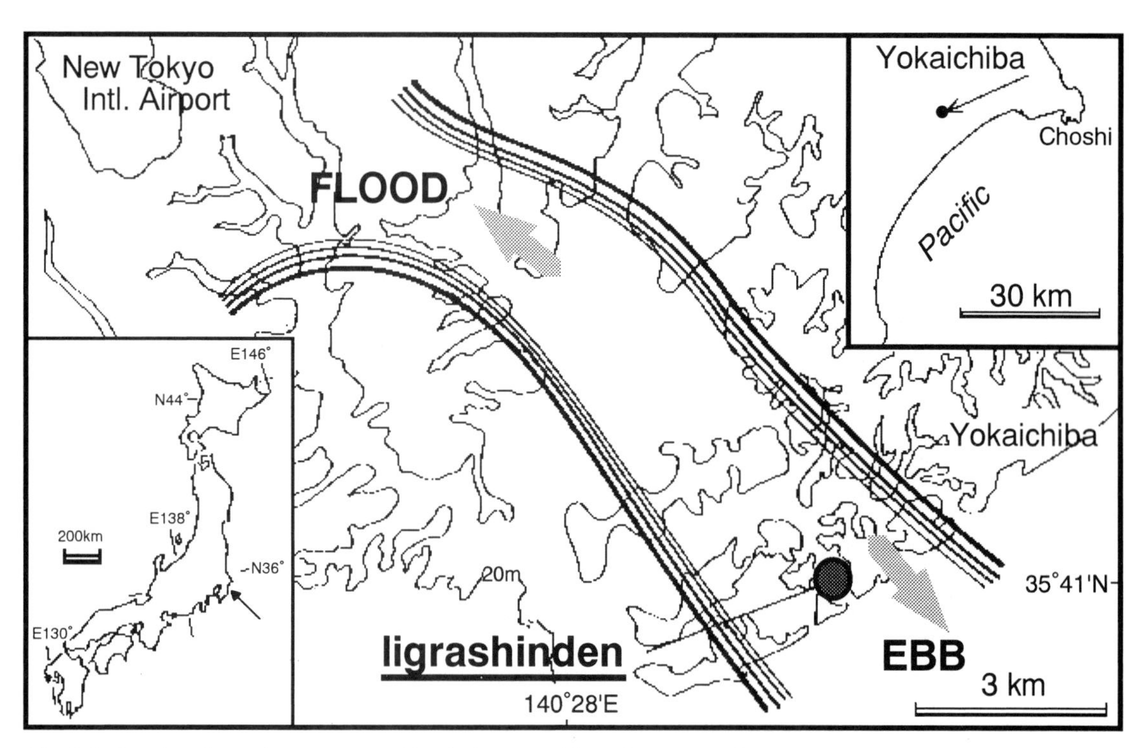

Fig. 1. Inferred position of the estuary valley during deposition of the basal Kioroshi Formation at Yokaichiba, Chiba. A flood (north-westward) and ebb (south-eastward) current system was reconstructed relative to the Pacific coast.

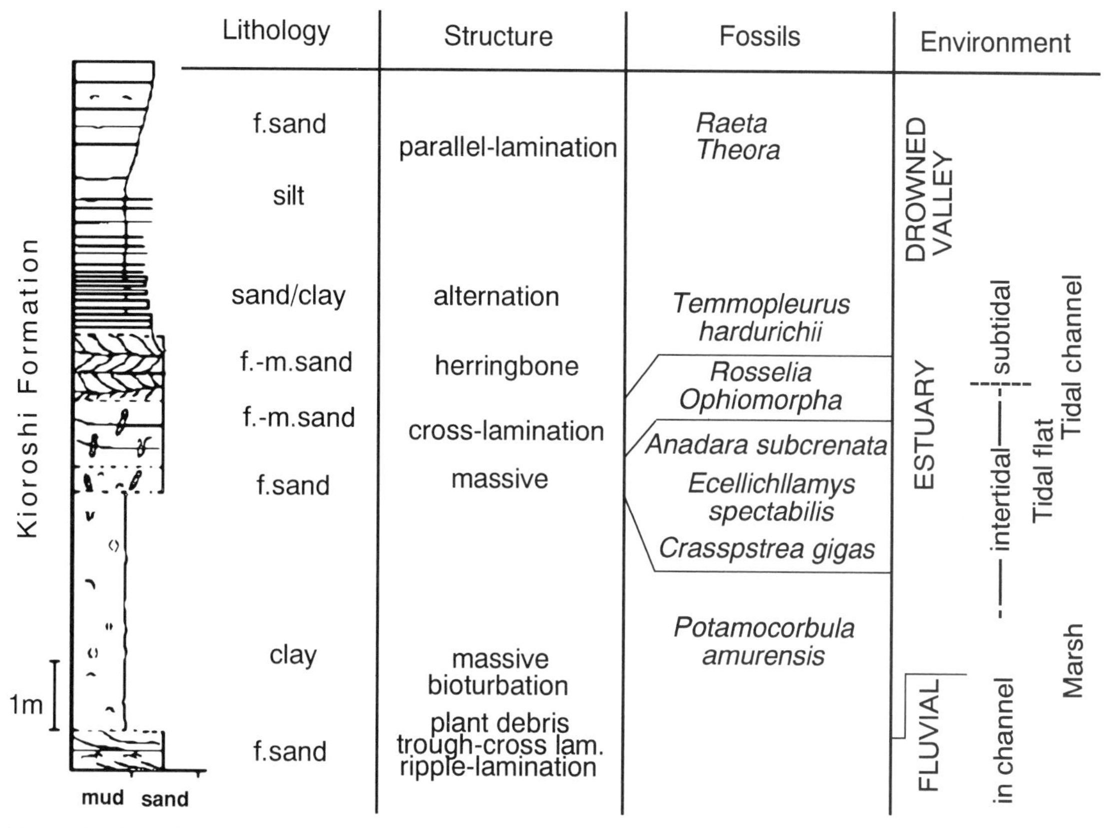

Fig. 2. Stratigraphical section of the estuarine deposits of the Kioroshi Formation showing lithology, sedimentary structures, fauna and environmental interpretation.

RECONSTRUCTION OF THE TIDAL REGIME

Analysis of sedimentary structures

Tidal bedding

Herringbone cross-bedding and tidal bundle sequences are exposed in road cuttings through Pleistocene strata at Yokaichiba (Figs 3 & 4). A north-eastward-facing stratum is exposed at the south-eastern end of a road cutting. The sediment consists of thin-bedded medium sands and the cross-beds suggest bidirectional tidal currents dominated by north-westward and south-eastward flow directions (Fig. 3b).

Mud drapes, a few millimetres in thickness, are only observed within the north-westward oriented cross-beds. Thin sand layers are intercalated between adjacent mud drapes, thus forming mud couplets. Within one cross-bedded set, parts of densely spaced mud drapes are preserved (Fig. 4).

The thickness of all tidal bundles in the north-westward oriented cross-bedded set were measured, bundle thickness being defined by adjacent mud drapes. In each case the measurement was taken at mid-level of a set in order to avoid inaccuracies owing to thickness variations along the vertical axis of individual bundles. The results, illustrated in Fig. 5, clearly show sequential changes in bundle thickness, manifesting themselves in thick–thin alternations. Moreover, a cyclicity of about 27 bundles is found in the overall thickness variation that is superimposed on the thick–thin alternations of successive bundles.

Interpretation

The analysis of palaeocurrent patterns relative to the palaeogeographic setting of the estuarine

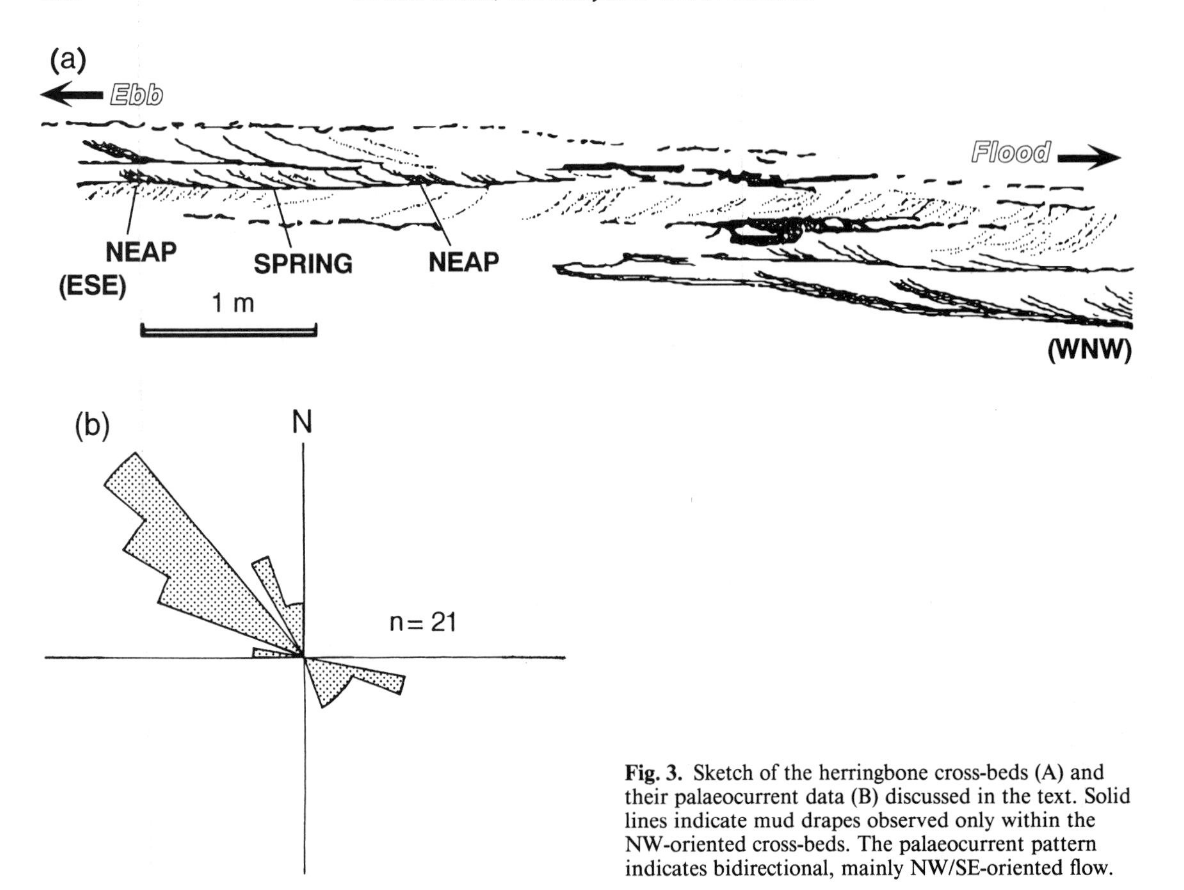

Fig. 3. Sketch of the herringbone cross-beds (A) and their palaeocurrent data (B) discussed in the text. Solid lines indicate mud drapes observed only within the NW-oriented cross-beds. The palaeocurrent pattern indicates bidirectional, mainly NW/SE-oriented flow.

environment along the Pacific coast suggests that the ebb current flowed south-eastwards and the flood current north-westwards (Fig. 1). Mud drapes are observed only within the flood-oriented cross-beds (Fig. 3a). No evidence for ebb flow (e.g. reactivation surfaces) could be detected in the cross-beds, which were evidently formed by sandwave migration during the dominant flood phase. The interbedded mud drapes were deposited at high water slacks. In the following ebb phase neither sand nor mud was deposited. At low water the bedforms may have been intertidally exposed.

Bedform progradation continued during the subsequent flood phase, following mud deposition at low water slacks. In this case, however, the secondary flood current was evidently weaker because each alternate bundle is thinner, reflecting a lower sand transport capacity of the flow. The observed thick–thin alternations of the bundles therefore suggest a strong diurnal inequality of the tide.

Besides the thick–thin periodicity of successive bundles, a larger-scale periodicity in thickness variation, comprising cycles of 26.5 tidal bundles on average, is observed. On this scale, thinner bundle sequences are produced around neap tide, when tidal ranges are smaller and hence tidal currents weaker, whereas the thicker bundle sequences result from higher current velocities around spring tide. If one bundle cycle were related to a neap–spring cycle, then more than 26.5 high-water events should occur per tidal cycle (neap–spring–neap). The results of this analysis therefore suggest that the sandwaves were generated in a semi-diurnal or mixed tidal system (Nio & Yang, 1989).

Reconstruction of tidal current regimes

The variation in bundle thickness reveals a clear and constant cyclic and sinusoidal pattern of about 26.5 bundles. The observed cycle of 26.5 bundles corresponds closely to the synodic neap–spring cycle, in the course of which the dominant flood

A

B

Fig. 4. Closely spaced mud drapes (A) indicating deposition during neap tide, and widely spaced mud drapes (B) produced during spring tide.

stage occurs 28 times in 14.76 days of a neap–spring cycle. The inferred tidal pattern thus has an almost semi-diurnal character.

The differences in thickness of adjacent bundles are very large and the bundles form an alternation of thick and thin pairs (Fig. 5B). These results indicate that the dominant flood current was alternatively stronger and weaker during consecutive flood tides. This pattern suggests a strong diurnal inequality.

Velocities of palaeotidal currents were calculated from the herringbone cross-beds (Fig. 6), based on the method of Allen & Homewood (1984) and Masuda & Nakayama (1988). For the estimation of the depth-mean velocity (U), the following equation is used:

$$U = u^*/\mathrm{k} \ln(h/\mathrm{e}\, z_0)$$

where u^* is shear velocity, k is the Karman constant (0.4), h is the estimated palaeodepth, e is the base of the natural logarithm (2.71), and z_0 is the roughness length. The mean palaeodepth was assumed to be 2 m in this case.

In the case of spring tides, the maximum tidal current velocity for the larger tidal excursion is estimated to have been 0.61 m s^{-1}, whereas the

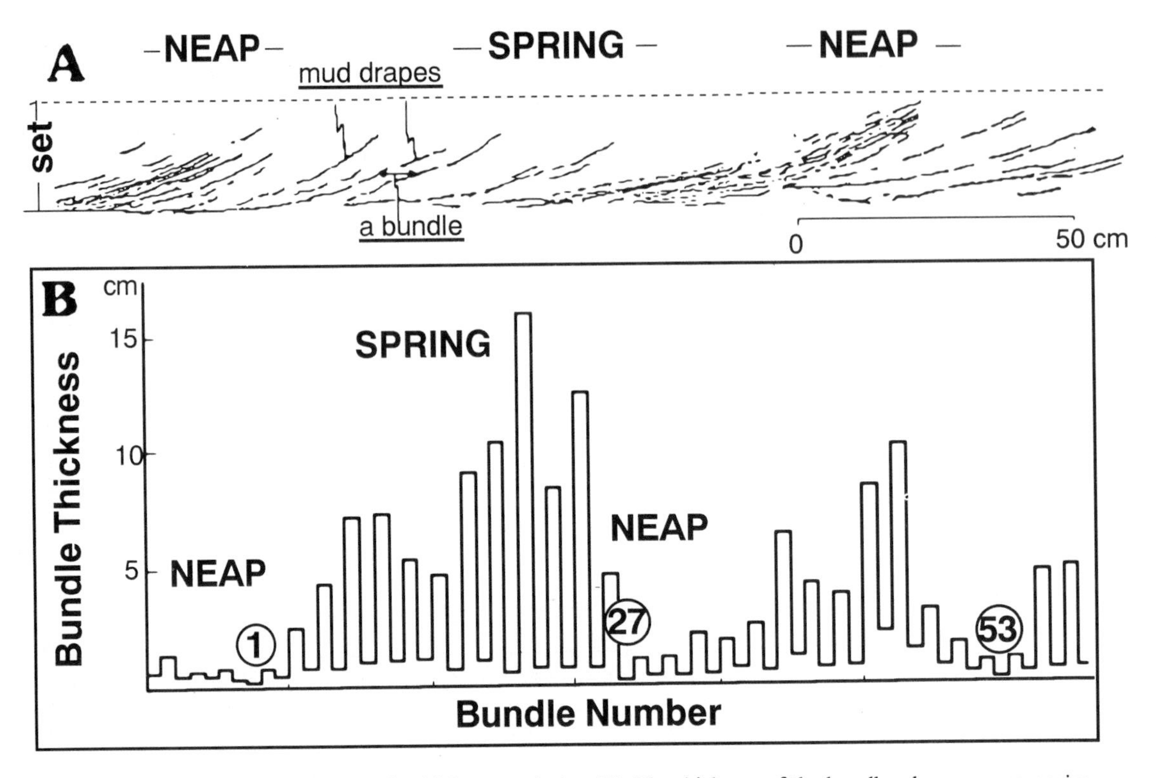

Fig. 5. Tidal bundle sequence (A) and its thickness variation (B). The thickness of the bundles shows a neap–spring variation and regular thick–thin pairs.

maximum velocity during the smaller tidal excursion is estimated at 0.52 m s^{-1}. Error bars indicate the variation of estimated velocities due to the difference of estimated palaeodepths ranging from 1 to 5 m. Since critical current velocities of 0.28 m s^{-1} and 0.24 m s^{-1} are required to initiate sediment transport in the mean grain sizes of consecutive bundles, a velocity variation of 0.24–0.61 m s^{-1} is suggested for the flood current at spring tide. By analogy, the maximum current velocity at neap tide was 0.59 m s^{-1}. The maximum velocity of ebb currents at spring tide is estimated at 0.61 m s^{-1}.

Cross-bedded sections indicating the influence of storm activity were also analysed. The tidal bundles in such cases show disturbed thick–thin sequences (Fig. 7). The disturbed parts include some very high thickness peaks in place of the common thick–thin sequences produced by the diurnal inequality of the tides. Such peaks reflect higher than average transport rates related to storm-amplified flood currents.

Analysis of fossil bivalve shells

Sample material

Fossil shells of the bivalve *Potamocorbula amurensis* Schrenck, an intertidal and brackish-water species (Fig. 8), were collected from the mud layer just beneath the sand with the herringbone cross-beds (Fig. 2). Based on facies analysis of the underlying and overlying deposits, the mud layer is inferred to have been deposited in an estuarine environment. The shells in the mud are autochthonous, having remained in life position. Other shells such as *Raeta yokohamensis* Pilsbry, *Raeta pellicula* Reeve, *Theora lubrica* Gould, *Anadara subcrenata* Lischke, *Excellichlamys spectabilis* Reeve and *Crassostrea gigas* Thunberg were also found in the intertidal facies, although these were not in life position.

Method

During shell growth, bivalves produce growth lines

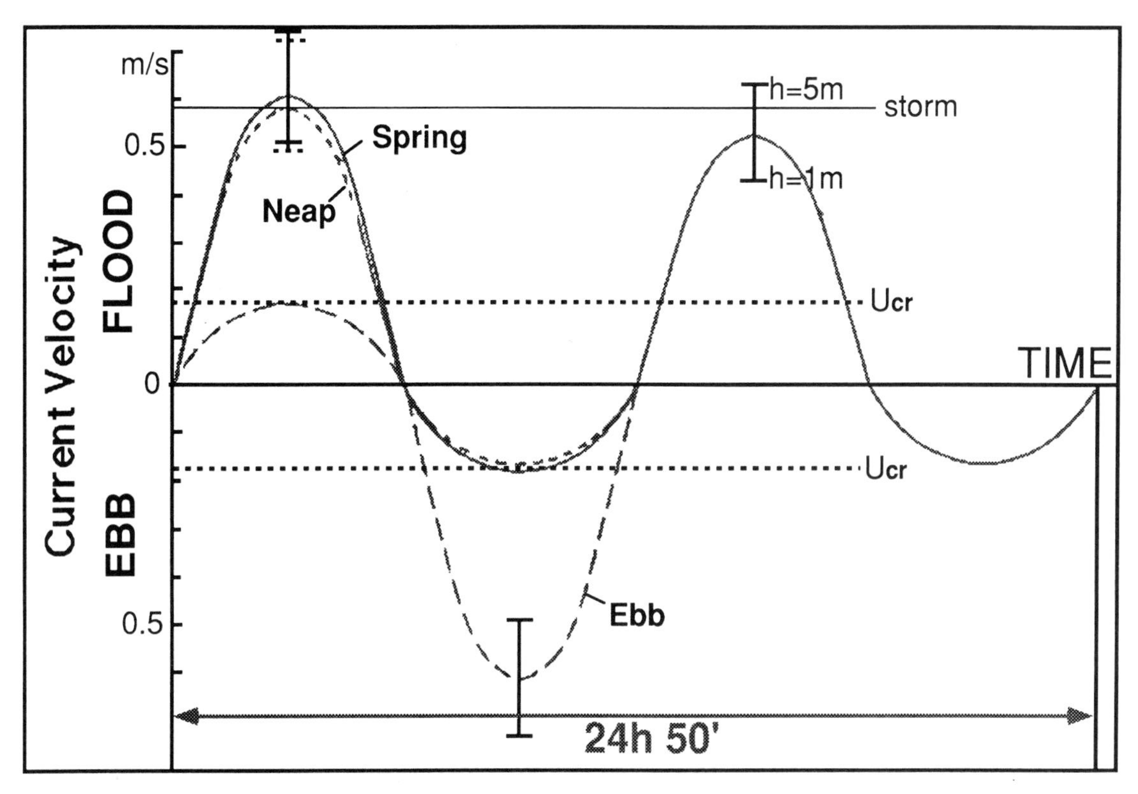

Fig. 6. Reconstructed variation of palaeocurrent velocity calculated from the herringbone cross beds. The solid line indicates spring tide, the dashed line neap tide. U_{cr} indicates the critical threshold velocity for each grain size. Error bars indicate the variation of estimated velocities owing to the difference in estimated palaeodepth ranging from 1 to 5 m. Note that only the velocity field for sediment deposition is indicated; the higher erosion velocity is not shown.

and growth increments. In shell cross-sections, the thinner and darker layers represent the growth lines, whereas the thicker layers between the growth lines are the growth increments. The growth lines correspond to shell portions consisting of ill-grown aragonite crystals associated with organic-rich

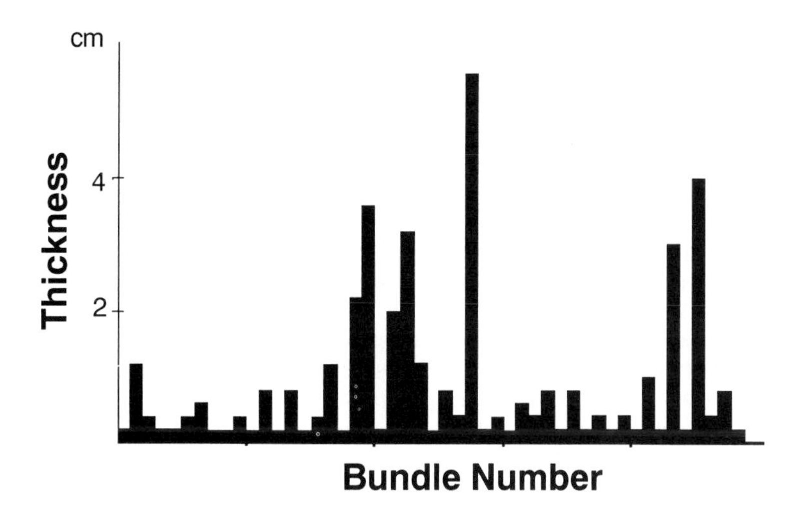

Fig. 7. Irregularities in the thickness variation of tidal bundles. Note that the regular thick–thin pattern is disturbed in some places with unexpected high peaks.

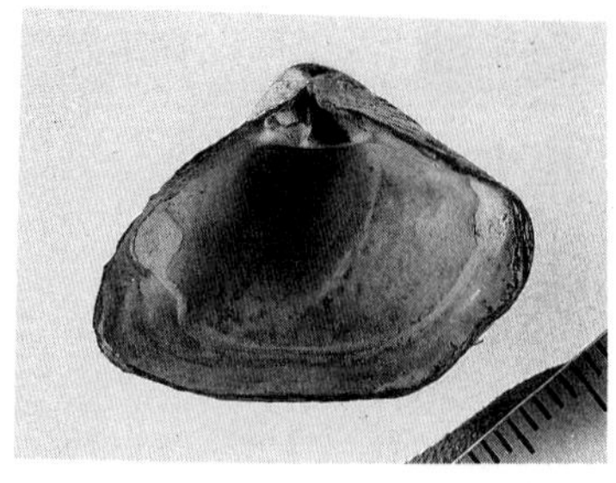

Fig. 8. Fossil shells of *Potamocorbula amurensis* Schrenck. Scales in mm.

matrices. The abundance of organic matrix relative to aragonite crystals is related to the speed of $CaCO_3$ precipitation (Iwasaki, 1977). In the case of the intertidal bivalve shells, distinct growth lines form during tidal exposure at low tides and growth increments form during submergence at high water. When bivalves are continuously submerged, they produce irregular growth bundles consisting of weak and irregularly spaced (non-tidal) growth lines.

By examining the pattern of growth lines and increments of recent and fossil bivalve shells, Pannella (1976) and Ohno (1989) established a method by which the characteristics of the tidal environment inhabited by the organisms can be estimated. The following relationships are suggested:

1 The width of each increment is proportional to the duration of submergence during high water. The wider increments are produced during higher high-water events, i.e. longer submergence periods, whereas the narrower parts relate to lower high-water events, i.e. shorter submergence periods. The alternating wide–narrow arrangement of growth increments reverses approximately every 26.4 increments, which relates to the number of 12.4 h tidal cycles per half tropical month. Moreover the widths of tidal growth increments are grouped into single- and double-spaced tidal growth lines. Single spacings between growth lines form during tidal exposure at low tide, approximately occurring every 12.4 h, whereas double spacings relate to the low tides occurring approximately every 24.8 h. The spacing of the latter is about twice as wide as that of the former.

2 Thick-thin growth-line alternations form by the interaction of semi-diurnal tidal exposure and 24-h day–night cycles. The thicker growth lines form during the day-time low tide, the thinner ones during the night-time low tide. The order of the thick–thin arrangement reverses approximately every 28.5 growth lines, which corresponds to the number of 12.4 h tidal cycles per fortnight.

The resulting succession of growth lines shows the following pattern (see Fig. 9): (i) double spacing of growth lines with alternation of regular and irregular bundles for diurnal tides; (ii) single spacing of growth lines with thickness alternation and frequent occurrence of double spacing for mixed tides with diurnal inequality at the heights of low water; (iii) single spacing of growth lines with thickness alternation and strong alternation in increment thickness for mixed tides with diurnal inequality at the heights of high water; and (iv) single spacing of growth lines with no thickness alternation for semi-diurnal tides with insignificant diurnal inequality.

Sample preparation

Samples of shells for growth analysis were set in synthetic resin and cut longitudinally from the umbo to the margin. The surfaces of the sections were ground with powders of Nos 2000 to 3000 and polished with diamond paste (where necessary) before being photographed using a reflection microscope. Some sections had to be etched chemically with diluted hydrochloric acid and then examined under a scanning electron microscope (SEM). Finally, photomosaics of each growth line sequence were compiled and evaluated.

Clear growth patterns are generally present in the outer shell layer. The thicknesses of growth increments and growth lines were measured from the assembled photomosaic to assess tidal influence on the growth of the shells.

Results

Growth patterns were examined in all of the fossil bivalve species. The clearest growth-line sequences

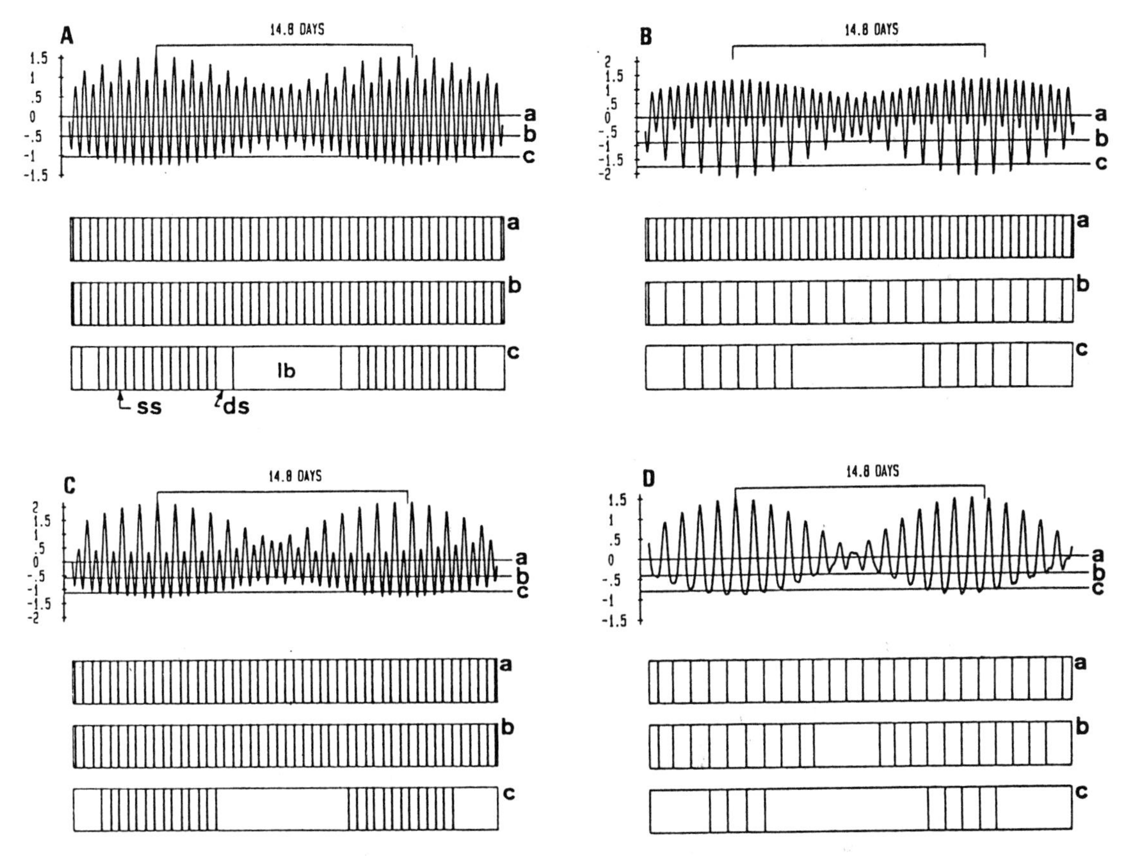

Fig. 9. Four typical tidal types with their associated potential tidal growth-line patterns (the scale shown on the left side of each diagram represents an arbitrary tidal range). A, Semi-diurnal tide with faint diurnal inequality. B, Mixed tide with strong diurnal inequality in the height of the low tides. C, Mixed tide with strong diurnal inequality in the heights of the high tides. D, Diurnal tide. Hypothetical patterns of growth line arrangements are shown at three different levels for each tidal type (**a**, mean tide level; **b**, between MTL and low water; **c**, near LW). Three stripes represent those parts of a shell cut parallel to the growth direction. The patterns are reconstructed assuming that one tidal growth line is formed at each low tide, when the shell is intertidally exposed. The boxes below each tidal curve represent a specific water level, and each vertical line in the boxes represents a tidal growth line corresponding to the associated low tide at which the shell is intertidally exposed. **ss**, Single spacing of tidal growth lines; **ds**, double spacing of tidal growth lines; **Ib**, irregular bundle. Possible variations in growth-line thickness and increment thickness, as well as non-tidal growth lines in irregular bundles, are not included in these hypothetical patterns, in order not to obscure the relationship between growth lines and tidal exposure (figure from Ohno, 1989).

were obtained from the shells of *Potamocorbula amurensis* (Fig. 10a–c) for which the longest sequence of 148 growth lines and increments was recorded. In the other fossil bivalve species it was not possible to detect any distinct tidal influence on shell growth.

In *P. amurensis*, the growth pattern has the following features: (i) a series of almost regularly spaced growth lines (except in the double-spaced part); (ii) reversal of growth-line thickness after 28 increments; (iii) double-spaced increments in some places, where thinner growth lines between two thicker lines have disappeared, thus producing double increments; (iv) in parts, a thick–thin alternation in growth-line thickness.

Interpretation

Using the criteria for reconstructing tidal type from shell growth patterns of intertidal bivalves (Fig. 9; see also table 2 in Ohno, 1989), the palaeotidal characteristics of the bivalve habitat were

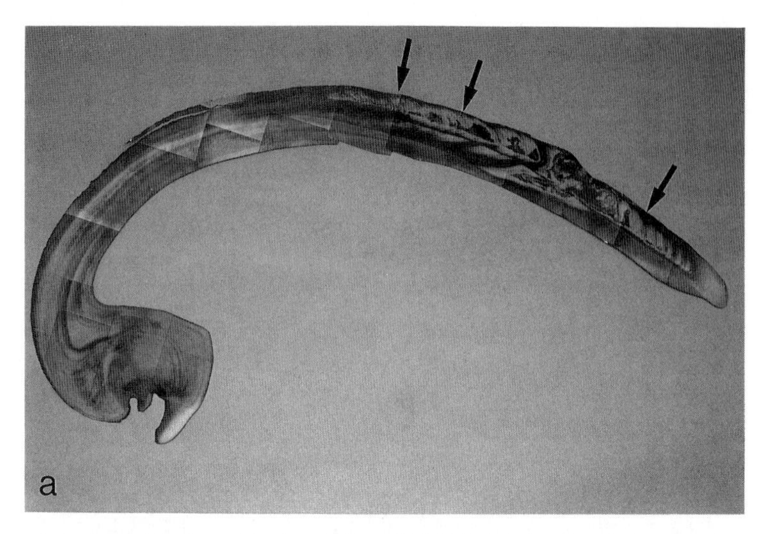

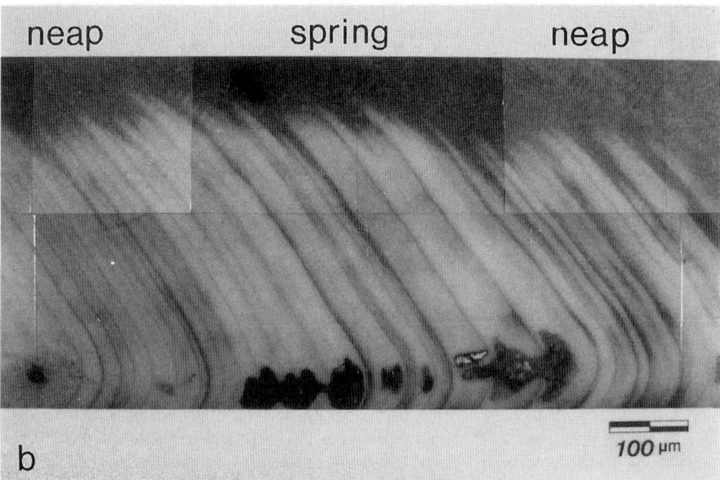

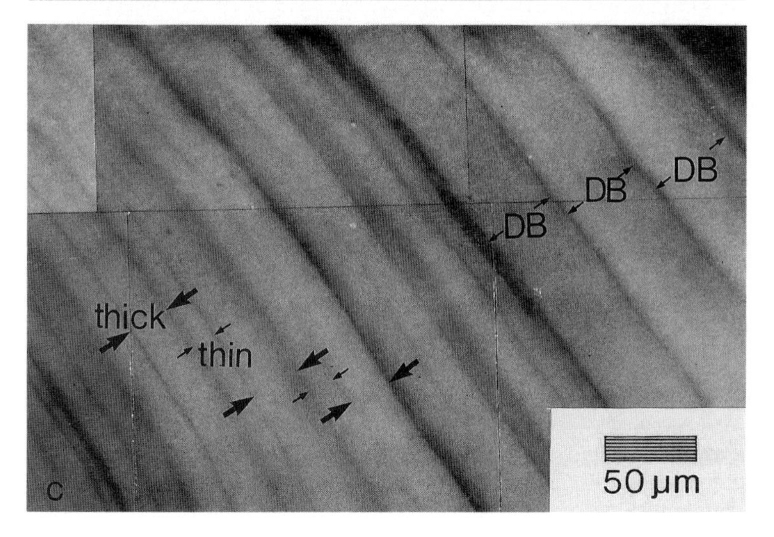

Fig. 10. (a) Photomosaic of a sectioned shell of *Potamocorbula amurensis* Schrenck. Growth lines appear in the outer layer of the shell (arrows). Growth line sequences are often interrupted by mud inclusion and shell erosion. (b) Close-up of a growth-line sequence showing the cyclicity pattern suggestive of a neap–spring cycle. (c) Close-up of a growth-line sequence showing growth increments in form of thick–thin pairs (arrows) that indicate diurnal inequality of high tides, and three double-spaced growth increments (DB) that indicate strong diurnal inequality of low tides.

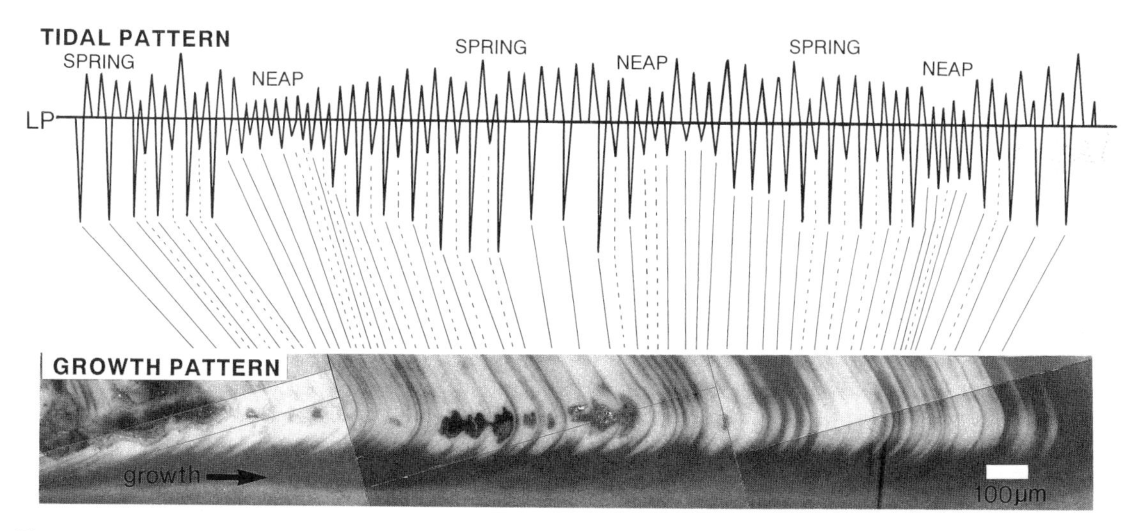

Fig. 11. The complete growth-line sequence with the inferred palaeotidal pattern reconstructed from a shell of *Potamocorbula amurensis*. The line marked **LP** designates the assumed life position of the shell. Relative high-tide levels are drawn proportional to the relative thickness of growth increments, whereas the relative low-tide levels are almost in proportion to the thickness of growth lines. The tidal range is arbitrary.

reconstructed from the growth patterns of examined shells.

Single spacing of growth lines with thickness alternation (features (i) and (ii) above) indicates semi-diurnal or mixed tides. The occurrence of double spacing of growth lines (feature (iii) above) suggests a mixed tide with strong diurnal inequality at the heights of low tide. The absence of growth lines in places where there are wider increments suggests that the bivalve occupied a tidal level that did not emerge at higher low-water events associated with the diurnal inequality of the tide. As a result of this, growth lines did not form and neighboring increments coalesced. Alternations in increment thickness (feature (iv) above) are produced if the duration of submergence events differs at successive high tides. This is a typical growth pattern observed in intertidal shells exposed to strong diurnal inequalities in tides (Fig. 11).

DISCUSSION

Palaeotidal patterns were reconstructed by both sedimentological and palaeontological approaches using tide-influenced sedimentary structures and intertidal bivalve shells recovered from the same strata of Pleistocene estuarine deposits. The results of both analyses showed good agreement in terms of detailed tidal characteristics such as tidal type and diurnal inequality. The results of both methods are comparable as well as complementary and are thus able to contribute towards a more precise reconstruction of palaeotidal patterns.

The results of this study are based on a rather limited number of examples. The main reason for this is the mostly discontinuous nature of data sets found in the field, the quality of which depends largely on site-specific conditions of exposure and preservation. Nevertheless, the approach can be applied to virtually any rock outcrop containing fossil shells.

An important tidal parameter that cannot be reconstructed accurately from growth patterns in shells is the palaeotidal range. High-tide levels are estimated from the relative width of growth increments, whereas low-tide levels are estimated from growth-line thickness, both tidal levels being related to the positions of live shells.

Whether a shell is suitable for growth pattern analysis depends on its composition and internal structure. The shell of *P. amurensis*, the only bivalve showing a clear tidal influence on shell growth in this study, consists of aragonite crystals. The other shells, such as scallops, are made up of calcitic material. From a microscopic and chemical point of view, the investigation of growth lines may also benefit from other shell characteristics, especially the form and size of carbonate crystals. Their relationship with growth patterns, however, has yet to be clarified.

CONCLUSIONS

Palaeotidal characteristics of Palaeo-Tokyo Bay were reconstructed from tide-influenced sedimentary structures and the growth lines of the fossil shells of intertidal bivalves. The analysis of herringbone cross-beds indicates: (i) a bimodal (NW–SE) tidal flow pattern of the palaeocurrents in each bed set; (ii) a neap–spring cyclicity in the thickness variation of tidal bundles; and (iii) a diurnal inequality pattern, revealed by thickness variations in alternate tidal bundles.

The analysis of growth lines of fossil bivalves, in particular of the intertidal bivalve *P. amurensis*, indicates (i) a neap–spring cyclicity revealed by the regular reversal of growth-line thickness; (ii) a mixed tidal pattern from double-spaced growth increments; and (iii) a strong diurnal inequality of both low and high water from the thickness variation of growth increments.

On the basis of comparable and complementary evidence derived from sedimentological and palaeontological data analysis, a tidal pattern of semi-diurnal to mixed character, with a strong diurnal inequality of low and high water, was reconstructed for Palaeo-Tokyo Bay.

ACKNOWLEDGEMENTS

We thank the Inoue Foundation for Science for providing financial support for N. Murakoshi to attend the Tidal Clastics 92 meeting in Germany. We also wish to thank the reviewers, Dr A. Archer and an anonymous referee, for constructive comments and suggestions.

REFERENCES

Allen, P.A. & Homewood, P. (1984) Evolution and mechanics of a Miocene tidal sandwave. *Sedimentology* **31**, 63–81.

Evans, J.W. (1972) Tidal growth increments in the cockle *Clinocardium nuttalli. Science* **176**, 416–417.

House, M.R. & Farrow, G.E. (1968) Daily growth banding in the shell of the cockle, *Cardium edule. Nature* **219**, 1384–1386.

Iwasaki, Y. (1977) Shell growth and growth lines. *Kumamoto J. Sci. (Geol.)* **10**, 41–54 (in Japanese with English abstract).

Makino, Y., Kimura, M., Terakado, N. & Yoshikawa, I. (1989) Tidal deposits in the Pleistocene Kamiizumi Formation, Shimosa Group, Japan. *Sci. Rept Fac. Educ. Ibaraki Univ.* **38**, 103–119 (in Japanese with English abstract).

Masuda, F. & Nakayama, N. (1988) Calculation of paleotidal current velocity. *J. Sediment. Soc. Japan* **29**, 1–8 (in Japanese with English abstract).

Masuda, F., Nakayama, N. & Ikehara, K. (1988) A cross-stratification produced by tidal current nine days in the Pleistocene at Uchijuku in Kitaura, Ibaraki. *Tsukuba Environ. Stud.* **11,** 91–105 (in Japanese with English abstract).

Murakoshi, N. & Masuda, F. (1992) Estuarine, barrier-island to strand-plain sequence and related ravinement surface developed during the last interglacial in the Paleo-Tokyo Bay, Japan. *Sediment. Geol.* **80,** 167–184.

Narita Research Group (1962) The Shimosueyoshi transgression and the Paleo-Tokyo Bay. *Earth Sci.* **60/61**, 8–15 (in Japanese with English abstract).

Nio, S.D. & Yang, C.S. (1989) *Recognition of Tidally-influenced Facies and Environments.* Short Course Note Ser. **1**, International Geoservices, BV, Leiderdorp, 230 pp.

Ohno, T. (1989) Palaeotidal characteristics determined by micro-growth patterns in bivalves. *Palaeontology,* **32**, 237–263.

Pannella, G. (1976) Tidal growth patterns in recent and fossil mollusc bivalve shells: a tool for the reconstruction of paleotides. *Naturwissenschaften* **63**, 539–543.

Richardson, C.A., Crisp, D.J. & Runham, N.W. (1981) Factors influencing shell deposition during a tidal cycle in the intertidal bivalve *Cerastoderma edule. J. Mar. Biol. Ass. U.K.* **61**, 465–476.

Yang, C.S. & Nio, S.D. (1985) The estimation of palaeohydrodynamic processes from subtidal deposits using time series analysis methods. *Sedimentology* **32**, 41–57.

Spec. Publs int. Ass. Sediment. (1995) **24**, 301–311

Climbing ripples recording the change of tidal current condition in the middle Pleistocene Shimosa Group, Japan

M. YOKOKAWA*, M. KISHI†, F. MASUDA* *and* M. YAMANAKA‡

**Department of Earth and Space Science, Faculty of Science, Osaka University, Toyonaka, Osaka 560, Japan;*
†Department of Geology, Faculty of Science, Shinsyu University, Matsumoto 390, Japan, and
‡Indonesian Oil Co. Ltd, Chiyoda, Tokyo 100, Japan

ABSTRACT

A lenticular sand body in the middle Pleistocene Shimosa Group in Japan, dominated by climbing ripple cross-lamination, is interpreted here as an estuarine or lagoonal deposit that accumulated under tidal influences. Based on the systematic change of sand-layer thickness in the alternation of sand and mud layers, two sets of four neap–spring tidal cycles are reconstructed from tidal flat deposits that occur just above the climbing-rippled sand;

The climbing ripple cross-laminae occur as four different types which signify variable current velocities and sediment supply. Isochrons can be drawn through the climbing-rippled sand along individual bedding surfaces on detailed tracings of the climbing ripple cross-laminae. The dominant types of climbing ripple cross-lamination vary along each isochron. Cyclic variations in current velocity during a single tidal cycle are demonstrated from the vertical evolution in the types of climbing ripple cross-lamination.

INTRODUCTION

Climbing ripple cross-lamination is common in fluvial and deltaic environments (e.g. Reineck & Singh, 1980; Ashley *et al.*, 1982). Subsequent to Sorby's (1859) early description of 'ripple-drift', new nomenclatures and classification schemes have been proposed (e.g. Jopling & Walker, 1968; Allen, 1982; Ashley *et al.*, 1982), and laboratory experiments have revealed the hydraulic conditions under which climbing ripple cross-lamination is produced (Jopling & Walker, 1968; Ashley *et al.*, 1982).

This paper describes climbing ripple cross-lamination in a Pleistocene tidal deposit. Although climbing ripple cross-lamination is rare in tidal environments, Wunderlich (1969) reported this structure from tidal flats with high sedimentation rates (Reineck & Singh, 1980). We examined a climbing-ripple laminated sand body that forms part of a tidal channel fill or crevasse splay near a river mouth. Based on previous experiments (Jopling & Walker, 1968; Ashley *et al.*, 1982), we were able to reconstruct details of the cyclic change of tidal current conditions from the observed vertical change in the type of climbing ripple cross-lamination.

GEOLOGICAL SETTING

The Pleistocene Shimosa Group lies beneath the Kanto Plain, central Japan (Fig. 1), in shallow marine deposits of Palaeo-Tokyo Bay. This bay has alternately expanded and contracted in response to changes in eustatic sea-level since about 0.45 Ma (e.g. Masuda, 1988).

A lenticular sand body dominated by climbing ripple cross-lamination (examined in this study) is intercalated in the Izumiyatsu Mud Member of the Jizodo Formation that forms the lowermost part of the Shimosa Group (Fig. 2). The Jizodo Formation is composed of a transgressive systems tract and a

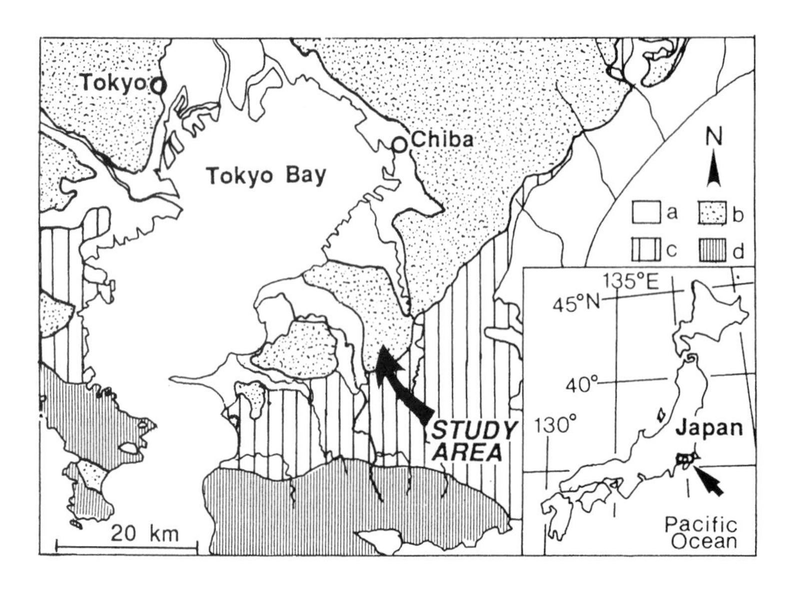

Fig. 1. Geological map of the study area, Chiba Prefecture, southern Kanto plain, central Japan. (a) Holocene; (b) Pleistocene Shimosa Group; (c) Pliocene Kazusa Group; (d) Pre-Pliocene.

highstand systems tract (Ito & O'Hara, 1990). The basal part of the Jizodo Formation rests on an erosional surface and consists of tidal channel deposits that infilled incised valleys during the transgression. The overlying Izumiyatsu Mud consists of mud, muddy silt and muddy fine to very fine sand in which current ripples are preserved. The Izumiyatsu Mud is highly bioturbated and includes many plant fragments and intercalated peat layers, and is thought to represent the deposits of tidal flats or lagoons (Tokuhashi & Endo, 1984; Tokuhashi & Kondo, 1989; Ito & O'Hara, 1990). The Izumiyatsu Mud Member is overlain by poorly-sorted muddy fine sand including a 1-m thick shell bed. The lowermost part of this sand has dense burrows of the decapod *Upogebia major*, suggesting an intertidal sand flat environment. The overlying molluscan fossil associations change from an inner-bay mud association (*Macoma tokyoensis* and *Mya arenaria*) to a sand–gravel association (*Callithaca adamsi* and *Saxidomus purpuratus*) (Kondo, 1989). This muddy sand is overlain by shoreface deposits consisting of hummocky cross-stratified very fine to fine sands which compose the main part of the upper Jizodo Formation.

TIDAL INDICATORS

The lenticular sand body, 19 m long and 0.8 m thick, is dominated by climbing ripple cross-lamination (Fig. 3) and occurs within the lower part of the Izumiyatsu Mud Member (Fig. 2). This fine-grained sand body is separated into two units by a 2–3 cm thick mud layer. The two units display opposing palaeocurrent directions (Fig. 4), indicating deposition in response to tidal currents, the Izumiyatsu Mud being a tidal flat or lagoon deposit. Based on the palaeocurrent indicators in this palaeogeographical setting (Tokuhashi & Endo, 1984; Ito & O'Hara, 1990), the lower part (Unit 1) was formed by an ebb current and the upper part (Unit 2) by a subsequent flood current. The mud layer is thought to be a slack-water drape.

A thinly bedded alternation of sand and mud layers (Fig. 5), about 60 cm thick, overlies the climbing-ripple laminated sand with a sharp boundary and a bioturbated, *c.* 2 cm thick, mud layer. Within this package, two sets of four upward-thickening cycles of sand beds are recognized (cf. Fig. 12a). Each cycle is interpreted to represent a single neap–spring tidal deposit as discussed below.

The lenticular shape of the sand body containing the climbing ripples is ascribed to deposition in a minor channel on the tidal flat of an estuary or a crevasse splay of a tidally influenced lagoon. In the former case, large amounts of fine sand were temporarily supplied by an ebb current, filling the existing tidal channel. In the latter case, a river that entered Palaeo-Tokyo Bay from the north or northwest was crevassed near its mouth, spilling its splay deposits into a lagoon during an ebb-tidal phase. In both cases, tidal influences played a major role in shaping the deposits. The overlying alternations of

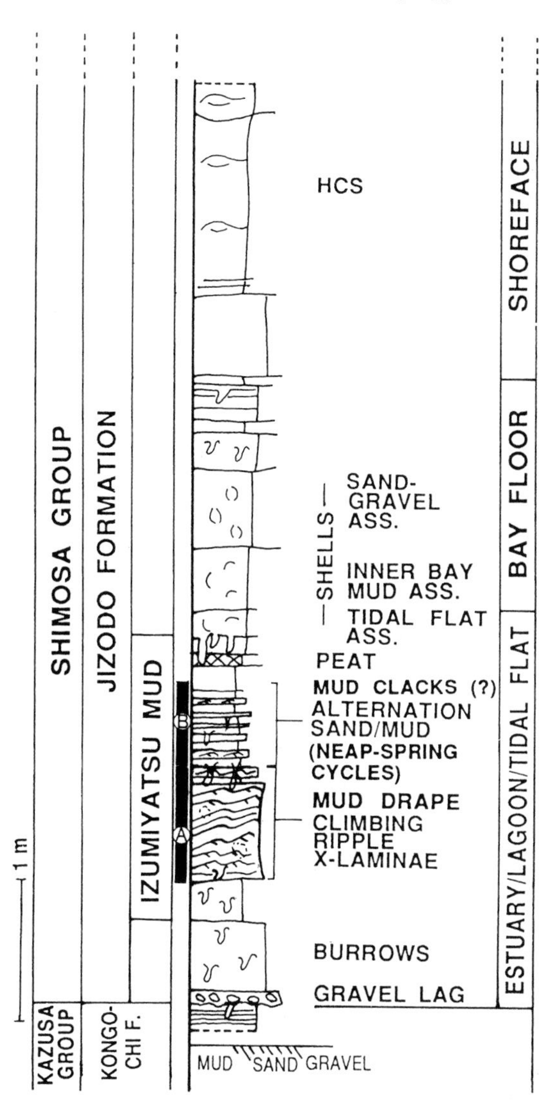

Fig. 2. Columnar section of the lower part of the Jizodo Formation in the study area. The black bars indicate the positions of the climbing-rippled sand (a) and the alternation of sand and mud layers (b).

sand and mud are considered to be tidal flat deposits in an estuarine or lagoonal environment. Because of the existence of mud-crack-like structures in the upper part of the alternation, this environment must have experienced prolonged exposures, as would be the case in higher intertidal or supratidal settings.

Mean diameters of both the climbing-rippled sand and the overlying sand range from *c.* 2.5 to 3.5 phi, standard deviations from *c.* 0.4 to 0.7 phi (Fig. 6). The sands are well to moderately well sorted (Folk & Ward, 1957). Sands within the thinly bedded alternations are relatively fine, this being consistent with higher intertidal or supratidal settings. Sand from just below the mud layer (which separates the climbing-ripple sand into two parts) is relatively coarse-grained. Physical sedimentary structures are not well preserved in this sand because of strong bioturbation. Flood and ebb deposits have similar grain sizes.

CLIMBING RIPPLE CROSS-LAMINAE

The climbing ripple cross-laminae occur in four different types (Figs 7 & 8). Type I is characterized by a low-angle climb and planar cross-laminae, corresponding to Allen's (1982) 'subcritical tabular' type. Type II has a higher angle of climb. This type is subdivided into Type IIa and Type IIb. In Type IIa the stoss slope is erosional, whereas in Type IIb it is depositional. Tangential and/or angular laminae dominate at the toe of foresets in Type IIa, whereas concave foreset toes dominate in Type IIb. The Type II climbing ripple cross-laminae are also subcritical (Allen, 1982) because the angles of climb are still smaller than the stoss slope angles of the bedforms. Type III climbing ripple cross-laminae, on the other hand, are supercritical (Allen, 1982) as the angle of climb exceeds the stoss slope angle of the bedform. In some cases the angles of climb decrease upwards (Type IIIa), whereas in others the angles of climb increase upwards (Type IIIb). Type IV has very low amplitude sinusoidal laminations formed by ripples that climb at angles exceeding the stoss slope angles of the ripples. This type has been called 'sinusoidal ripple lamination' (Jopling & Walker, 1968) or 'draped lamination' (Ashley *et al.*, 1982). Jopling & Walker (1968) and Ashley *et al.* (1982) divided climbing ripple cross-lamination into Type A and Type B. Our Type I and Type II correlate with their Type A, and our Type III corresponds to their Type B.

Increasing angles of climb indicate that the ratio of the vertical bed aggradation rate to the downstream ripple migration rate increases (Ashley *et al.*, 1982), i.e. the ratio of deposition between suspended bed material and traction bed load increases (Jopling & Walker, 1968). Ashley *et al.* (1982) have reported experimental work in which climbing ripple cross-laminations were generated in various

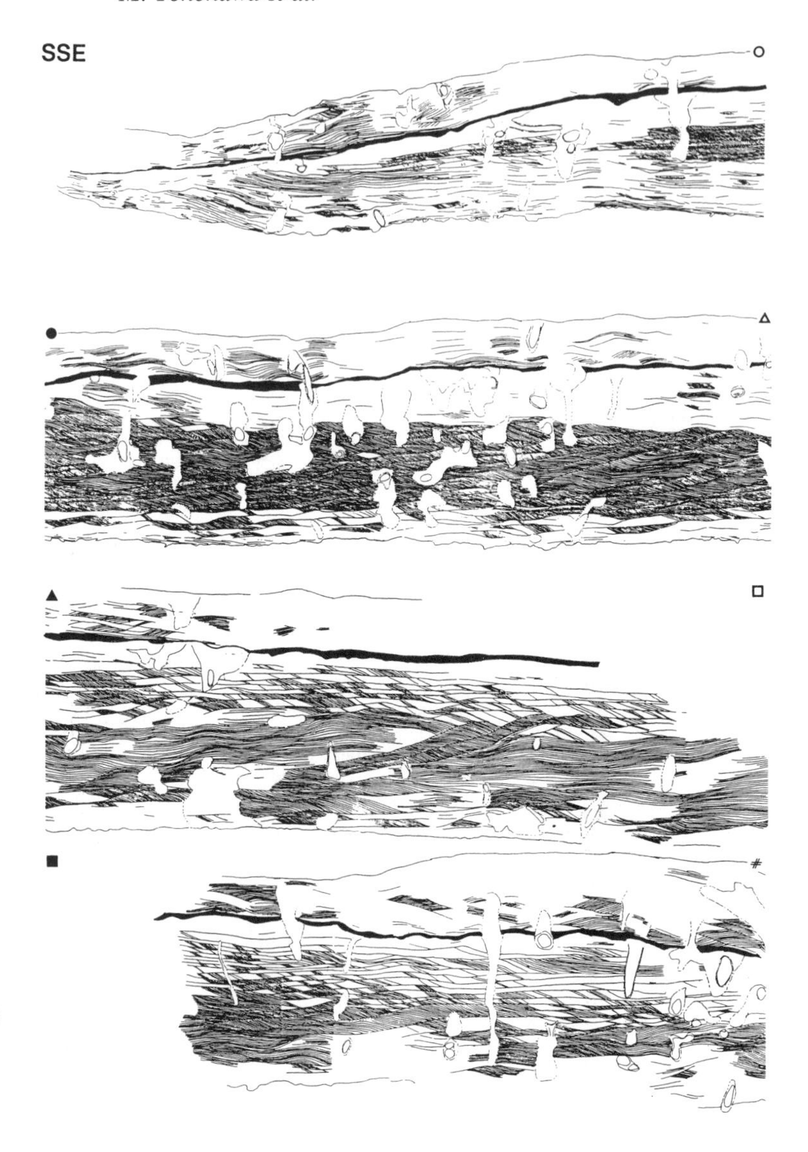

Fig. 3. Detailed sketch of climbing ripple cross-lamination developed in the lenticular sand body. The rectangles **a** and **b** correlate with those of Fig. 10.

combinations of current velocity and sediment supply. The most common succession of climbing ripple cross-lamination, i.e. from Type II to Type IV (Fig. 8b), was generated when the suspended load increased while the current velocity decreased.

RECONSTRUCTION OF TIDAL CURRENT CONDITIONS

Five isochrons labelled A to E have been drawn through the climbing-rippled sands of Unit 1 in order to highlight temporal variation in the type of cross-lamination. Isochrons were found to display three patterns (Fig. 9). The first coincides with one of the Type IV laminations which can be traced widely (Fig. 9a). The second consists of a boundary at which the profile of the ripple cross-lamination changes from Type II to Type IV (Fig. 9b). The third pattern consists of a boundary where adjacent foreset laminae of climbing ripples change from angular to concave in the downstream direction (Fig. 9c). These lines are thought to be depositional surfaces. By connecting short isochron segments

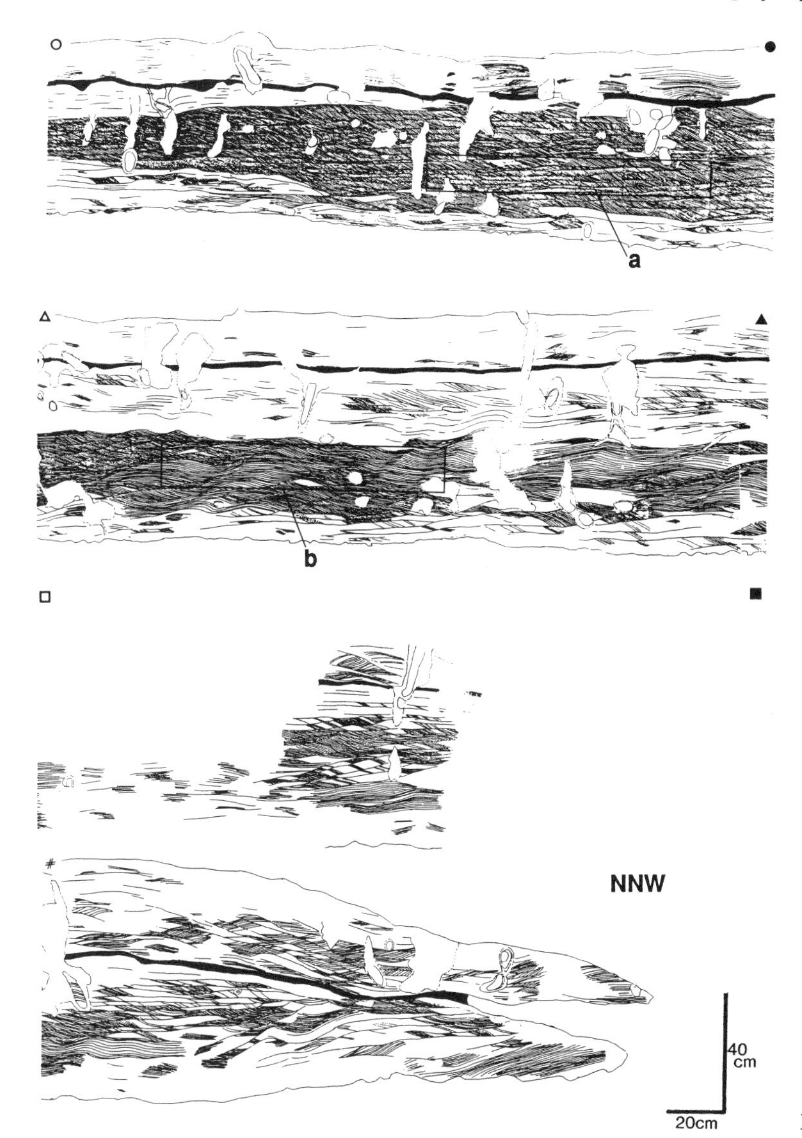

Fig. 3. *(Continued.)*

laterally it was possible to identify five isochrons in Unit 1: Line A, Line B, Line C, Line D and Line E (Fig. 10).

Along each isochron the ripple form may vary, although each isochron is dominated by one or more ripple types. Along Line A, Type I cross-lamination occupies about 87% and Type II cross-lamination occupies the remainder (Fig. 11). Types III and IV are absent. At the overlying Line B, Type I decreases sharply to 14%, Type II accounts for 43%, and Type III and IV account for the remaining 43%. Along Line C, Type I is rare (7%), Types III and IV form 53%, and Type II forms 40%. Along Line D, Type I increases to 43%, Type III and IV form 37%, and Type II contributes 20%. Along the uppermost Line E, Type I predominates as it did along Line A. Line E is overlain by a mud layer.

In Unit 2 only one isochron could be identified, comprising mainly Types III and IV cross-laminations (61%). The direction of ripple migration is reversed in Unit 2. Reactivation surfaces are absent in both Units 1 and 2.

The various types of cross-lamination are believed to indicate changing current conditions and

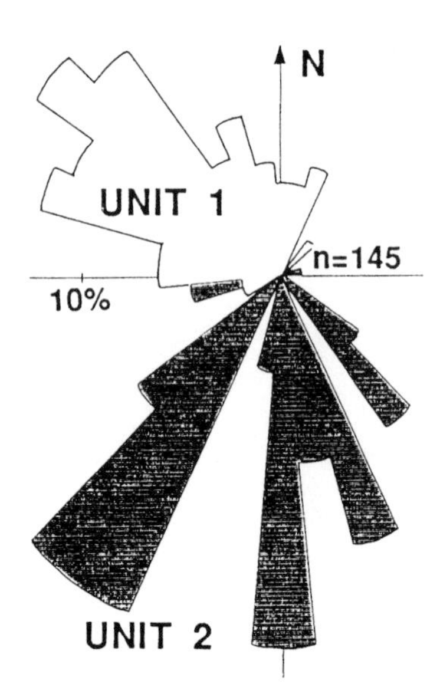

Fig. 4. Palaeocurrents of the cross-laminae in the lenticular sand body.

sediment supply, both being variably represented over individual tidal cycles in modern tidal environments (e.g. Carling, 1981; Collins *et al.*, 1981; Jago, 1981). In Unit 1 sediment supply is assumed to be constant so that the correlation of cross-lamination types with velocity is unproblematical. The lowest energy conditions are reflected by mud layers below and above Unit 1 (Figs 3 & 11), probably representing slack-water deposition. From the stage of mud deposition the current velocity at first increased to Line A. Thereafter it decreased gradually through Line B to Line C. Types III and IV cross-lamination occupy the greater part of Line C, indicating that the aggradation rate was much faster than the ripple migration rate. Current velocity then increased through Line D to Line E, before decreasing once more towards the mud layer that overlies Unit 1.

The progressive change of the dominant type of cross-lamination from Line A to Line E is asymmetrical. This is shown by the fact that Type I cross-

Fig. 5. Thinly bedded alternation of sand and mud layers that overlie Unit 2 of the climbing-rippled sand.

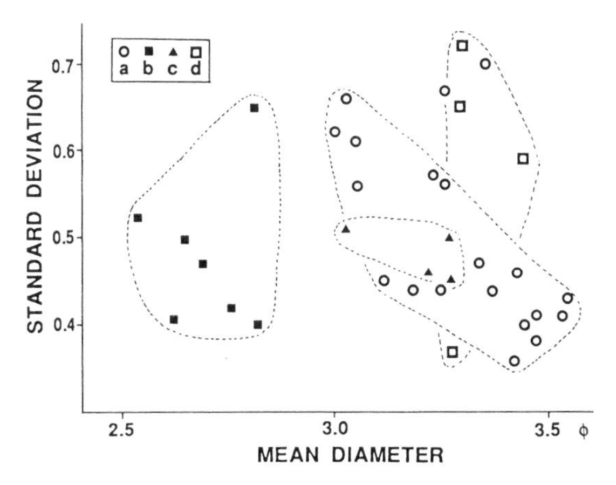

Fig. 6. Grain-size parameters of the climbing-rippled sand and sand layers in the overlying alternation sequence. (a) Unit 1 (ebb deposit); (b) uppermost part of Unit 1; (c) Unit 2 (flood deposit); (d) sand layers in the overlying alternation sequence.

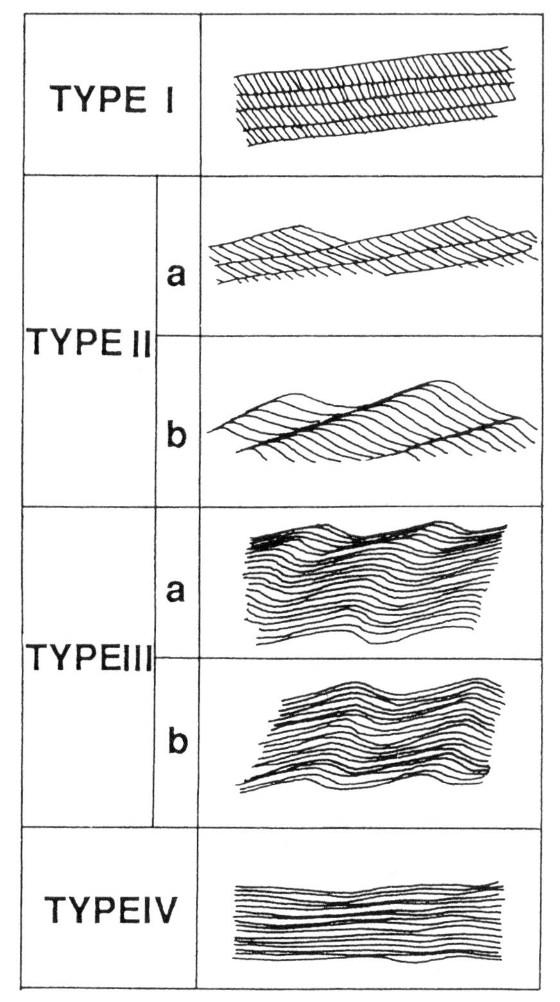

Fig. 7. Types of climbing ripple cross-laminae (for details see text).

lamination occupies a much greater part of Line D than of Line B. This can be explained by differences in sediment supply. If it is assumed that sediment supply initially increased from the lowermost mud layer to Line C before decreasing, then the maximum sediment supply probably existed between Line B and Line C. This combination of a sediment supply curve and a velocity curve can best explain the asymmetrical change in the dominant types of cross-lamination.

In Unit 2 only one isochron can be drawn. The current velocity curve during deposition of Unit 2 can therefore be reconstructed only as a unimodal curve. The isochron is occupied 61% by Types III and IV cross-lamination, the currents acting in the opposite direction to that of Unit 1.

Figure 11 summarizes the relationship between the flow and sediment transport conditions over the complete tidal cycle in the course of which the sand body was deposited. The absence of evidence of reactivation suggests that the channel fill (or the crevasse splay) represented by the lenticular sand body was deposited by continuous sedimentation over a single ebb-flood tidal cycle.

In Unit 1 the current velocity curve shows two peaks during the ebb period. Although many models explaining tidal current velocity use a single maximum (e.g. Allen, 1982), current investigations show that two or three maxima can occur (e.g. Carling, 1981). Carling (1981) reports two peaks during the ebb periods and three peaks during the flood periods in the macrotidal Burry Inlet of South Wales, UK. This bimodal ebb current velocity curve is very similar to the velocity curve reconstructed in the present study.

Earlier studies in modern tidal environments have shown that concentration levels of suspended sediment fluctuate with tidal current velocity (e.g. Collins, 1981; Jago, 1981). This relationship is incorporated in the model for the formation of climbing ripple cross-lamination in the Jizodo Formation presented in this paper (Fig. 11).

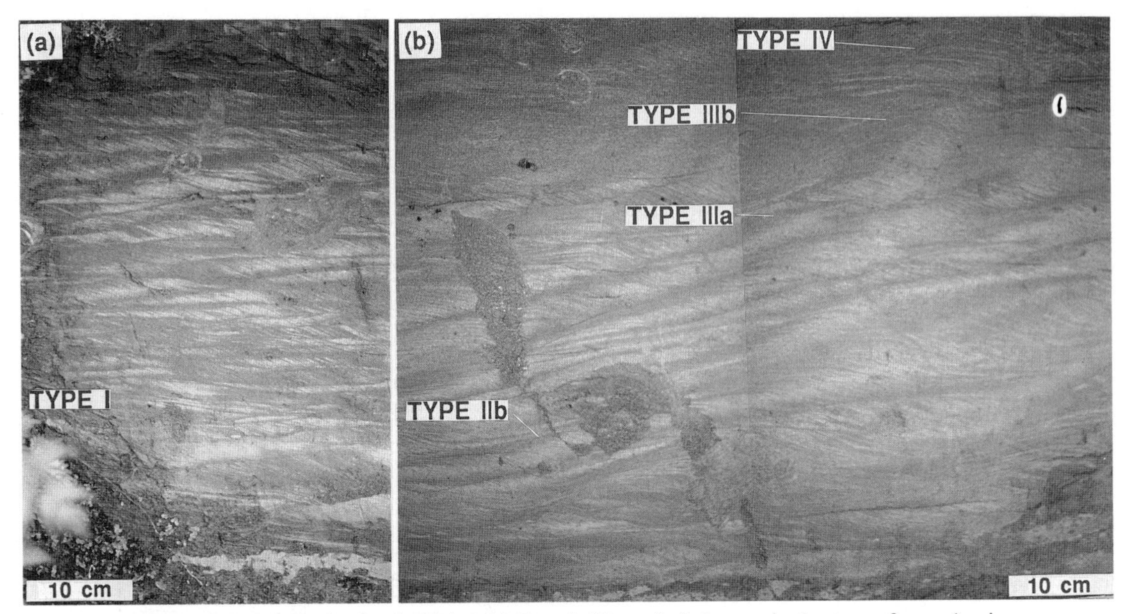

Fig. 8. Climbing ripple cross-lamination in Unit 1 (a) Type I; (b) vertical changes in the type of cross-laminae: lowermost Type IIb changes upward through Type IIIa, Type IIIb and Type IV.

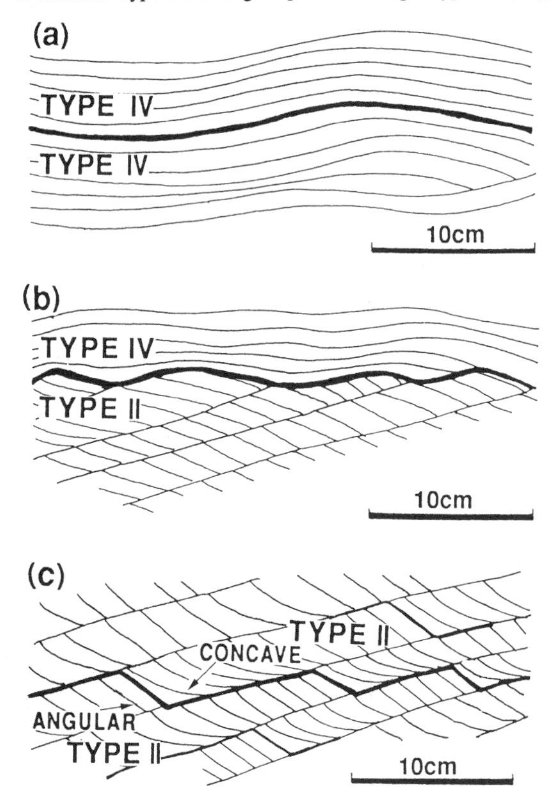

Fig. 9. Three patterns of isochrons in climbing-ripple laminae. (a) Coincidence with Type IV lamination; (b) boundaries marking Type II and Type IV; (c) boundaries where foreset laminae change from angular to concave.

RECONSTRUCTION OF NEAP–SPRING CYCLES

A thinly-bedded alternation of sand and mud layers, about 60 cm thick, overlies the climbing-ripple laminated sand (Fig. 5). This alternation contains 26 sand layers (Fig. 12a). Ten of these layers incorporate current ripple cross-lamination. One sand layer is composed of a single ripple train. Relatively thick sand layers contain one succession of climbing ripples. Each of these sand layers thus corresponds to one flood cycle. Only the lowermost sand layer was formed by an ebb current, all others having been formed by flood currents, as shown by the palaeocurrent pattern. Dalrymple & Makino (1989) have described similar flood-current dominated tidal deposits from the Bay of Fundy in Canada.

Four upward-thickening cycles are recognized in the sands of the lower half of the bed, each being interpreted as having deposited over a single neap–spring period. High-water levels become progressively higher during a neap–spring half cycle and progressively lower during a spring–neap half cycle. Tidal current energy is in general greater during spring tides than during neap tides, resulting in greater transportation and sedimentation during spring tides (Jago, 1981; Kohsiek *et al.*, 1988). Moreover, greater tidal ranges have greater tidal

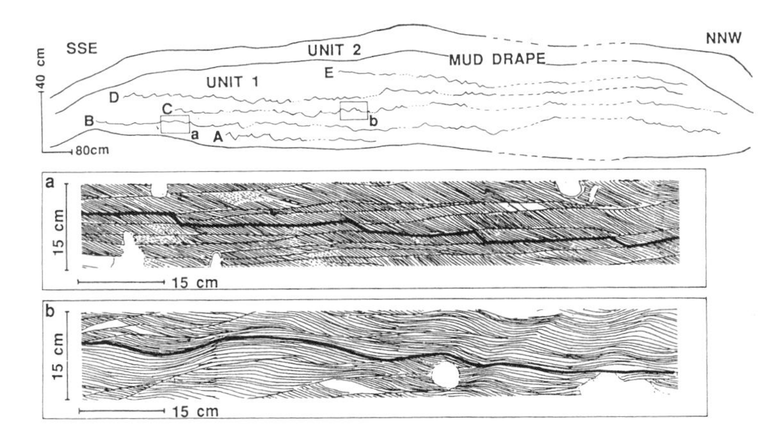

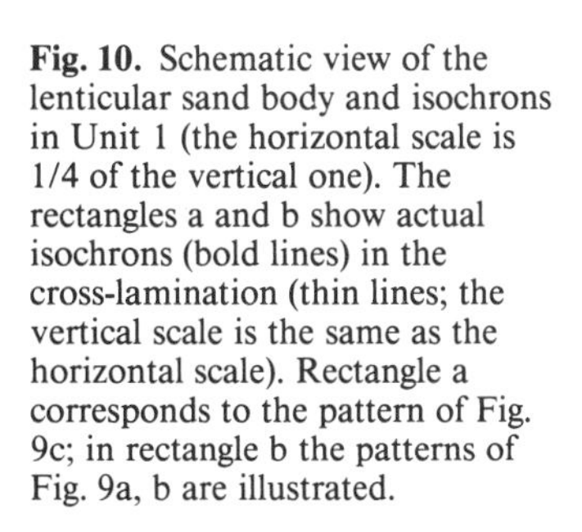

Fig. 10. Schematic view of the lenticular sand body and isochrons in Unit 1 (the horizontal scale is 1/4 of the vertical one). The rectangles a and b show actual isochrons (bold lines) in the cross-lamination (thin lines; the vertical scale is the same as the horizontal scale). Rectangle a corresponds to the pattern of Fig. 9c; in rectangle b the patterns of Fig. 9a, b are illustrated.

current energy. The upward-thickening sand layers are therefore believed to indicate increasing tidal range, one upward-thickening cycle representing one neap–spring half cycle. Along the modern Choshi coast (Pacific coast of central Japan), flood tidal ranges increase towards spring tide, the maximum flood range being recorded just after the highest high-water levels (Fig. 12b). At this maximum flood tidal range, the thickest sand layer is deposited. The thickest sand layer is followed by one of the thinnest sand layers and the intercalated mud layer is particularly strongly bioturbated. This indicates that there must have been a non-depositional period between the laying down of the thickest and thinnest layers, i.e. no sediment was deposited during the spring–neap half cycle. A decreasing number of sand layers in one thickening-upward cycle might indicate that the depositional surface was rising in response to sedimentation (through cycles II, III and IV; Fig. 12c). Another set of similar upward-thickening sand layers was deposited subsequently, following deposition of alternatively thick (*c.* 4 cm) mud layers. These mud layers are highly bioturbated, representing extended periods of low sedimentation.

SUMMARY

A lenticular sand body of the Pleistocene Shimosa Group in Japan, which is dominated by climbing ripple cross-laminations, was examined in detail. The sand body is thought to be a tidally influenced estuarine or lagoonal deposit, as inferred from palaeocurrents and associated facies. Based on the detailed examination of climbing ripple cross-

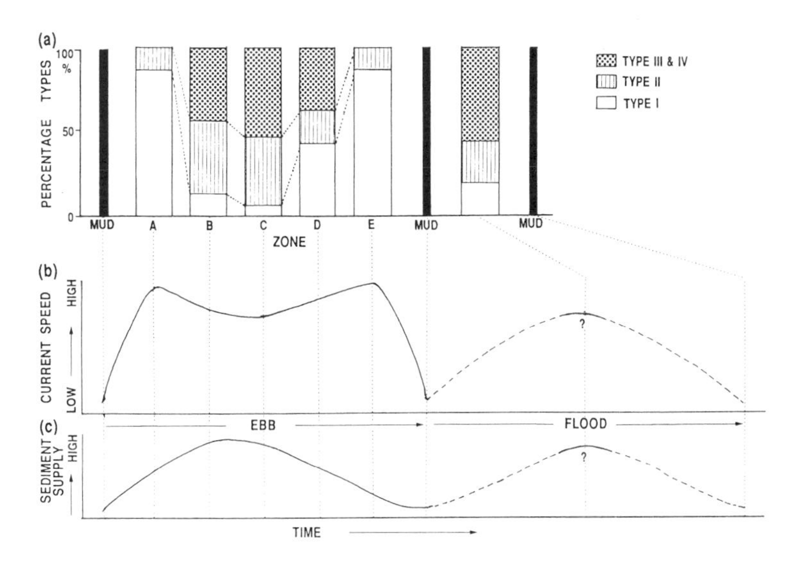

Fig. 11. (a) Percentage occurrence of the various types of cross-laminae along each isochron (letters A, B, C, D and E designate the different isochrons). (b) Current speed and (c) sediment supply are reconstructed from the variation of dominant types of cross-laminae along each isochron.

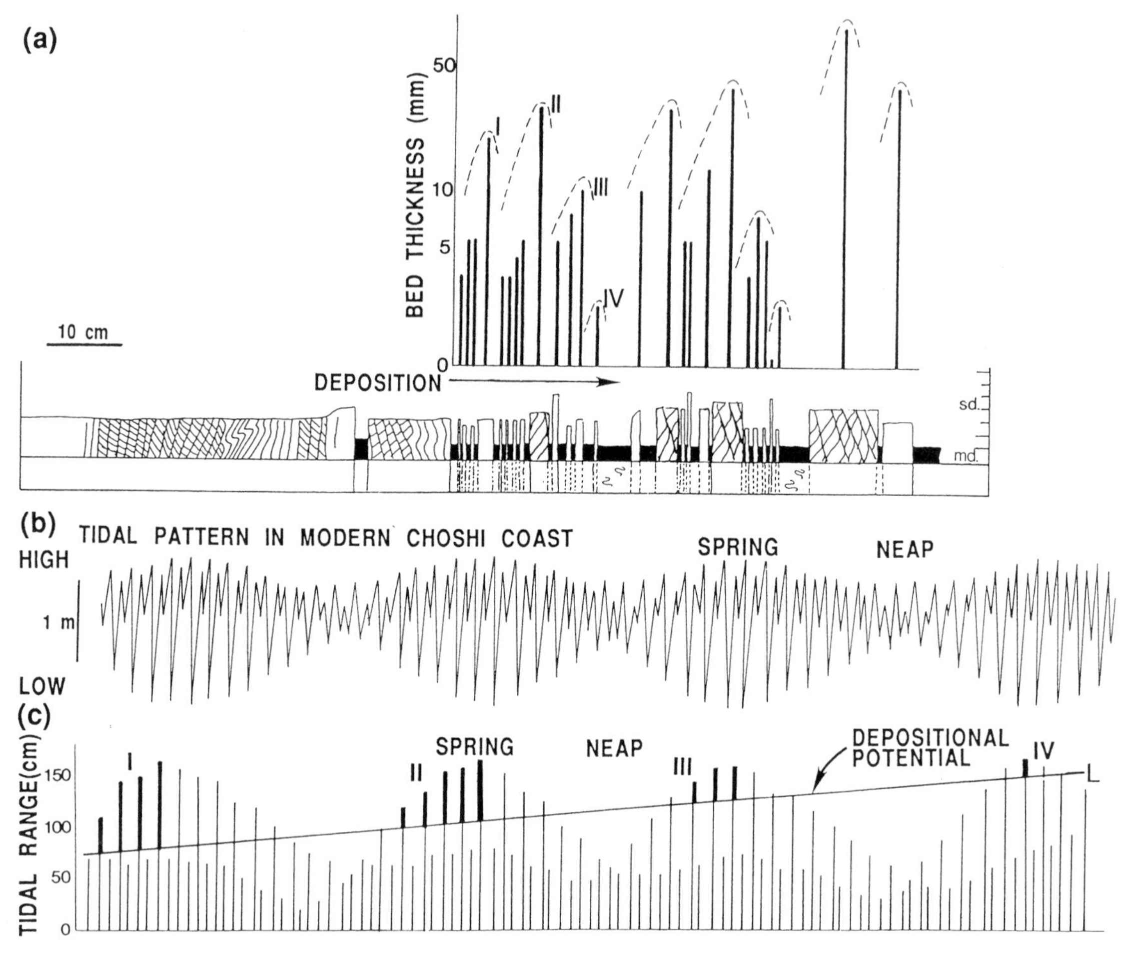

Fig. 12. (a) Thickness variation of the sand beds in the alternation sequence above the climbing-rippled sand. (b) Tidal pattern along the modern Choshi coast in Chiba Prefecture, Pacific coast of central Japan, June–July 1980. (c) Variation of tidal range. Assuming that the depositional surface (line L) rises gradually, the variation of sand layer thickness in (a) can be explained (for details see text).

laminations, isochrons have been drawn along individual bedding surfaces. Four different types of climbing ripple cross-lamination have been recognized and correlated with variable current velocities and sediment supply. Cyclic variations of both tidal current velocity and sediment supply during a single tidal cycle are inferred from the vertical evolution of dominant types of climbing-ripple lamination along each isochron. Similarly to some modern examples, the reconstructed current velocity curve shows two peaks during the ebb period. Moreover, neap–spring tidal cycles have been reconstructed from tidal flat deposits that occur just above the climbing-rippled sand.

ACKNOWLEDGEMENTS

We are much indebted to Dr Richard N. Hiscott, the Memorial University of Newfoundland, and Dr Richard J. Cheel, Brock University, for reading an earlier version of the paper.

REFERENCES

Allen, J.R.L. (1982) *Sedimentary Structures—their Character and Physical Basis* (Vol. I). Elsevier, Amsterdam, 593 pp.

Ashley, G.M., Southard, J.B. & Boothroyd, J.C. (1982)

Deposition of climbing-ripple beds: a flume simulation. *Sedimentology* **29**, 67–79.

CARLING, P.A. (1981) Sediment transport by tidal currents and waves: observations from a sandy intertidal zone (Burry Inlet, South Wales). In: *Holocene Marine Sedimentation in the North Sea Basin* (Eds Nio, S.D., Schüttenhelm, R.T.E. & Weering, Tj.C.E. van). *Spec. Publs int. Ass. Sediment.* **5**, 65–80.

COLLINS, M.B., AMOS, C.L. & EVANS, G. (1981) Observations of some sediment supply to the Wash, U.K. In: *Holocene Marine Sedimentation in the North Sea Basin* (Eds Nio, S.D., Schüttenhelm, R.T.E. & Weering, Tj.C.E. van). *Spec. Publs int. Ass. Sediment.* **5**, 81–98.

DALRYMPLE, R.W. & MAKINO, Y. (1989) Description and genesis of tidal bedding in the Cobequid Bays-Salmon River Estuary, Bay of Fundy, Canada. In: *Sedimentary Facies in the Active Plate Margin* (Eds Taira, A. & Masuda, F.), pp. 151–177. TERRAPUB, Tokyo.

FOLK, R.L. & WARD, W. (1957) Brazos River bar: A study in the significance of grain size parameters. *J. sediment. Petrol.* **27**, 3–26.

ITO, M. & O'HARA, S. (1990) Depositional sequences of the lower part of the Shimosa Group in Kimitsu and Futts Cities, Boso Peninsula, Japan. *Rept. Environ. Res. Organiz. Chiba Univ.* **16**, 1–8 (in Japanese with English abstract).

JAGO, C.F. (1981) Sediment transport measurements in Sizewell–Dunwich Banks area, East Anglia, U.K. In: *Holocene Marine Sedimentation in the North Sea Basin* (Eds Nio, S.D., Schüttenhelm, R.T.E. & Weering, Tj.C.E. van). *Spec. Publs int. Ass. Sediment.* **5**, 283–301.

JOPLING, A.V. & WALKER, R.G. (1968) Morphology and origin of ripple-drift cross-lamination, with examples from the Pleistocene of Massachusetts. *J. sediment. Petrol.* **38**, 971–984.

KOHSIEK, L.H.M., BUIST, H.J., BLOKS, P., MISDORF, R., BERG, J.H. VAN DEN & VISSER, J. (1988) Sedimentary processes on a sandy shoal in a mesotidal estuary (Oosterschelde, The Netherlands). In: *Tide-Influenced Sedimentary Environments and Facies* (Eds Boer, P.L. de, Gelder, A. van & Nio, S.D.), pp. 201–214. D. Reidel, Dordrecht.

KONDO, Y. (1989) Faunal condensation in early phases of glacio-eustatic sea-level rise, found in the middle to late Pleistocene Shimosa Group, Boso Peninsula, central Japan. In: *Sedimentary Facies in the Active Plate Margin* (Eds Taira, A. & Masuda, F.), pp. 197–212. TERRAPUB, Tokyo.

MASUDA, F. (1988) Dynamic stratigraphy—Facies analysis in Paleo-Tokyo Bay area. *J. Japan. Soc. Eng. Geol.* **29**, 312–321 (in Japanese).

REINECK, H.-E. & SINGH, I.B. (1980) *Depositional Sedimentary Environments*, 2nd edn. Springer, New York, 549 pp.

SORBY, H.C. (1859) On the structures produced by the currents present during the deposition of stratified rocks. *Quart. J. Geol. Soc. London* **64**, 171–233.

TOKUHASHI, S. & ENDO, H. (1984) *Geology of the Anegasaki District. Quadrangle Ser., scale 1:50,000.* Geol. Surv. Japan, 136 pp. (in Japanese with English abstract).

TOKUHASHI, S. & KONDO, Y. (1989) Sedimentary cycles and environments in the middle-late Pleistocene Shimosa Group, Boso Peninsula, central Japan. *J. Geol. Soc. Japan* **95**, 933–951 (in Japanese with English abstract).

WUNDERLICH, F. (1969) Studien zur Sedimentbewegung. 1. Transportformen und Schichtbildung im Gebiet der Jade. *Senckenbergiana marit.* **1**, 107–146.

Spec. Publs int. Ass. Sediment. (1995) **24**, 313–328

Internal geometry of ancient tidal bedforms revealed using ground penetrating radar

C.S. BRISTOW

Research School of Geological and Geophysical Sciences, Birkbeck College and University College London, Gower Street, London WC1E 6BT, UK

ABSTRACT

This paper represents the results of a ground penetrating radar (GPR) survey in the Woburn Sands Formation, Lower Greensand Group near Leighton Buzzard. Almost 4 km of survey lines were taken to investigate the internal structure of the sands. The radar survey achieved a resolution on the order of tens of centimetres with a penetration up to 10 m. It revealed a variety of horizontal, dipping and curved reflectors. The reflections are described in terms of radar facies by identifying repeated packages of reflections with similar character and geometry, following the approach for seismic facies analysis. On this basis nine different radar facies have been identified: (i) large sets of planar cross-stratification; (ii) cross-strata with asymptotic toesets; (iii) sets of trough cross-stratification; (iv) cosets of cross-stratification; (v) tidal bundles; (vi) clay drapes; (vii) erosional scours; (viii) diagenetic concretions or voids; and (ix) bioturbated sands.

A three-dimensional reconstruction revealed large bedforms with troughs resembling very large, curved-crested dunes. The troughs are up to 9 m deep and up to 60 m wide. The bedforms are strongly asymmetrical with steep lee slopes and low-angle stoss slopes, indicating a strong time–velocity asymmetry in the tidal currents. The internal erosion and reactivation surfaces are consistent with type II sandwaves of Allen (1980). Most of the bedforms observed appear to be truncated, suggesting that their original heights were substantially greater. Lateral changes in reflection character indicate lateral changes in bedform type and sedimentary structures. These are interpreted to have been formed by bedforms in the classes IA, IIA, IIIA and IVA of Allen (1980). The lateral variations in facies suggest that the Woburn Sands Formation is a complex sheet deposit consisting of very large curved-crested dunes and smaller superimposed dunes. The interpretations are compared with local outcrops in working sand pits.

INTRODUCTION

The Woburn Sands Formation is Aptian to Albian in age and forms part of the Lower Greensand Group which crops out in the Weald Basin, East Anglia, and the Isle of Wight. The thickness and distribution of the Woburn Sands Formation in the Leighton Buzzard area has recently been reviewed by Wyatt *et al.* (1988). In the Leighton Buzzard area the Lower Greensand Group lies with a minor unconformity on Upper Jurassic Kimmeridge Clay and is overlain disconformably by the Gault Clay. According to Bridges (1982) the sands accumulated in a narrow tidal seaway connecting the Boreal Sea to the Tethys Ocean and early North Atlantic Ocean. The strata are believed to be part of a transgressive sequence produced by the early Cretaceous global sea-level rise. Allen (1981a,b), Bridges (1982), Walker (1985) and Buck (1985) have interpreted the Lower Greensand Group as tidal sandwave deposits with local thickness and facies changes controlled by palaeotopography and palaeogeography. Eyers (1991) has suggested that there was a tectonic influence on sedimentation in the Leighton Buzzard area with faults controlling the thickness of the sands, but this has been disputed by Shepard-Thorn *et al.* (1994). The latter describe four informal units within the Woburn Sands: (i)

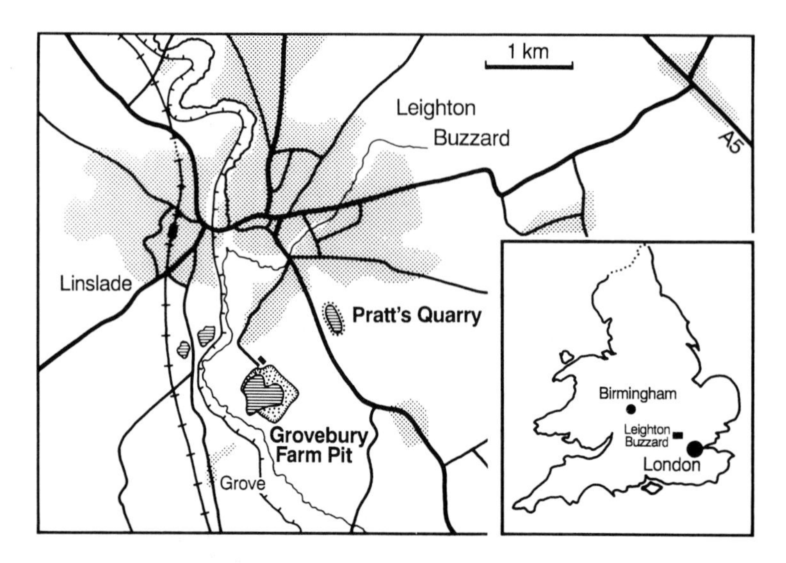

Fig. 1. Map showing the location of the Grovebury Farm and Pratt's sand pits near Leighton Buzzard, southern England.

Brown Sands; (ii) Silver Sands; (iii) Silty Beds; and (iv) Red Sands. The sediments described in this paper form part of the 'Red Sands'.

This paper provides new information on the geometry and internal architecture of the large sedimentary structures locally exposed in sand pits near Leighton Buzzard. The data have been collected from outcrop studies and ground penetrating radar (GPR) surveys in two sand pits, Pratt's Quarry (SP 240N 930E) and Grovebury Farm Pit (SP 230N 923E) (Fig. 1), which both expose sections in the 'Red Sands'. These localities were selected for the GPR survey because they contain coarse-grained, friable sandstones with low matrix content which favours good penetration by this method. The sandstones contain very large sedimentary structures which are clearly resolved by GPR.

STUDY METHODS

Survey method and procedure

In Grovebury Farm Pit, a survey consisting of 16 north–south oriented strike lines and 15 east–west oriented dip lines with two additional diagonal lines (Fig. 2) was made using a SIR System-3 ground penetrating radar (GPR) with a 300 MHz transducer. The survey was arranged in a grid with 10-m spacing. In this way almost 3250 m of GPR data were collected. In Pratt's Quarry, a smaller survey totalling approximately 720 m with four dip lines and 18 strike lines was carried out. Excellent penetration and resolution were obtained down to 140 ns two-way travel time (twt), which corresponds to a depth of approximately 8–9 m. Resolution of tens of centimetres was achieved in this interval. Beneath 140 ns resolution is limited by the influence of the water table.

Principles of radar interpretation

Gawthorpe *et al.* (1993) advocate a seismic stratigraphical approach to GPR data following the principles of seismic stratigraphy as described by Brown & Fisher (1980). This method relies upon the identification of reflector terminations on the radar image. These terminations indicate unconformity surfaces that define genetically related packages of strata. Gawthorpe *et al.* (1993) refer to the boundaries as *radar sequence boundaries* which define radar sequences as *fundamental stratigraphical units identifiable on radar data.* Accordingly, the radar sequence boundaries in the Madison River are interpreted as being equivalent to bedsets or parasequences (*sensu* Van Wagoner *et al.,* 1990).

A more detailed qualitative interpretation of GPR profiles may be achieved by identifying repeated packages of reflections with similar character and geometry following the approach for seismic facies analysis outlined by Mitchum *et al.* (1977) and Brown & Fisher (1977). Packages of reflections with similar configuration are termed *radar facies* (Gawthorpe *et al.*, 1993).

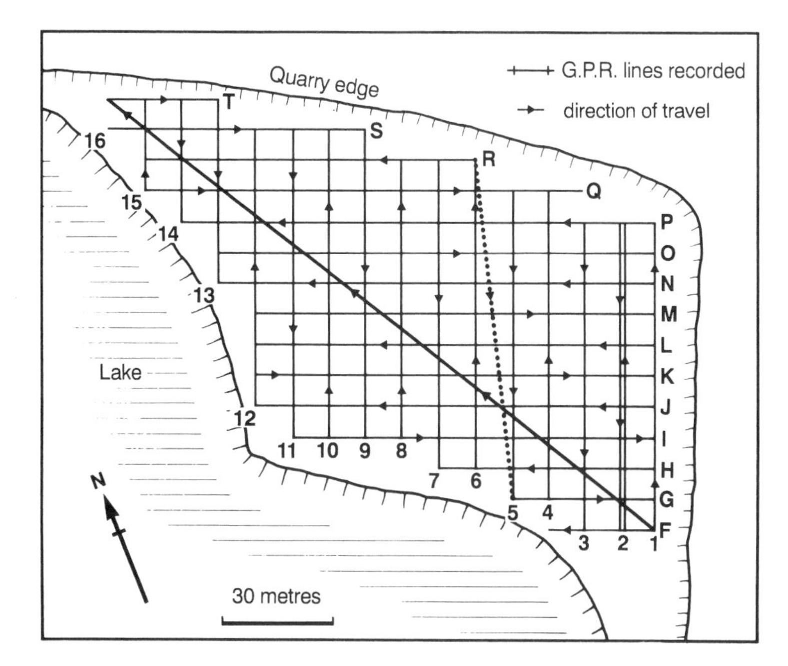

Fig. 2. Sketch map of the Grovebury Farm sand pit showing the location of the GPR survey lines.

The radar facies are controlled by the dielectric properties of the sediment and the arrangement of packages of sediment with similar or repeating dielectric properties. In sedimentary rocks the dielectric properties are primarily controlled by lithology and water content (McCann *et al.*, 1988). In the Woburn Sands Formation water content is a function of porosity and height above the water table. All profiles recorded in this study were taken above the water table.

RESULTS

Radar facies and mineralogical composition

In the Greensand nine different radar facies have been identified from their reflection character and geometry (Bristow, 1995). These are interpreted as: (i) large sets of planar cross-stratification; (ii) cross-strata with asymptotic toesets; (iii) sets of trough cross-stratification; (iv) cosets of cross-stratification; (v) erosional scours; (vi) tidal bundles; (vii) clay drapes; (viii) diagenetic concretions or voids; and (ix) possibly bioturbated sands (Fig. 3).

The Greensand is largely composed of detrital quartz with minor amounts of glauconite and limonite. Diagenetic minerals include goethite and rare carbonate concretions. Goethite occurs in the form of grain coatings and such grains appear to have been locally concentrated by hydraulic sorting. Colour changes at outcrop, being attributed to iron-bearing clay minerals, appear to follow primary sedimentary structures. The porosity of the sands is largely controlled by grain size and sorting, both of which are determined by the hydraulic conditions at the time of deposition. It may be concluded that changes in both composition and porosity in the Woburn Sands Formation are related to primary sedimentary structures. Changes in dielectric properties and reflection character in the Woburn sands should therefore correspond to these primary sedimentary structures. This is the principal tenet behind the interpretations of reflection profiles.

Line descriptions and interpretations

A representative sample of four lines from the survey grid in Grovebury Farm Pit are shown in Figs 4 & 5. Lines Q and K are NW–SE trending and lie perpendicular to lines 6 and 10 which trend NE–SW. The location of the survey lines is shown in Fig. 2. The vertical axes on the sections are scaled in nanosecond (ns) two-way travel time (twt) as recorded. For estimations of depth, a velocity of 15 ns m^{-1} twt is assumed. This corresponds to approximately 0.13 m ns^{-1}. The horizontal scale is

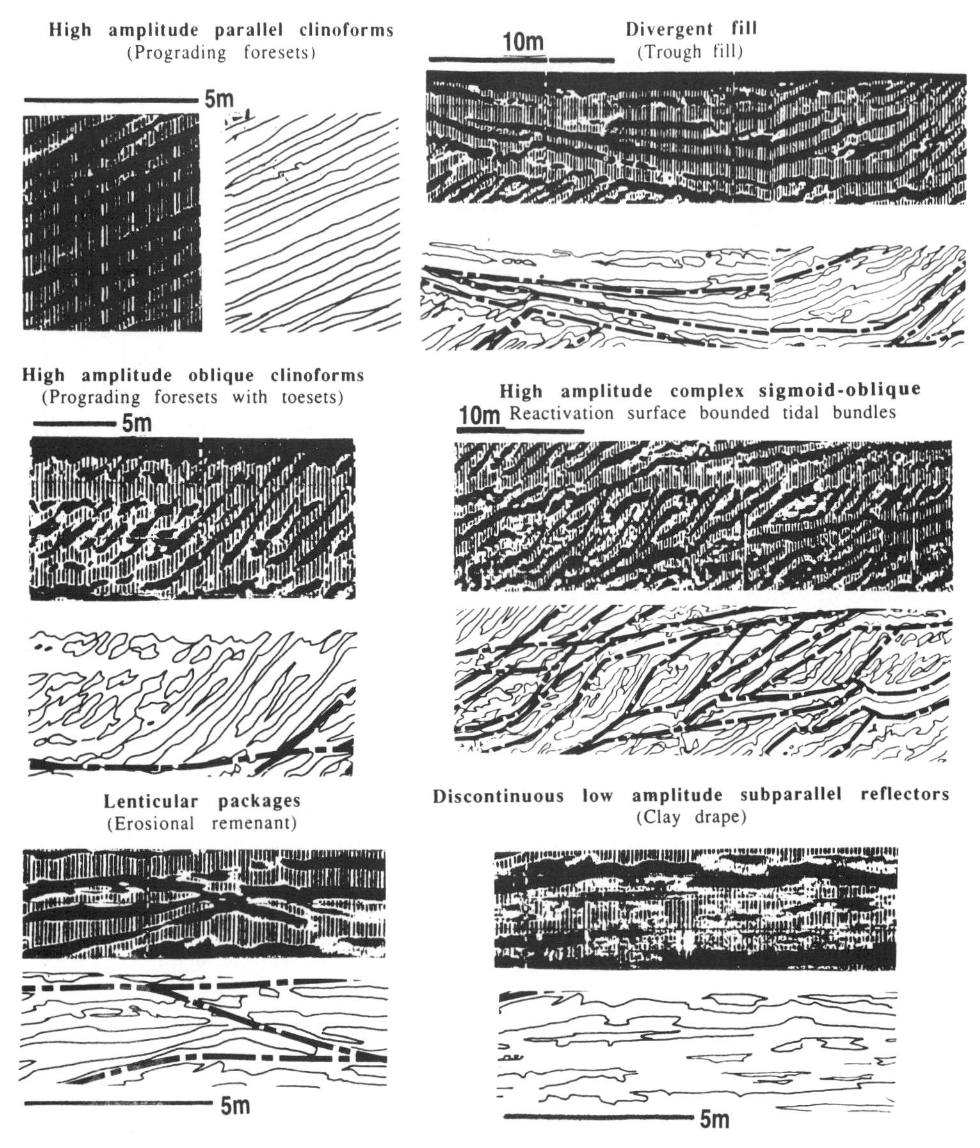

Fig. 3. Radar facies identified in the Woburn Sands Formation on the basis of reflection character and geometry, interpreted as a product of primary sedimentary structures within the sands.

shown in tens of metres measured by chain survey on the ground and marked at the intersection of the survey grid.

Line Q

Line Q is oriented approximately NW–SE, running slightly oblique to palaeoflow in Grovebury Farm pit. Its location is shown in Fig. 2, the cross-lines 15–3 being spaced at 10-m intervals. Between lines 15 and 11, line Q shows bundles of reflectors dipping left to right separated by low-angle reflectors which also dip left to right. Between lines 10 and 7, the upper part of the profile shows divergent reflectors. From 7 to 3 bundles of dipping clinoform reflectors can be seen climbing from left to right.

Line Q is interpreted as showing cut-and-fill structures including superimposed bedforms with

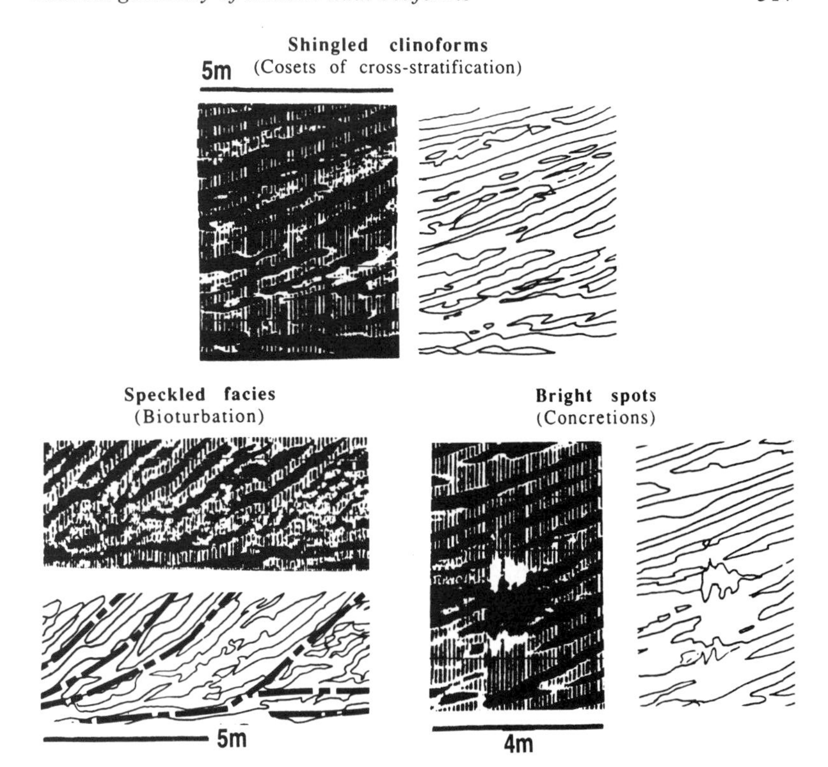

Fig. 3. *(Continued)*

both downstream and upstream accretion. The section between lines 15 and 11 is interpreted as showing cosets of cross-stratification dipping gently downcurrent as they prograded down the face of a larger bedform.

Between lines 11 and 7, the divergent reflectors at the top of the section are interpreted as an oblique section through the trough of a large 3D dune which has scoured down more than 4 m into the underlying sand. The top of the 3D dune is truncated so that only a part of the bedform has been preserved. The original height of the dune must therefore have been greater than 4 m.

Between lines 7 and 3 the dip of the low-angle bounding surfaces suggest that the bedforms accreted vertically as they migrated forward. This indicates either bedform climb, where the rate of sediment accumulation exceeds the rate of bedform migration, or upstream accretion of dunes on the back of a larger dune bedform.

The well-defined packages of reflectors between lines 5 and 3 are interpreted to represent tidal bundles similar to those shown in Fig. 6. The break in reflections at line 4 may indicate an hiatus between phases of sedimentation. Breaks in sedimentation can be clearly distinguished at bioturbated horizons at outcrop (Fig. 6). The top of the section is flat, all bedforms having been truncated by a submarine ravinement surface prior to deposition of the Passage Beds which pass up into the Gault Clay.

Overall, the profile shows a variety of bedform type and superposition with 3D dunes in excess of 4 m amplitude and tidal bundles migrating up and down larger bedforms.

Line K

Line K was recorded parallel to line Q, 60 m to the south (Fig. 2). Between lines 12 and 10 clinoform and oblique reflectors can be seen thickening to a maximum of 90 ns twt. From 10 to 8 these reflectors pass into shingled reflectors truncated by parallel and oblique reflectors between 8 and 7 which, in turn, pass into divergent reflectors. Beneath the divergent reflectors very large shingled clinoforms can be seen truncating against parallel reflectors at 90–100 ns twt. These very large shingled reflectors dominate the profile from lines 5 to 1.

The clinoform reflectors between lines 12 and 10

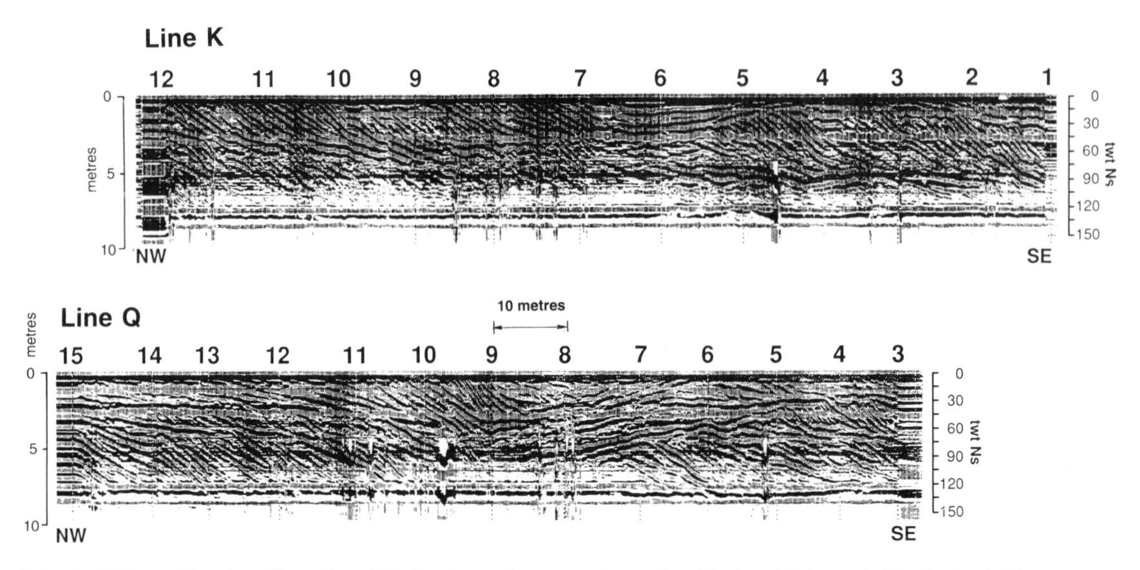

Fig. 4. GPR profiles along lines Q and K showing reflector packages described and interpreted in the text. The numbers across the top of the profile correspond to intersections with cross lines on the survey grid (Fig. 2), spaced at 10 m intervals. The vertical scale is in nanoseconds (ns) two-way travel time (twt).

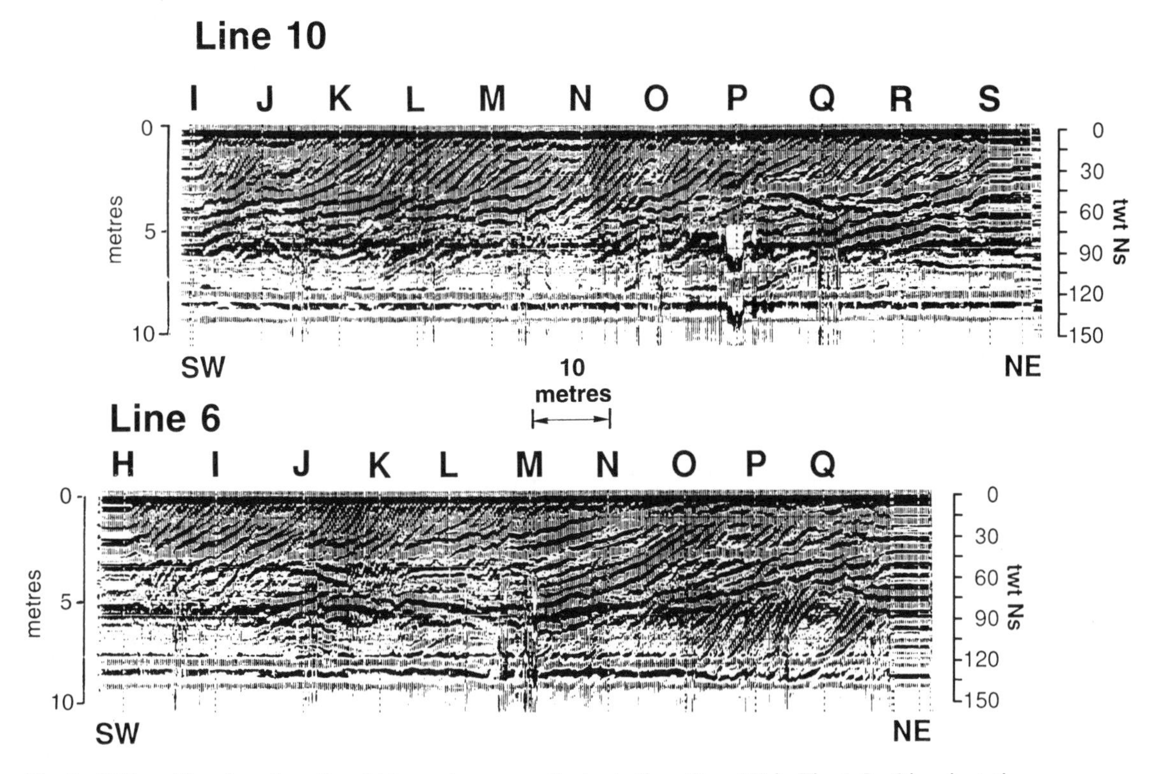

Fig. 5. CPR profiles along lines 6 and 10 running perpendicular to lines Q and K in Fig. 4. In this orientation bedforms can be seen migrating from right to left (NE–SW) in a section oblique to the palaeoflow. The numbers across the top of the profile correspond to intersections with cross lines on the survey grid (Fig. 2), spaced at 10 m intervals. The vertical scale is in nanoseconds (ns) two-way travel time (twt).

are interpreted as prograding foresets with tangential and oblique reflectors at reactivation surfaces marking small-scale breaks in sedimentation.

The shingled clinoforms between lines 10 and 8 are interpreted as bundles of cross-stratification with multiple reactivation surfaces. The regular spacing of the shingled reflectors suggests that these might be tidal bundles. The shingled reflectors, which are clearly truncated by the oblique and divergent reflectors between lines 8 and 7, pass back into tangential clinoforms around line 8.

The oblique and divergent reflectors are interpreted as a large set of trough cross-stratified sand, approximately 3 m thick and 20 m wide, which eroded down into the older sediments while prograding obliquely across and out of the line of section. This set of trough cross-stratification is correlated with the trough form between 11 and 6 in line Q. By comparing the two sections it is apparent that the trough filled and shallowed as it prograded.

The large, shingled progrades between lines 5 and 1 are interpreted as sets of cross-stratification with abundant reactivation surfaces. Some of the reactivation surfaces may be due to reversing tidal flows but outcrop studies indicate that many are marked by extensive bioturbation indicating a longer time gap. Alternatively pause planes could be due to bedform superposition on dunes which were at least 6 m high.

Line 10

Between lines S and P, the section down to 45 ns shows parallel and oblique reflectors above low-angle dipping reflectors. At the intersection with line P there is a bright spot at around 60 ns, reflectors beneath having been pushed down. Between lines P and Q, large oblique clinoforms down to 75 ns are developed, dominating the section to line K which shows parallel reflectors downlapping onto larger oblique clinoforms. Between lines K and I, the reflectors between the large oblique clinoforms appear to be shingled.

The clinoform and oblique reflectors at the top of the section, between lines S and P, are interpreted as prograding sets of cross-stratification deposits with abundant reactivation surfaces or pause planes at the larger oblique reflectors. The low-angle dipping reflectors beneath 45 ns are correlated with parallel clinoforms in line Q.

The two sets of apparent dips can be resolved as foresets within a southerly prograding dune. The combined parallel and oblique reflectors above 45 ns between lines L and P are interpreted as prograding, cross-stratified sets. The pushing down of reflectors beneath the bright spot at the intersection of line P indicates that the bright spot is a low velocity zone. In seismic interpretation this would indicate low density, in radar it probably represents low density and high resistance, characteristic of a void. The origin of the void is unknown.

Line 6

At the north-eastern end of line 6, low-angle dipping reflectors truncate dipping clinoforms down to 60 ns. At the intersection with line Q, small oblique clinoforms at the top of the sections occur, the low-angle reflectors correlating with shingled reflectors in line Q. The clinoform reflectors at the base of the section correlate with low-angle oblique clinoforms.

Large divergent reflectors start at line P and cut down to 75 ns at line M. From line N the top of the divergent clinoforms are filled by smaller oblique and parallel clinoform reflectors above 45 ns which extend laterally to line J. From line J these reflectors are truncated by larger oblique reflectors with many internal truncations.

Correlation of the oblique clinoforms at the top of the section, at the intersection of lines Q and 6, resolves into a true sedimentary dip towards the south. The low-angle dips in line 6 correlate with shingled reflectors in line Q, suggesting that the bedforms with apparent tidal bundles are of lower relief than the 3D dunes. This may indicate that increased flow asymmetry associated with tidal bundles leads to a more swept-out, lower amplitude bedform.

The very large divergent reflectors which extend down to 75 ns are interpreted as a very large, 5 m deep and 60 m wide, trough which is partially filled and truncated by overlying bedforms. These are correlated with a clear trough form shown in the perpendicular orientation at the intersection with line K. The regular truncation of oblique clinoform reflectors between L and H at the top of the section can be correlated with tidal bundles at outcrop.

Line C

Line C is a NW–SE trending line in Pratts Pit (Fig. 7a) which shows very large parallel clinoform reflectors with occasional truncations where reflectors downlap at minor unconformities. A photo-

Fig. 6. Outcrop photograph of the Woburn Sands Formation in Grovebury Farm Pit showing trough cross-stratified sets arranged in bundles and cosets picked out by iron cements. These cross-stratified sands are divided by packages of bioturbated sands which indicate reduced rates of deposition. The scale of sedimentary structures and packaging of sediments is comparable with the complex sigmoidal–oblique reflectors shown in Figs 5 & 7. The smaller-scale packaging of foresets into cosets may represent shingled clinoforms, whereas bioturbated sands might be represented by a speckled radar facies.

graph of a section subsequently excavated in this area shows very large cross-stratified sets with reactivation surfaces which correlate with the downlapping radar reflectors (Fig. 7b). The scale of radar resolution (a few dm) is in this case at the coset rather than the foreset level. The cosets are considered to be too large to represent tidal bundles and it is suggested that they indicate bedform superimposition as described by Rubin & McCulloch (1980). The 8–10 m high sets are thought to have been generated by an exceptionally strong storm-driven flow. The reactivation surfaces may represent disequilibrium conditions where smaller, 2–4 m high superimposed bedforms are migrating over the very large relict dunes, indicating a phase lag between flow and bedform.

An alternative hypothesis could incorporate the 'nested' boundary layer theory (see Ashley, 1990), where larger bedforms generate a boundary layer in which smaller bedforms are locally stable (Rubin & McCulloch, 1980). This theory could explain many of the features described above, except the dominance of cut-and-fill structures within the isometric diagram (Fig. 8).

DISCUSSION

Reconstruction of bedform scale and orientation

A line drawing of the reflectors recorded in Grovebury Farm pit from alternate dip and strike lines is shown in Fig. 8. This isometric reconstruction shows the form of the preserved cross-stratified sets. It is evident that the sedimentary structures are mostly trough shaped, indicating deposition from curved-crested 3D dunes (Fig. 9). Bounding surfaces between dunes can be seen to dip both upstream and downstream. It is clear that the cross-stratification, while very large, is essentially similar to smaller-scale sets of cross-stratification found in intertidal dunes (De Raaf & Boersma, 1971; Dalrymple, 1984; Nio & Yang, 1991a). The most striking feature is the exceptional size of the structures.

It should be noted that all of the bedforms described here appear to be truncated, suggesting that the original heights of the bedforms were substantially greater. Assuming a maximum bedform height of at least 10 m, this would correspond to a spacing of 200 m or more (Flemming, 1988; see also Ashley, 1990). Extrapolation from the depth/flow velocity diagram of Costello & Southard (1981) suggests a flow velocity of at least $1\ m\ s^{-1}$.

The bedforms evidently were strongly asymmetrical, with steep lee slopes and low-angle stoss slopes, indicating a high time/velocity asymmetry in the tidal currents (Allen, 1981b). The internal erosion and reactivation surfaces are consistent with type II sandwaves of Allen (1980). Lateral changes in reflection character indicate lateral changes in facies as shown in Fig. 4, where occasionally truncated clinoform reflectors increase in thickness up to 9 m high from right to left. This profile is interpreted to show large type I sandwaves eroding down and increasing in amplitude from right to left, which should indicate unidirectional flow (Allen, 1980). Fluctuations in flow velocity, superimposed bedforms and possible flow reversals are proposed to account for the reactivation surfaces. This would suggest a change from type I to type II bedforms. In Fig. 5 clinoform reflectors pass into complex sigmoidal to oblique reflectors and a divergent fill which overlies high-amplitude parallel and oblique clinoforms.

This lateral change in radar facies may represent the transformation of a single large sand wave with well-defined foresets into cross-stratified cosets and then oblique fill of the trough which overlies an older, partially preserved sand wave. The speckled facies in the toesets of the sand wave and the bioturbation seen at outcrop (Fig. 6) indicate phases of stillstand between migration events (type IIIA).

In line 6 (Fig. 5) complex sigmoidal reflectors, prograding from right to left, are overlain by a divergent trough fill. The complex sigmoidal to oblique reflectors are interpreted as pulsed accretion on a type IV A sandwave (Allen, 1980), overlain by an asymmetric trough fill from a large curved-crested dune. The preserved trough fill was approximately 6 m deep and 30 m wide. The underlying sand wave had a height of more than 3 m. However, because the top of the bedform has been eroded off, it is not possible to say exactly how high the bedform had originally been. Many sets of cross-stratification can be seen to scour down and then fill up (Fig. 8). This is interpreted to indicate bedform instability due to changes in flow conditions. It is suggested here that the very large bedforms were formed during exceptional events and then filled up during more normal conditions. The flat top of the Woburn Sands Formation is interpreted to represent a submarine ravinement surface.

The lateral variations in facies suggest that the Woburn Sands Formation was formed by a complex sheet of curved-crested dunes with extensive bedform superimposition, which was exposed to occasional very strong flow events and some tidal flux. Most descriptions of very large ancient cross-stratified deposits from shallow marine environments deal with straight-crested bedforms (e.g. Nio, 1976; Nio & Siegenthaler, 1978; Jerzykiewicz & Wojewoda, 1986; Kreisa *et al.*, 1986; Smith &

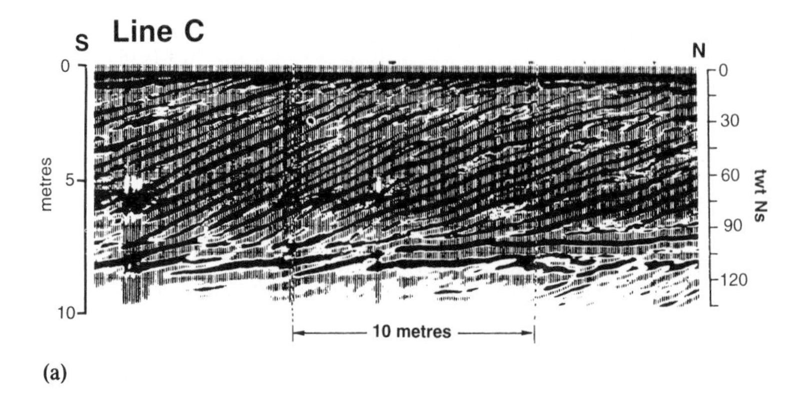

(a)

(b)

Fig. 7. (a) NW–SE trending radar profile in Pratt's Pit showing very large parallel clinoform reflectors with downlapping reflectors at reactivation surfaces. (b) Outcrop photograph of recently excavated very large, cross-stratified sets in Pratt's Pit which correlate with the reflectors seen in Fig. 7a.

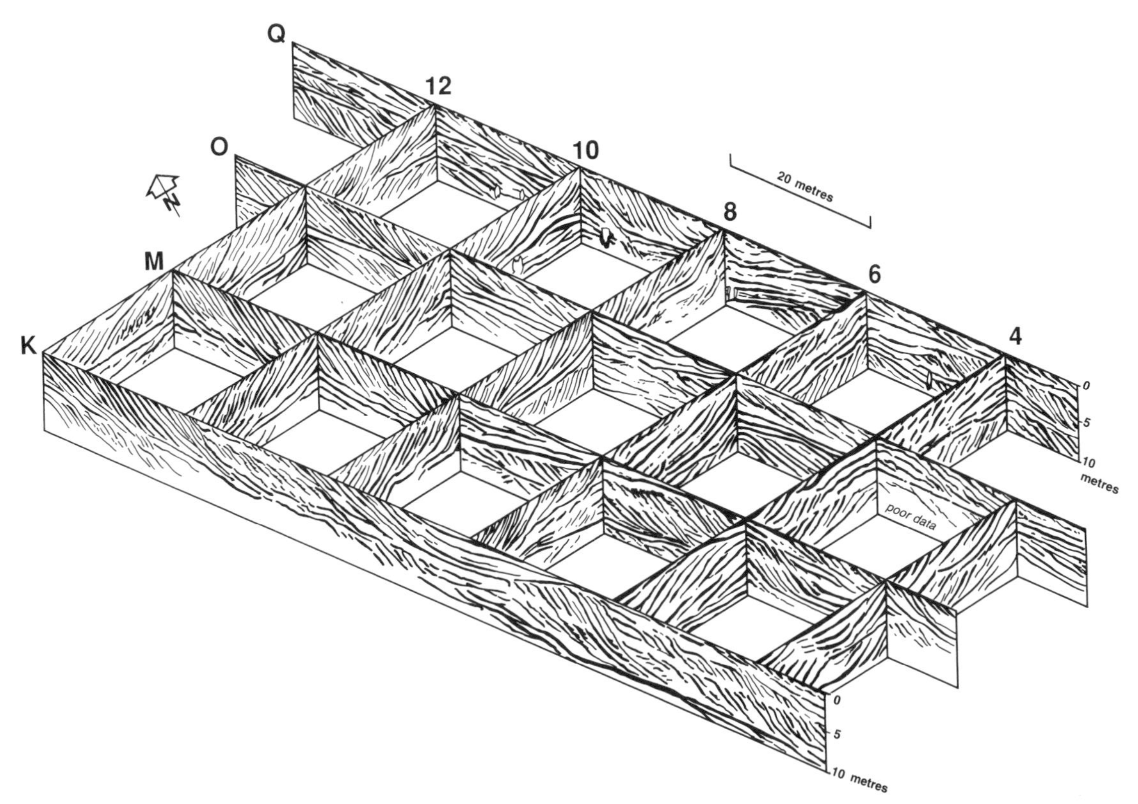

Fig. 8. Schematic reconstruction of bedforms in the Lower Greensand at Grovebury Farm Pit showing small 3D dunes superimposed on very large 3D dunes with 20–30 m wide slip faces, heights of at least 5 m and a spacings of approximately 100 m.

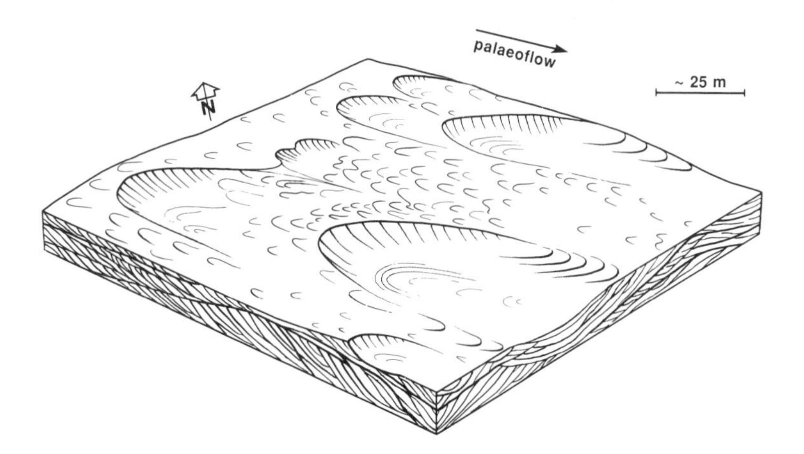

Fig. 9. Isometric diagram of radar profiles in Grovebury Farm Pit which shows the southerly trending very large trough cross-stratification in the Woburn Sands Formation.

Tavener-Smith, 1988; Sztano & Tari, 1992). The curved-crested 3D dunes described in this paper must have been formed under exceptionally high current velocities, flowing from north to south through the postulated palaeo-seaway. There is a very strong asymmetry to the bedforms and although tidal bundles are locally present, many of the reactivation surfaces and pause planes within the sands are suspected to indicate longer time breaks between dune activity. These longer time breaks may indicate a storm-driven component within the palaeoflow regime. Further evidence for breaks in sedimentation come from bioturbated horizons at outcrop. Pollard *et al.* (1993) have distinguished colonization windows between sedimentation events and have suggested a bedform migration rate of 20–70 m yr^{-1}.

It is proposed here that the tidal flow through the Cretaceous seaway may have been augmented by strong storm-driven currents flowing from the Boreal Sea to the Tethys Ocean. The southerly palaeoflow may indicate that the dominant storm winds blew from the north-east, which would be appropriate to a palaeolatitude of *c.* 35° N during the early Cretaceous.

Using the 'Bedforms' computer simulation package of Rubin (1987) it was possible to generate cross-stratified sets resembling those seen at outcrop in Leighton Buzzard (Fig. 10a & b). Figure 10a was generated by superimposing straight-crested bedforms on larger sinuous-crested ones migrating at a slower rate. In this model, cosets of cross-stratification resembling tidal bundles were generated without including current reversals. In Fig. 10b out-of-phase sinous-crested bedforms were superimposed on the large curved-crested ones. In this

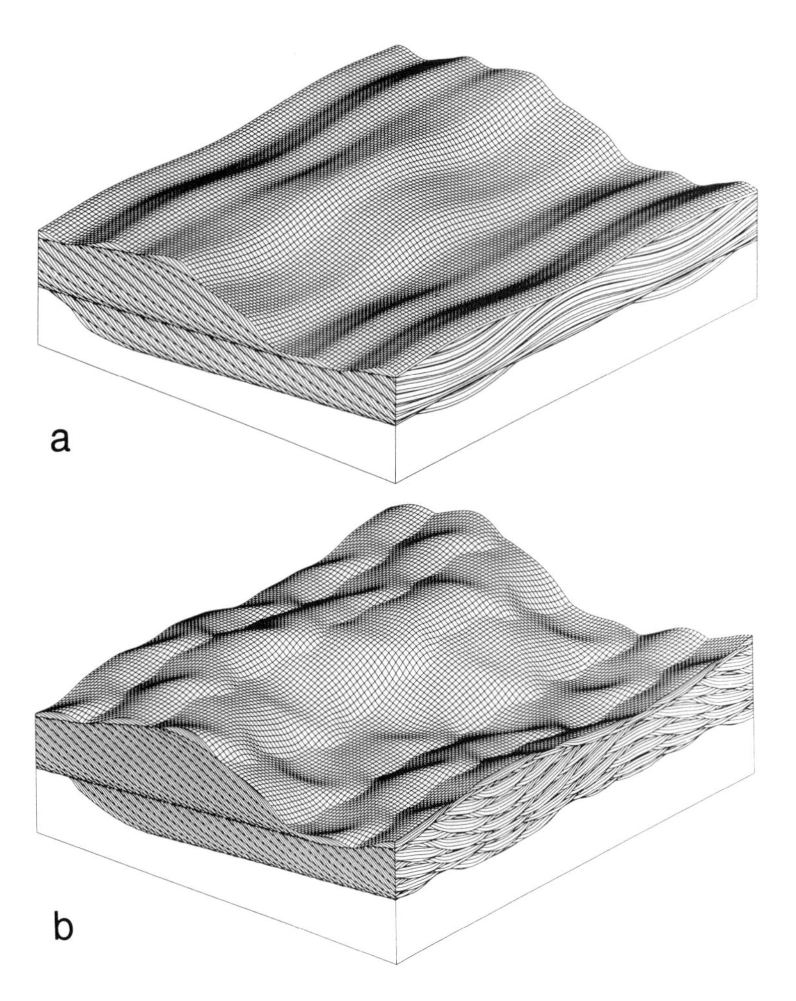

Fig. 10. (a) Cosets of cross-stratification simulated by superimposing straight-crested dunes on larger, slower moving curved-crested bedforms using the 'Bedforms' program of Rubin (1987). (b) Out-of-phase sinuous-crested bedforms superimposed on large curved-crested bedforms generates cosets of cross-stratification and 'nested' sets of trough cross-stratification. Although this program cannot accommodate changes in bedform geometry over time, many of the features found in the Red Sands are represented here without requiring current reversals.

manner cosets of cross-stratification were produced in the dip section and trough cross-stratification within troughs in the strike section.

Comparing these models with the isometric reconstruction in Fig. 8, it would appear that some combination of superimposed straight- and curved-crested bedforms is the most likely explanation for the sedimentary structures observed in the Woburn Sands Formation. Unfortunately it was neither possible to alter bedform geometry within a simulation run nor to model the tendency for bedforms to scour down and fill up as illustrated in Fig. 8. It is suggested that the large curved-crested bedforms were generated by exceptional flow events and subsequently overtaken by smaller straight-crested dunes. The very large bedforms were probably formed by storm-induced currents, whereas the smaller bedforms may represent normal tidal bedforms even though there is no clear evidence for tidal current reversals in the Red Sands.

Comparison with ancient analogues

Very large sets of cross-stratified sands have been described from a variety of modern and ancient sedimentary environments. These include: aeolian (McKee, 1979), fluvial (McCabe, 1977) and shallow marine (Smith & Travener-Smith, 1988). Previous studies of the Lower Greensand have clearly shown these to be of shallow marine origin (Allen, 1981a,b; Bridges, 1982; Buck, 1985; Walker, 1985). Within ancient shallow-marine sandstones, giant cross-stratified sets have been interpreted to represent shoreface-connected ridge deposits (Smith & Tavener-Smith, 1988). Surlyk & Noe-Nygaard (1992) have interpreted very large sets of cross-stratification in the Late Jurassic to Early Cretaceous Raukelv Formation as offshore prograding bar deposits, a model that is clearly not appropriate for the Woburn Sands. Giant foresets have been described from the Upper Cretaceous Radkow and Szcelinec sandstones in the Bohemian Basin (Jerzykiewicz & Wojewoda, 1986). These appear to be laterally persistent and straight-crested, being interpreted to represent tectonically controlled prograding foresets adjacent to a submarine fault scarp. Kreisa *et al.* (1986) describe giant cross-stratification in the Permian Rando Roja Sandstone in Arizona thought to have formed in an estuary or narrow seaway.

Nio (1976) and Nio & Siegenthaler (1978) attributed very large cross-stratified sets in the Roda Sandstone to an ancient sandwave complex, although these are now being interpreted as prograding delta deposits (Nio & Yang, 1991b). Phillips (1984) has described very large trough cross-stratified sets from the Miocene Santa Margarita Sandstones in California which he interpreted as tide-generated sedimentary structures in a shallow seaway. These display many features similar to those seen in the Woburn Sands, including abundant reactivation surfaces and bioturbated intervals, although there are very few conglomerates present in the Red Sands and less direct evidence of reversing currents. It was suggested that the giant bedforms had moved infrequently, possibly only during spring tides. However, in the case of the Woburn Sands, much longer breaks in sediment transport are postulated.

Comparison with modern analogues

Sandwaves or dunes with heights greater than 5 m have been described from the North Sea amongst others by Terwindt (1971) and McCave (1971). Lanckneus & de Moor (this volume, pp. 33–51) described straight-crested dunes on the Middelkerke Bank (southern North Sea). Bartholdy (1992) has described bedforms >5 m in height and up to 200 m in length from a tidal channel of the Danish Wadden Sea, where peak flow velocities are given as approximately 1.5 m s^{-1} at spring tide. The bedforms are straight-crested and have been observed to migrate 40 m yr^{-1}.

Flemming (1978) described very large 3D dunes from the southern African continental shelf. These appear to be supply-limited forms, a concept which is supported by Belderson *et al.* (1982) but which is clearly not applicable here.

Berné *et al.* (1988) describe the internal structure of subtidal sandwaves in the English Channel near Surtainville on the French coast. The bedforms were mapped by side-scan sonar and revealed a curved-crested barchanoid form with superimposed bedforms closely resembling the Leighton Buzzard reconstructions (Fig. 10a & b). The orientation of superimposed bedforms at Surtainville changes as the bedforms migrate over the larger dunes. This is likely to have occurred in the Woburn Sands, although this feature could not be incorporated into the model simulations. The internal structures of the dunes at Surtainville were revealed by high resolution seismic reflection profiling (Berné *et al.*, 1992) which, although having a lower resolution than GPR, showed many similar features to the Woburn Sands, including numerous internal erosion and reactivation surfaces. More recent profiles across the bedforms, reproduced in Berné *et al.* (1992), show troughs in cross-section. The bedforms have similar heights and lengths to those observed at Leighton Buzzard and the internal geometry is very similar. However, the bedforms described by Berné *et al.* (1988, 1992) appear to be supply-limited forms on a scoured shelf and are thus not quite comparable to those observed in the Woburn Sands. Berné *et al.* (1988) suggested that the sandwaves of Surtainville were relatively stable and that migration was slow with long-term phenomena, such as equinox cyclicity and storm events, generating the internal structures rather than lunar or other shorter period cyclicities.

The structures discussed in this study are attributed to very large dunes, the sinuous-crested 3D planforms reflecting exceptionally high palaeoflow velocities approaching 1–1.5 m s^{-1}. The dominance of unidirectional (southerly) palaeocurrents suggests either a high tidal asymmetry or storm-driven control, while the presence of bioturbated layers within the sands indicates that such strong currents were only intermittently active. During the early Cretaceous Leighton Buzzard lay in a palaeovalley which has been called the Bedforshire Straights (Bridges, 1982). It is possible that currents flowed from the Boreal Sea in the North to the Tethys Ocean in the South through the Bedfordshire Straights. It is suggested here that these southerly flowing, possibly storm-driven currents produced the very large bedforms in the Lower Greensand. Longer breaks in sedimentation are attributed to fair weather conditions between individual storm events.

CONCLUSIONS

Nine different radar facies are described based on the character and geometry of GPR reflections. These are interpreted to represent primary sedimentary structures within the Woburn Sands Formation and can be used to reconstruct the minimum sizes and geometries of ancient submarine bedforms. The top of the formation is flat, all bedforms apparently having been truncated by a submarine ravinement surface prior to deposition of the passage beds which pass up into the Gault Clay.

The profiles show a variety of bedform types and superpositions with 3D dunes in excess of 8 m amplitude occurring locally in Pratts Pit. The geometry of the reflectors indicates that the bedforms were curved-crested dunes with amplitudes up to 10 m, trough widths up to 60 m and wavelengths of at least 100 m. These exceptionally large bedforms contain many reactivation surfaces which can be resolved on radar. Outcrop studies show that the reactivation surfaces are often bioturbated, thereby indicating breaks in deposition. It is suggested that the reactivation surfaces are formed by the superimposition of smaller dunes, probably as a result of phase lags following exceptionally strong flows. Peak tidal/storm flow velocities are estimated at 1–1.5 m s^{-1}.

In Grovebury Farm Pit smaller-scale, complex oblique reflectors are interpreted as tidal bundles. Low-angle bounding surfaces, which are inclined upwards in the downstream direction, suggest that the bedforms scoured down and then filled up. The sedimentary structures indicate sand waves in classes IA, IIA, IIIA and IVA of Allen (1980). Although he did not discuss the 3D geometry of sand waves, a brief review of the literature has revealed that these forms are exceptional in having very large troughs indicative of 3D sinuous-crested dunes. Large curved-crested or barchanoid marine bedforms are often associated with a lack of sediment supply. However, it is clear that the very large dunes at Leighton Buzzard were not supply limited. Abundant reactivation surfaces within the cross-stratified sets indicate both superimposed bedforms and pulsed sediment transport. Tidal currents may be indicated by the presence of tidal bundles, while longer breaks in sedimentation are indicated by bioturbated pause planes. The larger-scale pulses are attributed to storm currents flowing south from the Boreal Sea towards the Tethys Ocean.

ACKNOWLEDGEMENTS

The author would like to thank Joseph Arnold and Sons Limited and ECC Quarries Limited for permission to work in Pratt's Quarry and Grovebury Farm Pit. Northumbrian Surveys and Addison Baxter kindly hired the equipment and demonstrated its application. Spectrum Energy and Information Technology Ltd kindly scanned the GPR data and converted it to Seg-Y format which was then interpreted on a LANDMARK seismic interpretation workstation at Phillips Petroleum Company United Kingdom Ltd with the help of Nigel Bramwell. Andreas Koeppen and Faz Jaffri are thanked for their assistance in the field. The manuscript has been improved by the comments of two unnamed referees.

REFERENCES

ALLEN, J.R.L. (1980) Sandwaves, a model of origin and internal structures. *Sediment. Geol.* **26**, 281–328.

ALLEN, J.R.L. (1981a) Lower Cretaceous tides revealed by cross-bedding with mud drapes. *Nature* **289**, 579–581.

ALLEN, J.R.L. (1981b) Palaeotidal speeds and ranges estimated from cross-bedding sets with mud drapes. *Nature* **293**, 394–396.

ASHLEY, G.M. (1990) Classification of large-scale subaqeuous bedforms: a new look at an old problem. *J. sediment. Petrol.* **60**, 160–172.

BARTHOLDY, J. (1992) Large-scale bedforms in a tidal inlet (Gradyb, Denmark). In: *Tidal Clastics 92, Abstract Volume* (Ed. Flemming, B.W.), *Cour. Forsch.-Inst. Senckenberg*, **152**, 7–8.

BELDERSON, R.H., JOHNSON, M.A. & KENYON, N.H. (1982) Bedforms. In: *Offshore Tidal Sands: Processes and Deposits* (Ed. Stride, A.H.), pp. 27–57. Chapman & Hall, London.

BERNE, S., AUFFRET, J-P. & WALKER, P. (1988) Internal structure of subtidal sandwaves revealed by high resolution seismic reflection. *Sedimentology* **35**, 5–20.

BERNE, S., DURAND, J. & WEBER, O. (1992) Architecture of modern tidal dunes (Sandwaves), Bay of Bourgneuf, France. In: *Three Dimensional Facies Architecture of Terrigenous Clastic Sediments and its Implications for Hydrocarbon Discovery and Recovery* (Ed. Miall, A.D. & Tyler, N.), pp. 245–260. S.E.P.M., Concepts in Sedimentology and Paleontology **3**.

BRIDGES, P. (1982) Ancient offshore tidal deposits. In: *Offshore Tidal Sands* (Ed. Stride, A.H.), pp. 172–192. Chapman & Hall, London.

BRISTOW, C.S. (1995) Facies analysis in the Lower Greensand using Ground Penetrating Radar. *J. Geol. Soc. London* (in press).

BROWN, L.F. & FISHER, W.L. (1977) Seismic stratigraphic interpretation of depositional systems: examples from Brazilian rift and pull-apart basins. In: *Seismic Stratigraphy—Applications to Hydrocarbon Exploration* (Ed. Payton, C.E.). *Am. Assoc. Petrol. Geol., Mem.* **26**, 218–248.

BROWN, L.F. & FISHER, W.L. (1980) Seismic stratigraphic interpretation and petroleum exploration. *Am. Assoc. Petrol. Geol. Educ. Course Note Ser.* **16**, 181 pp.

BUCK, S.G. (1985) Sand-flow cross-strata in tidal sands of the Lower Greensand (early Cretaceous): southern England. *J. sediment. Petrol.* **55**, 895–906.

COSTELLO, W.R. & SOUTHARD, J.B. (1981) Flume experiments on lower-flow-regime bed forms in coarse sand. *J. sediment. Petrol.* **51**, 849–864.

DALRYMPLE, R.W. (1984) Morphology and internal structure of sandwaves in the Bay of Fundy. *Sedimentology* **31**, 365–382.

DE RAAF, J.F.M. & BOERSMA, J.P. (1971) Tidal deposits and their sedimentary structures. *Geol. Mijnb.* **3**, 479–509.

EYERS, J. (1991) The influence of tectonics on early Cretaceous sedimentation in Bedfordshire, England. *J. Geol. Soc. Lond.* **148**, 405–414.

FLEMMING, B.W. (1978) Underwater sand dunes along the southeast African continental margin—observations and implications. *Mar. Geol.* **26**, 177–198.

FLEMMING, B.W. (1988) Zur Klassifikation subaquatischer, strömungstransversaler Transportkörper. *Boch. geol. u. geotechn. Arb.* **29**, 44–47.

GAWTHORPE, R.L., COLLIER, R.E.L., ALEXANDER, J., LEEDER, M. & BRIDGE, J.S. (1993) Ground penetrating radar: application to sandbody geometry and heterogeneity studies. In: *Characterisation of Fluvial and Aeolian Reservoirs* (Eds North, C.P. & Prosser, D.J.). Spec. Publ. Geol. Soc. Lond. **73**, 421–432.

JERZYKEWICZ, T., & WOJEWODA, J. (1986) The Radkow and Szceliniec sandstones: An example of giant foresets on a tectonically controlled shelf in the Bohemian Cretaceous basin (Central Europe). In: *Shelf Sands and Sandstones* (Eds Knight, R.J. & McLean, J.R.). *Can. Soc. Petrol. Geol. Mem.* **11**, 1–15.

KREISA, R.D., MOIOLA, R.J. & NOTTVED, A. (1986) Tidal sandwave facies, Rancho Rojo Sandstone (Permian), Arizona. In: *Shelf Sands and Sandstones* (Eds Knight, R.J. & McLean, J.R.). *Mem. Can. Soc. Petrol. Geol.* **11**, 227–291.

MCCABE, P.J. (1977) Deep distributary channels and giant bedforms in the Upper Carboniferous of the Central Pennines, northern England. *Sedimentology* **24**, 271–290.

MCCANN, D.M., JACKSON, P.D. & FENNING, P.J. (1988) Comparison of the seismic and ground probing radar methods in geological surveying. *Proc. Inst. Electr. Eng.* **135**, (F-4) **4**, 380–390.

MCCAVE, I.N. (1971) Sand waves in the North Sea off the coast of Holland. *Mar. Geol.* **10**, 199–225.

MCKEE, E.D. (1979) Sedimentary structures in dunes. In: *A Study of Global Sand Seas.* US Geol. Surv. Prof. Pap. **1052**, 87–134.

MITCHUM, R.M., VAIL, P.R. & SANGREE, J.B. (1977) Seismics stratigraphy and global changes of sea level, part 6: Stratigraphic interpretation of seismic reflection patterns in depositional sequences. In: *Seismic Stratigraphy—Application to Hydrocarbon Exploration* (Ed. Payton, C.E.). *Am. Assoc. Petrol. Geol. Mem.* **26**, 117–133.

NIO, S.D. (1976) Marine transgressions as a factor in the formation of sandwave complexes. *Geol. Mijnb.* **55**, 18–40.

NIO, S.D. & SIEGENTHALER, J.C. (1978) A Lower Eocene estuarine-shelf complex in the Isabena Valley. *State University of Utrecht, Sedimentology Group Report* **18**, 1–44.

NIO, S.D. & YANG, C.S. (1991a) Diagnostic attributes of clastic tidal deposits: a review. In: *Clastic Tidal Sedimentology* (Eds Smith, D.C., Reinson, G.E., Zaitlin, B.A. & Rahmani, R.A.). *Mem. Can. Soc. Petrol. Geol.* **16**, 3–28.

NIO, S.D. & YANG, C.S. (1991b) Sea-level fluctuations and

the geometric variability of tide dominated sandbodies. *Sediment. Geol.* **70**, 161–193.

PHILIPS, R.L. (1984) Depositional features of Late Miocene, Marine cross-bedded conglomerates, California. *Mem. Can. Soc. Petrol. Geol.* **10**, 345–358.

POLLARD, J.E., GOLDRING, R. & BUCK, S.G. (1993) Ichnofabrics containing Ophiomorpha: significance in shallow water facies interpretation. *J. Geol. Soc. London* **150**, 149–164.

RUBIN, D.M. (1987) *Cross-bedding, Bedforms, and Palaeocurrents*, S.E.P.M. Concepts in Sedimentology and Paleontology, **1**, 187 pp.

RUBIN, D.M. & MCCULLOCH, D.S. (1980) Single and superimposed bedforms: a synthesis of San Francisco bay and flume observations. *Sediment. Geol.* **26**, 207–231.

SHEPARD-THORN, E.R., MOORLOCK, B.S.P., COX, B.M., ALLSOP, J.M. & WOOD, C.J. (1994) Geology of the country around Leighton Buzzard. *British Geol. Surv. Memoir* **220**, 127 pp.

SMITH, A.M. & TAVENER-SMITH, R. (1988) Early Permian giant cross beds near Nqutu, South Africa, interpreted as part of a shoreface ridge. *Sediment. Geol.* **57**, 41–58.

SURLYK, F. & NOE-NYGAARD, N. (1992) Sand bank and dune facies architecture of a wide intracratonic seaway: Late Jurassic-Early Cretaceous Raukelv Formation, Jameson Land, East Greenland. In: *Three Dimensional Facies Architecture of Terrigenous Clastic Sediments and its Implications for Hydrocarbon Discovery and Recovery* (Eds Miall, A.D. & Tyler, N.), pp. 261–276. S.E.P.M. Concepts in Sedimentology and Paleontology **3**.

SZTANO, O. & TARI, G. (1992) Tide-influenced deposition after sea-level fall, early Miocene, northern Hungary. In: *Tidal Clastics 92, Abstr. Vol.* (Ed. Flemming, B.W.). *Cour. Forsch.-Inst. Senckenberg* **151**, 80–82.

TERWINDT, J.H.J. (1971) Sandwaves in the southern bight of the North Sea. *Mar. Geol.* **10**, 51–67.

WALKER, R.G. (1985) Ancient examples of tidal sand bodies formed in open, shallow seas. In: *Shelf Sand and Sandstone Reservoirs* (Eds Tillmann, R.W., Swift, D.J.P. & Walker, R.G.). *S.E.P.M. Short Course* **13**, 303–341.

WYATT, R.J., MOORLOCK, B.S.P., LAKE, R.D. & SHEPARD-THORN, E.R. (1988) Geology of the Leighton Buzzard to Ampthill district. *British Geol. Surv. Onshore Geology Series, Tech. Rept* **WA/88/1**.

VAN WAGONER, J.C., MITCHUM, R.M., CAMPION, K.M. & RAHMANIAN, V.D. (1990) Siliclastic sequence stratigraphy in well logs, cores and outcrops: Concept for high-resolution correlation of time and facies. *Am. Assoc. Petrol. Geol. Methods in Expl.* **7**, 55 pp.

Spec. Publs int. Ass. Sediment. (1995) **24**, 329–341

Tide-dominated sedimentation in the upper Tertiary succession of the Sitapahar anticline, Bangladesh

M.M. ALAM

Department of Geology, Dhaka University, Dhaka 1000, Bangladesh

ABSTRACT

The Baraichari Shale Formation (Late Miocene–Pliocene) in the Upper Tertiary clastic succession of the Sitapahar anticline, south-eastern fold belt of the Bengal Basin, displays a spectrum of tide-dominated facies. The fine-grained facies association (FFA), predominant in the upper and lower members of the formation, has mud- to sand-rich components and contains shale (sometimes sand/silt-streaked), lenticular-, wavy-, and flaser-bedded units and rippled siltstone. Fining-upward (FU) cycles and random intercalations are common, while coarsening-upward followed by fining-upward (CUFU) cycles are rare within the FFA. The medium-grained facies association (MFA), characterizing the middle member of the formation, comprises fine- and medium-grained, cross-bedded flat- and ripple-laminated sandstones with minor siltstone and shale intercalations.

The MFA is interpreted to represent migrating sandy shoals in a moderate energy, shallow-water subtidal environment. The shoals are developed in areas of most intensive and concentrated tidal energy. The FFA represents an intertidal environment, with the mud- and sand-rich components accumulating in the upper and lower intertidal subenvironments, respectively. The FU cycles have been developed by progradation of the tidal flats, their thicknesses probably being related to the tidal range at the time of deposition.

Large-scale alternations of MFA and FFA in the Baraichari Shale Formation are interpreted as resulting from minor transgressive and regressive phases during the overall regional regression in the south-eastern Bengal Basin. In the wake of subsequent tectonic upheavals, coupled with a regional regression, deposition of the coarse-grained facies association (CFA) of the Chandraghona Sandstone Formation took place in an essentially fluvial environment.

INTRODUCTION

The Bengal Basin, covering Bangladesh and part of eastern India, is known for extensive development of a thick Cretaceous–Tertiary sedimentary sequence. This sequence, which overlies the crystalline basement is usually concealed under an alluvial cover in most areas of the basin. However, a nearly complete Upper Tertiary (Neogene) succession crops out in south-eastern Bangladesh and eastern India in the form of a series of north–south trending folds of the Chittagong–Tripura Fold Belt (CTFB) (Fig. 1).

The Sitapahar anticline, located in the southern part of the CTFB, comprises a nearly 3-km thick succession of Upper Tertiary clastic rocks which are exposed along the Lichubagan–Kaptai road section (Fig. 2). The succession has been mapped and described by Rahman (1974), Chowdhury (1980) and Rahman (1980). Facies types and facies assemblages within the succession, particularly in the western flank of the structure, have been investigated by Ferdous (1990) and Alam & Ferdous (1991). A palyno-stratigraphical study was carried out by Chowdhury (1982).

The succession in the western flank of the Sitapahar anticline, between Lichubagan and Silchari Bazar (Fig. 2), comprises four conformable formations. In ascending order, these are: (i) the Sitapahar Sandstone–Shale (including the Silchari Shale and Wagga Sandstone Members); (ii) the Baraichari Shale; (iii) the Chandraghona Sandstone; and (iv)

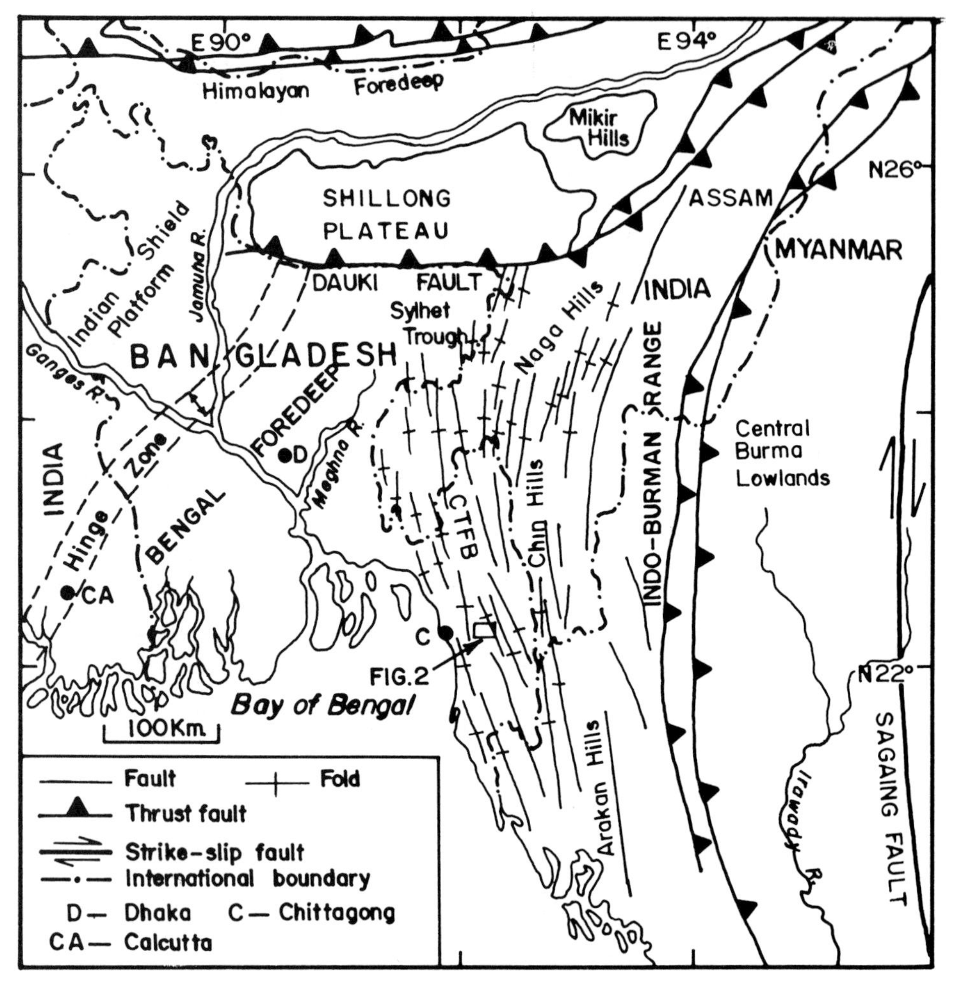

Fig. 1. Schematic map of the Bengal Basin showing the position of the Chittagong–Tripura Fold Belt (CTFB) in relation to the surrounding tectonic elements and the location of the study area (modified after Johnson & Alam, 1991).

the Lichubagan Sandstone Formations (Fig. 3) (Khan & Muminullah, 1988).

The present study is based on measured sections of the Baraichari Shale Formation (Late Miocene–Pliocene) which is 1050 m thick and is subdivided into a lower, a middle and an upper member (Fig. 3) on the basis of gross lithofacies.

REGIONAL TECTONIC SETTING

The Bengal Basin initially evolved as a NE–SW trending epicontinental basin along the eastern part of the Indian Precambrian Shield during Late Jurassic–Early Cretaceous times (Banerji, 1981). At that time a broad shelf zone extended from West Bengal north-eastwards up to Assam along the border of the crystalline massifs. Subsidence of the Bengal Basin probably began in the Late Cretaceous with the first Himalayan orogenic movement. It has been suggested that, throughout the period of its tectonic history, the basin has undergone several phases of transgressions, regressions, periodic uplifts and localized negative movements (Alam, 1972; Chowdhury, 1982; Banerji, 1984; Hiller, 1988) which were related to the Himalayan orogeny as well as upliftment of the Arakan–Chin–Naga hill ranges.

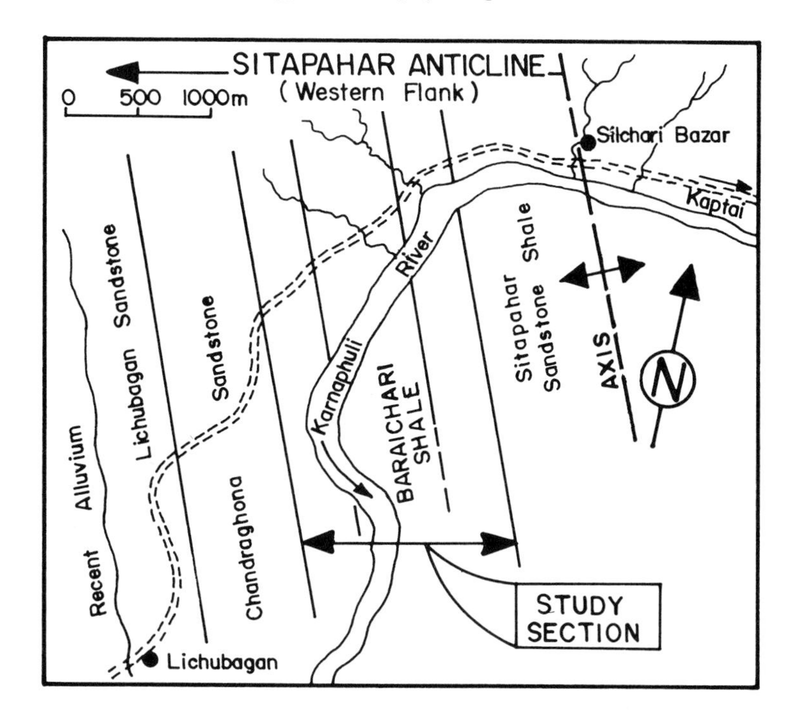

Fig. 2. Simplified geological map showing the Upper Tertiary clastic succession in the western flank of the Sitapahar anticline in south-eastern Bangladesh.

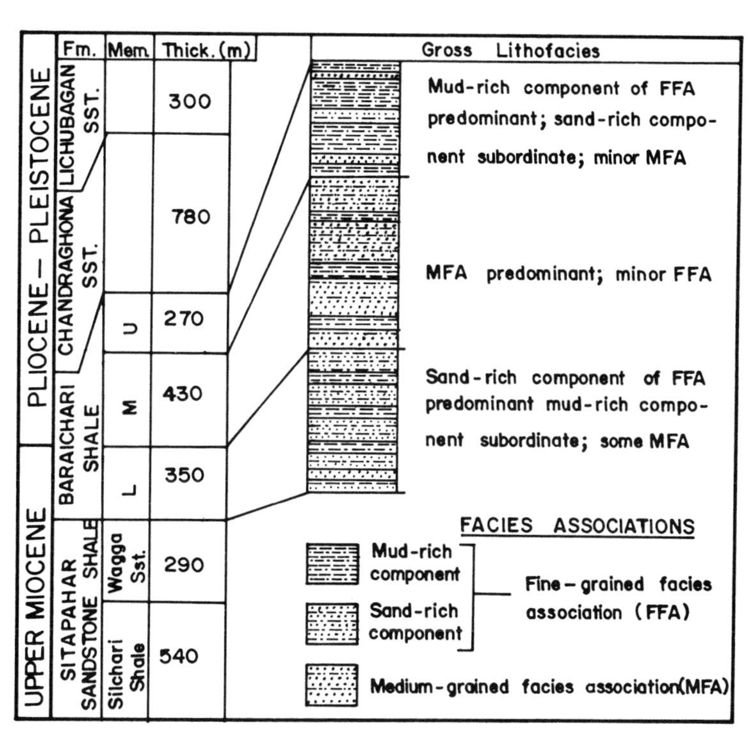

Fig. 3. Simplified stratigraphical column of the Upper Tertiary clastic succession and the composite lithostratigraphy of the Baraichari Shale Formation.

At present, the Bengal Basin is represented by a series of genetically related, but widely spaced tectonic geo-provinces (Fig. 1), each having its own geological and sedimentary history. The almost N–S striking folded belt of south-eastern Bangladesh is structurally a part of the CTFB, characterized by a series of long and narrow folds with echelon-like junctions representing the western extension of the Indo-Burman ranges lying to the east of the Bengal Basin. Salt *et al.* (1986) described the folded belt as a rifted passive margin that is slowly closing eastwards due to plate destruction in the subduction zone. The basin is hence gradually being encroached upon by the Indo-Burman orogenic belt (Hiller, 1988; Gupta *et al.*, 1990) associated with the eastward subduction of the Indian plate below Myanmar in Burma (Brunnschweiler, 1966; le Dain *et al.*, 1984; Sengupta *et al.*, 1990; Johnson & Alam, 1991).

The Sitapahar anticline, located in the south-central part of the CTFB (Fig. 1), represents an asymmetric anticline having a regional strike of N20°W–S160°E and a plunge of 4° towards S162°E.

FACIES ASSOCIATIONS

The Baraichari Shale Formation mainly comprises grey to bluish-grey shale, silty-shale, siltstone, sandy-siltstone and fine- to medium-grained sandstones of grey to brownish-grey colour. Within the formation, a fine-grained and a medium-grained facies association have been recognized. Variations in proportions of these facies associations form the basis for distinguishing between the lower, middle and upper members of the Baraichari Shale Formation (Figs 3 & 4).

Fine-grained facies association (FFA)

The fine-grained facies association (FFA) is charac-

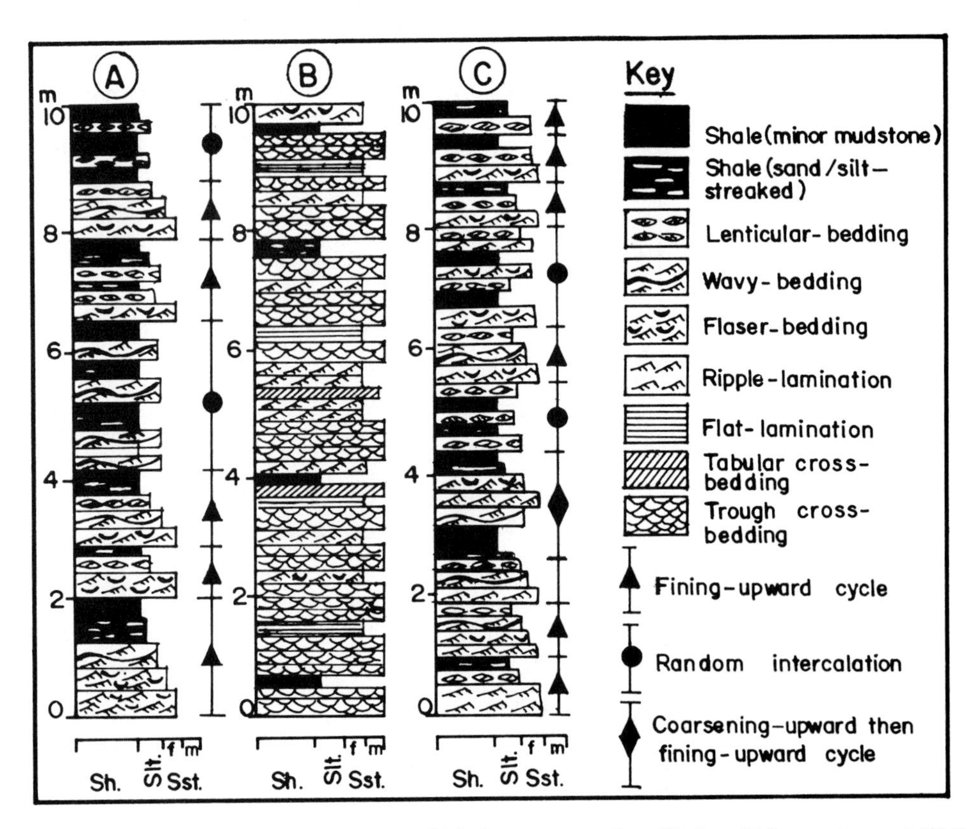

Fig. 4. Representative lithostratigraphical columns of (A) the upper member, (B) the middle member and (C) the lower member of the Baraichari Shale Formation.

Fig. 5. Outcrop section of the upper member of the Baraichari Shale Formation showing interbedding of mud-rich and sand-rich components of the FFA (note person for scale).

terized by a distinct facies spectrum, ranging from mud-rich to sand-rich components. The association consists of thinly interbedded shales, silty-shales, siltstones and sandy siltstones (Figs 4A, 4C & 5). Very fine-grained sandstones form a minor facies in this association. Individual facies in the FFA show a pattern of alternating textures, similar to the 'rhythmic sand-mud or alternate bedding' of Reineck (1967). The FFA also resembles the heterolithic strata of point bar deposits of tidally influenced meandering rivers described by Shanley *et al.* (1992). It ranges from very thinly laminated shale (sometimes sand/silt-streaked) and minor mudstone, through lenticular- and wavy-bedded units (Fig. 6), to continuously rippled sandy-siltstone with mud drapes (Fig. 7) and flaser-bedded units (Fig. 8) (Reineck & Wunderlich, 1968; Reineck & Singh, 1980).

The constituents of the mud-rich component include interbedded shale/sand/silt-streaked shale and lenticular-bedded units. The shale facies is evenly laminated, whereas the sand/silt-streaked shale is parallel-laminated with millimetre-thick siltstone/sandstone intercalations. These facies lack desiccation cracks, but are sometimes bioturbated and contain sand-filled retrusive burrows indicating rapid sedimentation. A gradation occurs from the shale facies into a lenticular-bedded rippled sandstone/siltstone facies with intercalated mudstone. The rippled lenses, commonly referred to as 'starved ripples' (de Raaf *et al.*, 1977), indicate weak currents in a predominantly low-energy environment. The marked asymmetry of most rippled lenses and the paucity of internally discordant cross-laminations indicate their formation by currents rather than wave oscillation.

The mud-rich component, although also present in the lower member, is most abundant in the upper member of the Baraichari Shale Formation (Fig. 4A). It rarely occurs in the middle member. In the

Fig. 6. Photo showing details of lenticular-bedding in the mud-rich component of the FFA (coin is 2 cm in diameter).

Fig. 7. Photo showing details of ripple-laminated sandy-siltstone facies with mud drapes in the sand-rich component of the FFA (scale is in mm).

lenticular-bedded units, ripples commonly have heights of 1–2 cm and sometimes contain very thin mud laminations along the foresets. Vertically adjacent rippled lenses showing opposed dips are common. Double-crested ripples are sometimes present.

Continuously rippled sandy-siltstones, commonly accompanied by mud drapes and flaser-bedded units, are the dominant facies in the sand-rich component of the FFA. Beds of this facies are usually 8–20 cm thick and grade upwards into wavy-bedded units. Ripple height varies from 1.5–2.5 cm and includes highly asymmetrical interference forms, symmetrical forms being rare. Reactivation surfaces and intersecting trough-shaped bundles of ripple laminations are apparent. Thin mud drapes commonly occur along foresets and set boundaries. Minor mud accumulation is restricted to the lee surfaces of ripples, resembling tidal ripple-drift cross-laminations described by Chandra & Bhattacharyya (1974). In exposed bedding surfaces the ripples are commonly almost straight-crested to highly sinuous (linguoid), sometimes associated with 'rib and furrow' structures that indicate deposition by currents. The sand-rich component of the FFA lacks burrows and desiccation cracks.

In the Baraichari Shale Formation, the sand-rich component characterizes the upper part of the upper member and middle part of the lower member. Palaeocurrent directions determined from ripple orientation are bimodal-bipolar with an overall NE–SW trend, although a south-westerly tidal current is predominant (Fig. 9A–C). The palaeocurrent pattern is essentially similar in all the members of the formation.

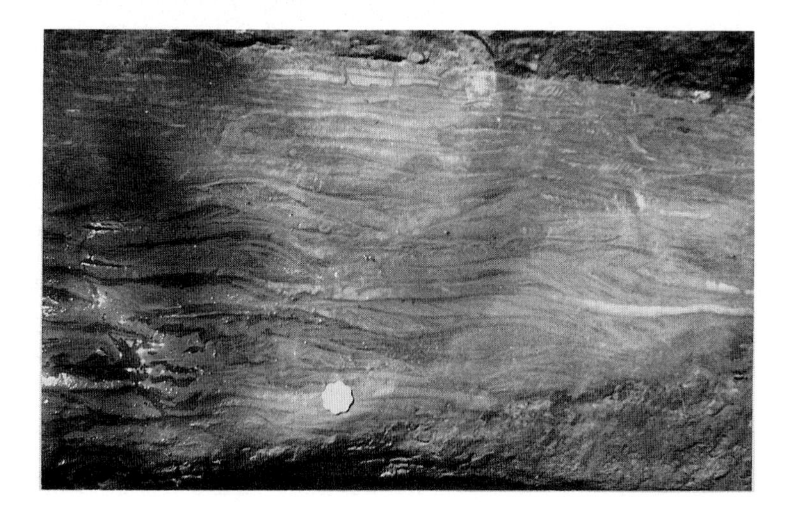

Fig. 8. Photo showing details of flaser-bedded facies passing upward into wavy-bedding (upper part of photo) in the sand-rich component of the FFA. Herring-bone structure is also visible. The photo is from the lower member (coin is 2 cm in diameter).

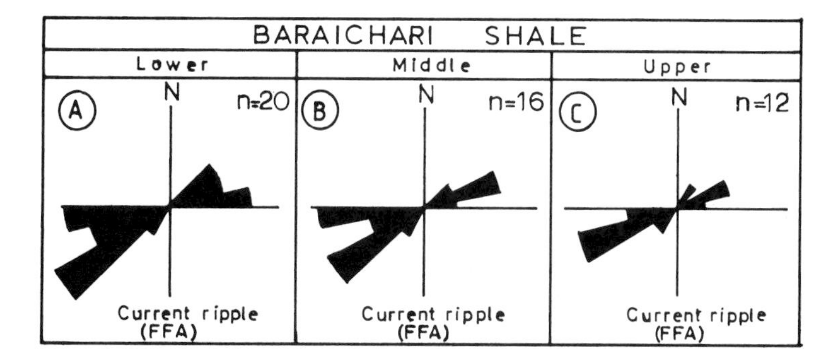

Fig. 9. Palaeocurrent rose diagram for current ripples in the FFA (interval 15°, *n* is number of readings).

Medium-grained facies association (MFA)

The medium-grained facies association (MFA) comprises mostly grey to brownish-grey, medium- to fine-grained litharenitic sandstones (Fig. 10) that are cross-bedded, flat- and ripple-laminated, and contain minor amounts of siltstone and shale. These facies are most abundant in the middle member of the Baraichari Shale Formation (Figs 3 & 4B) but also occur sporadically in the other two members. The uppermost 40 m of the upper member contain a few thick beds of the MFA, interbedded with sand-rich components of the FFA. This probably indicates a gradual transition of the Baraichari Shale Formation into the conformably overlying coarse-grained facies association (CFA) of the Chandraghona Sandstone Formation (Fig. 3) (Alam & Ferdous, 1991).

The most common facies of the MFA consists of fine- to medium-grained, moderately well sorted, trough cross-bedded sandstones (Fig. 11) and minor tabular sets. The trough cross-sets are low-angled, 10–25 cm in thickness and 0.5–1.5 m in width. Individual foresets are frequently separated by thin mud layers that tend to thicken towards the base of the cross-set (Fig. 12), resembling mud-draped cross-bedding and tidal bundles (Boersma & Terwindt, 1981; Singh & Singh, 1992). The cross-bedded sets are commonly arranged as multiple intersecting troughs in cosets up to 1.5 m thick. Internally the units may reveal some upward-fining trends, but generally they occur in a random pattern with intercalations of fine-grained facies elements.

Thicker sandstone units are occasionally convoluted and in many cases are loaded or may display small scours on the underlying fine-grained unit, but

Fig. 10. Outcrop section of the middle member of the Baraichari Shale Formation showing interbedding of various facies elements in the MFA (hammer at lower left is 32 cm long).

Fig. 11. Photo showing details of trough cross-bedded sandstone facies in the MFA. Note distinct herring-bone cross-stratification (pen is 13.5 cm long).

there is no evidence of major channelling. The palaeoflow direction is consistent within each cross-bedded set but adjacent sets sometimes show opposite dips and display herring-bone cross-stratification (Figs 11 & 12) (Reineck, 1967; Singh, 1969; Reineck & Singh, 1980; Alam *et al.,* 1985). The cross-bedded sandstone facies is generally overlain by the ripple-laminated, fine-grained sandstone facies, indicating a gradual reduction in the energy of depositional currents. Palaeocurrent directions indicated by the cross-beds are bipolar, approximating NE–SW for all members of the formation, although a predominance of the SW directed mode is evident (Fig. 13D–F).

The ripple-laminated sandstone facies is fine- to medium-grained, well to moderately sorted, and commonly overlies or is interbedded with the cross-bedded and flat-laminated sandstone facies. The ripple-laminated beds are rarely thicker than 15 cm and devoid of mudstone flasers, unlike the sand-rich component of the FFA. On bedding planes, ripples are typically linguoid and sometimes straight-crested with an average crest spacing of 8 cm and ripple heights of about 2–3 cm. Ripple orientations are invariably bimodal–bipolar having a general NE–SW direction with minor variations (Fig. 13A–C).

The flat-laminated sandstone facies is fine-grained and moderately sorted. Individual units vary in thickness from 10 to 20 cm, are interbedded with the ripple-laminated sandstone facies and sometimes with cross-bedded units. Sets of flat-laminated facies are separated by millimetre-thick mud laminae that occur in a rather regular manner (Fig. 14). On bedding planes, current lineations are not uncommon, their orientation coinciding with

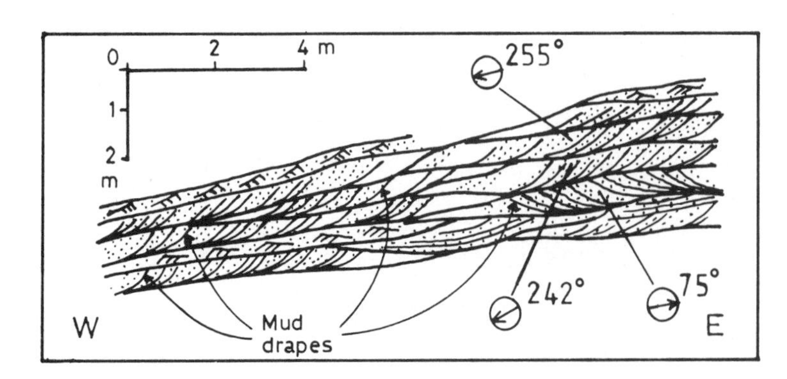

Fig. 12. Sketch of outcrop section showing cross-bedded sandstone facies with thin mud drapes along foresets. Note bipolar orientation of the cross-beds and interbedded ripple-laminated units.

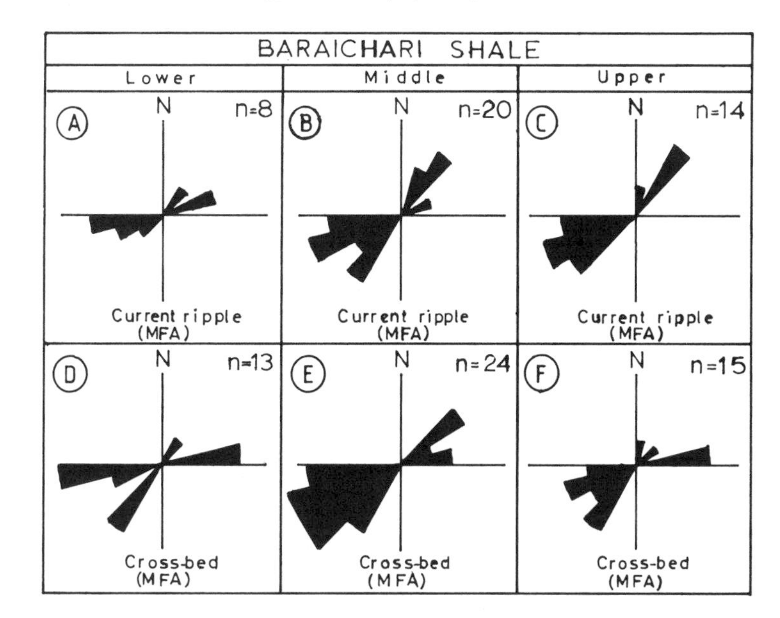

Fig. 13. Palaeocurrent rose diagram for current ripples (A–C) and cross-beds (D–F) in the MFA (interval 15°, *n* is number of readings).

those shown by current ripples and cross-beds. The flat-laminated sandstone facies show small-scale faults and folds (Fig. 14) that can be attributed to tectonic stress related to upliftment rather than to penecontemporaneous deformation.

INTERPRETATION OF PALAEO-ENVIRONMENTS

On the basis of gross lithofacies assemblages (Ferdous, 1990; Alam & Ferdous, 1991) and microfloral associations (Chowdhury, 1982), a generally marginal-marine depositional environment with possible tidal influence has been envisaged for the Baraichari Shale Formation. The facies associations (both FFA and MFA) presented in the present study reveal strong evidence for tide-influenced sedimentation of the Baraichari Shale Formation. Combinations of flaser-, wavy- and lenticular-bedded facies observed in the FFA occur in all members of the formation, resembling both modern and ancient

Fig. 14. Distortion in flat-laminated sandstone facies with thin mud laminae developing small-scale faults and folds probably caused by tectonic stress (lens cover is 6 cm in diameter).

tidal flat (intertidal) deposits (Evans, 1965; Ginsburg, 1975; Reineck & Singh, 1980; Klein, 1985).

The presence of mud drapes along foresets of the cross-bedded and ripple-laminated sandstone facies is also indicative of tidal environments (de Raff & Boersma, 1971; Klein, 1971; Boersma & Terwindt, 1981; Singh & Singh, 1992). The bimodal–bipolar orientation of cross-strata and the presence of herring-bone cross-stratification in sandstone units clearly indicate reversal of the depositional currents (Singh, 1969; Elliott, 1986) and therefore suggest a tidal environment.

Facies of the mud-rich and sand-rich components in the FFA are arranged in a parallel to wavy pattern throughout the Baraichari Shale Formation. The pattern is very similar to that described by Reineck (1967) as rhythmic sand–mud or alternate bedding, in which sand and mud alternate in various proportions. The rhythmic repetitions of facies clearly record transport of coarser sediments by ebb or flood tidal currents, with deposition of suspended fines during stagnant stages at high or low water. Similar tidal flat associations are known from other modern and ancient successions (Vilas *et al.*, 1988; Tessier & Gigot, 1989; Ramos & Galloway, 1990).

The spectrum of facies in the FFA is suggestive of an intertidal environment, the mud-rich facies components representing very low-energy upper intertidal (high tidal flats of Klein, 1971) to supratidal deposits, where sedimentation from suspension dominates and infrequent sandy incursions probably relate to unusually high tides or storms. The wavy-bedded facies is interpreted as a deposit of the middle part of an intertidal flat (mid-flat), where energy is moderate to low and bedload deposition alternates with suspension deposition. The sand-rich facies components with mud drapes and mud flakes in rippled siltstone and fine sandstone are interpreted as deposits of lower intertidal (low tidal flats) to shallow subtidal zones, where bedload transport occurs during moderate tidal activity.

The fining-upward cycles (FU), varying in thickness from 0.5 to 2.5 m, are most common in the FFA (Fig. 4). The cycles usually show sharp nonerosive bases and begin with units of ripple-laminated and/or flaser-bedded facies. These are overlain by thick wavy-bedded units that are topped by sand/silt-streaked shale, finally culminating in thinly laminated shale. The FU cycles are interpreted as 'tidalites' *sensu* Klein (1971) that may have developed either due to gradual progradation of the tidal flats or when conditions in the inner part of the tidal flats changed from those of lower intertidal to upper intertidal and possibly supratidal subenvironments during regressive phases.

Klein (1971) has suggested that thickness of the FU cycle is directly related to the palaeotidal range. However, since the thickness of the cycles would partly be controlled by the rate of basin subsidence as well as phases of transgression and regression during their development, it appears reasonable to suggest that the maximum of 2.5 m thickness of FU cycles in the Baraichari Shale Formation probably represents an absolute upper limit to the palaeotidal range at the time of deposition of the sediments.

The random intercalations in the FFA (Fig. 4) are interpreted to have developed when rates of sedimentation and subsidence were approximately balanced, or where a cycle of events was interrupted during random occurrence of, e.g. storms or unusually high tides with increasing tidal energy, augmented by wave energy or ocean current effects. Random intercalations may also be related to rapid influx of terrigenous clastics through runoff influencing tidal flat sedimentation. The CUFU cycles are rare in the Baraichari Shale Formation. It is therefore difficult to correlate such cycles with terrigenous influx related to tectonic activity which commonly give rise to abundant CUFU cycles (de Raaf *et al.*, 1977). Their presence can therefeore be explained either by lateral migration or construction and subsequent destruction of sandy shoals in the subtidal environment (de Raff *et al.*, 1977).

The trough cross-bedded sandstone facies in the MFA is thought to have been deposited by large migrating bedforms on sandy shoals in a broad and shallow subtidal environment near the inner shelf platform. Similar deposits are known to occur in a subtidal setting with water depths of up to 30 m (Reineck & Singh, 1980; Johnson & Baldwin, 1986). The shoals probably developed in areas where tidal energy was most intensive and concentrated.

The ripple-laminated sandstone facies were probably superimposed on the larger bedforms at times of weaker tidal currents. The minor intercalations of the ripple-laminated siltstone facies accumulated on the adjacent constricted low-energy parts of the tidal flats, while shales were deposited in more protected lower-energy parts of the flats due to frictional loss of currents and wave energy. Palaeocurrent data for cross-beds and current ripples

(Fig. 13) clearly indicate reversal in palaeoflow directions.

The flat-laminated sandstone facies with lineations, similar to those found in the Baraichari Shale Formation, have been reported from several ancient tidal successions (Sellwood, 1972; Hobday & Horne, 1977; Ireland *et al.,* 1978). This facies is interpreted as having been deposited under upper flow-regime conditions with high current velocities and highly turbulent flows in the tidal flat environment (Terwindt, 1971; Sellwood, 1972).

The Baraichari Shale sedimentation terminated with the advent of tectonic upheaval, which was probably the fourth phase of the main Himalayan orogeny in Pliocene times (Chowdhury, 1980), coupled with a regional regression. Banerji (1984) considered that by early Pliocene the geo-provinces in the eastern Bengal Basin (which includes the CTFB; Fig. 1) progressively came under tectonic control on account of the development of horizontal compressional forces. These forces were related to the great Himalayan orogeny and the crustal shortening due to the collision of the Indian and Asian plates. This resulted in extensive uplift and thrusting of older rocks in the east. In the CTFB, the sedimentary rocks generally grouped as the later molasse sequences were laid down in a dominantly non-marine fluviodeltaic environment during Plio–Pleistocene times (Chowdhury, 1982; Ferdous, 1990). In the Sitapahar anticline the coarse-grained facies association (CFA) of the Chandraghona Sandstone Formation (Figs 2 & 3) were conformably deposited on the Baraichari Shale Formation in an essentially continental–fluvial environment.

DISCUSSION AND CONCLUSIONS

The Baraichari Shale Formation flanking the south-eastern folded belt of the Bengal Basin shows an alternation of fine-grained and medium-grained facies associations (Figs 3 & 4) that record temporal changes in depositional environments within a tide-dominated shoreline. The FFA, which is predominant in the upper member of the formation, reflects lower intertidal to uppermost intertidal and possibly supratidal environments. The lack of desiccation cracks in most parts of the section is considered to be due either to a generally poor development of the supratidal zone or to the tendency of basinal subsidence to exceed the rate of sedimentation in the tidal flat environment at the time of deposition.

A notable feature of the MFA is the near absence of well-developed tidal channels in the cross-bedded sandstone facies which tend to be rather thick and laterally extensive units. The paucity of tidal channels has been considered by Walker & Harms (1975) as evidence of a low palaeotidal range, which is probably not true in the case of the Baraichari Shale sedimentation. It is more reasonable to suggest that the cross-bedded sandstone facies were deposited on a gentle palaeoslope, such as a low-gradient inner shelf, in which the tidal currents were not strong enough to develop deep tidal channels.

On the basis of the evidence for tidal influence in the Baraichari Shale Formation (reflected in the type of bedding structures and palaeocurrent orientations), it is suggested that the sediments were deposited by tidal currents that flowed perpendicular to an almost N–S striking coastline. The consistency of the bimodal–bipolar palaeocurrent distribution pattern throughout the study area strongly indicates that the NE–SW oriented tidal current system was maintained throughout the deposition of the formation. The paucity of symmetrical ripples and lack of wave-ripple cross-lamination as well as near absence of N–S directed palaeocurrents indicate that the wave activity and longshore currents were insignificant during deposition of the Baraichari Shale Formation.

Vertical relationships of the facies associations in the Baraichari Shale Formation document their development in a tide-dominated coastal setting with shoreline migrations. The regional marine regression (with minor intermittent transgressive and regressive phases) coupled with basinal subsidence were most probably responsible for the displacement of the shoreline, thereby producing the overall nature of large-scale alternations between the FFA and MFA in the formation. Some alternations may have been produced due to the shoaling of the outer part of the inner shelf. At present there is no conclusive evidence to suggest that the deposition of the Baraichari Shale was influenced by major tectonic activity in the source area to the east and north-east.

ACKNOWLEDGEMENTS

Funds for field work were provided by the Department of Geology, Dhaka University. I am grateful to my colleagues at the Department for their encouragement, valuable discussions and constructive suggestions in the field, especially Professor M. Haque, Dr B. Imam and Mr A Haque. Thanks are due to Mr M.H.S. Ferdous, presently a staff member of the Department, for some of the photographs. I also thank two anonymous referees for constructive and critical reviews of the manuscript.

REFERENCES

ALAM, M. (1972) Tectonic classification of Bengal Basin. *Geol. Soc. Am. Bull.* **83**, 519–522.

ALAM, M.M., CROOK, K.A.W. & TAYLOR, G. (1985) Fluvial herring-bone cross-stratification in a modern tributary mouth bar, Coonamble, New South Wales, Australia. *Sedimentology* **32**, 235–244.

ALAM, M.M. & FERDOUS, M.H.S. (1991) Marine-continental transition in the Upper Tertiary succession of the Sitapahar anticline, Kaptai, Rangamati District, Bangladesh. *Abstr. 16th Ann. Conf. Bangladesh Ass. Adv. Sci.*, pp. 2–3.

BANERJI, R.K. (1981) Cretaceous–Eocene sedimentation, tectonics and biofacies in the Bengal Basin, India. *Palaeogeogr. Palaeoclimat. Palaeoecol.* **94**, 57–85.

BANERJI, R.K. (1984) Post-Eocene biofacies, palaeoenvironments and palaeogeography of the Bengal Basin, India. *Palaeogeogr. Palaeoclimat. Palaeoecol.* **45**, 49–73.

BOERSMA, J.R. & TERWINDT, J.H.J. (1981) Neap–spring tide sequences of intertidal shoal deposits in a mesotidal estuary. *Sedimentology* **28**, 151–170.

BRUNNSCHWEILER, R.O. (1966) On the geology of the Indo-Burman ranges (Arakan coast and Yoma, Chin Hills, Naga Hills). *J. Geol. Soc. Australia* **13**, 137–194.

CHANDRA, S.K. & BHATTACHARYYA, A. (1974) Ripple-drift cross-lamination in tidal deposits: examples from the Precambrian Bhander Formation of Maikar, Satna District, Madhya Pradesh, India. *Geol. Soc. Am. Bull.* **85**, 1117–1122.

CHOWDHURY, S.Q. (1980) *Palynostratigraphy of the Neogene sediments as exposed in the western flank of the Sitapahar Anticline, Chittagong Hill Tracts.* MSc thesis, Univ. Dhaka.

CHOWDHURY, S.Q. (1982) Palynostratigraphy of the Neogene sediments of the Sitapahar Anticline (western flank), Chittagong Hill Tracts. *Bangladesh J. Geol.* **1**, 35–49.

DAIN, A.Y. LE, TARRONIER, P. & MOLNAR, P. (1984) Active faulting and tectonics of Burma and surrounding regions. *J. geophys. Res.* **89**, 453–472.

ELLIOTT, T. (1986) Siliciclastic shorelines. In: *Sedimentary Environments and Facies* (Ed. Reading, H.G.), pp. 155–188. Blackwell Scientific Publications, Oxford.

EVANS, G. (1965) Intertidal flat sediments and their environments of deposition in the Wash. *Quart. J. Geol. Soc. London,* **121**, 209–245.

FERDOUS, M.H.S. (1990) *Sedimentology of the sediments exposed in the Sitapahar Anticline, Kaptai–Chandraghona area, Rangamati, Bangladesh.* MSc thesis, Univ. Dhaka.

GINSBURG, R.N. (1975) *Tidal Deposits.* Springer-Verlag, New York, 428 pp.

GUPTA, H.K., FLEITOUT, L. & FROIDEVAUX, C. (1990) Lithospheric subduction beneath the Arakan-Yoma Fold Belt: Quantitative estimates using gravimetric and seismic analyses. *J. Geol. Soc. India* **35**, 235–250.

HILLER, K. (1988) On the petroleum geology of Bangladesh. *Geol. Jb.* **90**, 3–32.

HOBDAY, D.K. & HORNE, J.C. (1977) Tidally influenced barrier island and estuarine sedimentation in the Upper Carboniferous of southern West Virginia. *Sediment. Geol.* **18**, 97–122

IRELAND, R.J., POLLARD, J.E., STEEL, R.J. & THOMPSON, B.D. (1978) Intertidal sediments and trace fossils from the Waterstones (Scythian–Anisian?) at Daresbury, Cheshire. *Proc. Yorks. Geol. Soc.* **41**, 399–436.

JOHNSON, H.D. & BALDWIN, C.T. (1986) Shallow siliciclastic seas. In: *Sedimentary Environments and Facies* (Ed. Reading, H.G.), pp. 229–282. Blackwell Scientific Publications, Oxford.

JOHNSON, Y.J. & ALAM, A.M.N. (1991) Sedimentation and tectonics of the Sylhet trough, Bangladesh. *Geol. Soc. Am. Bull.* **103**, 1513–1527.

KHAN, M.R. & MUMINULLAH, M. (1988) Stratigraphic Lexicon of Bangladesh. *Records Geol. Surv. Bangladesh,* **5**, 70 pp.

KLEIN, G. DEVRIES (1971) A sedimentary model for determining palaeotidal range. *Geol. Soc. Am. Bull.* **82**, 2585–2592.

KLEIN, G. DEVRIES (1985) *Sandstone Depositional Models for Exploration of Fossil Fuels*, 3rd edn. Reidel, Dordrecht, 209 pp.

RAAF, J.F.M. DE & BOERSMA, J.R. (1971) Tidal deposits and their sedimentary structures. *Geol. Mijnb.* **50**, 479–503.

RAAF, J.F.M. DE, BOERSMA, J.R. & GELDER, A. VAN (1977) Wave generated structures and sequences from a shallow marine succession, Lower Carboniferous, county Cork, Ireland. *Sedimentology* **24**, 451–483.

RAHMAN, M. (1974) *Grain-size analysis and heavy mineral study of the Upper Tertiary sediments as exposed in the Kaptai–Chandraghona road section, Chittagong Hill Tracts.* MSc thesis, Univ. Dhaka.

RAHMAN, H. (1980) *Palaeostratigraphy of the Neogene sediments as exposed in the eastern flank of the Sitapahar Anticline along Silchari–Raptai road cut, Chittagong Hill Tracts.* MSc thesis, Univ. Dhaka.

RAMOS, A. & GALLOWAY, W.E. (1990) Facies and sandbody geometry of the Queen city (Eocene) tide-dominated delta-margin embayment, NW Gulf of Mexico Basin. *Sedimentology* **37**, 1079–1098.

REINECK, H.-E. (1967) Layered sediments of tidal flats, beaches and shelf bottoms of the North Sea. In: *Estuaries* (Ed. Lauff, G.D.), pp. 191–206. Am. Ass. Adv. Sci., Washington.

REINECK, H.-E. & SINGH, I.B. (1980) *Depositional Sedimentary Environments.* Springer-Verlag, New York, 549 pp.

REINECK, H.E. & WUNDERLICH, F. (1968) Classification and origin of flaser lenticular beddings. *Sedimentology* **11**, 99–104.

SALT, C.A., ALAM, M.M. & HOSSAIN, M.M. (1986) Bengal Basin—current exploration of the hinge zone of southwestern Bangladesh. *6th Offshore SE Asia Conf. Singapore*, pp. 55–67.

SELLWOOD, B.W. (1972) Tidal-flat sedimentation in the Lower Jurassic of Bornholm, Denmark. *Palaeogeogr. Palaeoclimat. Palaeoecol.* **11**, 93–106.

SENGUPTA, S., RAY, K.K., ACHARYA, S.K. & DESMETH, J.B. (1990) Nature of ophiolite occurrences along the easter margin of the Indian plate and their tectonic significance. *Geology* **18**, 439–443.

SHANLEY, K.W., MCCABE, P.J. & HITTINGER, R.D. (1992) Tidal influence in cretaceous fluvial strata from Utah, USA: a key to sequence stratigraphic interpretation. *Sedimentology* **39**, 905–930.

SINGH, I.B. (1969) Primary sedimentary structures in Precambrian quartzite of Telemark, southern Norway, and their environmental significance. *Norsk. Geol. Tidsskr.* **31**, 1–31.

SINGH, P.P. & SINGH, I.B. (1992) Cross-bedding with tidal bundles and mud drapes. Evidence for tidal influence in Bhuj Sandstone (Lower Cretaceous), Eastern Kachchh. *J. Geol. Soc. India* **39**, 487–493.

TERWINDT, J.H.J. (1971) Lithofacies of inshore estuarine and tidal inlet deposits. *Geol. Mijnb.* **50**, 515–526.

TESSIER, B. & GIGOT, P. (1989) A vertical record of different tidal cyclicities: an example from the Miocene marine molasse of Digne (Haute Province, France). *Sedimentology* **36**, 767–776.

VILAS, F., SOPENA, A., REY, L., RAMOS, A., NOMBELA, M.A. & ARCHE, A. (1988) The Corrubedo Tidal inlet, Galicia, N.W. Spain: sedimentary processes and facies. In: *Tide-influenced Sedimentary Environments and Facies* (Eds Boer, P.L. de, Gelder, A. van & Nio, S.D.), pp. 183–200. Reidel, Dordrecht.

WALKER, R.G. & HARMS, J.C. (1975) Shorelines of weak tidal activity: Upper Devonian Catskill Formation, Central Pennsylvania. In: *Tidal Deposits: A Case-book of Recent Examples and Fossil Counterparts* (Ed. Ginsburg, R.N.), pp. 103–108. Springer-Verlag, New York.

Spec. Publs int. Ass. Sediment. (1995) **24**, 343–351

Evidence of tidal influence in the Murree Group of rocks of the Jammu Himalaya, India

B.P. SINGH *and* H. SINGH

Department of Geology, University of Jammu, Jammu, India

ABSTRACT

The sediments of the Murree Group form a broad rim in the inner Tertiary belt of the Himalaya, extending from the Jhelum syntaxis to the Jammu foothills on the Indian subcontinent. The Murree foreland basin, in which the sediments from the Himalayan uplifts have accumulated, was formed as a result of the early Tertiary Himalayan Orogeny. Recent palaeomagnetic and palaeontologoic studies have revealed that these sediments were deposited in the late Eocene–early Miocene.

The present study is confined to the Murree Group of rocks exposed along the Jammu–Srinagar highway (north of Udhampur) and deals with the depositional environments in relation to lithofacies and their organization, sedimentary structures, trace fossils and palaeocurrent patterns. Lithologically, the Murree Group of rocks has been divided into a Lower Murree Formation, exhibiting a cyclic sandstone–siltstone–mudstone sequence, and an Upper Murree Formation with sandstone–mudstone cycles. The individual cycles in the Lower Murree start with an erosional base, followed by channel-lag conglomerates, cross-bedded sandstones with thin mud flasers and massive sandstones. The latter are overlain by laminated sandstones displaying ripples, massive sandstones and mudstone–siltstone intercalations. In the upper Murree Formation, the arrangement of the rock units changes with a general increase of sandstone facies as compared to mudstone facies. The absence of siltstone facies is a characteristic feature in the Upper Murree Formation. The presence of strongly bioturbated, pebbly conglomerates and cross-bedded sandstones with thin mud flasers indicates deposition in a subtidal environment. The sandstone–mudstone intercalations along with laminated siltstones represent a mixed tidal flat sequence deposited in the intertidal zone.

The presence of herringbone cross-bedding in the Lower Murree sandstones further enhances the evidence for tidal influence during the accumulation of these sediments. The sole marks (groove casts and flute casts) at the base of the cycles of the Lower Murree rocks indicate their formation in tidal channels. The bifurcated, straight-crested asymmetrical, linguoid starved ripples in the Lower as well as the Upper Murree rocks substantiate the tidal origin of these rocks. Wrinkle marks indicate intertidal emergence of the sediments. Vertical burrows point towards deposition in a shallow-water environment. Palaeocurrent patterns are bimodal and in part bipolar. On the basis of this evidence an estuarine depositional environment with strong tidal influence is suggested for the Murree Group of rocks.

INTRODUCTION

The Murree Group of sediments marks a substantial change in the sedimentation pattern of the rising Himalaya, the geological evolution of which commenced with continental sedimentation. It has a widespread distribution in north-western India, forming the eastward extension of the Murree belt in Pakistan. In Jammu and Kashmir State of India, these rocks are exposed for a distance of 50 km along the Jammu–Srinagar Highway between the towns of Udhampur and Batote.

The depositional environment of these clastic sediments has long been a point of controversy among geologists. Oldham (1892), Pilgrim (1910) and Wadia (1928) considered these rocks as brackish-water deposits in the absence of definite marine or non-marine fauna. Khan *et al.* (1971) considered the Murree Group of rocks as transitional deposits. Ranga Rao (1971) concluded that marine conditions ceased with the commencement of their sedimentation. Sharda & Verma (1977) have suggested a quiet-water, lagoonal environment for the Lower Murree and a fluviodeltaic environment for the Upper Murree sediments. Mehta & Jolly (1988) indicated a freshwater environment whereas Singh *et al.* (1990), on the basis of a petrochemical study, concluded that the sediments were deposited in a nearshore environment. In the present paper, another attempt is made to clarify the environments of deposition by investigating the lithofacies and their organization, sedimentary structures, nature of bioturbation and palaeocurrent patterns.

GEOLOGICAL SETTING

The Murree Group of clastic rocks of late Eocene–early Miocene age is disconformably underlain by the rocks of the Subathu Formation in Jammu and Kashmir State (India). The northern boundary is marked by the Main Boundary thrust whereas the younger rocks of the Siwalik Group are exposed in the southern part (Fig.1). The geological distribution of Tertiary rocks in Jammu and Kashmir State is controlled mainly by complicated tectonic features. Primary and secondary magnetization of the Murree Group red beds (Miocene to late Eocene) in the north-east of Jammu indicates a clockwise rotation of 45° with respect to the Indian shield since late Eocene/early Miocene times (Klootwijk *et al.*, 1986). There are numerous thrusts trending

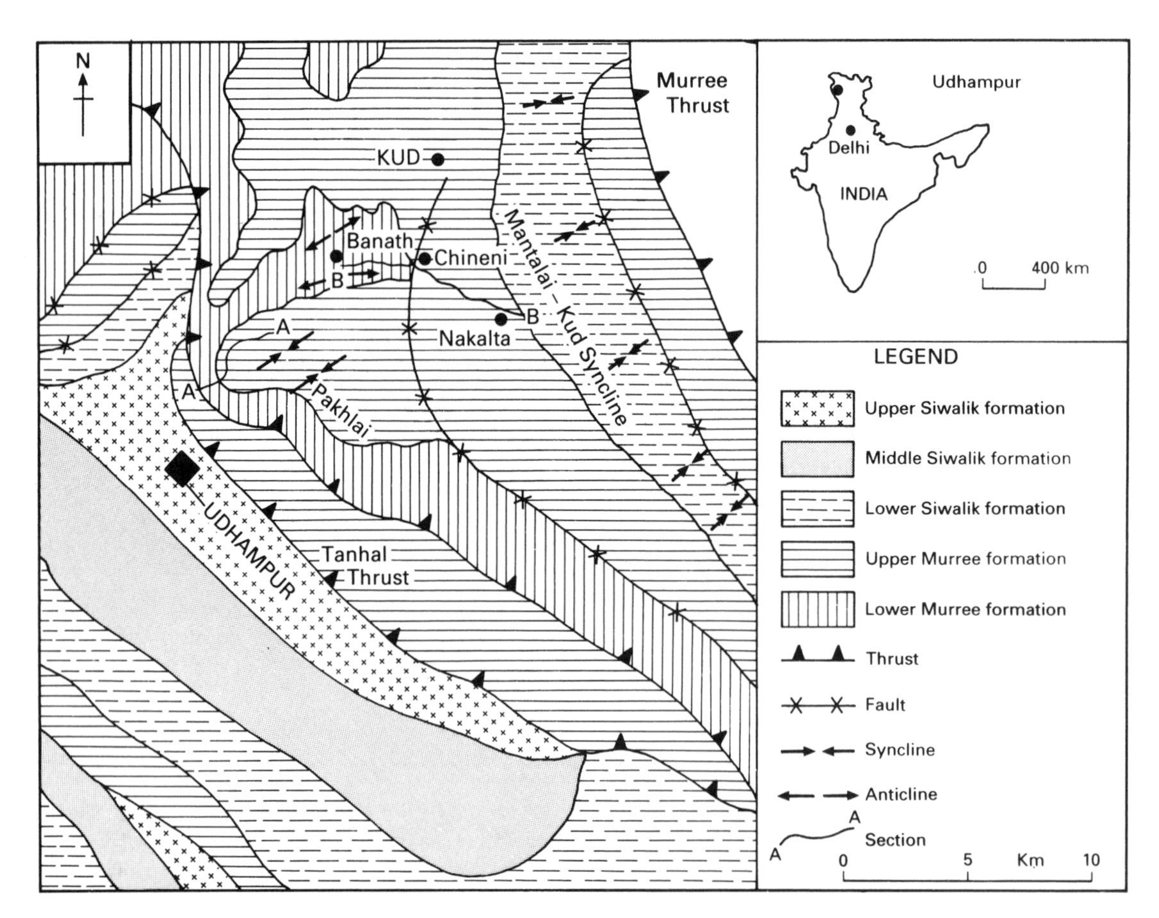

Fig. 1. Geological map of the study area in the vicinity of Udhampur, India.

north-west–south-east which have obliterated the systematic stratigraphical positions of the different formations. Other tectonic features are the Mantalai–Kud syncline, the Banath anticline, the Pakhlai syncline and the Udhampur syncline. The sections exposed between the Tanhal thrust and the Pakhlai syncline, and between the Banath anticline and the Lower Siwalik contact are illustrated in Fig. 2A & B.

LITHOFACIES

In accordance with lithological variations, the Murree Group of rocks has been divided into a Lower and an Upper Formation. The former exhibits sandstone–siltstone–mudstone cycles whereas the latter presents sandstone–mudstone cycles. Facies codes have been given to the different lithofacies in order to facilitate a better interpretation. Lithofacies codes **Cg**, **Sx** and **Sh** designate tidally influenced channel deposits, following McKie (1991). Lithofacies codes **Sma** and **Smb** are used by the authors to denote medium-grained massive sandstones and medium- to fine-grained massive sandstones respectively. Laminated siltstones with more than 70% silt-sized grains have been earmarked with the facies code **St**. Facies **Mb** follows the code of Johnson (1982) for laminated mudstones.

Lower Murree Formation

The Lower Murree Formation has a cyclic architecture, mainly comprising pebbly conglomerates, cross-bedded, laminated and massive sandstones, brown to purple siltstones and brown mudstones (Fig. 2A).

The pebbly conglomerate lithofacies (**Cg**) consists of mudstone pebbles 1–3 cm in diameter, embedded in a medium-grained sandy matrix. This facies starts with an erosional base exhibiting sole marks in most cases. The thickness of **Cg** varies from 0.15 to 0.30 m. In some of the cycles this lithofacies is absent. The conglomerates are overlain by medium- to coarse-grained cross-bedded sandstones (**Sx**). Cross-beddings are mostly trough-shaped, thicknesses of the cosets varying from 0.2 to 1.2 m. Herringbone cross-bedding suggests bidirectional flow. Tidal bundles with reactivation surfaces are found in the **Sx** facies. A remarkable feature of this lithofacies is the occurrence of thin mud flasers.

The **Sx** facies is followed by the medium-grained

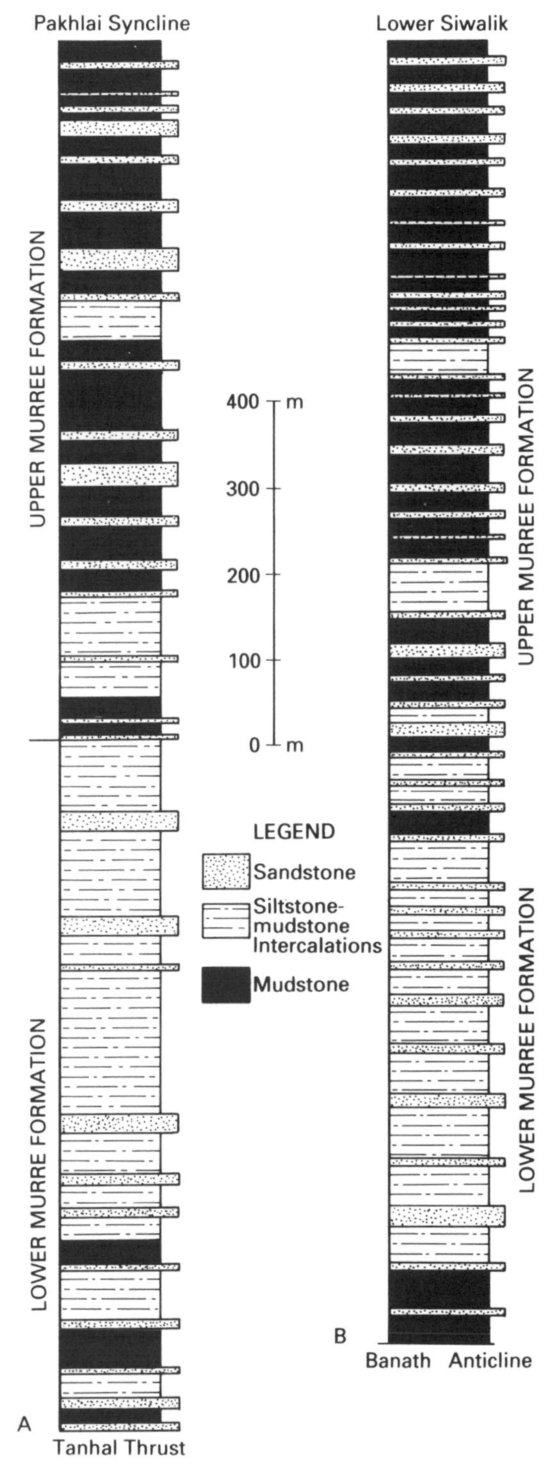

Fig. 2. Stratigraphic profile of the Murree Group of rocks. (A) Succession between the Tanhal thrust and the Pakhlai syncline. (B) Profile between Banath anticline and Nakalta.

massive sandstones facies (**Sma**). The sandstones comprise lithic arenites 0.5–1.0 m thick. The colour varies from grey to greenish grey. The **SMa** facies is separated from the overlying laminated sandstone facies (**Sh**) by a thin mud layer. The laminae in this facies vary in thickness from 0.1 to 3.0 cm, being characterized by alternations of ferruginous and non-ferruginous layers. The upper surface of this facies is marked by rippled sand sheets (wave as well as current ripples in many cycles). The **Sh** facies is again overlain by a massive, medium- to fine-grained sandstone (**Smb**) which is strongly bioturbated in places and 0.5–1.2 m thick. It is associated with 2–6 m thick, brown, laminated mudstones facies (**Mb**). These mudstones also contain calcrete layers at their upper parts. In most cases this facies is intercalated with 1.5–3.0 m thick, brown to purple, laminated siltstones facies (**St**).

Upper Murree Formation

The rocks of the Upper Murree Formation consist of conglomerates, cross-bedded, laminated and massive sandstones, and mudstones (Fig. 3B). Lithofacies **Cg** (conglomerates) is 0.2–0.5 m thick and consists of grey, greenish grey and yellow mudstone pebbles 1.5–6.0 cm in diameter. The subrounded and rounded mudstone pebbles are embedded in a coarse sandy matrix. The overlying **Sx** facies has an erosional contact with the **Cg** facies and comprises trough cross-bedded cosets displaying tidal bundles in many cases. The sandstones are medium- to coarse-grained. Thicknesses of invidual cosets vary from 0.5 to 2.0 m, whereas whole cycles vary in thickness from 0.5 to 2.3 m. Thin mudstone layers are observed between the sandstone beds.

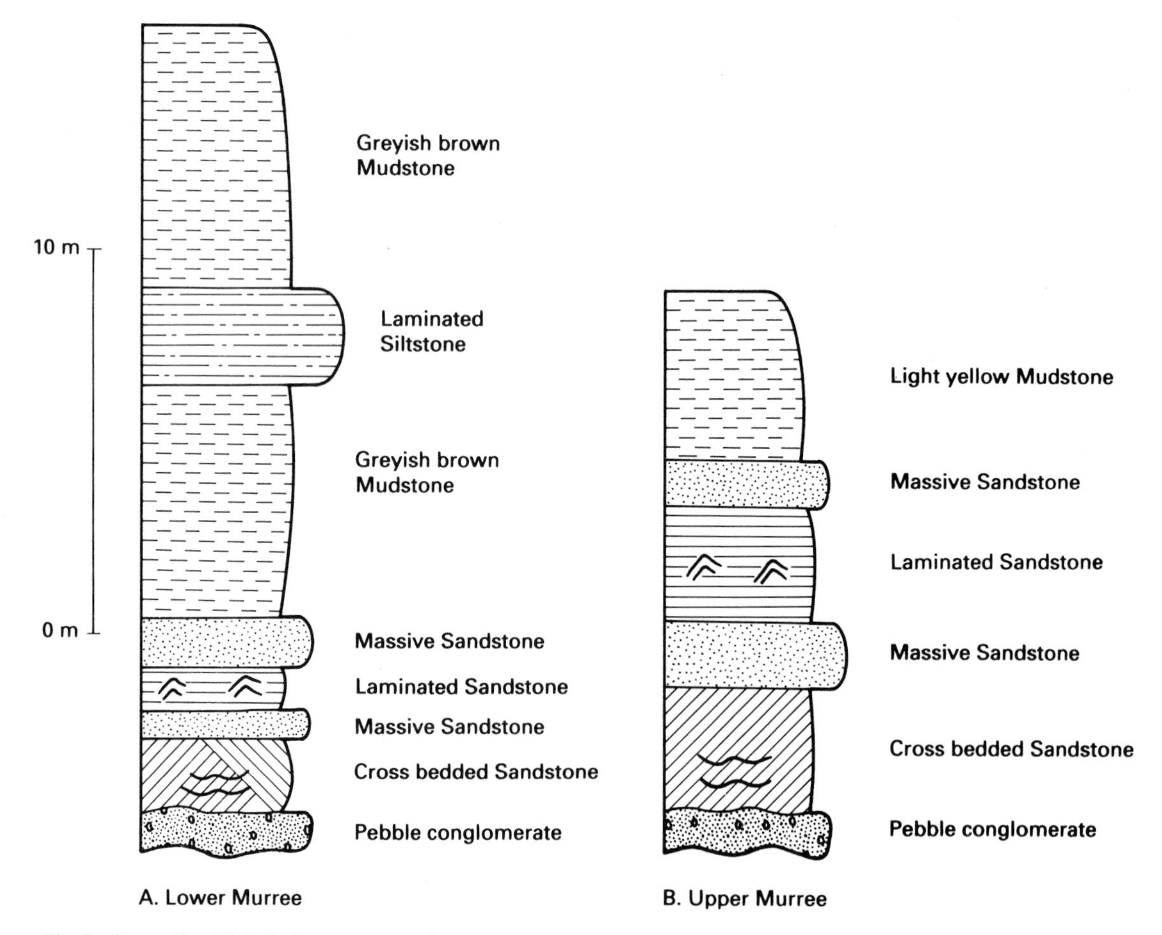

Fig. 3. Generalized lithofacies sequences of the Murree Group.

The **Sx** facies is overlain by a massive sandstone facies (**Sma**) of yellow to greenish yellow colour and medium to coarse grain sizes. The thickness of this facies varies from 0.5 to 1.2 m. The subrounded to rounded grains are cemented by calcite and Fe-calcite.

The **Sma** facies is overlain by an **Sh** facies. This facies is similar to the **Sh** facies of the Lower Murree Formation. Lamina thickness varies from 0.1 to 4.0 cm. The upper surface exhibits wave- and current-generated ripple sheets.

The **Sh** facies is followed by the massive sandstone facies (**Smb**) of greyish brown colour and fine grained nature. The thickness of the **Smb** facies varies from 0.3 to 1.5 m. Again the sandstones are cemented by a calcite and Fe-calcite cement.

In the Upper Murree Formation the **Smb** facies is in direct contact with a laminated mudstone facies (**Mb**) of brownish yellow and greenish yellow colour. Calcrete layers form a characteristic lithological element in this facies.

Sedimentary structures

The sedimentary structures in the Murree Group of rocks reflect the hydrodynamic conditions prevailing at the time of deposition. For the purpose of this study, all the primary sedimentary structures observed on the upper surfaces, within the sediment bodies and along the lower surfaces were evaluated.

Different types of ripple marks having varied geometry are observed in the Lower as well as the Upper Murree Group of rocks. These ripple marks have been identified and classified following the criteria and classification scheme of Reineck *et al.* (1971). Small asymmetrical and strongly undulating current ripples, forming large sheets on the sandstone surfaces, are found at Narsua and Samrauli (Fig. 4a). Wave ripples that are straight-crested, partly bifurcated and nearly symmetrical are observed near Maur in the Upper Murree sandstones (Fig. 4b). Starved ripple marks are found both in the Lower as well as the Upper Murree sandstones. Wrinkle marks are very prominent on the upper surfaces of the Lower Murree sandstones. These have elevations of 0.1–0.4 cm and occur in regular patterns (Fig. 4c).

Sole marks, in the form of flute casts and groove marks, appear in the Lower as well as the Upper Murree Formation. Flute casts with elongated and bulbous shapes are well developed at the interface of an overturned, folded sandstone–mudstone sequence. Lengths and widths of individual flutes approximate 15 and 10 cm respectively (Fig. 4b). Long and straight groove marks are present in several places.

Weakly to moderately developed cross-bedding and cross-lamination, mostly trough-shaped, are developed in these sandstones. Herringbone cross-bedding (Fig. 4e) with reactivation surfaces occurs mostly in the Lower Murree sandstones, while tidal bundles are also prominent. The sigmoidal tidal bundles display neap–spring cycles that reflect the regularly changing intensity of the tidal flow. In addition, mud flasers are observed in the sandstones, and lenticular sand bodies in the mudstones.

Bioturbation

In the absence of body fossils, trace fossils can also be successfully used to determine sedimentary environments. Seilacher (1953a) has distinguished five different types of trace fossils based on the life habits of the burrowing organisms. In the present study, bioturbation structures are observed near Dramthal and Narsua in the Lower Murree rocks and south of Bali Nalla and Chineni in the Upper Murree sandstones. Although the genera and species of the organisms could not be identified, the structures are indeed vertical burrows. In some places the physical surface structures have been destroyed by the formation of lebensspuren. In Fig. 4f, a thin sandy layer displaying a variety of colours covers a siltstone with a muddy bedding surface through which organisms have ploughed (internal lebensspuren—Seilacher, 1953b). This type of bioturbation may have been produced by vagile Hemiendobionta (Schäfer, 1956).

DEPOSITIONAL ENVIRONMENT

Lithofacies associations and sedimentary structures are the most significant features by which depositional environments can be reconstructed. Sandy tidal-channel environments are characterized by channel complexes separated by shallow-water shoals resulting from lateral channel migration. (Klein, 1970; Nio *et al.*, 1980; Boersma & Terwindt, 1981; van den Berg, 1982). Tidal-channel migration produces mud clasts by undercutting and collapse of channel banks. The channel lag conglomerates (**Cg**), cross-bedded sandstones (**Sx**), massive sandstones (**Sma**), and laminated sand-

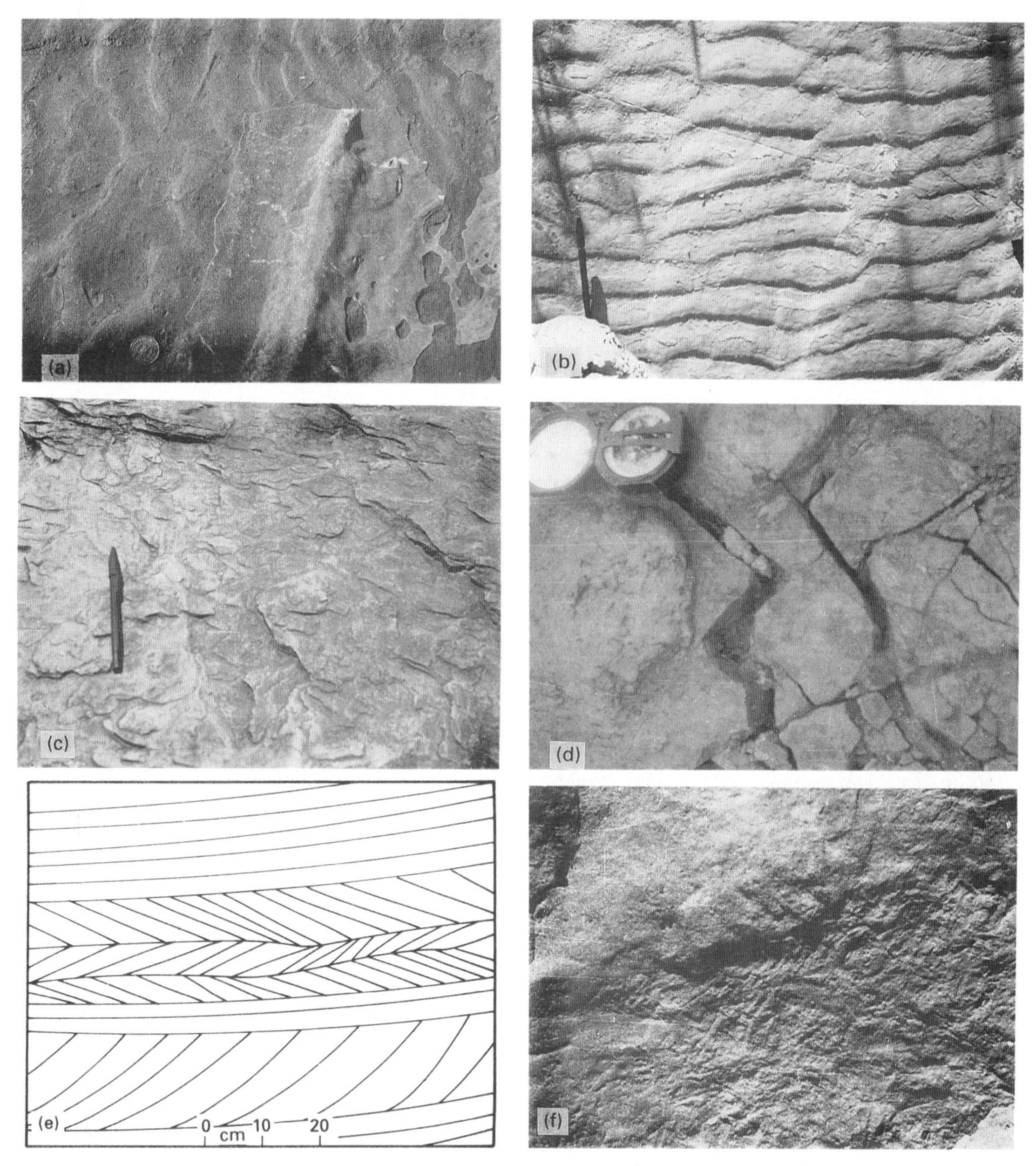

Fig. 4. Sedimentary structures. (A) Strongly undulating asymmetrical current ripples with a carbonate mat in the right corner on a bedding of the Lower Murree sandstone (scale: coin of 2.5 cm diameter). (B) Straight-crested, bifurcated asymmetrical wave ripples on a bedding surface of the Upper Murree sandstone (scale: pen of 15 cm length in the lower left corner). (C) Wrinkle marks in a regular pattern on the bedding surface of the Lower Murree sandstone (scale: 15 cm). (D) Bulbous and elongated flute casts on the base of a bed of the Lower Murree sandstone. Pronounced casts are on the right and below the compass (compass diameter is 8 cm). (E) Herringbone cross-bedding in the Lower Murree sandstone traced from a photograph. Trough cross-beds are also evident. (F) Surface lebensspuren with a spongy appearance at a sandstone–mudstone interface in the Upper Murree Formation.

stone facies (**Sh**) of the Murree Group have probably been deposited in migrating tidal-channel environments.

The intertidal zone can be divided into mud-flats, located near the high-water line, sand-flats near the low-water line, and mixed flats forming a transitional environment between mud-flats and sand-flats. The net sedimentation rate on intertidal flats is rather low and the sediments are therefore more or less strongly bioturbated (Reineck, 1975). The fine-grained massive sandstones (**Smb**), mudstones (**Mb**) and siltstone–mudstone intercalations are interpreted to have formed in an intertidal environment. The calcrete layers in the mudstones suggest supratidal sedimentation events in the Murree Group.

The rhythmites of the Lower Murree Formation, composed of mudstone, siltstone and sandstone, are the result of long-term changes in hydrodynamic conditions in the sedimentary basin. The sandstone layers were deposited in high-energy conditions by waves and currents. The mudstones, in turn, reflect periods of slack water or low-energy conditions. Deposition of siltstones may be attributed to intermediate conditions.

By comparison, the Upper Murree sediments were deposited under conditions of slightly higher energy, resulting in the formation of medium- and coarse-grained sandstones. However, low-energy conditions prevailed near the top, where mudstone layers reflect deposition in a macrotidal environment.

Green coloured sandstones occur mostly in the Upper Murree Formation and contain chloritic minerals indicative of shallow-water environments. Furthermore, the purple brown colours of many mudstones and siltstones suggest oxidizing conditions characteristic of nearshore environments, an observation that supplements the interpretation of the tidal flat origin of the Murree Group of rocks.

In some places of the Formation, carbonate-impregnated mud-cracked surfaces seem to have been eroded (or dissolved) and redeposited (or reprecipitated) in thin layers known as 'tidal flat paper' (Trusheim, 1936; Scott & Hayes, 1964).

Sand-flats are characterized by well-developed small-scale wave and current ripples. The most common bedding types found in mixed-flat environments are flaser and lenticular bedding which occur in many variations (Reineck, 1972, 1975). According to Reineck & Singh (1973), various types of ripple marks of different orientation and geometry are produced either simultaneously or sequentially in a hierarchical system that depends on morphology and hydrography. Increasing flow strength changes rippled bedforms from straight-crested into undulatory and finally cuspate forms (Hayes & Kana, 1976). In the case of the Lower Murree Group, strongly undulatory, small asymmetrical ripples indicate the transition from low-energy to high-energy conditions. Ripple spacing varies from 8 to 12 cm and heights reach 2 cm. These fossil ripple marks are very similar to those observed in modern tidal flat environments such as occur along the North Sea. Another form of commonly observed regular surface structure is straight-crested (longitudinal), partly bifurcated and asymmetrical low ridges with spacings of 7.5–12 cm. This type of surface structure results from selective erosion of longitudinal cross-bedding produced by lateral migration of tidal channels (Reineck, 1958, 1967a). Starved ripples are characteristic of tidal flat deposits where alternate layers of sand and mud occur, as is the case in the Murree Rocks. According to Allen (1985), wrinkle marks occur widely in intertidal environments, being formed soon after emergence of sedimentary surfaces. The presence of wrinkle marks in the fine-grained sandstones associated with mudstones suggests deposition in an intertidal environment.

Flute casts develop in the process of flow separation, whereby the current leaves the sediment surface at the upstream rims of flutes while small eddies or rollers, rotating in a horizontal plane, are trapped within the flutes (Allen, 1971). Although flute casts are particularly characteristic structures of turbidities, they are also reported from shallow-water deposits. Their presence in the Lower Murree Formation suggests shallow-water conditions in a turbulent environment. Grooves are sole marks produced by tools carried by the current along soft bottoms. In the Murree Group of rocks, groove casts are probably indicative of erosion along tidal channel beds.

Tidal flat environments are commonly characterized by herringbone cross-bedding (Singh, 1969). The presence of herringbone cross-bedding and flaser bedding in the Murree Group is further evidence for a tidal environment. Similarly, trough cross-bedding in the sandstone units is indicative of strong current action in the subtidal zone.

Reactivation surfaces form under flow conditions with a pronounced time-velocity asymmetry (Klein, 1970). Furthermore, tidal bundles are sedimentary structures first recognized in sandy shoals and large excavation pits associated with flood control structures in the tidal estuaries of the Dutch coast. These are described in detail by Visser (1980), Terwindt (1981), van den Berg (1982) and others. The regular increase and decrease of flow velocities in the course of a tidal cycle and the variations in tidal flow strength over the lunar month are physical parameters reflected in variable tidal bundle thickness (Visser, 1980). Since these features are formed only by tidal processes, they are a very powerful tool for the interpretation of shallow-water tidalites (Kreisa & Moiola, 1986). The presence of tidal bundles with spacings of 2–2.5 cm in the Lower as well as the Upper Murree sandstones is thus the most direct and unequivocal evidence for a tidal origin of the sediments.

Bioturbation is weakly to moderately developed in the Murree Group of rocks. The vertical burrows in the fine-grained sandstones suggest shallow-water deposits (Seilacher, 1967). Places of weaker bioturbation are attributed to rapid deposition (Reineck, 1967b).

The palaeocurrent directions in the Lower Murree sandstones show a NNE–SSW trend (Fig. 5). This pattern is certainly a product of bidirectional tidal currents. The Upper Murree sandstones reveal NNE–SSW flow directions with some subordinate modes. This is also evidence of bipolarity in the current system. Tidal flow is non-uniform not only along the channel axis but also across the tidal channel as shown by Boothroyd (1978). Variations in palaeocurrent direction in the Murree Group may be due to variations in current flow direction in tidal channels.

The present investigation thus clearly demonstrates that the Murree Group of rocks must have been deposited in a tidal environment. The individual cycles in the Lower as well as the Upper Murree Formation relate to short transgressive phases resulting in the preservation of subtidal, intertidal and supratidal facies. Similar conditions have previously been suggested by Singh (1978) and Casshyap & Srivastava (1983) for equivalent rocks of Himachal Pardesh in India. The lithofacies and palaeontological interpretations of the Murree Group of Pakistan by Bossart & Ottiger (1989) suggest deposition in a fluvial-tidal transitional system.

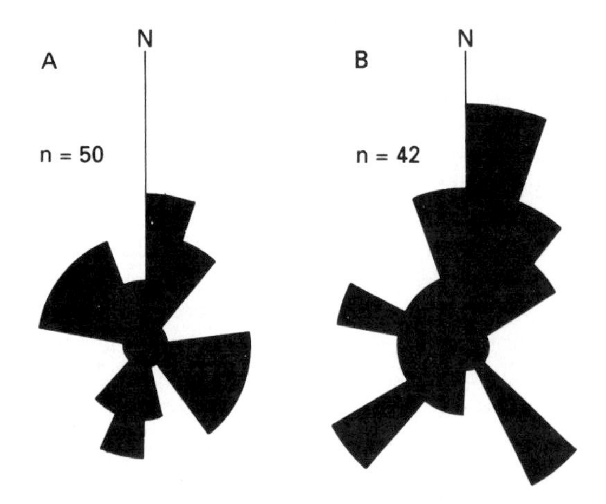

Fig. 5. Palaeocurrent patterns in the Murree Group. (A) Current rose diagram from the cross-bedded sandstones of the Lower Murree Formation. (B) Current rose diagram from the cross-bedded sandstones of the Upper Murree Formation.

ACKNOWLEDGEMENTS

The authors express their gratitude to Professor S.K. Shah, Head of Department of Geology, University of Jammu for providing facilities to work on the present topic. Thanks are further accorded to the Department of Science and Technology for providing funds to one of us (B.P.S.). The authors are also thankful to Professor B.W. Flemming for organizing Tidal Clastics '92 and providing the opportunity of publishing this work. The manuscript benefited from the critical comments of two anonymous reviewers.

REFERENCES

Allen. J.R.L. (1971) Transverse erosional marks of mud and rock: Their physical basis and geological significance. *Sediment. Geol.* **5**, 167–385.

Allen, J.R.L. (1985) Wrinkle marks: An intertidal sedimentary structure due to a seismic soft sediment loading. *Sediment. Geol.* **41**, 75–95.

Berg, J.H. van den (1982) Migration of large scale bedforms and preservation of crossbedded sets in highly accretional parts of tidal channels in the Oosterschelde, SW Netherlands. *Geol. Minjb.* **61**, 253–263.

Boersma, J.R. & Terwindt, J.H.J. (1981) Neap-spring tide sequences of intertidal shoal deposits in a mesotidal estuary. *Sedimentology* **28**, 151–170.

Boothroyd, J.C. (1978) Mesotidal inlets and estuaries. In: *Coastal Sedimentary Environments* (Ed. Davis, R.A., Jr), pp. 287–356. Springer, New York.

BOSSART, P. & OTTIGER, R. (1989) Rocks of the Murree Formation in northern Pakistan: Indicators of descending foreland basin of late Paleocene to mid Eocene age. *Eclog. Geol. Helv.* **82**, 133–165.

CASSHYAP, S.M. & SRIVASTAVA, V.K. (1983) Evolution of pre-Siwalik basin of Himachal Himalaya. *J. Geol. Soc. India* **24**, 134–147.

HAYES, M.O. & KANA, T.W. (1976) Terrigenous/clastic depositional environments—some modern examples. *Tech. Rept 11-CRD, Coast. Res. Div., Univ. South Carolina* **II-184**, 1–131.

JOHNSON, H.D. (1982) Shallow siliciclastic seas. In: *Sedimentary Environments and Facies* (Ed. Reading, H.G.), pp. 207–258. Blackwell, London.

KHAN, A., RAO, V.R., GANJU, J.L. & SANKARAN, V. (1971) Discovery of invertebrate and vertebrate fossils from Upper Murree Formation of Palkhai syncline near Udhampur, J & K State. *J. Pal. Soc. India* **16**, 16–21.

KLEIN, G.D. (1970) Depositional and dispersal dynamics of intertidal sand bars. *J. sediment. Petrol.* **40**, 1095–1127.

KLOOTWIJK, C.T., SHARMA, M.L., GERGAN, J., SHAH, S.K. & GUPTA, B.K. (1986) Rotational overthrusting of the northwestern Himalaya: further palaeomagnetic evidence from the Reasi Thrust sheet, Jammu foothills, India. *Earth & Planet. Sci. Letters* **80**, 375–393.

KREISA, R.D. & MOILA, R.J. (1986) Sigmoidal tidal bundles and other tide generated sedimentary structures of the Curtis Formation, Utah. *Geol. Soc. Am. Bull.* **97**, 381–387.

MCKIE, T. (1991) Tidal and storm influenced sedimentation from a Cambrian transgressive passive margin sequence. *J. Geol. Soc. London* **147**, 785–794.

MEHTA, S.K. & JOLLY, A. (1988) The Murree Group: Stratigraphic Limits. Correlations and Age implications from new fossil data. *Nat. Sem. Str. Bound. Pro. India*, 58–59.

NIO, S.D., BERG, J.H. VAN DEN, GOESTEN, M. & SMULDERS, F. (1980) Dynamics and sequential analysis of a mesotidal shoal and intershoal channel complex in the eastern Scheldt (Southwestern Netherlands). *Sediment. Geol.* **26**, 263–279.

OLDHAM, R.D. (1892) Report on the geology of Thal-Chotli and part of the Mari Country. *Rec. Geol. Surv. India* **25**, 18–29.

PILGRIM, G.E. (1910) Preliminary note on a revised classification of the Tertiary fresh water deposits of India. *Rec. Geol. Surv. India* **40**, 185–205.

RANGA RAO, A. (1971) New Mammals from Murree rocks of Kalakot zone of the Himalayan foothills, Kalakot, J & K State. *J. Geol. Soc. India* **12**, 126–134.

REINECK, H.-E. (1958) Longitudinale Schrägschichten im Watt. *Geol. Rundschau* **47**, 73–82.

REINECK, H.-E. (1967a) Layered sediments of tidal flat beaches and shelf bottoms of the North Sea. In: *Estuaries* (Ed. Lauff, G.H.), pp. 191–206. *Am. Assoc. Advance. Sci. Publ.* **83.**

REINECK, H.-E. (1967b) Parameter von Schichtung und Bioturbation. *Geol. Rundschau* **56**, 420–438.

REINECK, H.-E. (1972) Tidal flats. In *Recognition of Ancient Sedimentary Environments* (Eds Rigby, J.K. & Hamblin, W.K.) *S.E.P.M. Spec. Publ.* **17**, 146–159.

REINECK, H.-E. (1975) German North Sea tidal flats. In: *Tidal Deposits* (Ed. Ginsburg, R.N.), pp. 5–12. Springer, Heidelberg.

REINECK, H-E. & SINGH, I.B. (1973) *Depositional Sedimentary Environments.* Springer, Heidelberg, 551 pp.

REINECK, H.-E., SINGH, I.B. & WUNDERLICH, F. (1971) Einteilung der Rippeln und anderer mariner Sandkörper. *Senckenbergiana marit.* **3**, 93–101.

SCHÄFER, W. (1956) Wirkungen der Benthos-Organismen auf den jungen Schichtverband. *Senckenbergiana Lethaea* **37**, 183–263.

SCOTT, A.J. & HAYES, M.O. (1964) *Playa Lake and Clay Dunes. Depositional Environments of the South-Central Texas Coast.* Corpus Christi Geol. Soc., Univ. Texas, 38 pp.

SEILACHER, A.J. (1953a) Studien zur Palichnologie. *Neues Jahrb. Geol. Palaeontol. Abhd.* **96**, 421–452.

SEILACHER, A.J. (1953b) Die fossilen Ruhespuren (Cubichnia). *Neues Jahrb. Geol. Palaeontol. Abhd.* **98**, 87–124.

SEILACHER, A.J. (1967) Bathymetry of trace fossils. *Mar. Geol.* **5**, 413–428.

SHARDA, Y.P. & VERMA, V.K. (1977): Palaeoenvironments during Murree and Siwalik sedimentation around Udhampur, Jammu Himalaya. *Proc. Cent. Adv. Study Geol.*, Candigarth, 23A–35A.

SINGH, I.B. (1969) Primary sedimentary structures in Precambrian quartzites of Telemark, southern Norway, and their environmental significance. *Norsk Geol. Tidsskr.* **49**, 1–31.

SINGH, I.B. (1978) On some sedimentological and palaeoecological aspects of Subathu–Dagshai–Kasauli succession of Simla Hills. *J. Pal. Soc. India* **21**, 19–28.

SINGH, B.P., FOTEDAR, B.K. & RAO, A.S. (1990) Petrography and geochemistry of the sandstones of Murree Group around Laren, Udhampur, Jammu Himalaya. *J. Geol. Soc. India* **36**, 502–511.

TERWINDT, J.H.J. (1981) Origin and sequences of sedimentary structures in the inshore mesotidal of the North Sea. In: *Holocene Marine Sedimentation in the North Sea Basin* (Eds Nio, S.D., Shüttenhelm, R.T.E. & Weering, T.C.E. van). *Spec. Publ. int. Assoc. Sediment.* **5**, 4–26.

TRUSHEIM, F. (1936) Watten-Papier. *Natur und Volk* **66**, 103–106.

VISSER, M.J. (1980) Neap-spring cycles reflected in Holocene subtidal largescale bedform deposits: a preliminary note. *Geology* **8**, 543–546.

WADIA, D.N. (1928) The Geology of Poonch state and adjacent portions of the northern Punjab. *Mem. Geol. Surv. India* **55**, 185–370.

Index

Page numbers in *italic* indicate figures; those in **bold** indicate tables